REMOTE SENSING
OF LAND USE AND
LAND COVER

PRINCIPLES AND APPLICATIONS

Taylor & Francis Series in Remote Sensing Applications

Series Editor
Qihao Weng
Indiana State University
Terre Haute, Indiana, U.S.A.

Taylor & Francis Series in Remote Sensing Applications
Qihao Weng, Series Editor

REMOTE SENSING
OF LAND USE AND
LAND COVER

PRINCIPLES AND APPLICATIONS

EDITED BY
CHANDRA P. GIRI

CRC Press
Taylor & Francis Group
Boca Raton London New York

CRC Press is an imprint of the
Taylor & Francis Group, an **informa** business

CRC Press
Taylor & Francis Group
6000 Broken Sound Parkway NW, Suite 300
Boca Raton, FL 33487-2742

First issued in paperback 2019

ISBN-13: 978-1-4200-7074-3 (hbk)
ISBN-13: 978-0-367-86438-5 (pbk)

Library of Congress Cataloging-in-Publication Data

Remote sensing of land use and land cover : principles and applications / edited by Chandra P. Giri.
       p. cm. -- (Remote sensing applications series ; 8)
    Includes bibliographical references and index.
    ISBN 978-1-4200-7074-3
    1. Land use--Remote sensing. 2. Land use--Research--Methodology. 3. Remote sensing. I. Giri, Chandra P.

HD108.8.R47 2012
333.73'130285--dc23   2011052392

Visit the Taylor & Francis Web site at
http://www.taylorandfrancis.com

and the CRC Press Web site at
http://www.crcpress.com

Contents

SECTION I   Overview

SECTION II   Basic Principles

SECTION III   Application Examples

SECTION IV Looking Ahead

Series Preface

Land cover describes both natural and man-made coverings of the Earth's surface, including biota, soil, topography, surface and groundwater, and human structures. A related concept is land use, referring to the manner in which the biophysical attributes of the land are manipulated and the purpose for which the land is used. Remote sensing is a cost-effective technology for mapping land cover and land use and for monitoring and managing land resources. The remote sensing literature shows that a tremendous number of efforts has been made for mapping, monitoring, and modeling land cover and land use at the local, regional and global scales. However, a comprehensive book has not been published to specifically address the issues of land cover science, mapping techniques and applications, and future opportunities. *Remote Sensing of Land Cover: Principles and Applications* uniquely fills this niche.

I am pleased that Dr. Chandra Giri, a research physical scientist at the United States Geological Survey, has taken the initiative to compile this volume. Contributed by a group of leading and well-published scholars in the field, this book first discusses—following a nice overview chapter by Dr. Thomas Loveland—the principles of land cover mapping, monitoring, and modeling. The second part of the book deals with case studies, mostly examined at the continental scale, from all over the world. Last but not the least, land cover programs supported by NASA and GEO (Group on Earth Observation) are introduced, providing a prospect for future national and international efforts. Dr. Giri carefully selected and examined each contribution and created a well-structured volume in order to address the issues of land cover from the viewpoints of science, technology, practical application and future needs. This comprehensive approach presents the readers with both a systematic view of the field and a detailed knowledge of a particular topic.

Like other books in the Taylor & Francis Series in Remote Sensing Applications, this book is designed to serve as a guide or reference for professionals, researchers, and scientists, as well as a textbook or an important supplement for teachers and students. I hope that the publication of this book will further promote a better use of Earth observation data and technology and will facilitate the assessing, monitoring, and managing of land resources.

Qihao Weng, PhD
Hawthorn Woods, Indiana

Preface

Land-cover characterization, mapping, and monitoring are the most important and typical applications of remotely sensed data. The availability and accessibility of accurate and timely land-cover datasets play an important role in many global change studies. Several national and international programs have emphasized the increased need for better land-cover and land-cover change information at local, national, continental, and global scales. These programs, such as the International Geosphere–Biosphere Program (IGBP), U.S. Climate Change Science Program, Land Cover and Land Use Change (LCLUC) program of the National Aeronautics and Space Administration (NASA), Global Land Project, Global Observation of Forest and Land Cover Dynamics (GOFC/GOLD), and Group on Earth Observations (GEO), have been in the forefront of framing scientific research questions on land-change science.

Recent developments in earth-observing satellite technology, information technology, computer hardware and software, and infrastructure have helped produce land-cover datasets of better quality. As a result, such datasets are becoming increasingly available, the user base is ever widening, application areas are expanding, and the potential for many other applications is increasing. Despite such progress, a comprehensive book, such as *Remote Sensing of Land Use and Land Cover: Principles and Applications*, on this topic has not been available so far. This book aims at providing a synopsis of basic land-cover research questions and an overview of remote-sensing history. It also offers an overview of land-cover classification, data issues, preprocessing, change analysis, modeling, and validation of results.

Examples of application at global, continental, and national scales from around the world have been provided. Overall, the book highlights new frontiers in remote sensing of land use/land cover by integrating current knowledge and scientific understanding and provides an outlook for the future. Specific topics emphasize current and emerging concepts in land-use/land-cover mapping, an overview of advanced and automated land-cover interpretation methodologies, and a description and future projection of the major land-cover types of the world. The book offers a new perspective on the subject by integrating decades of research conducted by leading scientists in the field.

The book is expected to be a guide or handbook for resource planners, managers, researchers, and students at all levels and a valuable resource for those just starting out in this field or those with some experience in the area of land-use/land-cover characterization and mapping. The book also contains some advanced topics useful for seasoned professionals. It can also be used as a textbook or as reference material in universities and colleges.

Chandra P. Giri
Sioux Falls, South Dakota

Acknowledgments

I extend my heartfelt thanks to all the authors who have contributed to this book despite their busy schedule and workload. Besides thanking of my colleagues at work and throughout the world, who inspired me to work on this book, I express my deepest appreciation to the reviewers who offered critical comments and suggestions to improve the book. Finally, I thank my mother Rupa, wife Tejaswi, daughter Medhawi, and son Ash for their continued support and encouragement. I hope this book will help students and professionals who use remote-sensing technology for land-cover characterization, mapping, and monitoring.

The names of the reviewers are listed below alphabetically:

Thomas Adamson
Gulleid Artan
Roger F. Auch
Birendra Bajracharya
Marc Carroll
Robert B. Cook
Anna Cord
Pelilei Fan
Tejaswi Giri
Peng Gong
Mryka Hall-Beyer
Nazmul Hossain
Dave Johnson
Soe W. Myint
Ian Olthof
Md Shahrir Pervez
Bradley C. Reed
Laura Dingle Robertson
Ake Rosenquist
Angela Schwering
Ruth Swetnam
G. Gray Tappan
Larry L. Tieszen
George Z. Xian
Limin Yang
Raul Zurita-Milla

Editor

Chandra P. Giri received his BS in forest conservation from Tribhuvan University, Nepal, his MS in interdisciplinary natural resources planning and management, and his PhD in remote sensing and geographic information systems from the Asian Institute of Technology (AIT), Bangkok, Thailand. Currently a research physical scientist at the U.S. Geological Survey (USGS)/Earth Resources Observation and Science (EROS) Center, he is also a guest/adjunct faculty at South Dakota State University. Earlier, he had worked for Columbia University's Center for International Earth Science Information Network (CIESIN), United Nations Environment Programme (UNEP), AIT, and Department of Forests, Nepal. At EROS, he leads the International Land Cover and Biodiversity program. His work focuses on global and continental-scale land-use/land-cover characterization and mapping using remote sensing and geographic information systems (GIS). His recent research was on global mangrove forest-cover mapping and monitoring using earth-observation satellite data, and on studying the impact, vulnerability, and adaptation of sea-level rise to mangrove ecosystems, integrating both biophysical and socioeconomic data. He is also researching the development of remote-sensing–based state-of-the-art methodologies to monitor carbon stocks for the Reducing Emissions from Deforestation and Forest Degradation (REDD) initiative. He has experience working in the private sector, academia, government, and international organizations at national, continental, and global levels in different parts of the world. He serves as an expert in national and international working groups. He has more than 50 scientific publications to his credit and has received several awards from USGS, NASA, and other organizations.

Contributors

William Acevedo
Earth Resources Observation and Science
    Center
United States Geological Survey
Sioux Falls, South Dakota

Ola Ahlqvist
Department of Geography
Ohio State University
Columbus, Ohio

Olivier Arino
European Space Research Institute
European Space Agency
Frascati, Italy

Roger F. Auch
Earth Resources Observation and
    Science Center
United States Geological Survey
Sioux Falls, South Dakota

Patrice Bicheron
MEDIAS France
National Center for Space Studies (CNES)
Toulouse, France

Catherine Bodart
Global Environment Monitoring Unit
Institute for Environment and Sustainability
European Commission Joint Research Centre
Ispra, Italy

Sophie Bontemps
Earth and Life Institute
Catholic University of Louvain
Louvain-La-Neuve, Belgium

Andreas Bernhard Brink
Global Environment Monitoring Unit
Institute for Environment and Sustainability
European Commission Joint Research Centre
Ispra, Italy

Christelle Carsten Brockman
Brockmann Consulting
Geesthacht, Germany

Gyanesh Chander
Stinger Ghaffarian Technologies
Earth Resources Observation and Science
    Center
United States Geological Survey
Sioux Falls, South Dakota

Xuexia (Sherry) Chen
ASRC Research and Technology Solution
Earth Resources Observation and Science
    Center
United States Geological Survey
Sioux Falls, South Dakota

René R. Colditz
Directorate of Geomatics
National Commission for Knowledge and Use
    of Biodiversity (CONABIO)
Mexico City, Mexico

Pierre Defourny
Earth and Life Institute
Catholic University of Louvain
Louvain-La-Neuve, Belgium

Xiangzheng Deng
Institute of Geographic Sciences and Natural
    Resources Research
Chinese Academy of Sciences
Beijing, China

Antonio Di Gregorio
Global Land Cover Network
Land and Water Division
Food and Agriculture Organization of the
    United Nations
Rome, Italy

Mark A. Drummond
U.S. Geological Survey
Fort Collins, Colorado

Hugh Douglas Eva
Global Environment Monitoring Unit
Institute for Environment and Sustainability
European Commission Joint Research Centre
Ispra, Italy

Jan Feranec
Institute of Geography
Slovak Academy of Sciences
Bratislava, Slovak Republic

Robert Fraser
Canadian Centre for Remote Sensing
Natural Resources Canada
Ottawa, Ontario, Canada

Steffen Fritz
International Institute for Applied Systems
  Analysis
Laxenburg, Austria

Alisa L. Gallant
Earth Resources Observation and Science
  Center
United States Geological Survey
Sioux Falls, South Dakota

Sangram Ganguly
NASA Ames Research Center/BAERI
Moffett Field, California

Lorena Hojas Gascon
Global Environment Monitoring Unit
Institute for Environment and Sustainability
European Commission Joint Research Centre
Ispra, Italy

Chandra P. Giri
Earth Resources Observation and Science
  Center
United States Geological Survey
Sioux Falls, South Dakota

Nadine Gobron
Global Environment Monitoring Unit
Institute for Environment and Sustainability
European Commission Joint Research Centre
Ispra, Italy

Annemarie van Groenestijn
Laboratory of Geo-Information Science and
  Remote Sensing
Wageningen University and Research Centre
Wageningen, The Netherlands

Garik Gutman
Land-Cover/Land-Use Change Program
National Aeronautics and Space Administration
Washington, D.C.

Matthew C. Hansen
Department of Geography
University of Maryland
College Park, Maryland

Gerard Hazeu
Altera, Green World Research
Wageningen, The Netherlands

Martin Herold
Laboratory of Geo-Information Science and
  Remote Sensing
Wageningen University and Research Centre
Wageningen, The Netherlands

Colin Homer
Earth Resources Observation and Science
  Center
United States Geological Survey
Sioux Falls, South Dakota

Sheikh Nazmul Hossain
Stinger Ghaffarian Technologies
Earth Resources Observation and Science
  Center
United States Geological Survey
Sioux Falls, South Dakota

Chengquan Huang
Department of Geography
University of Maryland
College Park, Maryland

Gabriel Jaffrain
IGN France International
Paris, France

Chris Justice
Department of Geography
University of Maryland
College Park, Maryland

Vasileios Kalogirou
European Space Research Institute
European Space Agency
Frascati, Italy

LeeAnn King
Department of Geography
University of Maryland
College Park, Maryland

Lammert Kooistra
Laboratory of Geo-Information Science and
    Remote Sensing
Wageningen University and Research Centre
Wageningen, The Netherlands

Rasim Latifovic
Natural Resources Canada
Canadian Centre for Remote Sensing
Ottawa, Ontario, Canada

Marc Leroy
MEDIAS France
National Center for Space Studies (CNES)
Toulouse, France

Jiyuan Liu
Institute of Geographic Sciences and Natural
    Resources Research
Chinese Academy of Sciences
Beijing, China

Thomas R. Loveland
Earth Resources Observation and Science
    Center
United States Geological Survey
Sioux Falls, South Dakota

Landing Mane
Satellite Observatory of Central African
    Forests (OSFAC)
Kinshasa, Democratic Republic of Congo

Philippe Mayaux
Institute for Environment and Sustainability
Joint Research Centre—European Commission
Ispra, Italy

José Emilio Meroño De Larriva
Department of Graphic Engineering and
    Geomatics
University of Cordoba
Cordoba, Spain

Alberto Jesús Perea Moreno
Department of Applied Physics
University of Cordoba
Cordoba, Spain

Douglas O'Brien
IDON Technologies
Ottawa, Ontario, Canada

Ian Olthof
Natural Resources Canada
Canadian Centre for Remote Sensing
Ontario, Canada

Jean-François Pekel
Institute for Environment and Sustainability
Joint Research Centre—European Commission
Ispra, Italy

Darren Pouliot
Natural Resources Canada
Canadian Centre for Remote Sensing
Ottawa, Ontario, Canada

Rainer Ressl
Directorate of Geomatics
National Commission for Knowledge and Use
    of Biodiversity (CONABIO)
Mexico City, Mexico

Kristi L. Sayler
Earth Resources Observation and Science
   Center
United States Geological Survey
Sioux Falls, South Dakota

Chris Schmullius
Department for Earth Observation
University of Jena
Jena, Germany

Dario Simonetti
ReggianiSpA
Varese, Italy

Benjamin Sleeter
Western Geographic Science Center
United States Geological Survey
Menlo Park, California

Terry Sohl
Earth Resources Observation and Science
   Center
United States Geological Survey
Sioux Falls, South Dakota

Kuan Song
Department of Geography
University of Maryland
College Park, Maryland

Tomas Soukup
GISAT
Prague, Czech Republic

Hans-Jurgen Stibig
Institute for Environment and Sustainability
Joint Research Centre—European Commission
Ispra, Italy

Christelle Vancutsem
Monitoring Agricultural Resources Unit
European Commission Joint Research Centre
Ispra, Italy

Arturo Victoria
National Institute of Statistics and Geography
Aguascalientes, Mexico

James E. Vogelmann
Earth Resources Observation and Science
   Center
United States Geological Survey
Sioux Falls, South Dakota

Carlos de Wasseige
Satellite Observatory of Central African
   Forests (OSFAC)
Kinshasa, Democratic Republic of Congo

Section I

Overview

1 Brief Overview of Remote Sensing of Land Cover

Chandra P. Giri

CONTENTS

1.1 BACKGROUND

Land cover of the earth's land surface has been changing since time immemorial and is likely to continue to change in the future (Ramankutty and Foley, 1998). These changes are occurring at a range of spatial scales from local to global and at temporal frequencies of days to millennia (Townshend et al., 1991). Both natural and anthropogenic forces are responsible for the change. Natural forces such as continental drift, glaciation, flooding, and tsunamis and anthropogenic forces such as conversion of forest to agriculture, urban sprawl, and forest plantations have changed the dynamics of land-use/land-cover types throughout the world.

In recent decades, anthropogenic land-use/land-cover change has been proceeding much faster than natural change. This unprecedented rate of change has become a major environmental concern worldwide. As a result, almost all ecosystems of the world have been significantly altered or are being altered by humans, undermining the capacity of the planet's ecosystems to provide goods and services. Two main forces responsible for anthropogenic changes are technological development and the burgeoning human population (Lambin and Meyfroidt, 2011).

Land-cover changes play a significant role in the global carbon cycle, both as a source and a sink (Loveland and Belward, 1997a; Moore, 1998), and in the exchange of greenhouse gases between the land surface and the atmosphere. For example, deforestation releases carbon dioxide into the atmosphere and changes land-surface albedo, evapotranspiration, and cloud cover, which in turn affect climate change and variability. In contrast, afforestation and reforestation remove carbon from the atmosphere (sink). Recent evidence shows that human-induced changes in land use/land cover over the last 150 years have led to the release of an enormous amount of carbon into the atmosphere. Although combustion of fossil fuels is the dominant source of release of carbon into the atmosphere, land use still contributes a significant portion (~20%) of anthropogenic emission, particularly in tropical areas.

Land-cover and land-use changes may have positive or negative effects on human well-being and can also have intended or unintended consequences (DeFries and Belward, 2000; Hansen and DeFries, 2004). Conversion of forests to croplands had provided food, fiber, fuel, and a host of other products to an increasing human population throughout human history. At the same time, tropical deforestation has reduced biodiversity, degraded watersheds, increased soil erosion, and consequently raised the risk of unintended but devastating forest fire. Owing to the rapid and unprecedented land-use/land-cover change in recent years, negative consequences such as soil erosion, loss of biodiversity, water pollution, and air pollution have increased. The benefits and economic gains provided by ecosystems have started eroding because these benefits are derived at the expense of degradation of the ecosystem.

1.2   RESEARCH NEED, PRIORITIES, AND OPPORTUNITIES

Understanding the distribution and dynamics of land cover is crucial to the better understanding of the earth's fundamental characteristics and processes, including productivity of the land, the diversity of plant and animal species, and the biogeochemical and hydrological cycles. Assessing and monitoring the distribution and dynamics of the world's forests, shrublands, grasslands, croplands, barren lands, urban lands, and water resources are important priorities in studies on global environmental change as well as in daily planning and management. Information on land cover and land-cover change is needed to manage natural resources and monitor global environmental changes and their consequences (Loveland and Belward, 1997b).

Several national and international programs have emphasized the increased need for better land-cover and land-cover change information at local, national, continental, and global levels. These programs, such as International Geosphere Biosphere Program (IGBP), U.S. Climate Change Science Program, Land Cover and Land Use Change (LCLUC) program of the National Aeronautics and Space Administration (NASA), Global Land Project, Global Observation of Forest and Land Cover Dynamics (GOFC-GOLD), and Group on Earth Observations (GEO), have been in the forefront of scientific inquiry in land-change science. For example, GOFC-GOLD has provided detailed guidelines for land-cover products (Turner et al., 1994). Similarly, the GEO has identified key land-cover observations and desired products that are likely to contribute to specific areas of societal benefits (Figure 1.1). Land-cover observation and monitoring can provide critical information needed for several GEO areas of societal benefits (Table 1.1).

In essence, the GEO has (1) highlighted the societal needs and relevance of land observations, (2) provided a forum for advocating global land-cover and change observations as a key issue, (3) fostered integrated perspectives for continuity and consistency of land observations, (4) helped evolve and apply international standards for land-cover characterization and validation, (5) improved a shared vision within the land observation community and involved global actors, (6) advocated joint participation in ongoing global mapping activities, regional networking, and capacity building in developing countries, and (7) helped develop international partnership involving producers, users, and the scientific community to better produce and use existing datasets (http://www.geogr. uni-jena.de/~c5hema/telecon/geo_achievement_global_land_cover.pdf).

Similarly, the United States Global Change Research Program (USGCRP) have identified five strategic questions that are important for future research on land cover and land-cover change (http://www.usgcrp.gov/usgcrp/ProgramElements/land.htm).

1. What tools or methods are needed to better characterize historical and current land-use and land-cover attributes and dynamics?
2. What are the primary drivers of land-use and land-cover change?
3. What will land-use and land-cover patterns and characteristics be in 5–50 years?
4. How do climate variability and change affect land use and land cover, and what are the potential feedbacks of changes in land use and land cover to climate?
5. What are the environmental, social, economic, and human health consequences of current and potential land-use and land-cover change over the next 5–50 years?

Townshend et al. (2011) identified major stakeholders of global land observations that are relevant to land-cover observations and monitoring. They are as follows:

• National, regional, or local governments that need the information to assist them in developing and implementing their policies and to help them meet mandatory reporting requirements resulting from such policies

FIGURE 1.1   **(See color insert.)** Nine areas of societal benefit of the Group on Earth Observations (GEO).

- International initiatives to help develop and fund programs for countries that need the information to develop their policies and operational strategies
- Nongovernmental organizations
- Scientists who need the information to improve our understanding of the processes and uncertainties associated with the earth system
- The individual citizen who needs understandable and reliable information on global environmental trends
- The private sector that needs information to help partner and directly service the previous five stakeholders

With the recent advancement in remote sensing and geographic information systems (GIS) and computer technology, it is now possible to assess and monitor land-use/land-cover changes at multiple spatial and temporal scales (Hansen and DeFries, 2004). For example, the National Land Cover Database (NLCD) 2011 is an integrated database encompassing land-cover and land-cover change products at various thematic, spatial, and temporal resolutions (Figure 1.2).

Remote sensing offers several advantages. It is a relatively inexpensive and rapid method of acquiring up-to-date information over a large geographical area owing to its synoptic coverage and repetitive measurements. Remote-sensing data usually acquired in digital form are easier to manipulate and analyze; they can be acquired not only from visible but also from spectral ranges that are invisible to human eyes; they can be acquired from remote areas where accessibility is a concern; and they provide an unbiased view of land use/land cover. Similarly, historical data date back as early as the 1970s, and such data are becoming freely available. Several remotely sensed

TABLE 1.1

Linking the GEO Areas of Societal Benefits with Global Land-Cover Observation and User Requirements

GEO Areas of Societal Benefits	Key Land-Cover Observations and Desired Products
Disasters: reducing loss of life and property from natural and human-induced disasters	Fire monitoring (active + burn); surface-cover type changes and land degradation due to disasters; location of population and infrastructure
Health: understanding environmental factors affecting human-induced disasters	Land characteristics/change for disease vectors; land cover/change affecting environmental boundary conditions; demographics, socioeconomic conditions, and location and extent of settlement patterns
Energy: improving management of energy resources	Biofuel production sustainability; biomass yield estimates (forestry and agriculture); assessments for wind and hydropower generation and explorations
Climate: understanding, assessing, predicting, mitigating, and adapting to climate variability and change	Greenhouse gas emissions as the cause of land-cover change; land-cover dynamics forcing water and energy exchanges; location and extent of energy combustion
Water: improving water resources management through better understanding of the water cycle	Land-cover change affecting the dynamics of the hydrological systems; available water resources and quality distribution of water bodies and wetlands; water-use pattern (i.e., irrigation and vegetation stress) and infrastructure
Weather: improving weather information, forecasting, and warning	Land-cover change affecting radiation balance and sensible heat exchange; land surface roughness; biophysical vegetation characteristics and phenology
Ecosystems: improving the management and protection of terrestrial, coastal, and marine ecosystems	Changes in environmental conditions, conservation and provision of ecosystem services; land-cover and vegetation characteristics and changes; land-use dynamics and driving processes
Agriculture: supporting sustainable agriculture and combating desertification	Distribution and monitoring of cultivation practices and crop production; forest types and changes (e.g., logging); land degradations, and threats to terrestrial resources and productivity
Biodiversity: understanding, monitoring, and observing biodiversity	Ecosystem characterization and vegetation monitoring (types and species); habitat characteristics and fragmentation of invasive and protected species; changes in land cover and use affecting biodiversity

Source: Group on Earth Observations. Geo portal, http://www.geoportal.org.

data are available for assessing and monitoring land cover. A list of primary remote-sensing systems used for observing and monitoring land cover and land use is presented in Table 1.2.

Land use is difficult to observe because the intended use of the land may be different from the actual use. What we see are the physical artifacts of that use. For example, forest in many countries is defined as land designated as forest by the government regardless of whether the land is covered by trees or not. From a land-cover perspective, it could be barren land if the area is not covered by trees. Some land-use types such as industrial areas can be observed and measured using remotely sensed data, particularly with the help of very high-resolution satellite data, aerial photographs, ancillary data, and/or *a priori* knowledge. Certain land-use types can be derived from observed land-cover types because the realms of land use and land cover are interconnected. Observing land use using remotely sensed data becomes complicated when a single land-cover class is associated with multiple uses and multiple land-cover types are used for a single use. For example, a forest land cover can be used for timber production, fuel-wood production, recreation, biodiversity conservation, religious

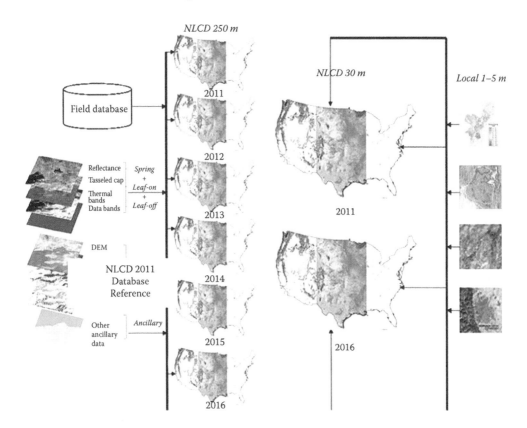

FIGURE 1.2 **(See color insert.)** A potential product framework proposed for NLCD 2011. (Adapted from Xian, G., Homer, C., and Yang, L., 2011. Development of the USGS National Land-Cover Database over two decades. In: Weng, Q. H., ed., *Advances in Environmental Remote Sensing—Sensors, Algorithms, and Applications.* CRC Press, Boca Raton, FL, 525–543.)

purposes, hunting/gathering, shifting cultivation, watershed protection, soil conservation, and carbon sequestration. Furthermore, several land-cover types such as croplands, grasslands, woodlots, and settlements can be used for a certain farming system (Meyer and Turner, 1992).

However, remote sensing of land cover may have many limitations. Data availability, accessibility, and cost of remotely sensed data may be an issue particularly in developing countries. However, since 2008, the U.S. Geological Survey/Earth Resources Observation and Science Center (USGS/EROS) has been providing free terrain-corrected and radiometrically calibrated Landsat data. Other space agencies and data providers are expected to follow suit. Much needs to be done to improve the preprocessing and classification accuracy of satellite imagery. Recently, the NASA-funded Web-Enabled Landsat Data (WELD) project demonstrated that large-scale (e.g., conterminous United States), cloud-free, and radiometrically and atmospherically corrected Landsat mosaics at 30-m resolution can be produced using the entire Landsat archive. The advantage is that "users do not need to apply the equations and spectral calibration coefficients and solar information to convert the Landsat digital number to reflectance and brightness temperature, and successive products are defined in the same coordinate system and align precisely, making them simple to use for multitemporal applications" (http://globalmonitoring.sdstate.edu/projects/weld/). The WELD product can then be used for land-cover characterization, mapping, and monitoring. At times, classification results may not be repeatable, and classification accuracy may be too low. Skilled manpower needed for the analysis may not be available. Incorporating field inventory data is critical for classification and validation.

Land-use/land-cover characterization and mapping is one of the most popular applications of remotely sensed data. Significant advances have also been made in the application of remote sensing

TABLE 1.2

List of Major Remote-Sensing Systems Used for Observing and Monitoring Land Cover and Land Use

Satellite	Web Site	Satellite	Web Site
ALOS/AVNIR/ PRISM	http://www.jaxa.jp/projects/sat/alos/ index_e.html	*MERIS (Envisat)*	http://envisat.esa.int/
ASTER	http://envisat.esa.int/	*MODIS*	http://modis.gsfc.nasa.gov/
CARTOSAT-1	http://www.isro.org/	*OrbView-3*	http://www.geoeye.com/
CBERS-1, 2, 2B	http://www.cbers.inpe.br/	*Quickbird*	http://www.digitalglobe.com
DMC	http://www.dmcii.com/	*RapidEye1–5*	http://www.rapideye.de/
EROS-A, EROS-B	http://www.imagesatintl.com	*SPOT 1–5*	http://www.spotimage.fr
FORMOSAT-2	http://www.spotimage.fr	*THEOS*	http://new.gistda.or.th/en/
GeoEye-1	http://launch.geoeye.com/LaunchSite/	*WorldView-1*	http://www.digitalglobe.com/
GOSAT	http://www.jaxa.jp/projects/sat/ gosat/index_e.html	*WorldView-2*	http://worldview2.digitalglobe.com/ about/
IKONOS	http://www.geoeye.com	*ASAR(Envisat)*	http://envisat.esa.int/
IRS-1A, 1B,1C, 1D	http://www.isro.org	*COSMO-SkyMed 1–3*	http://www.telespazio.it/cosmo.html
IRS-P2, P3, P4	http://www.isro.org	*ERS-1, ERS-2*	http://www.esa.int/esaCP/index.html
KOMPSAT-1	http://new.kari.re.kr/english/index.asp	*PALSAR*	http://www.jaxa.jp/index_e.html
KOMPSAT-2	http://earth.esa.int/object/index. cfm?fobjectid=5098	*RADARSAT-1, 2*	http://gs.mdacorporation.com/
Landsat 1–5, 7	http://landsat.gsfc.nasa.gov/	*TerraSAR-X*	http://www.astrium-geo.com/ en/228-terrasar-x-technical- documents

Source: Adapted from Remote sensing satellites. http://www.remotesensingworld.com/2010/06/16/remote-sensing-satel-lites/. With permission.
Note: This table is not intended to be complete.

for land-cover and land-use characterization, mapping, and monitoring to support global environmental studies and resource management. However, further work is needed not only for characterization and mapping but also for forecasting land-use/land-cover change for the future. Availability and accessibility of remotely sensed data are also critical. Scientific advancement in land-cover change analysis, accuracy assessment, use of multiscale data, addition of thematic richness (e.g., percent tree), and improved strategies for using land cover to more specifically infer land uses are needed (Loveland, 2004).

Looking ahead, the following were identified as the highest priority global land-cover issues (Townshend et al., 2011):

- Commitment to continuous 10–30-m resolution optical satellite systems with data acquisition strategies at least equivalent to that of the Landsat 7 mission.
- Development of *in situ* reference network for land-cover validation.
- Generation of annual products documenting global land-cover characteristics at resolutions between 250 m and 1 km, according to internationally agreed standards with statistical accuracy assessment.
- Generation of products that document global land cover at resolutions between 10 and 30 m at least every 5 years; a long-term goal is annual monitoring.
- Ensuring future continuity of mid-resolution multispectral SAR L-band data.
- Coordination of radar and optical data acquisitions so that radar data are usable to ensure regular monitoring of global land cover.
- Agreed upon internationally accepted land-cover and use classification systems.

The Ministry of Science and Technology of the People's Republic of China had approved the launching of a global land-cover mapping project to produce land-cover data products for 2000 and 2010, using Landsat, MODIS, and Chinese weather satellite data, with the minimum mapping unit of 30 m and the final product aggregated to 250 m. Similarly, the U.S. GEO announced the Global Land Cover Initiative at the Beijing GEO Ministerial Summit in November 2010, which aimed at the following:

1. Developing an initial global land-cover baseline for the 2010 period, using Landsat 30-m satellite data
2. Implementing an ongoing monitoring system that provides periodic (1, 2, 5 years) land-cover updates and land-cover change products from 2010 onwards
3. Improving the availability of 30-m class data (whenever possible)
4. Establishing the capability and capacity to develop historical land-change time series (1970s to present)

Significant progress in land-cover research has been made in the last two decades. With the development of remote sensing and computer technology, free availability of remotely sensed data, and availability of land-change expertise, a land-cover monitoring system is expected to be operational in the near future.

DEFINING LAND USE AND LAND COVER

Land use and land cover have often been confused and used interchangeably in the literature and also in daily practice. Thus, it is important to define and understand the meaning of these terms so that they can be used correctly, meaningfully, and to the best advantage. *Land cover* refers to the observed biotic and abiotic assemblage of the earth's surface and immediate subsurface (Meyer and Turner, 1992). Examples of major land-cover types are forests, shrublands, grasslands, croplands, barren lands, ice and snow, urban areas, and water bodies (including groundwater). As can be seen from the definitions and examples, the term now includes not only the vegetation that covers the land but also human structures, such as roads, built-up areas, and immediate subsurface features such as groundwater. *Land use* is defined as the way or manner in which the land is used or occupied by humans. In a nutshell, land cover represents the visible evidence of land use. A land covered by vegetation can be a forest as seen from the ground or through remote-sensing observations; however, the same tract of forest can be used for production, recreation, conservation, and religious purposes (Figure 1.3). In other words, land cover is the observed physical cover, whereas land use is based on function or the socioeconomic purpose for which the land is being used. A piece of land can have only one land cover (e.g., forests), but can have more than one land use (e.g., recreational, educational, and conservational).

LAND-COVER AND LAND-USE CHANGE

Land-cover change can be characterized as land-cover conversion and modification. Land-cover conversion is a change from one land-cover category to another, and modification is a change in condition within a land-cover category (Meyer and Turner, 1994). An example of the former is change from cropland to urban land, and an example of the latter is degradation of forests. Forest degradation may be due to change in phenology, biomass, forest density, canopy closure, insect infestation, flooding, and storm damage. Conversion is generally easier to measure and monitor than modification using remotely sensed data. Modification is usually a long-term process and may require multiyear and multiseasonal data for accurate quantification. Land-use change is a change in the use or management of land by humans. Land-use change may change without land-cover conversion or modification. For example, a production forest can be declared a protected area, and the number of visitors in a recreational forest may change without land-cover modification. On the contrary, land cover may change even if the land use remains unchanged; however, land-use change is likely to cause land-cover change.

Land cover = Forest

Land use = Recreational forest

FIGURE 1.3 Land cover and land use.

1.3 ORGANIZATION OF THE BOOK

The book is divided into four sections (Figure 1.4). Each chapter is organized around two basic themes: land cover and remote sensing; the chapters describe the salient issues in remote sensing and in land cover and their applications. Section I begins with a brief overview of remote sensing of land cover and the history of land-cover mapping. It provides a brief overview of key issues, opportunities, and recent advancements in the interpretation of remotely sensed data for land cover. Significant improvements have been made in land-cover research over the years, but many challenges remain for operational land-cover observation and monitoring (Giri et al., 2005). The second chapter in this section provides a comprehensive overview of the history of land-cover mapping. Historical perspective is needed to understand the data, classification system, infrastructure, and institutional issues and priorities better. Lessons learned from past experiences will be valuable for future land-cover initiatives.

Section II provides the basic principles of remote sensing for land-cover characterization, mapping, and monitoring. It highlights the fundamental mapping concepts that need to be considered during land-cover mapping using remote-sensing data. A land-cover classification system, including semantic issues and interoperability, is critical for evaluation, comparison, and change analysis of land-cover products. At present, no definitive universally accepted land-cover classification exists (Townshend et al., 2011). However, the Land Cover Classification System (LCCS) is currently the most comprehensive, internationally applied, and flexible framework for land-cover characterization. Thus, it is important to examine how LCCS is useful in evaluating land-cover legends. The section also highlights data records (e.g., AVHRR and MODIS) that can be routinely applied to

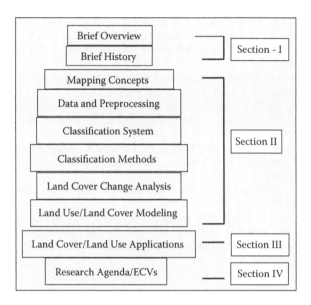

FIGURE 1.4 Main contents of the book.

study long-term changes in land-cover dynamics at multiple scales. Section II also addresses image-processing steps such as preprocessing, classification, change analysis, and validation of results. These chapters provide an overview of the science with examples. They also address the limitations and future possibilities of land-use/land-cover modeling in the United States.

Section III provides examples of land-cover application at global, continental, and national scales from around the world. Chapters in this section use multiple data sources and provide in-depth understanding of land cover and land-cover dynamics in multiple spatial, thematic, and temporal resolutions. Finally, Section IV highlights the research agendas for land-cover and land-use change and the importance of land cover as one of the major essential climate variables (ECVs). Recent research agendas and future research priorities from NASA's Land Cover and Land Use program are discussed. The final chapter also discusses how operational global and regional land-cover observations and monitoring are developed.

REFERENCES

DeFries, R.S. and Belward, A.S. 2000. Global and regional land cover characterization from satellite data: An introduction to the Special Issue. *International Journal of Remote Sensing*, 21, 1083–1092.

Giri, C., Zhu, Z.L., and Reed, B. 2005. A comparative analysis of the Global Land Cover 2000 and MODIS land cover data sets. *Remote Sensing of Environment*, 94, 123–132.

Hansen, M.C. and DeFries, R.S. 2004. Detecting long-term global forest change using continuous fields of tree-cover maps from 8-km advanced very high resolution radiometer (AVHRR) data for the years 1982–99. *Ecosystems*, 7, 695–716.

Lambin, E. and Meyfroidt, P. 2011. Global land use change, economic globalization, and the looming land scarcity. *Proceedings of the National Academy of Sciences*, 108, 3465–3472.

Loveland, T.R. (Ed.). 2004. *Observing and Monitoring Land Use and Land Cover*. Washington, DC: American Geophysical Union.

Loveland, T.R. and Belward, A.S. 1997a. The IGBP-DIS global 1 km land cover data set, DISCover: First results. *International Journal of Remote Sensing*, 18, 3291–3295.

Loveland, T.R. and Belward, A.S. 1997b. The International Geosphere Biosphere Programme Data and Information System global land cover data set (DISCover). *Acta Astronautica*, 41, 681–689.

Meyer, W.B. and Turner, B.L. 1992. Human-population growth and global land-use cover change. *Annual Review of Ecology and Systematics*, 23, 39–61.

Meyer, W.B. and Turner, B.L. 1994. *Changes in Land Use and Land Cover: A Global Perspective: Papers Arising from the 1991 OIES Global Change Institute*. Cambridge; New York: Cambridge University Press.

Moore, P.D. 1998. Climate change and the global harvest: Potential impacts of the greenhouse effect on agriculture. *Nature*, 393, 33–34.

Ramankutty, N. and Foley, J.A. 1998. Characterizing patterns of global land use: An analysis of global croplands data. *Global Biogeochemical Cycles*, 12, 667–685.

Townshend, J., Justice, C., Li, W., Gurney, C., and Mcmanus, J. 1991. Global land cover classification by remote-sensing—present capabilities and future possibilities. *Remote Sensing of Environment*, 35, 243–255.

Townshend, J.R., Latham, J., Justice, C.O., Janetos, A., Conant, R., Arino, O., Balstad, R., et al. (Eds.). 2011. *International Coordination of Satellite Land Observations: Integrated Observations of the Land* (pp. 835–856). New York: Springer.

Turner, B.L., Meyer, W.B., and Skole, D.L. 1994. Global land-use land-cover change—towards an integrated study. *Ambio*, 23, 91–95.

Xian, G., Homer, C., and Yang, L. (Eds.). 2011. Development of the USGS National Land-Cover Database over two decades. In: Weng, Q.H., ed., *Advances in Environmental Remote Sensing—Sensors, Algorithms, and Applications* (pp. 525–543). Boca Raton, FL: CRC Press.

2 History of Land-Cover Mapping

Thomas R. Loveland

CONTENTS

2.1 INTRODUCTION

The historical roots of land-cover mapping reside in the early history of aerial photography and applications spanning forestry, agriculture, urban planning, and water-resources management. Considering this long span of mapping, any attempt to provide an exhaustive treatment of the full history of land-cover mapping will necessarily be incomplete. For that reason, this chapter on the history of land-cover mapping emphasizes the "modern" era of land-cover mapping, which has been arbitrarily defined to begin in the early 1970s. This was when civil space-based remote sensing came of age, and intellectual efforts focused on strategies for using new observations in understanding the characteristics of, and the changes in, land use and land cover.

In an earlier perspective, Steiner (1965) provides an excellent summary of the state of land-use and land-cover mapping in the mid-1960s and identifies some of the pioneering work in land-cover mapping with aerial photos. Among the landmark efforts in the era of aerial photography is Marschner's (1958) "Major Land Uses in the United States" in which he used air photo index sheets to compile a map of general land-use types at a final scale of 1:5,000,000. Steiner also highlighted even earlier efforts such as the "Michigan Land Economic Survey" in which aerial photos were used to identify land uses needed to improve the conservation of previously cleared forests. The Michigan survey was initiated in the 1920s using field mapping, but aerial photos were used in the later phases of the survey (Foster, 1932). Other noteworthy early land-cover mapping examples include the "Land Use Categories in Pennsylvania," which was developed by the University of Pennsylvania Department of Geography (Klimm, 1958), and the Massachusetts Cooperative Wildlife Research Unit-led development, that is, the "Vegetative Cover Map of Massachusetts" using 1:20,000-scale aerial photographs (MacConnell and Garvin, 1956).

In this chapter, the history of land-cover mapping is reviewed for each of the four decades beginning with 1970. Each decade includes distinctive activity and emphasis, and subsequent decadal events build on the events of the previous decade. For example, the 1970s were the foundational period for space-based land-cover mapping, the 1980s saw the rapid growth of digital LC-mapping methods and projects, the 1990s represented the early stages in operational national to global land-cover mapping, and the first decade of the twenty-first century saw the maturation of operational mapping and a stronger emphasis on land-cover change studies.

2.2 THE DISTINCTION BETWEEN LAND-COVER MAPPING AND LAND-USE MAPPING

Before reviewing the modern history of land-cover mapping, it is important to note one contentious issue within the land-cover mapping community—the distinction between land cover and land use. Land cover refers to the vegetation and artificial constructions covering the land (Burley, 1961), and land use is the human activities on the land, which are directly related to the land (Clawson and Stewart, 1965). Both are clearly connected since changes in land use can change land cover, and changes in land cover can change land use. However, the connection is often complex since a given land use (e.g., grazing) may be associated with several different types of land cover (e.g., grassland and forestland), and a specific land cover (e.g., forestland) may have several different land uses (e.g., timber production, grazing, and recreation) (Loveland and DeFries, 2004).

Land-cover studies based on remote sensing often blur the distinction between land use and land cover and often interchange or mingle the terms. Land cover is often used as a surrogate for land use and vice versa. Some have attempted to clarify the differences between the two terms while rationalizing the necessary connections between them (Anderson et al., 1976), whereas others have concluded that the interchange of terms negatively affects the applications of both land-use and land-cover datasets (Comber, 2008).

Throughout the history of land-cover mapping using remote sensing, there has been an awkward but linked relationship between land use and land cover. Fundamentally, this is because some applications require land-cover data, whereas others need land-use inputs. Because land-cover and land use studies are relatively expensive, most datasets are designed to be multipurpose and attempt to satisfy both ends of the use spectrum. Remote sensing approaches are best suited for land cover investigations, but multispectral measures can provide context and patterns that can help understand and infer land use.

2.3 LAND-COVER MAPPING IN THE 1970s

Rapid maturation and growth in land-cover mapping capabilities was predicted in Steiner's mid-1960s characterization of the state of land-cover mapping (Steiner, 1965). Steiner noted that new forms of aerial imagery would be needed in the future to provide more detailed land-use information, and he suggested that infrared and color films and the application of multitype photography (e.g., multispectral) would become more commonplace. Steiner was correct.

The stage for the 1970s satellite era was set by two mid-1960 events. First, William Pecora, director of the U.S. Geological Survey (USGS), proposed the idea of a civilian remote-sensing satellite program to gather facts about the natural resources of the earth. Second, during the same time as Pecora's proposal, NASA initiated a series of remote-sensing investigations, using instruments mounted on an aircraft. Pecora's vision and NASA's growing interest in earth remote sensing resulted in NASA's launch of the Earth Resources Technology Satellite (later renamed Landsat) in July 1972. Landsats have operated continuously since 1972 and have been central to many land-cover mapping initiatives. Landsats 1–5 and 7 have acquired millions of images of the earth, which have been used in a wide range of scientific and operational applications. However, land-cover studies are a key driver of the Landsat mission. Landsats provide global, synoptic, and repetitive multispectral imagery coverage of the earth's land surfaces at a scale in which natural and human-induced changes can be detected, differentiated, characterized, and monitored over time.

In anticipation of the Landsat, in the early 1970s, NASA initiated a number of regional investigations using NASA research imagery (high-altitude aerial photography and multispectral scanner imagery) for regional studies of land-cover issues. Initiatives such as the Census Cities Project organized by NASA and the USGS were launched to test the viability of multispectral high-altitude photography for mapping urban lands within the 1970 census tracts (Barrett and Curtis, 1982). Regional studies, such as the Central Atlantic Regional Ecological Test Site (Alexander et al., 1975), were used to evaluate the potential for using a range of image sources for large-area land-cover assessments.

The early regional land-cover tests stimulated the establishment of a framework for using remotely sensed data for land-cover mapping. In 1972, Dr. James Anderson of the USGS and several colleagues introduced the first draft of what became the *de facto* standard for mapping land cover. The draft "A Land Use and Land Cover Classification System for Use with Remote Sensor Data" was published in final form in 1976 and provided a classification legend that defined land-use and land-cover categories that could be derived from remote-sensing sources (Anderson et al., 1976). The classification system was founded on a set of assumptions, of which three were particularly significant. First, the categories should permit vegetation and other types of land cover to be used as surrogates for land-use activity. The classes in the system generally corresponded to cover but included inferences to specific land uses. Second, the system was hierarchical with four levels, with each level designed for use with a specific scale or resolution of remotely sensed inputs (Table 2.1). The assumption is that the information at Levels I and II is of interest to users who need land-use and land-cover data for state and regional to national applications, whereas the data at Level III and IV apply to more localized places and regions. Anderson expected that Level III and IV categories would be defined by users to meet local requirements but that they could be aggregated to more general categories at Levels I and II for state to national reporting. Finally, the classes at each level of the hierarchy when mapped with the appropriate scale of remotely sensed data would have per-category interpretation accuracy of at least 85%. The 85% assumption became a *de facto* land-cover accuracy standard that is still used today.

The Anderson System was then applied to produce a national land-cover database often referred to as LUDA—Land Use Data Analysis (USGS 1990). Level II land cover was mapped using NASA high-altitude photography and other aerial sources, usually at a scale smaller than 1:60,000. The minimum mapping unit for developed classes was 4 ha, and the remaining classes were mapped at 16 ha. The maps were assembled by USGS 1:250,000 quadrangle and eventually digitized. The LUDA represented the first detailed land-cover map of the United States (Price et al., 2007).

Similar national mapping activities were carried out in other countries. For example, nationwide Mexico land-cover maps were also produced in the 1970s at a 1:250,000-scale using a classification system similar to that of Anderson (Velázquez et al., 2010). In Bolivia, the value of Landsats for

---

TABLE 2.1
Anderson's Classification System

Classification Level	Remote-Sensing Data	Example Classes
I	Landsat	1—Urban or built-up land
II	High-altitude aircraft data at 40,000 ft (12,400 m) or above (less than 1:80,000 scale)	11—Residential 12—Commercial and services 13—Industrial 14—Transportation, communications, and utilities 15—Industrial and commercial complexes 16—Mixed urban or built-up land 17—Other urban or built-up land
III	Medium-altitude aircraft data taken between 10,000 and 40,000 ft (3100 and 12,400 m) (1:20,000 to 1:80,000 scale)	TBD—User-defined subdivisions of Level II classes
IV	Low-altitude aircraft data taken below 10,000 ft (3100 m) (more than 1:20,000 scale)	TBD—User-defined subdivisions of Level III classes

land-use and land-cover mapping was discovered unexpectedly from a Bolivia Ministerio de Miner ia y Metalurgia (1977) study of geological resources.

The activities of the 1970s, most notably the launch of Landsat 1, 2, and 3, as well as the establishment of a framework for land-cover mapping, serve as the foundation for today's land-cover mapping. Though the early efforts were based on manual rather than computer-assisted methods, the basic tenets of the land-cover mapping of the 1970s continued through the remainder of the century.

2.4 LAND-COVER MAPPING IN THE 1980s

Three significant trends dominated the aforementioned decade—the development and acceptance of computer-assisted land-cover mapping techniques, the growth of land-cover mapping initiatives across the United States and other parts of the world, and the improvement in Landsat data quality due to Thematic Mapper instrument. Decisions to commercialize the Landsat data also had an impact on land-cover initiatives in the 1980s and 1990s.

The advantages and disadvantages of computer-assisted methods for land-cover mapping were relatively well known owing to the early pioneering research and development that took place at Purdue University's Laboratory for Applications of Remote Sensing (LARS) and other remote-sensing centers. Anderson of the USGS recognized the need to improve the level of spatial detail obtained from Landsats and gain efficiencies through the analysis of digital imagery. However, considering his team's initial research on digital land-cover classification, he was skeptical that sufficiently accurate land-cover maps could be produced using Landsat MSS data and automated classification methods owing to the complexity of the landscape in both urban and rural settings (Anderson, 1976). Methodological advances, improved Landsat imagery, access to relatively sophisticated image-processing software, and a growing cadre of remote-sensing experts, however, contributed to the acceptance of the computer approaches.

Intellectual contributions by Robinove (1981) helped frame the physical relationships between land-cover surface properties and electromagnetic physics, and advances in multispectral classification methods such as Bryant (1989) and Landgrebe (1980) spatial-spectral classification algorithm and texture-based classification algorithms by Swain et al. (1981) improved image processing. Better understanding of the supervised and unsupervised classification strategies allowed analysts to make intelligent choices regarding classification strategies (Fleming et al., 1975; Justice and Townshend, 1981) and to recognize the role of ancillary data to improve land-cover mapping accuracy (Hutchinson, 1982; Strahler et al., 1978). Software systems such as LARSYS developed by Purdue University (Lindenlaub, 1973), ELAS—Earth Resources Land Analysis System—developed by NASA (Stennis, 1989), the Land Analysis System developed by NASA Goddard and the USGS (Wharton et al., 1988), and VICAR-IBIS—Video Image Communication and Retrieval/ Image-Based Information System—developed at NASA's Jet Propulsion Lab enabled the processing of digital Landsat imagery to create land-cover products (Bryant and Zobrist, 1982). Sophisticated, integrated commercial systems such as the Interactive Digital Image Manipulation System (IDIMS) (Fleming, 1981) and the Earth Resources Data Analysis System (ERDAS) also contributed to the maturing of land-cover mapping (ERDAS, 1994).

Several statewide land-cover mapping programs were initiated during the late 1970s and 1980s in Arizona, Kansas, Nebraska, North Dakota, South Dakota, and Texas—see Cornwell's (1982) review of state land-cover and geographic information system (GIS) activities. Most states used computer-assisted analysis of Landsat data state programs that typically incorporated land-cover mapping functions within larger GIS offices. Although most were envisioned to provide ongoing mapping and monitoring of change, the survival rate of state land-cover initiatives was low owing to the expenses and the complexity of the mapping activities. The commercialization of the Landsat and the subsequent higher prices had a particularly negative impact on state mapping programs (Lamm, 1980).

Land-cover mapping received a significant boost when Landsats 4 and 5 were launched in 1982 and 1984, respectively. A new sensor, the Thematic Mapper (TM), offered improved spatial and multispectral capabilities that had major advantages for land-cover mapping and monitoring. The improved ground resolution (from 79 m by 79 m to 30 m by 30 m) and the addition of short-wave infrared spectral measurements increased the ability to identify land cover in complex settings and detect vegetation conditions that better correlated to land-cover features of interest. Similarly, the French Satellite Pour l'Observation de la Terre (SPOT) mission with its high resolution visible (HRV) instrument also provided high-resolution data (10 m and 20 m pixels) and multispectral content that helped land-cover mapping studies. The advantages of the TM sensor are lasting, and there is a widespread acceptance of the value of 30-m resolution and of the TM instrument specifications for land-cover investigations (Wulder et al., 2008).

While Landsat and SPOT data were the mainstay sources for land-cover applications, Tucker et al. (1985) provided evidence of the value of coarse-resolution space-based imagery with their use of advanced very high resolution radiometer (AVHRR) data to map land cover for the African continent. Their research has long-term implications for the next 20 years of land-cover mapping.

2.5   LAND-COVER MAPPING IN THE 1990s

The aforementioned decade can be best characterized as the start of the large-area operational era for land-cover mapping. AVHRR-based land-cover projects ushered in the global land-cover mapping era, and the end of the commercial era contributed to the growth of national-scale Landsat land-cover activities.

The merits of AVHRR for large-area land-cover characterization, which were clearly demonstrated by Tucker et al. (1985), spawned land-cover investigations at national scales, such as Frederiksen and Lawesson's (1992) Senegal study, Gaston et al.'s (1994) mapping of the land cover of former Soviet Union, Cihlar et al.'s (1996) study of Canada, Zhu and Evan's (1994) assessment of the forest cover of the United States, and Loveland et al.'s (1995) characterization of the land cover of the United States. Tateishi and Kajiwara's (1991) Asia land-cover demonstration, Stone et al.'s (1994) land-cover map of South America, and Achard and Estreguil's (1995) study of the land cover of Southeast Asia provided evidence of the potential for mapping land cover with AVHRR data for multinational areas. These studies used AVHRR data at different resolutions and formats. Some used single data scenes, whereas others used multiple maximum-greenness composites.

Those and other studies set the stage for the first global land-cover products—all developed from AVHRR data. Initially, DeFries et al. (1995) used a series of seasonal metrics (e.g., length of the growing season) to produce a global land-cover map with 1° by 1° resolution. DeFries et al. (1998) later improved on that map with a global land-cover dataset with 8-km resolution. In response to the needs of the International Geosphere-Biosphere Program, Loveland et al. (1999) generated the first 1-km global land-cover map. The IGBP map also represented the first global product with accuracy validated using a statistical sampling design (Scepan, 1999).

Collectively, the AVHRR studies demonstrated several innovations and advantages. The use of seasonal and annual time series datasets, including derived seasonal metrics, based on the normalized difference vegetation index (NDVI), added significant information content that permitted overcoming the limitations of coarse-resolution inputs. The use of seamless datasets reduced the impacts of scene boundaries. The role of ancillary data and stratification to improve classification accuracy, as well as more sophiscated classification strategies emerged from these global studies.

In parallel with the global AVHRR studies, large-area land-cover initiatives based on high-resolution imagery, such as Landsat and SPOT, also grew during the 1990s. In the United States, a consortium of Federal agencies organized the Multi-Resolution Land Characterization (MRLC) consortium to facilitate the development of land-cover products needed by their respective agencies (Loveland and Shaw, 1996). Initially motivated by the need to pool resources to acquire still-commercial Landsat data, the MRLC group contributed to expanding national land-cover mapping capabilities as well as to reducing duplication and increasing product consistency between programs with land-cover data needs. An outcome of the MRLC Landsat data purchase was the development of the USGS National Land Cover

Database (Vogelmann et al., 1998) and Gap Analysis Project natural vegetation map (see Scott et al. [1993] for a program description) and the National Oceanic and Atmospheric Administration (NOAA) Coastal Change Analysis Project coastal land-cover change dataset (Dobson et al., 1995). Although these projects had different goals and mapping objectives, they were all based on the recognition of the value of collaboration, use of common standards, and continuing innovation.

Landsat-class national-scale land-cover initiatives continue to grow around the world. The European Union CORINE (Coordination of Information on the Environment) land-cover program represents a comprehensive approach to providing ongoing land-cover products for most of the European Union (Bossard et al., 2000). Landsat and SPOT data were used to map detailed land-cover categories. CORINE is updated on a regular cycle and now includes over 38 European countries.

The United Nations Food and Agriculture (FAO) Africover project was started late in this decade (Africover, 2002). The goal of the Africover project was to provide consistent, high-resolution land cover for all areas of Africa. The activity was based on the manual interpretation of Landsat and other similar resolution data, and country maps were developed using in-country teams. Africover stresses capacity building and improving national and subregional capabilities for establishing, updating, and using Africover and cover maps and databases.

A key element of Africover is the Land-Cover Classification System (LCCS) developed by Di Gregorio and Jansen (2000). LCCS represents the first significant land-cover legend advance since the Anderson system in the 1970s. LCCS is an *a priori* classification system that provides the flexibility to meet unique user requirements while maintaining consistency in language and definitions. The system uses a set of independent criteria that allow correlation with existing classifications and legends. With LCCS, land covers are defined by sets of diagnostic criteria. By standardizing language and definitions, LCCS provided a powerful tool for developing flexible, yet consistent and comparable land-cover products. LCCS has evolved into an international standard and is now used around the world.

The land-cover accomplishments of the 1990s arguably exceeded the advances of the earlier decades. Technical advances in classification methods, including artificial neural networks (Hepner et al., 1990; Gopal and Woodcock, 1994) and regression and decision trees (Friedl and Brodley, 1997; DeFries et al., 1998; Hansen et al., 2000) improved the accuracy and repeatability of land cover mapping in projects spanning all scales and geographic venues. The availability of data from new missions (e.g., India Remote Sensing Satellite and others) and the end of the commercial Landsat era were also important factors in ensuring investments in land-cover programs and enabling innovation.

2.6 LAND-COVER MAPPING IN THE 2000s

Large-area land-cover mapping continued to mature with more innovative and quantitative land-cover product characteristics, emphasis on change, methodological advances, and significant growth in quasi-operational land-cover mapping programs. Global land-cover monitoring matured owing to the launch of the NASA Terra and Aqua satellites with the Moderate Resolution Imaging Spectroradiometer (MODIS). In addition, the availability of other coarse-resolution imagery from the SPOT Vegetation instrument and the European MERIS (Medium-Spectral Resolution Imaging Spectrometer) onboard ENVISAT also were important sources of remotely sensed data for global land-cover mapping.

As part of the NASA MODIS Land (MODLand) team activity, Friedl et al. (2002) developed a method to periodically produce 500-m resolution land cover based on the IGBP classification system and a supervised classification approach. An additional MODIS land-cover product, vegetation continuous fields, was developed, which estimates basic land-cover fractions, including forest, grassland, and bare ground (Hansen et al., 2002), at 500-m subpixel level. Both MODIS land-cover products continue to be updated on a cyclic schedule in order to contribute to the studies that address global land-cover dynamics.

The European Space Agency (ESA) sponsored two global mapping projects during this decade. The initial ESA global land-cover project was Global Land Cover 2000 (GLC2000), which was led by the European Commission's Joint Research Centre but which involved nearly 30 other groups (Bartholomé and Belward, 2005). The land-cover map is based on daily data from the VEGETATION sensor on-board SPOT 4. This effort produced both detailed, regionally optimized land-cover legends for each continent and a less thematically detailed global product. The follow-on to GLC2000, Globcover 2006, uses 300-m MERIS to create global products consistent with the FAO LCCS (Arino et al., 2007). This dataset represents the highest resolution global land-cover product currently available.

Landsat-scale national land-cover initiatives continue and keep improving. The USGS NLCD continues to be updated, and additional land-cover data layers have been added, including imperviousness and tree cover continuous fields (Homer et al., 2007) and land-cover change layers (Xian et al., 2009). The European Union CORINE Project is continuing, as is the FAO Africover activity. The FAO has also initiated a similar project—Asiacover—that provides similar land-cover capabilities for Southeast Asian countries. Canada mapped 2000-era forest cover using Landsat data and produced the most detailed nationwide forest cover map ever (Wulder et al., 2003), and the Mexican Instituto Nacional de Estadística, Geografía, e Informatics (INEGI) used Landsat data to map 2002–2005 land cover across the nation. Increasingly, land-cover programs are moving beyond baseline mapping and are focusing on change analysis.

The establishment of accuracy standards for land-cover products has matured significantly. Although the early USGS LUDA land-cover products included accuracy assessments, generally, formal accuracy assessment of large-area land-cover maps was less common in the earlier decades. Because of research in the 1990s—see, for example, Congalton (1991, 2001), Stehman (1999), Stehman and Czaplewski (1998), and Foody (2002)—more land-cover projects include validation as a standard practice. The international community has recently published standards for global land-cover accuracy assessments (Strahler et al., 2006).

2.7   LOOKING AHEAD

It is not appropriate to write the future history of land cover, but it is clear that history has set the stage for upcoming innovations. A significant boost in these land cover mapping capabilities has come from the USGS Landsat Data Policy decision in 2008, which makes all USGS-managed global Landsat data available at no cost to users via the Internet. By eliminating the relatively high cost of Landsat data, studies spanning longer temporal periods and covering larger geographic areas are possible, funds are freed to expand or improve land-cover study capabilities, and new methodological innovations are possible. Because land-cover analysts have access now to all of the data they need rather than being restricted to the data they can afford to, advances in multitemporal analysis should result in improved land-cover products and broader application of these products. Global land-cover initiatives based on Landsat are now being planned (Stone, 2010).

The second significant trend is the shift from baseline land-cover mapping to land-cover change mapping and monitoring. To better understand environmental dynamics and the impacts of land change on natural and human systems, land-cover change data are critical. A clear need, methodological improvements, and better access to appropriate remotely sensed data are driving this emphasis.

REFERENCES

Achard, F. and Estreguil, C. 1995. Forest classification of Southeast Asia using NOAA AVHRR data. *Remote Sensing of Environment*, 54, 198–208.

Africover. 2002. Africover--Eastern Africa module, land cover mapping based on satellite remote sensing. Rome: Food and Agriculture Organization of the United Nations. http://www.sciencedirect.com/science/article/pii/S0016718507000528; www.africover.org/download/documents/Short_Project_description_en.pdf.

Alexander, R.H., Fitzpatrick, K., Lins, H.F., Jr., and McGinty, H.K., III. 1975. Land use and environmental assessment in the central Atlantic region. In *NASA Earth Resources Survey Symposium*, 1-C (pp. 1683–1727), NASA Johnson Space Center.

Anderson, J.R., Hardy, E.E., and Roach, J.T. 1972. A land-use classification system for use with remote-sensor data. Reston VA, U.S. Geological Survey Circular 671, 16 p.

Anderson, J.R. 1976. Land use and land cover map and data compilation in the U.S. Geological Survey. In *Proceedings, Second Annual Pecora Memorial Symposium: Mapping with Remote Sensing Data*, (pp. 2–12), Falls Church, VA: American Society of Photogrammetry.

Anderson, J.R., Hardy, E.E., Roach J.T., and Witmer R.E. 1976. A land use and land cover classification system for use with remote sensor data. Reston VA, U.S. Geological Survey Professional Paper 964, 28 p.

Arino, O., Gross, D., Ranera, F., Bourg, L., Leroy, M., Bicheron, P., Latham, J., et al. 2007. GlobCover: ESA service for global land cover from MERIS. *Proceedings IEEE Geoscience and Remote Sensing Symposium*, 2412–2415.

Barrett, E.C. and Curtis, L.F. 1982. *Introduction to Environmental Remote Sensing*, 3rd edition. London: Chapman and Hall.

Bartholomé, E. and Belward, A.S. 2005. GLC2000: A new approach to global land cover mapping from Earth observation data. *International Journal of Remote Sensing*, 26(9), 1959–1977.

Bolivia Ministerio de Miner ey Metalurgia, Servicio Geologico de Bolivia. 1977. Programa del Satilite Tecnologico de Recursos Naturales ERTS—Bolivia: Procesamien to digital de datos multiespectrdes proyecto experimental. *La Paz, GEOBOL*, 1, 58.

Bossard, M., Feranec, J., and Otahel, J. 2000. CORINE land cover technical guide, addendum 2000. Copenhagen, European Environment Agency Technical Report N. 40, 105 p.

Bryant, J. 1989. A fast classifier for image data. *Pattern Recognition* 22 45–48.

Bryant, N.A. and Zobrist, A.L. 1982. Some technical considerations on the evolution of the IBIS system (Image Based Information System). *Proceedings, Seventh Pecora Symposium*, Sioux Falls, SD, (pp. 465–475).

Burley, T.M. 1961. Land use or land utilization. *Professional Geographer*, 13, 18–20.

Cihlar, J., Ly, H., and Xiao, Q. 1996. Land cover classification with AVHRR multichannel composites in northern environments. *Remote Sensing of Environment*, 58, 36–51.

Clawson, M. and Stewart, C.L. 1965. *Land Use Information: A Critical Survey of U.S. Statistics Including Possibilities for Greater Uniformity*. Baltimore, MD: The Johns Hopkins Press for Resources for the Future, 402 p.

Comber, A.J. 2008. Land use or land cover? *Journal of Land Use Science*, 3(4), 199–201.

Congalton, R.G. 1991. A review of assessing the accuracy of classifications of remotely sensed data. *Remote Sensing of Environment*, 37, 35–46.

Congalton, R. 2001. Accuracy assessment and validation of remotely sensed and other spatial information. *International Journal of Wildland Fire,* 10, 321–328.

Cornwell, S.B. 1982. History and status of state natural resource systems. *Computers, Environment and Urban Systems*, 7(4), 253–260.

DeFries, R., Hansen, M., and Townshend, J. 1995. Global discrimination of land cover types from metrics derived from AVHRR pathfinder data. *Remote Sensing of Environment*, 54, 209–222.

DeFries, R.S., Hansen, M., Townshend, J.R.G., and Sohlberg, R. 1998. Global land cover classifications at 8 km spatial resolution: The use of training data derived from Landsat imagery in decision tree classifiers. *International Journal of Remote Sensing*, 19(16), 3141–3168.

Di Gregorio, A. and Jansen, L.J.M. 2000. *Land Cover Classification System: Classification Concepts and User Manual*. Rome, Italy: UN FAO, 179 p.

Dobson, J.E., Bright, E.A., Ferguson, R.L., Field, D.W., Wood, L.L., Haddad, K.D., Iredale, H., et al. 1995. NOAA Coastal Change Analysis Program (CCAP): Guidance for regional implementation. Seattle, WA: U.S. Department of Commerce, NOAA Technical Report NMFS 123.

ERDAS. 1994. *ERDAS Field Guide*. Atlanta, GA: ERDAS Inc. 628 p.

Fleming, M.D. 1981. Interactive digital image manipulation system (IDIMS). In: *Proceedings, NASA Ames Research Center Western Regional Remote Sensing Conference* (SEE N82-22546), (pp. 13–43, 160–162).

Fleming, M.D., Berkebile, J.S., and Hoffer, R.M. 1975. *Computer-Aided Analysis of LANDSAT-1 MSS Data: A Comparison of Three Approaches, Including a "Modified Clustering Approach," LARS Technical Reports. Paper 96*. Richmond, IN: Purdue University.

Foody, G.M. 2002. Status of land cover classification accuracy assessment. *Remote Sensing of Environment*, 80(1), 185–201.

Foster, Z.C. 1932. The use of aerial photographs in the Michigan Land Economic Survey. *Bulletin of the American Soil Survey Association*, 13, 86–88.

Frederiksen, P. and Lawesson, J.E. 1992. Vegetation types and patterns in Senegal based on multivariate analysis of field and NOAA-AVHRR satellite data. *Journal of Vegetation Science*, 3, 535–544.

Friedl, M.A. and Brodley, C.E. 1997. Decision tree classification of land cover from remotely sensed data. *Remote Sensing of Environment*, 61(3), 399–409.

Friedl, M.A., McIver, D.K., Hodges, J.C.F., Zhang, X.Y., Muchoney, D., Strahler, A.H., Woodcock, C.E., et al. 2002. Global land cover mapping from MODIS: Algorithms and early results. *Remote Sensing of Environment*, 83, 287–302.

Gaston, G.G., Jackson, P.L., Vinson, T.S., Kolchugina, T.P., Botch, M., and Kobak, K. 1994. Identification of carbon quantifiable regions in the former Soviet Union using unsupervised classification of AVHRR global vegetation index images. *International Journal of Remote Sensing*, 15(16), 3199–3221.

Gopal, S. and Woodcock, C.E. 1994. Theory and methods for accuracy assessment of thematic maps using fuzzy sets. Photogrammetric Engineering and Remote Sensing 60(2), 181–188.

Hansen, M.C., DeFries, R.S., Townshend, J.R.G., Sohlberg, R., Dimiceli, C., and Carroll, M., 2002. Towards an operational MODIS continuous field of percent tree cover algorithm: Examples using AVHRR and MODIS data. *Remote Sensing of Environment*, 83, 303–319.

Hansen, M.C., DeFries, R.S., Townshend, J.R.G., and Sohlberg, R. 2000. Global land cover classification at 1 km spatial resolution using a classification tree approach. *International Journal of Remote Sensing*, 21(6– 7), 1331–1364.

Hepner, G.F., Logan, T., Ritter, N., and Bryant, N. 1990. Artificial neural network classification using a minimal training set: Comparison to conventional supervised classification. *Photogrammetric Engineering and Remote Sensing*, 56(4), 469–473.

Homer, C., Dewitz, J., Fry, J., Coan, M., Hossain, N., Larson, C., Herold, N., McKerrow, A., VanDriel J.N., and Wickham J. 2007. Completion of the 2001 national land cover database for the conterminous United States. *Photogrammetric Engineering and Remote Sensing*, 73(4), 337–341.

Hutchinson, C.F. 1982. Techniques for combining Landsat and ancillary data for digital classification improvement. *Photogrammetric Engineering and Remote Sensing*, 48, 123–130.

Justice, C.O. and Townshend, J.R.G. 1981. A comparison of unsupervised classification procedures of Landsat MSS data for an area of complex surface conditions in Basilicata, S. Italy. *Remote Sensing of the Environment*, 12, 407–420.

Klimm, L.E. 1958. Description of a land use map of Pennsylvania. Technical Report 2. Philadelphia, PA.

Lamm, R.D. 1980. Recommendations of the National Governor's Association, National Conference of State Legislatures, Intergovernmental Science, Engineering and Technology Advisory Panel, National Resources and Environment Task Force, for the final transition plan for the National Civil Operating Remote Sensing Program. Denver Colorado, Office of the Governor.

Landgrebe, D.A. 1980. The development of a spectral-spatial classifier for Earth observational data. *Pattern Recognition*, 12(3), 165–175.

Lindenlaub, J.C. 1973. *Guide to multispectral data analysis using LARSYS. LARS Information Note 062873*, Purdue University.

Loveland, T.R. and DeFries, R. 2004. Observing and monitoring land use and land cover change. In R. DeFries, G. Asner, and R. Houghton (Eds.), *Ecosystem Interactions and Land Use Change* (pp. 231–248). Washington, DC: AGU Publications Geophysical Monograph 153.

Loveland, T.R., Merchant, J.W., Reed, B.C., Brown, J.F., and Ohlen, D.O. 1995. Seasonal land cover regions of the United States. *Annals of the Association of American Geographers*, 85(2), 339–355.

Loveland, T.R. and Shaw, D.M. 1996. Multiresolution land characterization: Building collaborative partnerships. In T. Tear and M. Scott (Eds.), *Gap Analysis, a Landscape Approach to Biodiversity Planning* (pp. 17–25). Bethesda, MD: American Society of Photogrammetry and Remote Sensing.

Loveland, T.R., Zhu, Z., Ohlen, D.O., Brown, J.F., Reed, B.C., and Yang, L. 1999. An analysis of the global land cover characterization process. *Photogrammetric Engineering and Remote Sensing*, 65(9), 1021–1032.

MacConnell, W.P. and Garvin, L.E. 1956. Cover mapping a state from aerial photographs. *Photogrammetric Engineering*, 22(4), 702–707.

Marschner, F.J. 1958. *Land Use and Its Patterns in the United States*. Washington, D.C.: U.S. Department of Agriculture, Agricultural Handbook, 153–277.

NASA 1989. *ELAS User Reference Manual*, v. 2. NASA John C. Stennis Space Center, Science and Technology Laboratory, Report 18, 126 p.

Price, C.V., Nakagaki, N., Hitt, K.J., and Clawges, R.M. 2007. Enhanced historical land-use and land-cover data sets of the U.S. Geological Survey.

Reston, VA: U.S. Geological Survey, Online Data Series 240. http://pubs.usgs.gov/ds/2006/240/.

Robinove, C.J. 1981. The logic of multispectral classification and mapping of land. *Remote Sensing of Environment*, 11, 231–244.

Scepan, J. 1999. Thematic validation of high-resolution global land-cover data sets. *Photogrammetric Engineering and Remote Sensing*, 65(9), 1051–1060.

Scott, J.M., Davis, F., Csuti, B., Noss, R., Butterfield, B., Groves, C., Anderson, H., et al. 1993. Gap analysis: A geographic approach to protection of biological diversity. *Journal of Wildlife Management Wildlife Monographs*, 123 p.

Stehman, S.V. 1999. Basic probability sampling designs for thematic map accuracy assessment. *International Journal of Remote Sensing*, 20(12), 2423–2441.

Stehman, S.V. and Czaplewski, R.L. 1998. Design and analysis for thematic map accuracy assessment: Fundamental principles. *Remote Sensing of Environment*, 64(3), 331–344.

Steiner, D. 1965. Use of air photographs for interpreting and mapping rural land use in the United States. *Photogrammetria*, 20(2), 65–80.

Stone, R. 2010. Earth-observing summit endorses global data sharing. *Science*, 330, 902.

Stone, T.A., Schlesinger, P., Houghton, R.A., and Woodwell, G.M. 1994. A map of the vegetation of South America based on satellite imagery. *Photogrametry Engineering and Remote Sensing*, 60, 541–551.

Strahler, A.H., Logan, T.L., and Bryant, N.A. 1978. Improving forest cover classification accuracy from Landsat by incorporating topographic information. In *Proceedings of 12th International Symposium on Remote Sensing of Environment*, 20–26 April, Manila, Philippines, (pp. 927–942).

Strahler, A.H., Boschetti, L., Foody, G.M., Friedl, M.A., Hansen, M.C., Herold, M., Mayaux, P., Morisette, J.T., Stehman, S.V., and Woodcock, C.E. 2006. Global land cover validation: Recommendations for evaluation and accuracy assessment of global land cover maps. Edmonton, Alberta Canada, GOFC-GOLD Report No. 25, 60 p.

Swain, P.H., Vardeman, S.B., and Tilton, J.C. 1981. Contextual classification of multispectral image data. *Pattern Recognition*, 13(6), 429–441.

Tateishi, R. and Kajiwara, K. 1991. Land cover monitoring in Asia by NOAA GVI data. *Geocarto International*, 6(4), 53–64.

Tucker, C.J., Townshend, J.R.G., and Goff, T.E. 1985. African land cover classification using satellite data. *Science*, 227, 369–375.

USGS. 1990. Land use and land cover digital data from 1:250,000- and 1:100,000-scale maps. Reston, VA: U.S. Geological Survey, Data user guide 4, 25 p.

Velázquez, A., Mas, J.-F., Bocco, G., and Palacio-Prieto, J.L. 2010. Mapping land cover changes in Mexico, 1976–2000 and applications for guiding environmental management policy. *Singapore Journal of Tropical Geography*, 31(2), 152–162.

Vogelmann, J.E., Sohl, T., and Howard, S.M. 1998. Regional characterization of land cover using multiple sources of data. *Photogrammetric Engineering and Remote Sensing*, 64(1), 45–47.

Wharton, S.W., Lu, Y.-C., Quirk, B.K., Oleson, L.R., Newcomer, J.A., and Irani, F.M. 1988. The land analysis system (LAS) for multispectral image processing. *IEEE Transactions on Geoscience and Remote Sensing*, 26(5), 693–697.

Wulder, M.A., Dechka, J.A., Gillis, M.A., Luther, J.E., Hall, R.J., Beaudoin, A., and Franklin, S.E. 2003. Operational mapping of the land cover of the forested area of Canada with Landsat data: EOSD land cover program. *Forestry Chronicle*, 79(6), 1075–1083.

Wulder, M.A., White, J.C., Goward, S.N., Masek, J.G., Irons, J.R., Herold, M., Cohen, W.B., Loveland, T.R., and Woodcock, C.E. 2008. Landsat continuity: Issues and opportunities for land cover monitoring. *Remote Sensing of Environment*, 112, 955–969.

Xian, G. Homer, C. and Fry, J. 2009. Updating the 2001 National Land Cover Database land cover classification to 2006 by using Landsat imagery change detection methods. *Remote Sensing of Environment*, 113, 1133–1147.

Zhu, Z. and Evans, D.L. 1994. U.S. forest types and predicted percent forest cover from AVHRR data. *Photogrammetric Engineering and Remote Sensing*, 60(5), 525–531.

Section II

Basic Principles

3 Semantic Issues in Land-Cover Analysis
Representation, Analysis, and Visualization

Ola Ahlqvist

CONTENTS

3.1 INTRODUCTION

The use of categorical data in computer-based land analysis is a significant challenge because it usually leads to a binary treatment of the information in a subsequent analysis. Cognitive science suggests that humans need categorical data to process experiences, form memories, analyze, or summarize and communicate knowledge (Lakoff, 1987; Rosch, 1978). Similar reasons underlie the common practice of measuring and storing land-cover information as categorical data. We find it intuitive to talk about "forest cover," "grassland," and "sand dunes," but despite the inherently experiential and subjective nature of these terms, we are able to effectively communicate ideas using them.

Land-cover data also serve as a rich and generic resource as they are often used for purposes other than just finding out what the land cover is at a location; examples are climate modeling, monitoring of biodiversity, and simulation of urban expansion. Many of these uses call for a deeper understanding of the categories to repurpose the data. As more and more land-cover datasets have been developed, there is greater recognition that variation in nomenclature and class definitions poses significant hurdles to effective and synergistic use of these resources. A frequently proposed solution to these issues, and one of the recurring themes in land-use/land-cover monitoring initiatives, is the effort to harmonize classification systems for landscape analysis. The idea is that the use of standardized taxonomies will create homogeneous information sources that can be merged across space into comprehensive datasets with regional, national, or global coverage, which will also make possible comparisons over time. Some examples of datasets that use standardized nomenclatures are the CORINE Land Cover (CEC, 1995), AFRICOVER (Kalensky, 1998), and Global Land Cover 2000 (Bartholomé and Belward, 2005). Despite the availability of standardized classification systems, problems of category semantics have lingered in remote-sensing literature for a long time (Fisher and Pathirana, 1990; Gopal and Woodcock, 1994; Robbins, 2001).

Engineering and other domains have used formal ontologies to address category heterogeneity, and about 10 years ago, ontology began to be suggested as a way of addressing taxonomy heterogeneity and improving geographic data interoperability (Fonseca et al., 2002). There are now several examples of promising land-cover ontology-matching methods. For example, Kavouras and Kokla (2002b) used formal concept lattices to integrate the European CORINE land-cover taxonomy with a Greek National Cadastre Land Use Classification system, and later Kavouras et al. (2005) used a syntactic analysis of natural language definitions to compare CORINE land-cover categories with MEGRIN's hydrology categories. These efforts have provided important building blocks for addressing semantic heterogeneity in harmonizing land-cover data. However, many implementations of ontology use a formal logic founded on crisp representations of objects and relations. This is somewhat surprising since land-cover classes in different taxonomies often only partially correspond rather than have direct, one-to-one matches. This noncrisp nature of mental categories is well known in the cognitive sciences:

> …The gradation of properties in the world means that our smallish number of categories will never map perfectly onto all objects: The distinction between member and nonmembers will always be difficult to draw or will even be arbitrary in some cases […]; if the world consists of shadings and gradations and a rich mixture of different kinds of properties, then a limited number of concepts would almost have to be fuzzy. (Murphy, 2004)

The graded and fuzzy nature of land-cover categories has been recognized for a long time by the remote-sensing community (cf. Foody, 2002; Gopal and Woodcock, 1994), but no overarching framework for incorporating semantic uncertainty in land-cover studies has been proposed. The following sections aim at outlining such a framework by summarizing existing studies on the representation and analysis of the semantics of land-use and land-cover categories.

3.2   REPRESENTATION

The issue of representing semantic information about categories in general is an active research area. In geographic information sciences, notable contributions to this research started to emerge in the late 1990s (Bishr, 1998; Harvey et al., 1999; Rodriguez et al., 1999), followed by a surge of work in the past decade. Useful summaries can be found in the works of Agarwal (2005) and Schwering (2008).

Philosophy and science have primarily defined categories using summary definitions such that every object is either part of a category or not, and all members of a category are equally good examples of it. This "classical view" (Murphy, 2004) of categories forms the basis for many common knowledge-representation theories and logic in use today (see Sowa [2000] for an overview). Nevertheless, many deem the classical view to be insufficient to deal with the several semantically imprecise and vague notions that are so pervasive in geography (Bennett, 2001; Couclelis, 1992; Fisher, 2000; Fisher and Wood, 1998).

Three main theories have replaced the classical view on concepts: prototype, exemplar, and knowledge theories (Murphy, 2004). All three accept that categories will have gradations of typicality and that there will be borderline cases. Although this is intuitive and familiar to most people who have worked with land-cover and land-use data, this is a big step away from how geographic information systems were designed to model real-world concepts and objects in a spatial database. Several methods now proposed to formally represent and handle category gradations in geographic data can be found in the works of Ahlqvist (2004), Comber et al. (2004), Feng and Flewelling (2004), and Kavouras et al. (2005). From these and other examples, the emerging picture seems to be that we cannot expect to see one general representation of category semantics but more likely a collection of complementary semantic assessment frameworks. Recently, Schwering (2008) reviewed five different models to represent and measure semantic similarity: geometric, feature, network, alignment, and transformational. Of the five models, the geometric, feature, and network models have generated most interest in the GIScience literature. Some example formalizations can be found in the works of Ahlqvist (2004), Feng and Flewelling (2004), Kavouras and Kokla (2002b), Rodriguez and Egenhofer (2003), and Song and Bruza (2003).

Geometric and feature models use a collection of characteristics to define a category. A geometric model defines a characteristic as a value along some attribute dimension, whereas the feature model uses a list of Boolean characteristics. As an illustrative example, the IGBP-DIS land-cover classification scheme (Loveland and Belward, 1997) characterizes forest classes as lands dominated (>60% cover) by woody vegetation. The geometric model can represent this by defining a dimension called "percent vegetation cover" and specify the corresponding interval, 60%–100%, that characterizes the forest classes. A feature model can instead add "forest dominated" to a list of characteristic features for the forest classes (Figure 3.1).

The evaluation of similarity will then be based on comparing vegetation cover for two objects or classes of interest. In the geometric case, the 60%–100% interval can be compared, for example, with the 10%–100% crown closure criterion for forestlands in the widely used USGS (Anderson, 1976) classification system. Using some form of interval or other difference-based metric will give a quantitative estimate of the semantic similarity. In the feature model, the object of interest would be compared to the criteria of being "forest dominated." If that binary evaluation comes out true, it will indicate semantic similarity. One important problem with these models is the need to identify a common set of characteristic features/dimensions. In the above example, the descriptive features may be hard to reconcile; for example, if the USGS formalization uses a feature called *closed forest cover*, it is not clear how that will be matched with the "forest dominated" feature selected for the IGBP class. The proposed solution to this issue is to seek a similar set of descriptive characteristics (Di Gregorio, 2004; Jansen and Di Gregorio, 2002).

Network models focus on evaluating semantic relationships between categories in an existing taxonomy or other types of networks made up of links and nodes, where the nodes represent concepts, objects, or properties and the links represent some form of a relationship. Arguing that the knowledge embedded in such concept networks can be the basis for a similarity assessment, Rada et al. (1989) developed various distance metrics to measure "conceptual distance" between 15,000 biomedical categories such as "anatomy," "organism," and "disease." Although this line of semantic evaluations has garnered significant attention in many fields, the direct application of this evaluation is problematic in the land-cover domain mainly because of the relatively small size of land-cover hierarchies. For examples of works that have used a network representation as a foundation, refer to Kavouras and Kokla (2002a) and Rodriguez and Egenhofer (2003).

While there are significant differences in these methods to represent category semantics, it is important to remember that just recognizing similarity is of limited value. A semantic similarity assessment is usually only a first step in some type of targeted analysis. Land-cover data analysis is often of a spatiotemporal nature where we may be interested, for example, in land-cover change over time, pattern analysis across space, or accuracy assessment for descriptions of data quality. For these and many other questions, we can use semantic knowledge and derived similarity metrics to quantitatively evaluate the similarity between any two land-cover classes. In this manner, nominal land-cover data, which can otherwise be restrictive in terms of possible analysis methods, can apply numerical methods through the semantic similarity metrics. This opens a possibility for

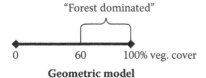

| Geometric model | Feature model |

FIGURE 3.1 Schematic of the geometric (left) and feature (right) model of representing land-cover classes.

more nuanced assessments rather than resorting to, for example, a binary "change or no change" assessment. The semantic formalizations can also be subjected to scrutiny by others, allowing for alternative interpretations of a dataset based on other classification criteria or for other purposes of analysis. The following section demonstrates the use of formalized semantic information to make more nuanced land-cover data analysis.

3.3   ANALYSIS

As a foundation for the analysis examples below, we assume that the representation of land-cover category semantics uses any of the above techniques that can produce a semantic relationship measure for pair-wise land-cover class comparisons. With a total of n classes, it is possible to generate a cross-product of a semantic relationship for all pair-wise combinations of categories. The result is a semantic relationship matrix. The term "relationship" here is meant to be generic since, for example, similarity is only one of many semantic relations of potential interest. Other relationships could be inclusion, resemblance, dissimilarity (Bouchon-Meunier et al., 1996), and various spatial relationships (Schwering and Raubal, 2005). Using the semantic relationship matrix as a foundation enables us to look beyond the particulars of any one representational model and focus on how these metrics can potentially be used in various land-cover analyses. The use of a semantic cross-product matrix has many similarities to the use of the contingency matrices that are frequently used in accuracy assessment and change analysis of land-cover data. Indeed, those applications are also exemplified below, but the semantic relationship matrix is also used here in ways more akin to how distance matrices are used in various clustering and geostatistical techniques.

3.4   LAND-COVER ACCURACY ASSESSMENT

One of the first steps in making a land-cover product useful is to evaluate its quality. Data uncertainty is an inseparable companion of almost any type of land-cover product, and today there are many techniques to handle uncertainty representation and analysis for remote sensing and Geographic information system (GIS) (cf. Foody, 2002; Zhang and Goodchild, 2002). A standard method to describe thematic uncertainty in land-cover data is using an error or confusion matrix (Card, 1982). This matrix is used for many different measures of agreement between data estimates and ground truth conditions. In addition, scholars have presented ways to expand on the traditional use of an error matrix to compare map data with various types of associated uncertainty (Ahlqvist, 2000; Gopal and Woodcock, 1994; Pontius and Cheuk, 2006; Woodcock and Gopal, 2000). Ahlqvist and Gahegan (2005) followed the soft-accuracy assessment ideas, specifically related to semantic analysis, described by Congalton and Green (1999) to generate a semantic similarity matrix that could identify land-cover classes that were easily confused because of their similarities. For example, when gathering data for an accuracy assessment, a reference site could be labeled a "mixed forest" although it is very similar to a "coniferous forest," so a dataset could be almost right even if it classified that object as a "coniferous forest." Intuitively, a "mixed forest" is much more similar to a "coniferous forest" than to "open water," so the two forest types would probably be harder to distinguish and would, more often, result in some classification confusion. In their study, Ahlqvist and Gahegan (2005) found a significant correlation between semantic similarity metrics based on the class definitions and empirical estimates of ambiguity between classes. These results supported the hypothesis that a semantic similarity matrix can predict those land-cover classes that are more prone to confusion and those that are not. This can help data producers during taxonomy formation as a means to test out likely uncertainties before gathering expensive field data. Using it, we can decide that when two categories overlap too much, including both categories in the classification would lead to unacceptable error rates in the resulting maps. The alternative could be either to collapse the categories into one or to consider how the category definitions could be modified so that the attributes make a firmer separation between them.

3.5 LAND-COVER CHANGE ANALYSIS

One of the prime uses of land-cover data is for change analysis, and many different methods have been devised to assess change in the landscape (Lu et al., 2004; Mas, 1999; Singh, 1989). Among these, the postclassification method is frequently used because of (1) the detailed information that can be gained from the produced change matrix, (2) the limited impact that image calibration and atmospheric and environmental differences will have on the multitemporal image comparison, and (3) its intuitive interpretation as opposed to numerically based image analysis methods that need careful interpretation to assess what the identified changes mean (Lu et al., 2004). Part of the appeal is then closely tied to the fact that classification into land-cover categories embeds rich semantic information with the class labels that allow for interpretation and use in many different application contexts. However, as I have already noted before, these semantics are also problematic owing to the sometimes-limited descriptions of what the land-cover labels exactly represent (Comber et al., 2005). This problem becomes particularly vexing when certain types of land-cover change are of importance, but where the original classes need to be reclassified into these more relevant categories or the data on land cover from different times are classified using different classification systems (Comber et al., 2004). Traditional postclassification change analysis typically uses a binary image overlay logic where areas are classified either as change/no-change or as change from class A to class B. Because a subtle change from "row crops" to "pasture" is treated equally as a drastic change from "row crops" to "strip mine," researchers have suggested alternative ways of understanding and analyzing the content of a pixel (Foody, 2007). Alternative, "soft" land-cover classification methods have been suggested (Fisher and Pathirana, 1990; Foody and Cox, 1994; Pontius and Cheuk, 2006), but these have mostly addressed the vague or "fuzzy" relationship between an observation and a target category. In contrast, the notion of land-cover semantics and similarity introduced in the previous section is concerned with relations between categories, and these are particularly relevant where already available data use heterogeneous classification systems. In a study of land-cover change from 1992 to 2001, Ahlqvist (2008) used data from the U.S. Geological Survey (USGS, 2006a, 2006b) to demonstrate the use of semantic similarity metrics as a measure of land-cover change. In that study, two different semantic relationship measures were used on a geometric representation of land-cover semantics: the class *distance* and the class *overlap*. These two measures are illustrated schematically in Figure 3.2, where two hypothetical land-cover classes "park" and "forest" are formally defined to have a tree cover of 30%–80% and 60%–100%, respectively. These intervals are partially overlapping, and this can be measured using an overlap metric.

The two intervals are also partially separated, and that aspect of the formal semantics is measured by a distance metric. It is also important to realize that many categories use several attribute dimensions in their definition, and in these cases, the metrics can easily be extended to provide

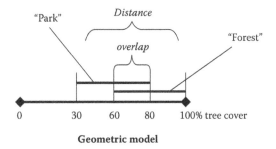

FIGURE 3.2 Graphic illustration of the overlap and distance metric for measuring semantic relations in a geometric model.

a summary metric for all dimensions as well as for one dimension at a time. Please see Ahlqvist (2004) for more details on the specific implementation of these metrics.

Combinations of these metrics take on four main interpretations, as illustrated by Figure 3.3.

When overlap is very small and distance is also small, this is interpreted as "similar but disjoint classes." When overlap is small and distance is large, the classes are "very different." On the other hand, when distance is small and overlap is large, the classes are "very similar." Finally, when distance is large and overlap is also large, this is interpreted as a "class/subclass relationship." Clearly, these qualitative interpretations are based on quantitative evaluations, and the distance and overlap values can vary continuously from zero to a maximum determined by the scaling of the metric. This means that two categorical land-cover data layers can be compared, class by class, in a continuous fashion such that the final change map can visualize and distinguish between dramatic changes, such as a change from "row crops" to "strip mine," and more subtle changes such as one from "row crops" to "pasture." In addition, the semantic assessments can be done between completely incompatible classification systems, and we can use a combination of semantic similarity metrics to make detailed and spatially explicit interpretations of the detected changes. Using the above example again, the two metrics—Distance and Overlap—can be used to construct a bivariate graded color scheme that enables the above interpretations across an entire change map (Figure 3.4).

Here, the blue-orange color scale follows the same overlap-distance combinations outlined in Figure 3.3 and should therefore be interpreted as follows: significant land-cover changes ("very different classes") will show as dark gray, intermediate changes will show as either more or less saturated blue ("similar but disjoint classes") or orange tones ("class/subclass relationship"), and little or no-change situations ("very similar classes") will show as very pale colors or no color at all. The change map in Figure 3.4 thus illustrates "semantic change," and the most significant change is the larger dark gray area in the upper left part of the map, representing change from the 1992 class "quarries/strip mines/gravel pits" to the 2001 "developed" classes. Other distinct patterns are the concentric bands of orange gray and blue-colored areas around the city of West Chester. These represent changes related to actual land cover and also changes in class definitions. Separating these and the ability to recognize major changes from more subtle ones are obviously of value to an in-depth landscape change analysis.

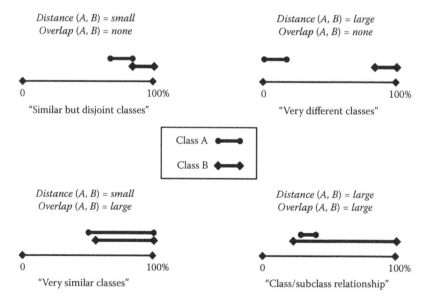

FIGURE 3.3 Graphic illustration of four different concept relationships in one attribute dimension. The two concepts are represented by horizontal lines, specifying numerical intervals on the dimension.

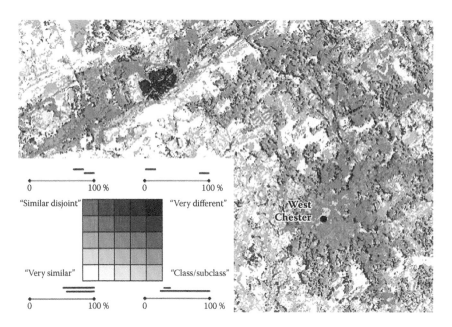

FIGURE 3.4   (See color insert.) An example of semantic land-cover change map using a bivariate color scheme to represent different combinations of the semantic distance and overlap metric.

3.6   LANDSCAPE PATTERN ANALYSIS

This final example will address the use of semantic similarity measures to analyze the spatial structure of land-cover data. Investigating spatial patterns of land cover is a significant theme of spatial statistics and has many applications in the environmental and social sciences, including, for example, climate variability (DeFries et al., 1995), urban sprawl (Wu and Webster, 1998), and habitat loss (Lambin et al., 2001). Patch and pixel-based pattern metrics, such as contagion and fractal dimension, and numerous other metrics that quantify spatial autocorrelation such as join-count statistics, Moran's I, Geary's C, and semivariogram techniques are frequently applied in landscape research (Gustafson, 1998). These metrics are well established, but their use is also known to be sensitive to the spatial scale of analysis (Lam and Quattrochi, 1992) as well as the level of detail in the categorical classification system (Li and Wu, 2004). As an example, the Contagion index (O'Neill et al., 1988) is frequently used to measure the spatial pattern of land-cover data, but because the spatial dimension is captured only through binary adjacency of areas and the attribute difference is only measured as same or different classes (binary), it is very sensitive to changes, for example, in the number of land-cover classes and the pixel resolution.

In a recent study, Ahlqvist and Shortridge (2010) developed a conceptual typology of autocorrelation metrics for categorical data and demonstrated that a semantic approach to the indicator semivariogram technique (Goovaerts, 1997) could overcome some of the mentioned shortcomings of other existing pattern metrics. Arguing that autocorrelation patterns are measured as some form of cross-product of a spatial relation metric (usually spatial distance) and an attribute relation metric (usually the difference between recorded attributes), Ahlqvist and Shortridge (2010) suggested a semantic variogram to capture both attribute and spatial relations simultaneously.

The basis for a semantic variogram is the regular variogram that measures semivariance (the sum of squared differences between all data points separated by a distance l) for a sample dataset and estimates a function that describes how semivariance changes with different values of l. If compared points are increasingly different as distance increases, the semivariance will increase, and by looking at the shape of the semivariogram, important information about the degree and range of spatial autocorrelation can be derived.

Regular variogram uses numerical attribute differences to measure the semivariance, but this is conceptually analogous to measuring the difference in attribute as a semantic difference, for example, between recorded land-cover classes at two locations. Details of the definition of semantic variogram can be found in the work of Ahlqvist and Shortridge (2006). Figure 3.5 illustrates the utility of this method on the Land Cover 2005 map from The North American Land Change Monitoring System (http://landcover.usgs.gov/nalcms.php).

In Figure 3.5, 110 sample points are distributed randomly across the entire land-cover dataset, and each point is assigned the land-cover class at that location. All the 110 points are then pair-wise cross-tabulated so that every point pair has information about the distance between the two points and the land-cover classes at the two points. In a normal variogram analysis, the attribute values would be numbers, but in the semantic variogram analysis we replace the numerical attribute difference with a semantic relationship metric calculated through one of the methods described in the sections 3.3 through 3.5. The "semantic variogram" scatterplot (Figure 3.5, top-right) shows all the 12,100 spatial-semantic distance pairs. Because the general tendency of a plot like this can be hard to distinguish, a summary graph is provided below it (Figure 3.5, bottom-right). In this graph, a series of box plots gives a summary of the distribution of semantic distance values for specific spatial distance intervals. As we can see from these plots, there is a general tendency for close observations to be more similar than distant observations, which is one of the signals of spatial autocorrelation. We can also estimate at which distance (between 50 and 74 km) this effect ceases to be noticeable.

In a detailed analysis of U.S. National Land Cover Data (NLCD) from 1992 (USGS, 2006a) for portions of Ohio, Michigan, and Massachusetts, Ahlqvist and Shortridge (2010) also demonstrated that the semantic variogram is relatively robust to class aggregation compared with other comparable pattern metrics. The 21 NLCD Level 2 land-cover classes were aggregated hierarchically to nine Level 1 classes, and semantic similarity values were calculated for the Level 1 classes. While other metrics, such as the Contagion index and the regular indicator variogram, were substantially affected by the class aggregation, the semantic variogram showed only limited change.

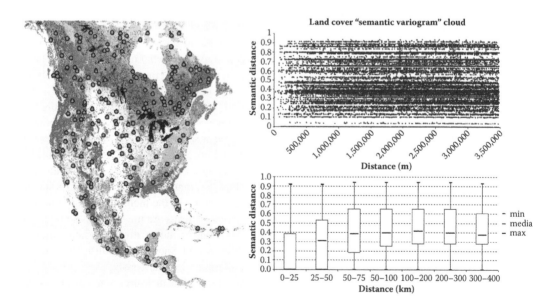

FIGURE 3.5 **(See color insert.)** Random sample points ($n = 110$) across a North American land-cover dataset are plotted in a semantic variogram according to pair-wise comparison of spatial distance of the points and the semantic difference between the land-cover classes at these points. Box plots summarize the points for specified spatial distance intervals.

3.7 CONCLUSION

To summarize, the use of semantic relationship metrics on land-cover data offers many interesting opportunities to build upon existing quantitative methods for land-cover analysis and to develop new ones. Many existing methods with attractive analytical capabilities have been restricted to interval and ratio data and are typically not applicable to categorical land-cover data. Although the semantic methods described above open exciting prospects, a few words of caution are also in place. Much research is still needed to validate the development of formal semantic descriptions of land-cover data. Whether developed from automated methods such as natural language processing of category definitions (Jensen and Binot, 1987) or through manual elicitation from domain experts (Feng and Flewelling, 2004), carefully evaluating the validity of derived specifications will need to be a collaborative process where many users and experts contribute their own understanding of the data. Another issue is related to algebraic evaluation of image characteristics in general. Although many problems associated with radiometric and atmospheric correction of images are mitigated by postclassification methods, it is still problematic to establish at what level the image differences separate actual change from apparent change. In the context of semantic difference, we may find that a category has changed a lot in its definition, but it is still a subcategory of the original class, and the actual object may not have changed at all. In that case, both the proposed semantically based methods and the traditional binary methodology would indicate change, but the semantic indication would offer a more nuanced estimation of the magnitude of that change, essentially reducing the error of the change estimate. An example of this situation can be seen in Figure 3.4. Still, the use of semantic relationship measures should provide researchers and organizations with a finer instrument to understand land cover and associated analysis.

REFERENCES

Agarwal, P. 2005. Ontological considerations in GIScience. *International Journal of Geographical Information Science*, 19(5), 501–536.

Ahlqvist, O. 2000. Rough classification and accuracy assessment. *International Journal of Geographical Information Science*, 14(5), 475–496.

Ahlqvist, O. 2004. A parameterized representation of uncertain conceptual spaces. *Transactions in GIS*, 8(4), 493–514.

Ahlqvist, O. 2008. Extending post-classification change detection using semantic similarity metrics to overcome class heterogeneity: A study of 1992 and 2001 U.S. National Land Cover Database changes. *Remote Sensing of Environment*, 112(3), 1226–1241.

Ahlqvist, O. and Gahegan, M. 2005. Probing the relationship between classification error and class similarity. *Photogrammetric Engineering and Remote Sensing*, 71(12), 1365–1373.

Ahlqvist, O. and Shortridge, A. 2006. Characterizing land cover structure with semantic variograms. In *Progress in Spatial Data Handling—12th International Symposium on Spatial Data Handling* (pp. 401–415). Springer-Verlag.

Ahlqvist, O. and Shortridge, A. 2010. Spatial and semantic dimensions of landscape heterogeneity. *Landscape Ecology*, 25(4), 573–590.

Anderson, J.R. 1976. *A Land Use and Land Cover Classification System for Use with Remote Sensor Data*. US Government Print Office.

Bartholomé, E. and Belward, A.S. 2005. GLC2000: A new approach to global land cover mapping from Earth observation data. *International Journal of Remote Sensing*, 26(9), 1959–1977.

Bennett, B. 2001. What is a forest? On the vagueness of certain geographic concepts. *Topoi*, 20(2), 189–201.

Bishr, Y. 1998. Overcoming the semantic and other barriers to GIS interoperability. *International Journal of Geographical Information Science*, 12(4), 299–314.

Bouchon-Meunier, B., Rifqi, M., and Bothorel, S. 1996. Towards general measures of comparison of objects. *Fuzzy Sets and Systems*, 84(2), 143–153.

Card, D.H. 1982. Using known map category marginal frequencies to improve estimates of thematic map accuracy. *Photogrammetric Engineering and Remote Sensing*, 48(3), 431–439.

CEC. 1995. *CORINE Land Cover*. Luxembourg: Commission of the European Communities. Available at: http://reports.eea.europa.eu/COR0-landcover/en/land_cover.pdf

Comber, A., Fisher, P., and Wadsworth, R. 2004. Integrating land-cover data with different ontologies: Identifying change from inconsistency. *International Journal of Geographical Information Science*, 18(7), 691–708.

Comber, A., Fisher, P., and Wadsworth, R. 2005. What is land cover. *Environment and Planning B: Planning and Design*, 32, 199–209.

Comber, A., Fisher, P.F., and Wadsworth, R. 2004. Assessment of a semantic statistical approach to detecting land cover change using inconsistent data sets. *Photogrammetric Engineering and Remote Sensing*, 70(8), 931–938.

Congalton, R.G. and Green, K. 1999. *Assessing the Accuracy of Remotely Sensed Data: Principles and Practices*. Boca Raton, FL Lewis Publications.

Couclelis, H. 1992. People manipulate objects (but cultivate fields): Beyond the Raster-Vector debate in GIS. In A.U. Frank, I. Campari, and U. Formentini (Eds.), *Theories and Methods of Spatio-Temporal Reasoning in Geographic Space*, Lecture notes in computer science (Vol. 639, pp. 65–77). Berlin, Heidelberg, New York: Springer-Verlag.

DeFries, R.S., Field, C.B., Fung, I., Justice, C.O., Los, S., Matson, P.A., Matthews, E., et al. 1995. Mapping the land surface for global atmosphere-biosphere models: Toward continuous distributions of vegetation's functional properties. *Journal of Geophysical Research*, 100(D10), 20867–20882.

Di Gregorio, A. 2004. *Land Cover Classification System (LCCS), Version 2: Classification Concepts and User Manual*, Rome: FAO.

Feng, C.C. and Flewelling, D.M. 2004. Assessment of semantic similarity between land use/land cover classification systems. *Computers, Environment and Urban Systems*, 28(3), 229–246.

Fisher, P. 2000. Sorites paradox and vague geographies. *Fuzzy Sets and Systems*, 113(1), 7–18.

Fisher, P. and Wood, J. 1998. What is a mountain? or the Englishman who went up a Boolean geographical concept but realised it was Fuzzy. *Geography*, 83(3), 247–256.

Fisher, P.F. and Pathirana, S. 1990. The evaluation of fuzzy membership of land cover classes in the suburban zone. *Remote Sensing of Environment*, 34(2), 121–132.

Fonseca, F.T., Egenhofer, M.J., Agouris, P., and Camara, G. 2002. Using ontologies for integrated geographic information systems. *Transactions in GIS*, 6(3), 231–257.

Foody, G. 2007. Map comparison in GIS. *Progress in Physical Geography*, 31(4), 439–445. doi:10.1177/0309133307081294.

Foody, G.M. 2002. Status of land cover classification accuracy assessment. *Remote Sensing of Environment*, 80(1), 185–201.

Foody, G.M. and Cox, D.P. 1994. Sub-pixel land cover composition estimation using a linear mixture model and fuzzy membership functions. *International Journal of Remote Sensing*, 15(3), 619–631.

Goovaerts, P. 1997. *Geostatistics for Natural Resources Evaluation*. New York: Oxford University Press.

Gopal, S. and Woodcock, C. 1994. Theory and methods for accuracy assessment of thematic maps using fuzzy sets. *Photogrammetric Engineering and Remote Sensing*, 60(2), 181–188.

Gustafson, E.J. 1998. Quantifying landscape spatial pattern: What is the state of the art? *Ecosystems*, 1(2), 143–156.

Harvey, F., Kuhn, W., Pundt, H., and Bishr, Y. 1999. Semantic interoperability: A central issue for sharing geographic information. *Annals of Regional Science*, 33(2), 213–232.

Jansen, L.J.M. and Di Gregorio, A.D. 2002. Parametric land cover and land-use classifications as tools for environmental change detection. *Agriculture, Ecosystems and Environment*, 91(1), 89–100.

Jensen, K. and Binot, J.L. 1987. Disambiguating prepositional phrase attachments by using on-line dictionary definitions. *Computational Linguistics*, 13(3–4), 251–260.

Kalensky, Z.D. 1998. AFRICOVER: Land cover database and map of Africa. *Canadian Journal of Remote Sensing*, 24(3), 292–297.

Kavouras, M. and Kokla, M. 2002a. A method for the formalization and integration of geographical categorizations. *International Journal of Geographical Information Science*, 16(5), 439.

Kavouras, M. and Kokla, M. 2002b. A method for the formalization and integration of geographical categorizations. *International Journal of Geographical Information Science*, 16(5), 439–453. doi:10.1080/13658810210129120.

Kavouras, M., Kokla, M., and Tomai, E. 2005. Comparing categories among geographic ontologies. *Computers & Geosciences*, 31(2), 145–154.

Lakoff, G. 1987. *Women, Fire, and Dangerous Things*. Chicago: University of Chicago Press.

Lam, N.S. and Quattrochi, D.A. 1992. On the issues of scale, resolution, and fractal analysis in the mapping sciences. *Professional Geographer*, 44(1), 88–98.

Lambin, E.F., Turner, B.L., Geist, H.J., Agbola, S.B., Angelsen, A., Bruce, J.W., Coomes, O.T., et al. 2001. The causes of land-use and land-cover change: Moving beyond the myths. *Global Environmental Change, Part A: Human and Policy Dimensions*, 11(4), 261–269.

Li, H. and Wu, J. 2004. Use and misuse of landscape indices. *Landscape Ecology*, 19(4), 389–399.

Loveland, T.R. and Belward, A.S. 1997. The IGBP-DIS global 1km land cover data set, DISCover: First results. *International Journal of Remote Sensing*, 18, 3289–3295.

Lu, D., Mausel, P., Brondízio, E., and Moran, E. 2004. Change detection techniques. *International Journal of Remote Sensing*, 25(12), 2365–2407.

Mas, J.F. 1999. Monitoring land-cover changes: A comparison of change detection techniques. *International Journal of Remote Sensing*, 20(1), 139–152.

Murphy, G.L. 2004. *The Big Book of Concepts*. Cambridge, MA: MIT Press.

O'Neill, R.V., Krummel, J.R., Gardner, R.H., Sugihara, G., Jackson, B., DeAngelis, D.L., Milne, B.T., et al. 1988. Indices of landscape pattern. *Landscape Ecology*, 1(3), 153–162.

Pontius, R.G., and Cheuk, M.L. (2006). A generalized cross-tabulation matrix to compare soft-classified maps at multiple resolutions. *International Journal of Geographical Information Science*, 20(1), 1–30.

Rada, R., Mili, H., Bicknell, E., and Blettner, M. 1989. Development and application of a metric on semantic nets. *IEEE Transactions on Systems, Man and Cybernetics*, 19(1), 17–30.

Robbins, P. 2001. Fixed categories in a portable landscape: The causes and consequences of land-cover categorization. *Environment and Planning A*, 33, 161–179.

Rodríguez, M.A., Egenhofer, M., & Rugg, R. (1999). Assessing semantic similarities among geospatial feature class definitions. In A. Vckovski, K. Brassel, & H.-J. Schek (Eds.), *Interoperating Geographic Information Systems*—Second International Conference, INTEROP'99, Zurich, Switzerland, March 10–12, 1999. Proceedings, Lecture Notes in Computer Science (Vol. 1580, pp. 189–202). Berlin/ Heidelberg: Springer.

Rodríguez, M. and Egenhofer, M. 2003. Determining semantic similarity among entity classes from different ontologies. *IEEE Transactions on Knowledge and Data Engineering*, 15(2), 442–456.

Rosch, E. 1978. Principles of categorization. In E. Rosch and B.B. Loyd (Eds.), *Cognition and Categorization* (pp. 27–48). Hillsdale, NJ: Lawrence Erlbaum Associates.

Schwering, A. 2008. Approaches to semantic similarity measurement for geo-spatial data: A survey. *Transactions in GIS*, 12(1), 5–29.

Schwering, A. and Raubal, M. 2005. Spatial relations for semantic similarity measurement. In J. Akoka (Ed.), *Perspectives in Conceptual Modeling, ER 2005 Workshops CAOIS, BP-UML, CoMoGIS, eCOMO, and QoIS*, Lecture Notes in Computer Science (Vol. 3770, pp. 259–269). Klagenfurt, Austria: Springer.

Singh, A. 1989. Digital change detection techniques using remotely-sensed data. *International Journal of Remote Sensing*, 10, 989–1003.

Song, D. and Bruza, P. 2003. Towards context sensitive information inference. *Journal of the American Society for Information Science and Technology*, 54(4), 321–334.

Sowa, J.F. 2000. *Knowledge Representation: Logical, Philosophical, and Computational Foundations* (P. 594, xiv). Pacific Grove, CA: Brooks Cole Publishing Co.

USGS. 2006a, March. National Landcover Dataset 1992. Available at: http://landcover.usgs.gov/natllandcover.php

USGS. 2006b, September 13. National Landcover Dataset 2001. Available at: http://www.mrlc.gov/mrlc2k_nlcd.asp

Woodcock, C.E. and Gopal, S. 2000. Fuzzy set theory and thematic maps: Accuracy assessment and area estimation. *International Journal of Geographical Information Science*, 14(2), 153–172.

Wu, F. and Webster, C.J. 1998. Simulation of land development through the integration of cellular automata and multicriteria evaluation. *Environment and Planning B*, 25, 103–126.

Zhang, J. and Goodchild, M.F. 2002. *Uncertainty in Geographical Information*. London; New York: Taylor & Francis.

4 Overview of Land-Cover Classifications and Their Interoperability

Antonio Di Gregorio and Douglas O'Brien

CONTENTS

4.1 INTRODUCTION

There is a great need for data harmonization as there is a huge problem of compatibility and comparability between different land-cover (LC) products. Harmonization should be the process whereby differences among existing definitions of land characterization are identified and clarified and inconsistencies are reduced. However, this is not the reality, since current maps exist mostly as independent and incompatible datasets. This lack of harmonization can be explained by the poor compatibility of LC classifications or legends, which is often an arcane "black box" to anyone outside the immediate group involved in the preparation of legends. By its nature, mapping is a local activity, thus facilitating the tendency to establish unique classification systems to fit local environmental conditions. However, these incompatibilities hamper the aggregation toward broader regional and global datasets. To be able to integrate data from multiple sources, there is a strong need for semantic interoperability.

Semantic interoperability is one of the major unsolved problems in the modern use of LC data. Uncertainty is an inescapable element in all types of geographical information because truth as a distinct and indubitable fact cannot exist in a derived representation. Information is thus always relative to context. However, in some disciplines (like LC), the level of semantic vagueness and the relative misuse of data are far too high, and the practical use of semantic interoperability in many applications entails risk. Diffuse use of geographical information systems (GIS) and spatial analysis has further exacerbated this problem, creating a vicious circle of vagueness and ambiguity in the LC semantic, which propagates constantly and is strengthened through the interoperability issues encountered when using different datasets.

LC is one of the most easily detectable indicators of human intervention on land; therefore, information on LC is critical in any geographical database. In modern maps, LC has become a sort of "boundary object" between different disciplines. This development, on the one hand, enhances the intrinsic value of LC information, but on the other hand, it poses new challenges for its harmonization and correct use by further enlarging the base of potential users. Any land surface is heterogeneous, and the mapping standards to acquire, represent, and generalize land characteristics are about as diverse as the land surface itself.

In addition, there has been an explosion of LC datasets in the world, coupled with the growing use of new technologies and the rapid changes in how information can converge across previously disparate families of disciplines. Hence, fostering discussions and reviews for developing internationally accepted LC standards is a crucial task in minimizing current inadequacies and responding to the requests and needs of the international community.

4.2   LC AND LC MAPPING

LC can be defined as the observed (bio)-physical cover of the earth's surface. It can be considered a geographically explicit feature that other disciplines can use as a geographical reference (e.g., for land-use, climatic, or ecological studies).

Any LC-mapping activity can be defined as a process of information extraction governed by a process of generalization. As a matter of course, this implies a loss of several levels of detail in the abstract representation of the real world. The degree of generalization—and thus the efficiency of a database to represent the real world in two-dimensional form—is, at one level, linked to cartographic standards (cartographic scale and the minimum mapping unit [MMU]) and the way the "interpretation" process has been conducted. However, it is also strongly related to the thematic content of the map, how exhaustive is the formalization of the meaning of this thematic content, and how it can be understood by a large user community.

Flexibility and semantic interoperability of datasets are key elements when considering a multitude of potential users and applications. In the past, LC was not a stand-alone subject but was subsumed in many disciplines, so the same geographical areas could have been mapped several times for different purposes with different discipline-specific legends. However, in those times, the tools for data integration were absent or limited; thus, exchange of environmental data and their integrated use were hampered. Today, although technologies such as GIS have drastically increased the potential of flexibility and exchangeability of different datasets, there has been little progress in the effective integrated use of LC information. This primarily reflects the large heterogeneity of LC ontologies and the poor or absent formalization of the meaning of their semantics. For those reasons, deriving efficient maps that are interoperable and that satisfy the requirements of diverse end-user communities is still challenging. It should always be kept in mind that GIS is a functional tool for data integration; it cannot solve the problem of harmonization and standardization at the semantic level.

4.3   CLASSIFICATION, LEGENDS, AND STANDARDS

To classify is a human activity. Classification is the means whereby we put knowledge into order. Our lives are surrounded by systems of classification, limned by standards, formats, etc. The oldest method of communicating knowledge was, no doubt, human language and conversation, where specific language elements or specialized terms were created to exchange particular types of information. A body of shared knowledge as a basis for communication is, therefore, part of most sciences, and historically we find ample evidence of specialized terminology, hierarchical thinking, and classifications established within those disciplines. Each discipline has its own jargon.

Bjelland (2004: 2) proposes two distinct classification processes: cognitive and logical.

> … in the cognitive sense, classification is concerned with how people conceptualize the world in the form of mental representation and operations. In the logical sense, classification is concerned with the definition of terms in order to concretise concepts. The main difference is that in the cognitive sense concepts are subjective and private, while in the logical sense concepts are public and hence made inter-subjectively available by intentional definitions. It appears that classification in the cognitive sense is the justification for classification in the logical sense. Research within cognitive science has repeatedly demonstrated that concepts in general are subjective and vague and liable to change both between individuals and over time within the same individual. It is exactly the vagueness, instability, and subjectivity of mental concepts that cognitive theories of classification attempt to explain and that logical theory attempts to overcome.

Categorization can therefore be associated with a cognitive process, whereas classification as a social process can be linked to a logical process.

In the case of spatial information, classification is an abstract representation of features of the real world (Figure 4.1), using classes or terms derived through a mental process. Sokal (1974) defines it as "the ordering or arrangement of objects into groups or sets on the basis of their relationships," and Bowker and Star (1999) as "a spatial, temporal or spatio-temporal segmentation of the world." They define a "classification system" as "a set of boxes (metaphorical or literal) into which things can be put in order to then do some kind of work bureaucratic or knowledge production."

In the case of spatial information, as for LC, a classification describes the systematic framework, with the names of the classes, the criteria used to distinguish them, and the relationship between classes themselves. Classification thus requires the definition of class boundaries, which should be clear, precise, possibly quantitative, and based on objective criteria.

In an abstract, ideally a classification system should thus exhibit the following properties:

- Use of consistent, unique, and systematically applied classificatory principles
- Adapted to describing fully the whole gamut of the types of features

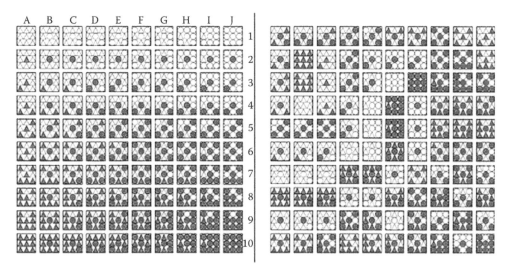

FIGURE 4.1   Abstract presentation of a classification consisting of a continuum with two gradients (left), in comparison with a concrete field situation (right). Triangles and circles represent the two elements being considered. (From Kuchler, A.W. and Zonneveld, I.S. (Eds.), *Vegetation Mapping. Handbook of Vegetation Science*, vol. 10, Kluwer Academic, Dordecht, The Netherlands, 1988. With permission.)

- Completeness, meaning total coverage of the area it describes
- Unique, mutually exclusive, and unambiguous classes

In addition, they should include some key characteristics to support evolving standards and, in general, the dynamics of science:

- Be potentially applicable as a common reference system or be able to converse with other systems.
- Recognize the balancing act inherent in classifying (Bowker and Star, 1999).
- Render voice retrieval (Bowker and Star, 1999) by allowing users to detail and compare classes using the detailed class description (systematically organized with a list of explicit measurable diagnostic attributes), thus avoiding the risk of systems being impermeable to the end user.

For LC mapping and all other disciplines producing two-dimensional representations of a certain portion of the land, the classification scheme appears in a specific database in the form of a legend. A legend can, therefore, be defined as the application of certain classification criteria (classification rules or classes) in a specific geographical area, using a defined mapping scale and a specific dataset. A legend may, therefore, contain only a proportion or a subset of all possible classes of the reference classification system.

Classification can be done in two ways: either *a priori* or *a posteriori*. In an *a priori* classification system, the classes are an abstract of the types expected to occur. The approach is based on a definition of classes before any data collection takes place. Thus, all possible combinations of classification criteria must be dealt with beforehand in the system. Basically, in a field (or with remote-sensing data), each sample plot (or polygon) is identified and labeled according to its similarity or compatibility with the predefined set of classes. This method is used extensively in soil science, such as *The Revised Legend of the Soil Map of the World* (FAO, 1988) and the *USDA Soil Taxonomy* (USDA, 1999). The main advantage of the *a priori* classification is that the classes being created independently from the study area predispose class definitions to a certain level of homogeneity and standardization among different users.

In contrast, the *a posteriori* classification differs fundamentally by its direct approach and its freedom from preconceived notions. The approach is based on defining the classes after clustering the field samples collected. An example is the Braun-Blanquet method used in vegetation science. This is a floristic classification approach, which uses the total species richness and composition to cluster samples in sociological groups (Kuchler and Zonneveld, 1988). The advantage of this type of classification is its flexibility and adaptability compared with the implicit rigidity of the *a priori* classification. Further, the *a posteriori* approach implies a minimum of generalization, and thus it better fits the collected field observations in a specific area. At the same time, because the *a posteriori* classification depends on the specific area described and is adapted to local conditions, it is not qualified to define standardized classes. Clustering of samples to define the classes can be done only after data collection, and the relevance of certain criteria in a certain area may be limited when the criteria are used elsewhere or in ecologically different regions.

A third way to organize spatial information is by establishing a "feature catalogue." In some types of maps, a set of feature types is identified for inclusion in the spatial dataset (map). For example, a road map may include roads, some rivers, and other significant landmarks and selected features. The main difference here is that the feature types are selected. Further, the road map schema may not comprise all road types. For example, only major roads might be included. Creating an information model by selecting particular features according to a collection criterion is a valid approach that is distinct from classification. Classification endeavors to address the entire information domain and subdivide it according to a set of rules to produce a set of classes and subclasses, allowing for all of the possibilities in the logical space.

4.4   LC CLASSIFICATIONS—A HISTORICAL BACKGROUND

The origin of the concept of systematic classification of vegetation can be traced to the ideas of Carl Linnaeus in the early eighteenth century in Sweden. The development of pure LC classification systems started with the use of aerial photographs at the beginning of the twentieth century, in 1920, in Canada. In this case, the study was focused mainly on forest mapping. In the mid-1940s, mapping of major land-use associations for the entire United States began using aerial photographs taken during the late 1930s and the early 1940s. The project produced a set of state-level land-use maps at a scale of 1:1,000,000 from mosaics of the aerial photographs, and later a map of major land uses at 1:5,000,000 was derived. Most of the LC classifications in the early maps were based on classification of vegetation and focused more on land use than on LC; however, they were very rudimentary, single-purpose-oriented, and unsystematic.

The real introduction of land-use-cum-LC classification systems and related concepts took place in the 1950s. The systems were all based on aerial photographs and related to the production of a particular map or a particular single exercise, not aiming at producing a reference system. At that time, LC was mostly understood as a variation of the dominant land-use classifications or was a feature in forestry maps. It was in 1972, with the launch of the first civilian-accessible satellite, ERTS 1, that a new satellite-image-based era began. LC started to be intermixed officially with land use in both the title and the purpose of many classification systems. It was at that time that the first official definitions of LC were made (Anderson et al., 1976; Burley, 1961).

This early work resulted in the development of several legends and nomenclatures to serve specific single-mapping exercises. In many cases, these legends and nomenclatures seem to be more of an adaptation of a specific nomenclature to the results of an automated classification of digital satellite images rather than a real coherent system. Consequently, and as result of the first appearance of spatial-modeling techniques, the problem of harmonization and comparability of different classifications and legends became evident. Initial efforts at harmonization started in the 1990s in parallel with the increasing use of GIS.

An important step toward LC standardization was taken by the Food and Agriculture Organization of the United Nations (FAO) and the United Nations Environment Programme (UNEP). In 1994, these organizations launched a joint initiative on standardizing LC and land-use terminology. An important result has been the suggestion to clearly separate land use from LC in current systems.

4.5   VEGETATION CLASSIFICATIONS AS A BASIS FOR DERIVING LC CATEGORIZATIONS

Vegetation is one of the major features of almost all parts of the earth's surface. Apart from the Arctic and Antarctic landscapes and deserts, most of the terrestrial surfaces beyond human constructions are covered by vegetation. Therefore, it is not surprising that LC derives directly from vegetation science, especially structural and physiognomic categorization studies.

Plant communities can be classified according to many different criteria, depending on which of their properties are emphasized:

1. Properties of the vegetation itself:
   A. Physiognomic and structural criteria
   B. Floristic criteria
   C. Numerical relation criteria (community coefficients)
2. Properties external to the vegetation:
   A. The presumed final stage in vegetation succession (climax)
   B. The habitat or environment
   C. Geographical location of communities

3. Properties combining vegetation and environment:
   A. By independent analysis of vegetation and independent analysis of environment
   B. By combined analysis of vegetation and environment

Danserau (1961) defines vegetation structure as "the organization in space of the individuals that form a stand," and he states that "the primary elements of structure are growth form, stratification, and coverage." Fosberg (1961) defines vegetation physiognomy as the external appearance of vegetation. Physiognomy, in this sense, is defined as the biomass structure, functional phenomena (such as leaf fall), and gross compositional characteristics (such as luxuriance or relative xeromorphy).

Several structural physiognomic or structuro-physiognomic vegetation systems exist; some of them have deeply influenced the development of the most common LC systems in use. The structural classification scheme of Danserau (1961) and Kuchler and Zonneveld (1988) is a well-known scheme that employs six categories:

- Plant life form
- Plant size
- Coverage
- Function (in the sense of deciduous or evergreen)
- Leaf shape and size
- Leaf texture

The categories are then subdivided into subcategories. Thus, "plant life form," for instance, is subdivided into five subtypes:

- 1a Trees
- 1b Shrubs
- 1c Herbs
- 1d Bryophytes
- 1e Epiphytes and lianas

Fosberg's structural formation system (Fosberg, 1961) was adopted as a guide for mapping vegetation for the International Biological Program (IBP). The Fosberg system is similar to the Danserau and Kuchler system, which is based on actual vegetation and which purposely avoids incorporating environmental criteria. This system has the advantage that the vegetation units established in this manner can be easily detected using remote-sensing satellite imagery. The criteria proposed by Fosberg are applicable on a global scale. Fosberg makes a distinction between physiognomy and structure, where physiognomy refers to the external appearance of vegetation and to its gross compositional features, implying broad units such as forests, grasslands, savannahs, and deserts. Structure relates more specifically to the arrangements in space of the plant biomass. In addition, Fosberg uses *function* in the sense of seasonal leaf shedding versus *retention* (Mueller-Dombois and Ellenberg, 1974).

UNESCO's Structural-Ecological Formation System (UNESCO, 1973) intends to serve as a basis for mapping world vegetation at a scale of 1:1 million. As in Fosberg's system, structure forms the main separating criterion for the three main levels. The top two levels are *formation class* and *subclass*:

Formation class (spacing and height of dominant growth form)
    Closed forests
    Woodland or open forests
    Shrubland
    Dwarf shrubland

Terrestrial herbaceous communities
Deserts and other sparsely vegetated areas
Aquatic plant formations
Subclass: (leaf phenology)
Evergreen
Deciduous

At the lower levels (formation groups, formation, and subformation), the criteria are macroclimatic and floristic aspects. Both the UNESCO classification and Fosberg's scheme can be applied to categorize vegetation in the field and on the maps in comparative terms within each scheme and also between them.

Eiten (1968) proposed a system based on five main vegetation groups—forest, woodland, shrubland, savannah, and herbaceous field—characterized by the presence or absence of major growth forms (trees, shrubs, and herbs). These growth forms are differentiated into vegetation subgroups according to extra characteristics, such as cover, height, and leaf phenology. The system resulted in 31 distinct final structural vegetation categories. The Yangambi vegetation nomenclature is the result of a meeting of experts in tropical vegetation, which was held in 1956 in Yangambi (former Congo). The nomenclature was intended to be a descriptive system for vegetal formations of tropical Africa and was proposed to resolve the extreme confusion of vegetation terms in Africa. It encompasses 7 main vegetation groups and 24 subgroups. Its structure is somewhat unsystematic, with the main separation criterion between the different vegetation formations being mainly physiognomic, coupled with climatic and altitudinal conditions.

Among the above-mentioned vegetation classification systems, those most used for reference are the UNESCO (at the first two levels: formation class and subclass) and the Fosberg systems. The others, however, have had a certain influence in specific continents or geographical areas, such as the Eiten in Latin America and the Yangambi in West Africa.

4.6   MAIN CURRENT LC CLASSIFICATIONS AND NOMENCLATURES

An internationally accepted reference LC classification does not really exist; however, there exist major classifications and legends that in the past have played a major role in specific geographical areas. The most famous and widely applied is the Anderson land-use and land-cover classification system (Anderson et al., 1976), a revision of the land-use classification system presented by Anderson, Hardy, and Roach (1972). The classification was first developed to meet the needs of the U.S. federal and state agencies to have an up-to-date overview of land use and LC throughout the country. In this context, the system was the final result of several efforts to generate a common land-use and LC system for the whole country. One effort was the Land Use Information and Classification Conference held in Washington, D.C., in June 1971. The conference was attended by more than 150 representatives of federal, state, and local government agencies, universities, etc. One of the results of the conference was a proposal for developing a land-use and LC classification system that could be used with remote-sensing data. The Anderson system was developed at two levels, with 9 major classes in Level 1 and 37 in Level 2. It has been left open-ended with the specific objective that other levels can be added to satisfy more detailed user needs. The system has been developed to be used mainly with remote-sensing data, and even the land-use classes present in the system are directly interpreted using LC as the principal surrogate. Despite its intended use for local studies only at the national level in the United States, the system has also been applied in other parts of the world and in global initiatives (e.g., Earthsat Geocover).

Another widely applied system is the Coordination of Information on the Environment (CORINE) system. In 1985, the European Commission launched the CORINE program to produce a consistent LC database for the whole of Europe. For this purpose, a three-level hierarchical nomenclature was developed. The CORINE system was defined as a "physical and physiognomic land cover nomenclature." Despite the intention to generate a pure LC system, CORINE includes several combinations of LC and land-use terms in the 44 classes of the third level. The whole of Western Europe has been mapped several times with this system at a scale of 1:100,000 based on Landsat data, thereby producing three regional databases in 1990, 2000, and 2006. The CORINE nomenclature has been enlarged to a fourth level, creating 97 final classes in the framework of the MURBANDY and MOLAND projects that were initiated in 1998 to monitor the development of urban areas in Europe.

Today, many other national or single-project-oriented LC classifications and legends exist. Each country has at least a national LC nomenclature. In Europe, despite the use of CORINE nomenclature at the regional level, countries have continued to develop their national systems, which were then converted into the CORINE system to fulfill European Union obligations. These national systems, in countries such as Norway, Sweden, UK, and Germany, are often more detailed and tailored to local requirements. Among the long list of existing national LC nomenclatures, it is worth mentioning the extension of the area mapped according to the LC legend of China, formed by 23 classes, which is the basis for a 5-year cycle of mapping activity that started in 1990. Also worth noting is the system in India, with more than 40 LC classes, with the aim of mapping India at 1:50,000 scale.

4.7 SHORTCOMINGS AND PROBLEMS OF SEMANTIC INTEROPERABILITY WITH CURRENT SYSTEMS

Categorization has always been a useful method to minimize the complexity of the real world. However, the use of a single ontology system (a class name with class description) with a predefined list of categories implies important constraints that increase the fuzziness of the data and create huge interoperability problems. Categories (classes) are usually limited in number. This forces the map producer to drastically generalize reality. Such generalization does not necessarily correspond to the needs of many studies, which ask for more and more detailed information on natural resources. The result is an explosion in the number of classes, which can be unsystematic (an expansion of classes limited to only particular aspects of LC due to the specific needs of a particular project) and which, therefore, is difficult to manage in a GIS system.

Generalization, as well as the creation of the class itself, is often an arbitrary process. Reality is a continuum, and any division of the continuum into categories often reflects specific needs of the data producer and not necessarily the varied needs of individual end users. Threshold parameters, for instance, produce arbitrary and artificial differences in values in the real world. For most LC classification systems, class definitions are imprecise, ambiguous, or absent. The composition of class definitions in the form of a narrative text is unsystematic (many diagnostic criteria forming the system are not always applied in a consistent way) and in any case do not always reflect the full extent of the information.

Generalization into categories where meaning is very often limited to the class name, or has only an unclear class description, implies rigidity in the transfer of information from the data producer to the end-user community. End users have a limited possibility, if any, of interacting with the data, and they must therefore accept them "as is." Representation of the granularity of the aspects summarizing a specific feature of the real world is drastically reduced or lost. Often some vagueness in the class definition is artificially included by the map producer to hide some "technical anomalies" when reproducing a certain feature on the map. Moreover, vagueness or extreme complexity in the class definition makes it difficult to assess correctly the accuracy of the dataset. Further, the structure of the data with just a name and a corresponding separate text description often hampers data management with modern GIS techniques.

Semantic interoperability is actually the main challenge in spatial data infrastructures (SDIs). Interoperability is defined as "the ability of systems to operate in conjunction on the exchange or re-use of available resources according to the intended use of their providers" (Kavouras and Kokla,

2002). In the case of "semantic interoperability," we refer to the understanding of the "meanings" of different classes and relations among concepts.

On these aspects, current classifications and legends show severe limitations that bear the risk of affecting the practical use of LC information. The list below shows the most common problems encountered when dealing with semantic interoperability of classification systems:

- Different terms used for concepts (*synonymy*)
- Different understandings of homonymous concepts (*polysemy*); for example, the various meanings of the term "forest" for forestry environmental modeling
- Different understandings of the relationship of common concepts
- Common instances across databases assigned to different concepts in different ontologies
- Common instances allocated to a more general concept in one hierarchy than in the other
- Equivalent concepts formalized differently
- Equivalent concepts explicated differently

4.8 THE FAO LCCS

In 1996, FAO tried to remedy this situation by developing a new way to approach the problem. A new set of classification concepts was elaborated, discussed, and endorsed at the meeting of the International Africover Working Group on Classification and Legend in Senegal in July 1996 (Di Gregorio and Jansen, 1996, 1997a, 1997b). The system was developed in collaboration with other international initiatives on classification of LC, such as the U.S. Federal Geographic Data Committee (FGCD)—Vegetation Subcommittee and Earth Cover Working Group (ECWG); the South African National Land Cover Database Project (Thompson, 1996); and the International Geosphere-Biosphere Programme (IGBP)—Data and Information System (DIS) Land Cover Working Group, and Land Use Land Cover Change (LUCC) Core Project.

After a test period in the FAO, the Africover project (1997–1999), the first official release of LCCS (v.1), was published in 2000 (Di Gregorio and Jansen, 2000). A second version was developed based on an international feedback involving a large global community and published in 2005 (LCCS v.2) (Di Gregorio, 2005). A new version (v.3) was released in 2011.

The LCCS adheres to the concept that it is deemed more important to standardize the attribute terminology rather than the final categories. The LCCS works by creating a set of standard diagnostic attributes (called "classifiers") to create or describe different LC classes. The classifiers act as standardized building blocks and can be combined to describe the more complex semantics of each LC class in any separate application ontology (= classification system) (Ahlqvist, 2008).

The creation of or an increase in detail in conceptualizing and describing an LC feature is not linked to a text description of the classifier (as in most other systems) but to the choice of clearly defined diagnostic attributes. Hence, the emphasis is no longer on the class name but on the set of clearly quantifiable attributes. This follows the idea of a hybrid ontology approach, with standardized descriptors allowing for heterogeneous user conceptualization (Ahlqvist, 2008). The LCCS approach thus differs from most other examples of standardized LC systems (e.g., Anderson or CORINE) that follow a single ontology approach where all semantic descriptions available have been created with a very similar view on a domain and have to be shared by all users (Lutz and Klein, 2006).

During the practical use of the LCCS in recent years, there has been an unexpected trend in the utilization of the system by the international user community. In addition to the creation of legends for specific applications, the system has also been used as a reference bridging system to compare classes belonging to other existing classifications. An example is the GOFC-GOLD report no. 43 (Translating and evaluating land cover legends using the UN Land Cover Classification System—LCCS).

In 2003, FAO submitted the LCCS to the ISO Technical Committee 211 on Geographic Information as a contribution toward establishing an international standard for LC classification systems. This was the first time that the ISO committee had addressed a standard for a particular community of interest within the general field of geographical information. All of its previous standards had been high-level or abstract standards that established rules for application schema, spatial schema, or similar concepts. There was some initial difficulty in initiating the standardization activity owing to this more specific focus. The result was that a standard was first developed to address classification systems in general (ISO 19144-1 Classification Systems) and then one to address LC (ISO 19144-2 Land Cover Meta-Language). The first one, ISO 19144-1, has already become an ISO standard; the second one, ISO 19144-2, has already passed the stage of FDIS (Final Draft International Standard) and is in the final approval stage.

There are many LC systems in different countries (with differing levels of detail), and there is a large volume of legacy information that must be maintained. Some of these requirements to maintain information are, in fact, linked to the environmental and forestry laws in those countries and cannot be changed in any way. It was not the intent of the LCCS to establish a new standard that would displace all others. The intent was to bridge between the different (e.g., national) systems, using the concept of linking to a set of clearly quantifiable attributes. This would allow the description of different LC systems using a common set of elements so that they could be compared and—even more importantly—a semantic bridge could be built between them to integrate different national datasets into regional or global datasets.

Difficulties were encountered because the LCCS itself is a classification system, and it was unclear how the aggregation and bridging processes among systems would work. To make this clear, the standard's focus was shifted. A high-level meta-language called LCML (Land Cover Meta-Language) was developed for version 3. This meta-language uses the same concept of building on a set of clearly quantifiable attributes as the basis elements. The difference is that LCML is intended only to model an LC classification system and is not a system in itself. The LCML standard contains a large number of examples that clearly show that the meta-language is capable of representing classes from a large proportion of the existing LC classification systems in use in the world. Since the representation is in terms of comment elements, these elements can be used to define bridges between classification systems. LCML thus enables us to (1) aggregate data from multiple sources into global sets and (2) produce explicit and precise LC class definitions.

REFERENCES

Ahlqvist, O. 2008. In search of classification that supports the dynamics of science—the FAO Land Cover Classification System and proposed modifications. *Environment and Planning B: Planning and Design*, 35(1), 169–1996.

Anderson, J.R., Hardy, E.E., and Roach, J.T. 1972–1976. A land-use classification system for use with remote sensor data. U.S. Geological Survey Circular, 671. Washington D.C.: USGS.

Anderson, J.R., Hardy, E.E., Roach, J.T., and Witmer, R.E. 1976. A land use and land cover classification system for use with remote sensor data. A revision of the land use classification system presented in U.S. Geological Survey Circular 671 by Anderson, Hardy and Roach (1972). U.S. Geological Survey Professional Paper, No. 964. Washington D.C.: USGS.

Bjelland, T.K. 2004. Classification: Assumptions and implications for conceptual modelling. Dissertation in Information Science. Department of Information Sciences and Media Studies, Faculty of Social Science, University of Bergen, Norway. 240 p.

Bowker, G.C. and Star, S.L. 1999. *Sorting Things Out: Classification and Its Consequences*. Cambridge, MA: MIT Press, 377 pp.

Burley, T.M. 1961. Land use or land utilization? *Professional Geographer*, 13(6), 18–20.

Danserau, P. 1961. Essai de representation cartographique des elements structuraux de la vegetation. In H. Gaussen (Ed.), *Metodes de la cartographie de la vegetation* (pp. 233–255). Centre National de la Recherche Scientifique. 97th International Colloqium. Toulouse, France 1960.

Di Gregorio, A. 2005. Land Cover Classification System—Classification concepts and user manual for Software version 2. *[FAO] Environment and Natural Resources Series*, No. 8, 190 pp.

Di Gregorio, A. and Jansen, L.J.M. 1996. FAO Land cover classification system: A dichotomous, modular-hierarchical approach. Paper presented at the Federal Geographic Data Committee Meeting—Vegetation Subcommittee and Earth Cover Working Group. Washington D.C., USA.

Di Gregorio, A. and Jansen, L.J.M. 1997a. Part I—Technical document on the Africover Land Cover Classification Scheme. In *FAO Africover Land Cover Classification* (pp. 4–33; 63–76). [FAO] Remote Sensing Centre Series, No. 70. Rome: FAO.

Di Gregorio, A. and Jansen, L.J.M. 1997b. A new concept for a land cover classification system. In *Proceedings of the Earth Observation and Environmental Information 1997 Conference* (pp. 13–16). Alexandria, Egypt, October 1997.

Di Gregorio, A. and Jansen, L.J.M. 2000. Land Cover Classification System (LCCS). Classification concepts and user manual for software version 1.0. Rome: FAO, 179 pp.

Eiten, G. 1968. Vegetation forms. A classification of stands of vegetation based on structure, growth form of the components, and vegetative periodicity. *Boletim do Instituto de Botanica (San Paulo)*, No. 4.

FAO. 1988. FAO-UNESCO Soil Map of the World. Revised Legend. FAO/UNESCO/ISRIC World Soil Resources Reports No. 60 (Reprinted 1990).

Fosberg, F.R. 1961. A classification of vegetation for general purposes. *Tropical Ecology*, 2, 1–28.

Kavouras, M. and Kokla, M. 2002. A method for the formalization and integration of geographical categorization. *International Journal of Geographical Information Science*, 16(5), 439–453.

Kuchler, A.W. and Zonneveld, I.S. (Eds.). 1988. *Vegetation Mapping. Handbook of Vegetation Science*, Vol. 10. Dordecht, The Netherlands: Kluwer Academic.

Lutz, M. and Klein, E. 2006. Ontology-based retrieval of geographic information. *International Journal of Geographical Information Science*, 20, 233–260.

Mueller-Dombois, D. and Ellenberg, J.H. 1974. *Aims and Methods of Vegetation Ecology*. New York; London: John Wiley.

Sokal, R. 1974. Classification: Purposes, principles, progress, prospects. *Science*, 185(4157), 111–123.

Thompson, M. 1996. A standard land-cover classification for remote-sensing applications in South Africa. *South African Journal of Science*, 92, 34–42.

UNESCO. 1973. *International Classification and Mapping of Vegetation*. Paris: UNESCO.

USDA [United States Department of Agriculture]. 1999. Soil Taxonomy. *A Basic System of Soil Classification for Making and Interpreting Soil Surveys*, 2nd ed. Prepared by Soil Survey Staff of the Natural Resources Conservation Service. *USDA/NRCS Agriculture Handbook*, No. 436.

5 Revisiting Land-Cover Mapping Concepts

Pierre Defourny and Sophie Bontemps

CONTENTS

5.1 INTRODUCTION

Historically, the terrestrial surface has been studied from a disciplinary perspective driven by each discipline's own specifications and well-defined objectives. Vegetation mapping by ecologists can be traced back to a century-long tradition, describing the surface in terms of presence and abundance of specific plant species. Early work in terrain classification systems focused on the physiographic description of the land forms and of the plant physiognomic types. As for geographers, they were more concerned with land-use information gathered manually through field and socioeconomic observations. Similarly, public administrations and agencies relied on their own data specifications and data collection and categorization methodologies for defining and recording land-based features of interest. However, such a disciplinary perspective for terrestrial surface characterization is no longer affordable. In addition, the scientific agenda of these disciplines has shifted from land inventory to process understanding and numerical modeling. Meanwhile, specific disciplines dealing with georeferenced information to study the terrestrial surface have emerged, supported by technological development in earth observation (EO), geographical information systems, and image processing.

There are many ways of describing and representing land-surface features. Historically, land use has been considered more relevant for many applications, and the overriding trend has been to focus on this information (Fisher et al., 2005). Recording of land cover is a relatively recent phenomenon and is closely linked to the availability of satellite imagery. Recently, the need for interdisciplinary approaches to fully understand the interactions within the land system (Verburg et al., 2009) has been widely recognized. Land cover has been transformed into a universal panacea for land inventory and has been adopted by a wide range of disciplines (Comber et al., 2005). Today, an appropriate land-cover map is increasingly required by a broad spectrum of scientific, economic, and governmental applications as an essential input to assess ecosystem status and biogeochemical cycling, understand spatial patterns of biodiversity, parameterize the land surface for modeling (e.g., water, climate, and carbon), and develop land management policy.

Thanks to information technology development and methodological advances, the remote-sensing community is becoming an undisputed provider of land information to a very wide range of users at all geographical scales. This has been recently illustrated by the tens of thousands of users registered for the GlobCover products. The split between land information producers and users has drastically increased and is enhanced by spatial-data infrastructure development making widely available "relevant, harmonized and quality geographic information to support formulation, monitoring and evaluation of policies" (EU INSPIRE directive available at http://inspire.jrc.ec.europa.eu).

Land cover is being used as a surrogate to describe the landscape structure and character by an increasing number of users who may be unaware or ignorant of the origin and semantics of land-cover information. Comber et al. (2004) demonstrated that land cover is perceived differently according to the discipline. If users do not fully understand the meaning of land cover and the assumptions behind it, then they impose their own interpretations of what land cover should encapsulate relative to their constraints, focus, and objectives, which may affect their assessment of the data and their subsequent analyses. Indeed, the literature on remote sensing addresses very well the land-cover variations related to data source type (e.g., Atkinson and Aplin, 2004) and the image-processing methods (e.g., Fritz et al., 2008), but it rarely discusses the assumptions and paradigms related to land-cover information. Similarly, metadata standards are adequate for assessing technical constraints, but they convey nothing about the organizational or epistemological context that gave rise to the data in the first place (Comber et al., 2005).

In this context, this chapter discusses the land-cover mapping practices and proposes to revisit the land-cover concept to address current shortcomings and describe the land surface better. This investigation supports the land-cover component of the European Space Agency (ESA) Climate Change Initiative and specifically focuses on the global scale. First, the current practices in the remote-sensing community are discussed. Common ideas about land-cover classification are then presented with some major examples. Finally, the conceptualization of land cover is reviewed and revisited to facilitate a better land-cover description.

5.2 CURRENT CHALLENGES FOR GLOBAL LAND-COVER PRODUCTS

Building on the increasing availability of EO satellite data, land-cover mapping from spectral and temporal signatures has progressively become one of the most popular approaches to describing land surface. The land surface in different regions of the world has been mapped and characterized several times. A number of global land-cover mapping activities have emerged and evolved with the availability of global satellite observations of moderate spatial resolution since the early 1990s. These efforts have yielded several products in the 300-m to 1-km spatial resolution range, all based on a "single-sensor" approach.

More recently, the accumulation of global multiyear time series of EO data has allowed the delivery of several and/or successive global land-cover products derived from the same sensor. This capacity to produce successive maps based on data acquired by a single sensor is certainly a major advance, but it has also raised new issues. For instance, Friedl et al. (2010) illustrated how significant the differences are between collection 4 and collection 5 land-cover products, both based on MODIS time series. Owing to the improvement in collection 5 product, new annual 500-m spatial resolution maps for 2001–2007 were released. Yet, significant year-to-year variations in land-cover labels not associated with land-cover change are observed (Friedl et al., 2010). This problem seems to be partly explained by the fact that many landscapes include mixtures of classes at a 500-m

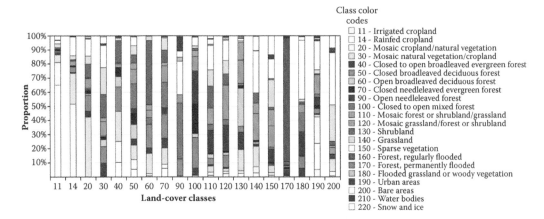

Class color codes
☐ 11 - Irrigated cropland
☐ 14 - Rainfed cropland
☐ 20 - Mosaic cropland/natural vegetation
☐ 30 - Mosaic natural vegetation/cropland
■ 40 - Closed to open broadleaved evergreen forest
■ 50 - Closed broadleaved deciduous forest
☐ 60 - Open broadleaved deciduous forest
■ 70 - Closed needleleaved evergreen forest
■ 90 - Open needleleaved forest
■ 100 - Closed to open mixed forest
■ 110 - Mosaic forest or shrubland/grassland
■ 120 - Mosaic grassland/forest or shrubland
■ 130 - Shrubland
■ 140 - Grassland
☐ 150 - Sparse vegetation
■ 160 - Forest, regularly flooded
■ 170 - Forest, permanently flooded
■ 180 - Flooded grassland or woody vegetation
■ 190 - Urban areas
☐ 200 - Bare areas
■ 210 - Water bodies
☐ 220 - Snow and ice

FIGURE 5.1 (See color insert.) Classification trajectories of the pixels that are not identically classified in the GlobCover 2005 and 2009 land-cover products. (From Bontemps, S. et al., GlobCover 2009—Products description and validation report, version 2.0, 17/02/2011. Available at: http://ionia1.esrin.esa.int/. With permission.)

spatial resolution. Further, year-to-year variability in phenology and disturbances such as fire, drought, and insect infestations make a consistent annual characterization rather difficult. Similarly, the comparison between the GlobCover 2005 and 2009 maps as well as between the GlobCorine 2005 and 2009 maps highlighted discrepancies between products even though they were based on the same sensor and the same methods (Bontemps et al., 2011; Defourny et al., 2010). Even if this issue is mostly observed between classes that are ecologically proximate, as illustrated in Figure 5.1 for the GlobCover products, there is a need for reducing the amount of spurious year-to-year change in the maps (Friedl et al., 2010).

In the context of the ESA land-cover project in the framework of its Climate Change Initiative (ESA, 2009), preliminary tests completed from daily SPOT-Vegetation time series also clearly highlighted this issue. Figure 5.2 displays consecutive but slightly different land-cover products obtained for the years 2007 and 2008 over Africa.

These results were produced by the automated GlobCover processing chain from the same type and same amount of SPOT-Vegetation data and using the same legend. At this resolution of 1 km, very few land-cover changes are expected to be visible for a 1-year interval. Therefore, the difference between these products can probably be related to the random component of the classification error and to the interannual variability of the seasonality observed for the different biomes.

As for many land-cover mapping activities, including those based on high spatial resolution data like the CORINE land-cover program, the expected stability of the product over time is not easy to reach. In the literature (Jung et al., 2006; McCallum et al., 2006), the discrepancy between several and/or successive land-cover products is often explained by the incompatibility between the land-cover typology and the limited accuracy of the classification outputs, that is, around 85% in general and around 75% for global products. As a result, land-cover change information cannot be derived from the direct comparison of such products. Clearly, the land-cover instability across products calls for alternative approaches or alternative concepts.

5.3 LAND-COVER CLASSIFICATION ISSUES

The real world is infinitely complex, and any interpretation of EO data involves processes such as abstraction, classification, aggregation, and simplification. For a long time, there has been some diversity of opinion about what land cover is and how it is distinct from land use. As there is no

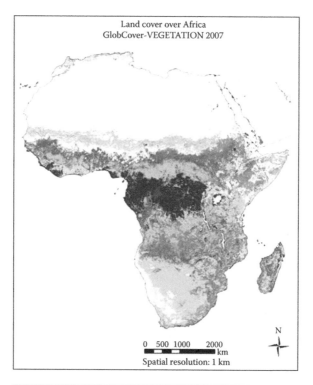

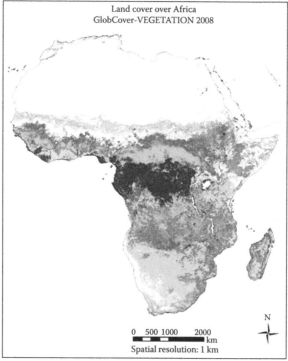

FIGURE 5.2   (See color insert.) Land-cover results obtained by the automated GlobCover classification chain from 2007 to 2008 daily SPOT-Vegetation time series. (From Moreau, I., Méthode de cartographie globale de l'occupation du sol par télédétection spatiale: Analyse de la stabilité interannuelle de la chaîne de traitement GlobCover, mémoire de fin d'études, Université Catholique de Louvain, Faculté d'ingénierie biologique, agronomique et environnementale, 2009. With permission.)

agreed fundamental unit for land observation, land-cover mapping must be understood as a process of information extraction governed by rules grounded in individual or institutional objectives.

Most of the major land-cover mapping initiatives have created their own classification systems and described them in great detail. At the very beginning of the satellite observation era, the U.S. Geological Survey (USGS) had established a standardized land-use and land-cover classification system based on 40 years of mapping experience using aerial photographs (Anderson et al., 1976). This is considered one of the most influential works in the area of development of national standards to serve various agencies.

With the increasing expectations of users and the ever-growing data availability, this kind of documentation effort is still going on all over the world. To support its land-use typology, the European CORINE (Coordinating Information on the European Environment) classification (European Commission, 2001) had to redefine what it considered to be land:

A delineable area of the Earth's terrestrial surface, embracing all attributes of the biosphere immediately above or below this surface, including those of the near surface climate, the soil and terrain forms, the surface hydrology including shallow lakes, rivers, marshes and swamps, the near-surface sedimentary layers and associated groundwater and geohydrological reserves, the plant and animal populations, the human settlement pattern and physical results of past and present human activity (terracing, water storage or drainage structures, roads, buildings, etc.).

In this case, owing to the difficulty in establishing clear thresholds between land and water (e.g., for wetlands), the concept of land was extended to inland water areas and tidal flats. This definition, however, is to be clearly separated from the concept of land area used for statistical purposes (e.g., by EUROSTAT [1998]), which excludes lakes, rivers, and coastal areas.

To ensure full interoperability between typologies and provide a common ground for land assessment, the AFRICOVER program led by the Food and Agriculture Organization of the United Nations (FAO) developed Land Cover Classification System (LCCS) as a conceptual tool for legend definition. Through a dichotomous, modular hierarchical system based on several sets of descriptors, namely the classifiers, this FAO-LCCS tool aims at explicitly clarifying each land-cover class and therefore allows translation from one typology to another (Di Gregorio and Jansen, 2000). This system is based on independent and universally valid land-cover diagnostic criteria rather than on a predefined set of land-cover classes. Its output is a comprehensive land-cover characterization, regardless of mapping scale, land-cover type, data collection method, or geographic location (Di Gregorio, 2005). As there was no internationally accepted LCCS, the FAO, jointly with the United Nations Environment Programme (UNEP), submitted the LCCS for approval in 2006 to make it an international standard through the technical committee of the International Organization for Standardization (ISO).

However, the objective of land-cover scheme standardization was challenged by Comber et al. (2008), who argued that land cover is in essence a socially constructed concept and that data producers use a classification scheme that is appropriate for their own context and related to their specific sociopolitical and technical setting. Meanwhile, Ahlqvist (2008) proposed a set of modifications to improve the flexibility of the LCCS, such as unbounded classifiers and a richer class description. According to this author, the LCCS imposes a view of land-cover categorization that is strictly and precisely hierarchical and that often imposes crisp univariate distinctions.

While the land-cover community is still debating the diversity of conceptualizations in how to represent land cover and the drawbacks of the incompatibility of typologies, other scientific communities such as the climate and vegetation modelers (Bonan et al., 2002) are expressing their land-information needs. For many years, these communities required maps expressed in Plant Functional Types (PFT), including C3/C4 plant discrimination, for their dynamic global vegetation models. This categorization of vegetation into a limited set of discrete types is based on morphological and physiological traits. Although the suite of MODIS land-cover products already includes a PFT map,

Poulter et al. (2011) recently converted the LCCS legends of existing global land-cover products in PFT to make them readily usable to the modelers. On the other hand, to move toward viewing vegetation as continuum rather than as discrete classes, Ustin and Roberts (2010) proposed the concept of optically distinguishable functional types, namely the "optical types," as a unique way of addressing the scale dependence of the vegetation description. Because plants are essentially solar energy factories, remote sensing directly assesses key structural and physiological features of plants. This new concept of optical types would be based on fundamental physical principles (e.g., radiative transfer theory and principles of spectroscopy) that interact with the vegetation structure, phenology, and biochemistry and physiology. These variables are related in predictable ways according to the functional convergence theory (Ustin and Roberts, 2010).

5.4 ALTERNATIVE STRATEGY FOR LAND-COVER INFORMATION

To reduce the constraints in classical land-cover classification schemes, different alternatives to land characterization have been developed.

The first alternative is the development of fusion methods from several existing land-cover products, which have been proposed to derive a better map—with reduced uncertainties and the desired classification legend—for specific applications (Jung et al., 2006).

Other initiatives, driven by well-targeted objectives, focus on the delivery of single land-cover class products or binary masks. This can be achieved by compiling all the available information about a single land-cover or land-use class from multiple sources, as is the case for the global croplands map at 10-km spatial resolution (e.g., Thenkabail et al., 2009). More recently, the global croplands extent has been directly derived from multiyear 250-m MODIS time series (Pittman et al., 2010). A set of 39 multiyear MODIS metrics was employed to depict cropland phenology and to derive a global per-pixel cropland probability layer using global classification tree algorithms. This study also resulted in a discrete cropland/non-cropland indicator. Hansen et al. (2005) also processed a large amount of data to obtain a forest/no-forest map at a global scale. Looking very specifically at the global urban extent, Schneider et al. (2010) developed a decision-tree classification algorithm based on temporal and spectral information in 1 year of MODIS observations and on a global training database. To overcome the confusion between urban areas and other land-cover types, stratification based on climate, vegetation, and urban topology was *a priori* applied. Such a class-specific approach also allowed working at the global scale, based on high spatial resolution data, as demonstrated by Giri et al. (2010) with the production of a mangrove atlas. All these initiatives offer the advantage of providing an extended description of the land-cover class of interest. Conversely, a major drawback is the absence of any concern for complementarities between products, thus possibly leading to significant spatial incompatibility or semantic inconsistency.

A third type of alternative strategy sets out to describe the vegetation in terms of continuous fields (DeFries et al., 1995; Smith et al., 1990). The MODIS continuous-field products are subpixel layers representing the percentage of bare ground, herbaceous, and tree cover and, for tree cover, the proportions of evergreen, deciduous, needle-leaved, and broadleaved species (Hansen et al., 2002). Continuous fields were obtained from a regression tree algorithm using (1) a continuous training dataset covering the whole range of vegetation cover and (2) multitemporal metrics based on a full year of coarse spatial resolution satellite data. The regression tree algorithm used the multitemporal metrics as independent variables to recursively split the tree cover (amounting to the dependent variable in this case) into subsets, which maximize the reduction in the residual sum of squares. Continuous fields of vegetation properties offer advantages over traditional discrete classifications since they allow better representation of areas of heterogeneity by depicting each pixel as a percent coverage. In this respect, this approach seems to be very appealing and relevant for many natural and seminatural landscapes. On the other hand, it is quite difficult to validate it because of the lack of a reference dataset.

Unlike the other approaches described here, the retrieval of biophysical variables from satellite time series should result in a quantitative description of the land surface in all dimensions—thanks to a physically based algorithm. For instance, remote-sensing products for leaf area index (LAI), fraction of absorbed photosynthetically active radiation (fAPAR), albedo, burnt areas, and soil moisture provide direct estimates of variables that can also be measured on the ground. The combination of all these biophysical variables is expected to fully characterize the land surface, which could then possibly be converted to land cover (if this information is still needed). However, the comparison of the current global biophysical products has highlighted some significant discrepancies depending on the sensors and the methods used. Furthermore, except for the method developed by Pinty et al. (2010) encompassing some specific areas, the current retrieval algorithms run separately for each variable and do not consider the necessary consistency across variables.

5.5 TOWARD A MORE DYNAMIC LAND-COVER DESCRIPTION

Despite the increasing use of land-cover information for scientific and policy purposes, land-cover data collection and interpretation processes are not operational in comparison to other major EO domains such as oceans and atmosphere (Chapter 26). The current state of the art in land cover, the increasing availability of remote-sensing data, and the shortcomings of the current approaches call for revisiting the land-cover concept even while capitalizing on all the experiences acquired in various contexts of land-cover mapping.

In a global-scale mapping perspective, which would be primarily supported by multisensor and multiannual remote-sensing datasets, the proposed land-cover description aims at integrating the advantages of some existing approaches into a more dynamic land-cover conceptualization. The objective of this section is to present the interactions between the epistemology of land-cover mapping and the ontology of the derived land-cover information in order to fully introduce this more dynamic type of land-cover information.

Ontology was originally the branch of metaphysics that dealt with the nature of being. As recalled by Ahlqvist (2008), the term has, during the last 10 years or so, been used in the literature on geographic information science, where its meaning ranges from the metaphysical science of being to a more computer-oriented concept. In this latter case, ontology is a formal specification of a common terminology by which shared knowledge can be represented. Therefore, it describes what land cover actually means in a wider sense that includes the epistemology of data collection, preprocessing and processing, and the ontological aspects of determining what features are to be included in each class.

5.6 LAND-COVER MAPPING EPISTEMOLOGY

The extraction of land-cover information from remotely sensed data relies on a series of complex processes, as the radiance measured in $W/m^2 \cdot str$ by the sensors does not allow for directly inferring the land features (owing to the seemingly inherent high levels of variance for a given feature type and between feature types).

Geographic data necessarily abstract from the reality or perception of reality from space. The abstraction process is deeply entrenched in the social and political context of the operatives (Comber et al., 2005) and results in relativistic measures of reality. Clearly, the production and use of land-cover data cannot be divorced from social experience (interest, constraint, context, etc.). This implication is clearly illustrated by the fact that different agencies have developed their own view of the world because of their particular mandates.

As reported by Comber et al. (2005), Jones suggests a middle ground in the realism–relativism debate: accept epistemological relativism (which assumes that we can never know reality exactly as it is) while rejecting ontological relativism (according to which our accounts of the world are not constrained by nature). This middle-ground position accepts diverse interpretations of a common reality as "meanings" rather than "truths" and sees the real world as being culturally filtered as meanings are constructed (Jones, 2002), thus avoiding both the naivety of "pure" realism and the impracticality of "pure" relativism. Defining a set of descriptive primitives (standing for building blocks that can be aggregated according to needs, thus allowing to deal with the social construction of the land cover) refers to this position.

5.7   FROM PIXEL TO OBJECT

Although the use of units or objects is self-evident in many scientific fields, it is not so in land-cover/land-use fields. Rasters made of pixels and vectors made of objects are the two main conceptual models designed to describe the spatial dimension of the world. The land is discretized in pixels by satellite imagery. When the pixel size is close to or larger than the land-cover features to map, land-cover information is generally presented as pixels. For very high spatial resolution imagery providing pixels much smaller than the land-cover features, the vector model is usually preferred, and land-cover objects are delineated.

The meaning of an object is a complex problem since the description of a part of the earth's surface presupposes that the area is clearly defined in space (Duhamel and Vidal, 1998). Although many objects are easily identifiable and have boundaries corresponding to physical discontinuities (e.g., plots of farmland or built-up areas), these boundaries typically become blurred in natural landscapes. In this case, the approach by continuous field is much more suitable for depicting natural gradient over space. Along these lines, an interesting concept is a hybrid of fiat and bona fide boundaries according to the properties of the geographic objects (Smith and Mark, 2001; Smith and Varzi, 2000).

The proposed land-cover information model obviously tries to take the best of these two worlds by using a pixel structure where pixel clusters would be handled as objects described by attributes and by supporting continuous fields for objects showing some gradient.

5.8   FROM CRISP CLASSIFICATION TO RICH DESCRIPTION

Most of the land-cover classification problems come from the attempt to classify the infinite variety of landscapes into a limited number of closed classes. Any classification system may be subject to controversy and discussion, all the more so if they are fixed and precise. Indeed, as explained here, land-cover classification systems are socially constructed from a specific cultural and technical context. However, as discussed by Ahlqvist (2008), classification is the necessity to structure a specific knowledge domain in order to create consistency and stability in communication between individuals.

The maximum of flexibility in a classification system can be preserved by defining a minimum set of descriptive primitives that act as building blocks. The land-cover features of the real world can then be classified, starting from a very simple group of elements (the descriptive primitives) and assembling them in different ways to describe the more complex semantic in any separate application ontology (legends).

A number of attempts to use descriptive primitives can be found in the literature. A literature review done by Comber et al. (2008) to better distinguish land cover from land use identified the following list of 14 primitives:

1. Naturalness, the extent to which the class was a naturally occurring feature or was directly the result of anthropogenic activity
2. Vegetation height, indicating the minimum height of the vegetation
3. Vegetation canopy coverage, indicating the minimum percentage of vegetation coverage
4. Homogeneity of appearance
5. Seasonality, the extent to which the class is seasonal or perennial

6. Structure, indicating complexity of vegetation structure
7. Wetness, specifying the dependency on specific wetness conditions (e.g., soil, growing medium, and climate)
8. Biomass production, related to the amount of energy fixed through photosynthesis by the class
9. Human activity, indicating the amount of human-related activity in the class
10. Human disturbance, defining the extent to which the existence and nature of this class reflects anthropogenic activity
11. Economic value, the economic importance of this class—how much money can be earned or how much it is worth
12 Production of crop-related food
13. Production of animal-related food
14. Artificiality, the extent to which the surface has been artificially created

Building on the classifiers' experience, the LCCS team (http://www.glcn.org/ont_2_en.jsp) is aiming at developing a Land Cover Meta Language (LCML), which should work as a "boundary object" to mediate and support negotiations of different ways to represent land cover. This means that classes derived by this LCML could be customized to user requirements but should have common identities between users. Such an LCML approach should also allow extension of similarity assessment and semantic distance expression as requested by Ahlqvist (2008). The challenge of such an approach is to define appropriate "building blocks," which provide a common ground to all users and thus guarantee a global standardization, and at the same time, limit the number of these blocks as much as possible to open the possibilities of representing distinctive land-cover situations.

Beyond the descriptive primitives, the object-based approach allows enrichment of this pre-defined set of land-cover basic blocks on their semantic significance with external qualities and attributes. These qualities and attributes can vary according to the descriptive primitive values and can be optional in some cases. In this way, their essential purpose is to describe the land cover with the best knowledge available rather than to merely identify the corresponding land-cover class from a predefined legend.

Such an approach allowing a richer description of land cover is expected to bring significant enhancement of the land characterization. This was already reported based on the new object-oriented data model of the Spanish mapping agency. Facing the same limitations and shortcomings of the hierarchical classification conceptual models used in CORINE, the Spanish mapping agency developed a data model concept to describe rather than classify each map polygon. Arozarena et al. (2006) explained that each polygon could have one or more covers and that each cover could be qualified by one or more attributes of biophysical or socioeconomic nature. Designed according to the main INSPIRE principles and ISO TC/211 standards, this concept moves from previous hierarchical land-cover databases toward a land-cover feature data model that allows deriving as many "land-cover views" as users require. This Land Cover and Use Information System of Spain (SIOSE for Sistema de Información sobre Ocupación del Suelo de España) was successfully demonstrated between 2006 and 2009, based on 2.5-m spatial resolution SPOT 5 imagery. Furthermore, it can integrate differ-ent datasets from various administrations (Environmental Ministry, Agriculture Ministry, Housing Ministry, Economy and Treasury Ministry, and Education and Science Ministry) into a decentralized and cooperative production model by the Spanish national and regional administrations (SIOSE 2011).

5.9   A NEW LAND-COVER ONTOLOGY

The absence of agreed-upon fundamental units for land-cover observation probably prevents any definitive standardization. However, it is recognized that at any time and place, there is a land cover to some level of observable granularity. This is why the most appropriate land-cover definition is "the observed bio-physical cover on the Earth's surface" as proposed by Di Gregorio and Jansen (1997).

Indeed, Burley ([1961] in Anderson et al. [1976]) first defined land cover as the vegetation and the artificial constructions covering the land. In the context of LCCS, land cover refers to the physical and biological cover over the surface of land, including water, vegetation, bare soil, and/or artificial structures (Di Gregorio, 2005). The Integrated Global Observation for Land (IGOL) theme also reported land-cover definition as "the observed bio-physical cover on the earth's surface" while recognizing the confusion between land cover and land use in current practices (Townshend et al., 2008). Land use characterizes the arrangements, socioeconomic activities, and inputs people are undertaking on a certain land-cover type. It includes both space and time dimensions, and theoretically it should be considered separately from land-cover type to ensure internal and external consistency and comparability (GLP, 2005).

Such a land-cover definition—much related to the observation process—is somewhat incompatible with the basic requirement of temporal stability expressed by users. According to the Global Climate Observing System (GCOS) community (GCOS, 2004, 2010), for instance, the stability of land-cover information between compatible products is of higher priority than the accuracy of the respective products. However, the most recent series of global land-cover products specifically point out this inconsistency issue as a quite difficult one to tackle (Bontemps et al., 2011; Friedl et al., 2010). Yet it must be recognized that land cover cannot, at the same time, be defined as the physical and biological cover on the earth's surface (Di Gregorio, 2005; Herold et al., 2009) and remain stable and consistent over time as expected by most users.

This conclusion calls for the development of a new land-cover ontology, which explicitly addresses the issue of inconsistency between annual land-cover products and/or of the sensitivity of the products to the observation period. The proposed land-cover ontology assumes that the land cover is organized along a continuum of temporal and spatial scales and that each land-cover type is defined by a characteristic scale, that is, by typical spatial extent and time period over which its physical traits are observed (Miller, 1994). This twofold assumption requires introduction of the time dimension in land-cover characterization, which contributes to defining land cover in a more integrative way. This conceptualization, detailed below, still attempts to build on most of the past experiences in the field, including the recent developments around the LCCS.

5.10   LAND-COVER FEATURES AND CONDITIONS

Accounting for the time dimension allows us to distinguish between the stable and the dynamic component of land cover. The stable component, named as "land-cover features," refers to the set of land elements that remain stable over time and thus define the land cover independently of any sources of temporary or natural variability. Conversely, the dynamic component is directly related to this temporary or natural variability that can induce some variation in land observation over time but without changing the land-cover feature in its essence. This second component is referred to as "land-cover conditions."

Land-cover features and land-cover conditions can be mapped through the use of *descriptive primitives*, corresponding to the building blocks of any landscape.

Land-cover features are defined by an ensemble of descriptive primitives depicting the most permanent aspect or stable elements of the landscape. They are characterized at least by the following:

1. The *nature* of the observed features, such as tree, shrub, herbaceous vegetation, moss/lichen vegetation, terrestrial or aquatic vegetation, inland water, built-up areas, and permanent snow/ice
2. The *structure* of the observed features, which refers to vegetation height, vegetation cover, and building density according to nature
3. The *naturalness* of the observed features, such as the level of artificiality, species information, and the number of cropping cycles
4. The *homogeneity* of the observed features at the level of observation, leading to a pure or mosaic object

The land-cover features could still be described using the LCCS classifiers if compatibility with the existing products is required. The anthropogenic dimension, included in the "level of artificiality" of the features' naturalness, refers to the land use and not the land cover and should not be mixed up from a conceptual point of view. However, the typical uses of land-cover products need to include this kind of simple surrogate for land use. The main argument for including it at the production level is that most users would anyway attempt to convert some land-cover information into this level of land-use information.

The *land-cover conditions* encompass the interannual processes modifying temporally the land surface throughout the year. Typically driven by biogeophysical processes, they correspond to an annual time series mode of "instantaneous observations" of the land-cover features. The land-cover conditions are described by different observable variables:

1. The *green vegetation phenology* through vegetation index (e.g., the normalized difference vegetation index—NDVI) profiles
2. The *snow coverage* allowing users to derive the snow-cover period
3. The *open water presence* related to floods, water extent dynamics, or irrigation
4. The *fire occurrence* and the associated *burn scars*

The land-cover condition can be described in a relevant way through an interpolation between "instantaneous" observations of the land-cover features. This can take the form of time profiles in the case of continuous variables (e.g., NDVI) or of temporal distribution of occurrence probabilities in the case of discrete variables (e.g., snow or water). In this way, the land-cover condition provides reference information depicting the land-cover seasonal pattern, which is not related to a given year. Ideally, this information should be obtained on a multiyear basis. In the case of continuous variables, mean time profiles are associated with standard deviation values, which then convey the interannual variability.

Table 5.1 illustrates this new land-cover concept (features and conditions) with two distinct illustrations, the first one referring to artificial urban areas and the second to a dense tropical forest.

Using this new concept of land cover made up of features and conditions offers the opportunity to characterize land cover in a more integrative way than as just categories (forest or open water) or as continuous variable classifiers (fraction of tree canopy cover). This new concept helps address the critical requirements of stability between successive annual products while integrating the dynamic dimension at the intraannual and seasonal levels. Of course, such land-cover ontology calls for specific methods to extract these different land-cover components appropriately and efficiently. On the other hand, validation of this land-cover information appears to be more compatible with expert knowledge, often used as a reference source.

As a result of this revisited definition of land cover, *land-cover change* must be referred to as a permanent modification of the land-cover features, not of the land-cover conditions—compared to a baseline status. Indeed, in the broadest sense of the term, change can be defined as the process of passing from one status to another. Applying this generic definition of change to the new land-cover concept introduced here would mean that the land cover has changed when its features (i.e., its permanent aspect or stable elements) have been modified, over time and/or in space, in such proportions that other values of descriptive primitives are required to describe them.

Such conceptualization comprises three peculiarities of land-cover change that would have to be considered to set up monitoring activities. First, a change is not an intrinsic event of the land cover but is related to some of its features. According to the features' descriptive primitives that are modified and the intensity of the modification, the land cover can be *transformed* in essence (i.e., become radically different by losing its original-feature nature) or *altered* in some particular way (i.e., become different while retaining the same-feature nature). Second, since each land-cover type is defined by a characteristic spatiotemporal scale, any change needs to be appreciated along spatial and temporal scales. If scales significantly higher or smaller than the characteristic

TABLE 5.1 (See color insert.)

Illustration of the Proposed Concepts of Land-Cover Features and Land-Cover Conditions

Land-cover features (permanent aspect or stable elements of the landscape)

Features' nature: built-up
Features' structure: high density of building
Features' naturalness: artificial
Features' homogeneity: urban patterns made of a mixture of green areas, buildings, houses, and water channels

Land-cover condition (dynamic component of land cover)

Seasonal behavior of the *green vegetation* (NDVI profile)
Snow cover usually from December 15 to January 15
No *flooding* dynamic
No *fire* dynamic

Possible denomination of this land cover according to the following:
A land-cover typology A: Urban area
A land-cover typology B: Residential area
A land-cover typology C: Impervious surface area

Land-cover features (permanent aspect or stable elements of the landscape)

Features' nature: tree cover
Features' structure: high tree density (canopy cover of 92%)
Features' naturalness: natural broadleaved, evergreen vegetation
Features' homogeneity: homogeneous canopy (few clearings)

Land-cover condition (dynamic component of the land cover)

Slight seasonal behavior of the *green vegetation* (NDVI profile)
No *snow* dynamic
No *flooding* dynamic
No *fire* dynamic

Possible denomination of this land cover according to the following:
A land-cover typology A: Closed evergreen forest
A land-cover typology B: Natural woody vegetation
A land-cover typology C: Dense broadleaved forest

scale are used in the monitoring activities, there is a high risk of misinterpretation of the land-cover type, because land-cover features observed at one scale are not automatically relevant at another scale. Third, change is a relational difference between statuses (more precisely, between the status before and the status after the event inducing the change): the land cover has changed compared to baseline requirements. The specification of the baseline requirements (i.e., of the change thresholds) is directly linked to the descriptive primitives relevant for the land-cover features. Accordingly, coupling a new land-cover concept, which allows us to distinguish between the stable and the dynamic component of land cover, with a more flexible classification system based on a limited number of descriptive primitives also opens up new possibilities in the field of land-cover change ontology.

5.11   CONCLUSION

Satellite remote sensing measures land-surface properties in the spectral domain—thanks to the radiative transfer—and in the temporal domain through time series of observations. Both measurements allow the recognizing and mapping of terrestrial surface features. The availability of multiannual time series from instruments of coarse to medium spatial resolution and the increasing processing capability have made feasible the production of regular or annual land-cover maps, even at the global scale. However, the current land-cover classification products are found to be very sensitive to the timing of the observations and to the content of the annual time series, with any variation in one of them inducing various discrepancies between successive annual products. This issue partly results from a rather ambiguous land-cover definition.

To enhance the land-cover description and address this stability issue, a new land-cover ontology based on few descriptive primitives has been proposed, in which the land-cover features (standing for the stable elements of the landscape) are explicitly separated from the land-cover conditions (standing for its dynamic component). The proposed approach remains fully compatible with the standardized LCCS while being much more supported by the Media Center Markup Language (MCML) ontology. In the context of the ESA land-cover project in the framework of its Climate Change Initiative, information extraction processes to characterize both the land-cover features and the land-cover conditions will be tested and possibly implemented at the global scale.

Major steps toward land-cover characterization are still to come with the future availability of high spatial resolution time series, such as those announced from Sentinel 2 missions. More original processes are also expected from light detection and ranging (LIDAR) imagers providing information in the vertical domain to build 3-D land-surface descriptions. Furthermore, collaborative data collection voluntarily by various stakeholders, shared through geowiki interfaces, may completely change the epistemology of land-cover mapping but would still support well the revisited land-cover concept.

REFERENCES

Ahlqvist, O. 2008. In search of classification that supports the dynamics of science: The FAO Land Cover Classification System and proposed modifications. *Environment and Planning B: Planning and Design*, 35, 169–186.

Anderson, J.R., Hardy, E.E., Roach, J.T., and Witmer, R.E. 1976. A Land Use and Land Cover Classification System for use with remote sensor data. U.S. Geological Survey Paper 964. Washington, D.C.: USGS. Available at: http://landcover.usgs.gov/pdf/anderson.pdf

Arozarena, A., Villa, G., Valcárcel, N., Peces, J.J., Domenech, E., and Porcuna, A. 2006. New concept on land cover/land use information system in Spain. Design and production. *In Proceedings of the 2nd Workshop of the EARSeL SIG on Land Use and Land Cover*. Bonn, Germany: Centre for Remote Sensing of Land Surfaces, September 28–30, 2006.

Atkinson, P.M. and Aplin, P. 2004. Spatial variation in land cover and choice of spatial resolution for remote sensing. *International Journal of Remote Sensing*, 18, 3687–3702.

Bonan, G.B., Levis, S., Kergoat, L., and Oleson, K.W. 2002. Landscapes as plant functional types: An integrating concept for climate and ecosystem models. *Global Biogeochemical Cycles*, 16, 5–17.

Bontemps, S., Defourny, P., Van Bogaert, E., Kalogirou, V., and Arino, O. 2011. GlobCover 2009—Products description and validation report, version 2.0, February 17, 2011. Available at: http://ionia1.esrin.esa.int/

Comber, A.J., Fisher, P., and Wadsworth, R. 2004. Integrating land-cover data with different ontologies: Identifying change from inconsistency. *International Journal of Geographical Information Science*, 18, 691–708.

Comber, A.J., Fisher, P., and Wadsworth, R. 2005. What is land cover? *Environment and Planning B: Planning and Design*, 32, 199–209.

Comber, A.J., Fisher, P., and Wadsworth, R. 2008. The separation of land cover from land use using data primitives. *Journal of Land Use Science*, 3, 215–229.

Defourny, P., Bontemps, S., Van Bogaert, E., Weber, J.L., Steenmans, C., Brodsky, L., Kalogirou, V., and Arino, O. 2010. GlobCorine 2009—Description and validation report, version 2.2, December 3, 2010. Available at: http://ionia1.esrin.esa.int/

DeFries, R., Field, C.R., Fung, I., Justice, C.O., Los, S., Matson, M.A., Matthews, E.A., et al. 1995. Mapping the land surface for global atmosphere-biosphere models: Towards continuous distributions of vegetation's functional properties. *Journal of Geophysical Research*, 100(D10), 20867–20882.

Di Gregorio, A. 2005. UN Land Cover Classification System (LCCS)—Classification concepts and user manual for software version 2. Rome: FAO. Available at: http://www.fao.org/docrep/003/X0596E/X0596e00.HTM

Di Gregorio, A. and Jansen, L.J.M. 2000. Land cover classification system (LCCS): Classification concepts and user manual. GCP/RAF/287/ITA Africover-East Africa Project and Soil Resources, Management and Conservation Service, Food and Agriculture Organization.

Di Gregorio, A. and Jansen, L.J.M. 1997. A new concept for a land cover classification system. In *Earth Observation and Environmental Classification*, Conference Proceedings, October 13–16, 1997, Alexandria, Egypt.

Duhamel, C. and Vidal, C. 1998. Objectives, tools and nomenclatures. In *Eurostat (1998): Land Cover and Land Use Information Systems for European Union Policy Needs. Proceedings of the Seminar*. Luxembourg, January 21–23, 1998.

ESA—European Space Agency. 2009. ESA Climate Change Initiative. Description. Reference: SEP/TN/0030-09/SP. Available at: http://www.esa-cci.org

European Commission. 2001. *Manual of Concepts on Land Cover and Land Use Information Systems*. Luxembourg: Office for Official Publications of the European Communities, ISBN 92–894-0432-9. Available at: http://ec.europa.eu/eurostat/ramon/statmanuals/files/KS-34-00-407-__-I-EN.pdf

EUROSTAT, 1998. European landscapes: Farmers maintain more than half of the territory. Statistics in Focus—Agriculture. Available at http://ec.europa.eu/agriculture/publi/landscape/ch2.htm (accessed on February 2, 2012).

Fisher, P.F., Comber, A.J., and Wadsworth, R.A. 2005. Land use and land cover: Contradiction or complement. In P. Fisher and D. Unwin (Eds.), *Re-Presenting GIS* (pp. 85–98). Chichester: Wiley.

Friedl, M.A., Sulla-Menashe, D., Tan, B., Schneider, A., Ramankutty, N., Sibley, A., and Huang, X. 2010. MODIS Collection 5 global land cover: Algorithm refinements and characterization of new datasets. *Remote Sensing of Environment*, 114, 168–182.

Fritz, S., Scholes, R.J., Obersteiner, M., Bouma, J., and Reyers, B. 2008. A conceptual framework for assessing the benefits of a global earth observation system of systems. *IEEE Systems Journal*, 2, 3, 338–348.

GCOS—Global Climate Observing System. 2004. Implementation plan for the Global Observing System for Climate in Support of the UNFCCC, World Meteorological Institute. Available at: http://www.wmo.int/pages/prog/gcos/Publications/gcos-92_GIP.pdf

GCOS—Global Climate Observing System. 2010. Implementation plan for the Global Observing System for Climate in Support of the UNFCCC, August 2010 (update), World Meteorological Organisation. Available at: http://www.wmo.int/pages/prog/gcos/Publications/gcos-138.pdf

Giri, C., Ochieng, E., Tieszen, L.L., Zhu, Z., Singh, A., Loveland, T., Masek, J., and Duke, N. 2010. Status and distribution of mangrove forests of the world using earth observation satellite data. *Global Ecology and Biogeography*, 20, 154–159.

GLP. 2005. Science plan and implementation strategy. IGBP Report No. 53/IHDP Report No. 19. Stockholm: IGBP Secretariat, 64 pp.

Hansen, M.C., Townshend, J.R.G., DeFries, R.S., and Carroll, M. 2005. Estimation of tree cover using MODIS data at global, continental and regional/local scales. *International Journal of Remote Sensing*, 26, 4359–4380.

Hansen, M., DeFries, R., Townshend, J.R.G., Sohlberg, R., Dimiceli, C., and Carrol, M. 2002. Towards an operational MODIS continuous field of percent tree cover algorithm: Examples using AVHRR and MODIS. *Remote Sensing of Environment*, 83, 303–319.

Herold, M., Woodcock, C., Wulder, M., Arino, O., Achard, F., Hansen, M., Olsson, H., et al. 2009. GTOS ECV T9: Land Cover—Assessment of the status of the development of standards for the Terrestrial Essential Climate Variables. Available at: http://www.fao.org/gtos/doc/ECVs/T09/T09.pdf

Jones, S. 2002. Social constructionism and the environment: Through the quagmire. *Global Environmental Change*, 12, 247–251.

Jung, M., Henkel, K., Herold, M., and Churkina, G. 2006. Exploiting synergies of global land cover products for carbon cycle modelling. *Remote Sensing of Environment*, 101, 534–553.

McCallum, I., Obersteiner, M., Nilsson, S., and Shvidenko, A. 2006. A spatial comparison of four satellite derived 1 km global land cover datasets. *International Journal of Applied Earth Observation and Geoinformation*, 8, 246–255.

Miller, R.I. 1994. *Mapping the Diversity of the Nature*. London; New York: Chapman & Hall.

Moreau, I. 2009. Méthode de cartographie globale de l'occupation du sol par télédétection spatiale: Analyse de la stabilité interannuelle de la chaîne de traitement GlobCover, mémoire de fin d'études, Université Catholique de Louvain, Faculté d'ingénierie biologique, agronomique et environnementale.

Pinty, B., Andredakis, I., Clerici, M., Kaminski, T., Taberner, M., and Plummer, S. 2010. Exploiting surface albedo products to bridge the gap between remote sensing information and climate models. In *Proceedings of the Earth Observation for Land-Atmosphere Interaction Science*. Frascati, Italy, November 3–5, 2010 (ESA SP-688, January 2011).

Pittman, K., Hansen, M.C., Becker-Reshef, I., Potapov, P.V., and Justice, C.O. 2010. Estimating global crop-land extent with multi-year MODIS data. *Remote Sensing*, 2, 1844–1863.

Poulter, B., Ciais, P., Hodson, E., Lischke, E., Maignan, F., Plummer, S., and Zimmermann, N.E. 2011. Plant functional type mapping for Earth System Models. *Geoscientific Model Development*, 4, 1–18.

Schneider, A., Friedl, M., and Potere, D. 2010. Mapping global urban areas using MODIS 500-m data: New methods and datasets based on "urban ecoregions." *Remote Sensing of Environment*, 114, 1733–1746.

SIOSE—Sistema de Información de Ocupación del Suelo en España, Equipo Technico Nacional. 2011. Documento Técnico SIOSE 2005, version 2. Available at: http://www.ign.es/siose/

Smith, B. and Mark, D.M. 2001. Geographical categories: An ontological investigation. *International Journal of Geographical Information Science*, 15, 591–612.

Smith, M.O., Ustin, S.L., Adams, J.B., and Gillespie, A.R. 1990. Vegetation in deserts: 1. A regional measure of abundance from multispectral images. *Remote Sensing of Environment*, 31, 1–26.

Smith, B. and Varzi, A.C. 2000. Fiat and bona fide boundaries. *Philosophy and Phenomenological Research*, 60, 401–420.

Thenkabail, P.S., Biradar, C.M., Noojipady, P., Dheeravath, V., Li, Y.J., Velpuri, M., Gumma, M., et al. 2009. Global irrigated area map (GIAM) derived from remote sensing for the end of the last millennium. *International Journal of Remote Sensing*, 30, 3679–3733.

Townshend, J.R., Latham, J., Arino, O., Balstad, R., Belward, A., Conant, R., Elvidge, C., et al. 2008. Integrated Global Observation of the Land: An IGOS-P Theme, IGOL Report No. 8.

Ustin, S.L. and Roberts, D.A. 2010. Remote sensing of plant functional types. *New Phytologist*, 186, 795–816.

Verburg, P.H., Van de Steeg, J., Veldkamp, A., and Willemen, L. 2009. From land cover change to land function dynamics: A major challenge to improve land characterization. *Journal of Environmental Management*, 90, 1327–1335.

6 Evaluating Land-Cover Legends Using the UN Land-Cover Classification System

Martin Herold and Antonio Di Gregorio

CONTENTS

6.1 TOWARD HARMONIZED LAND-COVER MAPPING

A number of global and regional land-cover datasets, classification systems, and legends have been developed with the use of satellite remote sensing for large-scale land monitoring. Monitoring initiatives have different interests, objectives, methodologies, and mapping standards, which limit the capacity of compatibility and comparability of land-cover data. A large and growing user community and a variety of applications require consistency and continuity in land observations, which can be achieved by harmonizing the multitude of datasets. In particular, harmonizing can improve change analysis, cross-comparison, and validation; derive an advanced product by aggregating or integrating datasets and different levels of information; and improve the monitoring of standardized land cover in future efforts.

Harmonization is the process whereby similarities between existing definitions of land characterization are enhanced and inconsistencies are reduced. Beginning from a state of divergence in land-cover datasets, harmonization seeks compatibility and comparability; however, it does not necessarily eliminate all differences. Ideally, harmonization should be guided by existing or evolving standards, and therefore, it has to use a common language for reference. Specific existing legends often lack a consistent way of formalizing the meaning of the classes they propose. The UN Land Cover Classification System (LCCS) currently provides the most comprehensive, the most flexible, and the most internationally accepted approach to land-cover characterization. The first step toward harmonization is the translation of existing legends in a common language provided by the LCCS to improve land-cover monitoring in the future.

This chapter presents the translation results of the Anderson Classification System (ACS), the European Coordination of Information on the Environment (CORINE), International Geosphere–Biosphere Program (IGBP), and University of Maryland (UMD) land-cover legend. The translations were developed through cooperation between the Land Cover Topic Centre (LCTC) of the UN Global Land Cover Network (GLCN) (http://www.glcn-lccs.org) and the GTOS/GOFC-GOLD (Global Terrestrial Observing System/Global Observation of Forest and Land Cover Dynamics) Land Cover Implementation Team Project Office (http://www.gofc-gold.uni-jena.de/; Herold et al., 2006b; Townsend and Brady, 2006). The translations and suggestions in this report are open for discussions and comments by the international community.

6.2 UN LAND-COVER CLASSIFICATION SYSTEM

6.2.1 THE LCCS CONCEPT

The LCCS (Di Gregorio, 2005) was developed by the Food and Agriculture Organization (FAO) and the United Nations Environment Programme (UNEP) to meet the need for a standardized global reference classification system. It is a classification system, not a land-cover legend that has distinct differences (Di Gregorio, 2005; McConnell and Moran, 2001). A single standardized legend significantly reduces the relevance of application of land-cover datasets (Wyatt et al., 1994). The principal characteristics of the LCCS are as follows:

- Flexibility: mapping at different scales and at different levels of detail, allowing cross-reference from local to global maps without loss of information
- Consistency: systematic class description with clearly defined land-cover criteria unambiguously delimited from environmental and technical attributes
- Comprehensiveness: allows the description of a complete range of land-cover features
- Comprehensibility: an essential set of classifiers minimizes possible errors and validation efforts
- Applicability: multipurpose land-cover classification that can be adapted to user needs

The LCCS provides a system of common diagnostic criteria (land-cover classifiers) that are in no particular hierarchy, thus providing a standardization of terminology, not categories. The LCCS was created to ensure fundamental rules of unambiguous definition of each class, avoid overlap on class boundaries, provide consistency in class description, and clearly define class relationships (possibly with mathematical parameters). Existing "classifications" usually fail to meet these rules, since many of them are often geographically limited "legends." The LCCS approach is therefore, in this way, different from most other examples (like CORINE and IGBP) of standardized land-cover systems (Ahlqvist, 2008). It can be considered a "boundary object" to evaluate and mediate different approaches to represent land-cover features around which similarities, differences, and internal consistencies can be understood and expressed in a rigorous way.

The LCCS classification concepts were endorsed in 1996. The initiative developed an internationally accepted reference base for land cover. The LCCS was used for the first time with FAO's Africover project (Di Gregorio and Jansen, 1996a, 1996b). Based on that experience, a second version of the software was developed. Currently, version 2.4 is in use, and version 3 is available as a prototype. In addition, the LCCS concept is a form of the Land Cover Data Macro Language, which would become an ISO standard for land-cover classification.

To facilitate collection of data coming from different land-cover projects, GLCN LCTC provides a translation form (see GOFC-GOLD, 2009) designed according to LCCS methodology/translation concepts (Herold et al., 2006a, 2006b; Jansen, 2004). This form is filled with information coming from the original legend and LCCS translation data. Furthermore, users can add notes, and GLCN LCTC staff members can evaluate the translation.

6.2.2 Classification with the LCCS

The LCCS is an *a priori* classification system, meaning that all classes have to be defined in advance of data collection and land-cover classification. Usually, *a priori* classification systems have a disadvantage, since a large amount of classes have to be defined to describe land cover all over the world in a consistent way. However, instead of predefined classes, the LCCS offers a set of predefined classification criteria—preventing inconsistencies while simultaneously providing standardization. This is an independent diagnostic criterion where the classifiers are hierarchically arranged, and they differ depending on the land-cover type—different land covers demand suitable sets of classifiers. Hence, the classification process with the LCCS goes through two main phases: first the dichotomous phase (Figure 6.1) and later the modular-hierarchical phase (Figure 6.2).

The dichotomous phase distinguishes eight major land-cover types. The appropriate set of classifiers in the modular-hierarchical phase (Figure 6.2) ensures certainty, standardization, and comprehensibility of the classification. Higher levels of detail can be achieved by using optional modifiers and attributes. These involve environmental (e.g., climate, lithology) as well as technical properties (e.g., crop type, salinity of water bodies), which go beyond the use of "pure" land-cover classifiers.

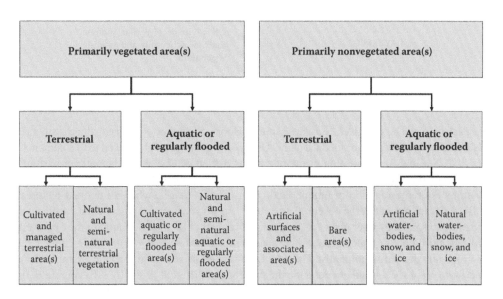

FIGURE 6.1 The initial dichotomous phase in the LCCS as used in the LCCS-2 software.

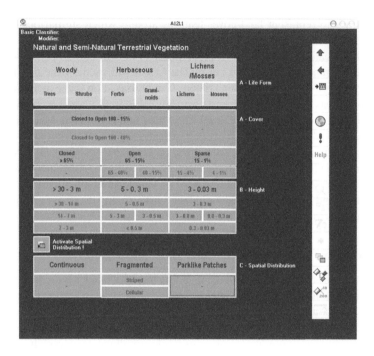

FIGURE 6.2 The modular-hierarchical phase (first level of "natural and seminatural terrestrial vegetation").

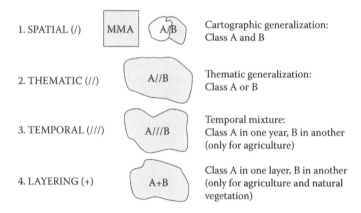

FIGURE 6.3 Mixed unit concept within the LCCS (MMA = minimum mapable area).

For each defined class, the LCCS creates a unique Boolean formula (comprising the classifiers used), a unique numerical code, and a standard name. User-defined names can be linked to this nomenclature.

The LCCS allows the definition of mixed classes, which can be either thematic or cartographic (spatial and/or time-related) mixes. The first case can be applied if the scale (minimum mapable area) limits the representation of unique land-cover classes, that is, when all defined features ("A" and "B", A/B) are present in the observed area. In the second case, no unique thematic information is provided; that is, a land-cover class "A" or a land-cover class "B" may be found in the observed area (A//B). The third one is a special case where spatial mixed coding may occur within cultivated areas when crops are alternating annually. Then the time-related mixed coding applies (temporal, A///B). Furthermore, the LCCS is able to describe the presence of different layers (A + B) (Figure 6.3).

Besides Di Gregorio (2005), Di Costanzo and Ongaro (2004) first presented a detailed description of LCCS v.2 as a classification language. The authors define language syntax and semantics building as a complete description of LCCS v.2 rules, which will be the basis for developing new tools that can integrate the LCCS into existing applications of GIS or remote sensing, thus benefiting both software developers and researchers. From these efforts to create a formal language to share the meaning of different ontologies, FAO has developed a new version of the LCCS. LCCS v.3 will be reflected by a UML and an XML to better share within the user community the conceptual bases of the system. LCCS v.3 can be considered a metalanguage containing a logical general framework of rules to describe land-cover features.

6.3   OVERVIEW OF LEGENDS

Four legends included in the translation are described in the following sections. Some global legends developed using LCCS do not require translation. These include the legend for Global Land Cover 2000 (Bartholomé and Belward, 2005; GLC2000; http://bioval.jrc.ec.europa.eu/products/glc2000/legend.php) and the new GlobCover 2005 product (Arino et al., 2007).

6.3.1   Anderson Classification System

The ACS, essentially developed by Anderson et al. (1976), was designed for national use in the United States, aimed at categorizing remote-sensing information (Table 6.1). The classification system itself offers four levels of increasing detail from level I to level IV, being adaptable to user demands by defining categories that are more detailed and simultaneously compatible for generalizations up to the smaller scales at the national level. Level II was intended for statewide and interstate regional land-use/land-cover compilation and mapping. The level II class, in this work, has been translated into LCCS (Anderson et al., 1976).

A modified version of the ACS was used by the USGS Land Cover Institute in its Landsat TM-based National Land Cover Data (NLCD) classification scheme (see http://landcover.usgs.gov/classes.php).

6.3.2   CORINE Land Cover

CORINE Land Cover, CLC, is jointly managed by the European Environment Agency (EEA) and the Joint Research Center (JRC). The priority of CLC is to provide a land-cover dataset for the European environmental policy, which is comparable across Europe. Initiated in the mid-1980s, the first dataset (Table 6.2) shows the land cover of the 15 EC member states around 1990 (CLC90), whereas the exact date differs mainly between 1986 and 1995. It uses a three-level nomenclature with 5 classes on the first, 15 classes on the second, and 44 classes on the third level. The mapping scale is 1:100,000. Of late, an updated database, CORINE Land Cover 2000 (CLC2000), is available with the reference year 2000 (±1 year). This new version also includes information about CLC changes between the reference years 1990 and 2000. Updates are intended to come out every 10 years; that is, the next update is expected in 2010. Major data sources of CLC2000 are orthocorrected Landsat-7 Enhanced Thematic Mapper (ETM) satellite images (<25-m root mean square error [RMSE]) with a spatial resolution of 25 m or rather 12.5 m for multispectral and panchromatic bands, respectively. The minimum mapping unit (MMU) is 25 ha; changes are accounted for areas of at least 5 ha (Büttner et al., 2004; JRC-IES, 2005).

6.3.3   IGBP Discover

On behalf of the Land Cover Working Group of the International Geosphere-Biosphere Programme Data and Information System (IGBP-DIS), the U.S. Geological Service guided the development of the Discover dataset to meet the demands of various IGBP initiatives for global land-cover data, since existing datasets proved unsuitable for upcoming IGBP core projects (IGBP, 1990). Data of 1 km resolution from the advanced very high resolution radiometer (AVHRR) were considered the

TABLE 6.1
The Anderson Classification System (ACS)

Level 1	Level 2
1 Urban or built-up	11 Residential
	12 Commercial and services
	13 Industrial
	14 Transportation, communications, and utilities
	15 Industrial and commercial complexes
	16 Mixed urban or built-up land
	17 Other urban or built-up land
2 Agricultural land	21 Cropland and pasture
	22 Orchards, groves, vineyards, nurseries, and ornamental horticultural areas
	23 Confined feeding operations
	24 Other agricultural land
3 Rangeland	31 Herbaceous rangeland
	32 Shrub and brush rangeland
	33 Mixed rangeland
4 Forestland	41 Deciduous forestland
	42 Evergreen forestland
	43 Mixed forestland
5 Water	51 Streams and canals
	52 Lakes
	53 Reservoirs
	54 Bays and estuaries
6 Wetland	61 Forested wetland
	62 Nonforested wetland
7 Barren land	71 Dry salt flats
	72 Beaches
	73 Sandy areas other than beaches
	74 Bare exposed rock
	75 Strip mines, quarries, and gravel pits
	76 Transitional areas
	77 Mixed barren land
8 Tundra	81 Shrub and brush tundra
	82 Herbaceous tundra
	83 Bare ground tundra
	84 Wet tundra
	85 Mixed tundra
9 Perennial snow or ice	91 Perennial snowfields
	92 Glaciers

adequate basis for the Discover dataset. The legend of the dataset comprises 17 classes (Table 6.3), designed to provide a consistent and exhaustive characterization of global land cover. More detailed specifications of the Discover dataset can be found in the work of Belward (1996).

The dataset is based on unsupervised classification of multitemporal monthly maximum NDVI composites collected from April 1992 to March 1993. For final class assignment, ancillary datasets were used during postclassification processing. Primary intentions of use targeted the environmental modeling community, especially for global-scale applications (e.g., climate) (Hansen and Reed, 2000; Loveland et al., 2000). The Discover dataset is available through the Global Land Cover Characteristics database via the World Wide Web (http://edc2.usgs.gov/glcc/glcc.php).

TABLE 6.2

The Three-Level Nomenclature of CLC

Level 1	Level 2	Level 3
1 Artificial surfaces	1.1 Urban fabric	1.1.1 Continuous urban fabric
		1.1.2 Discontinuous urban fabric
	1.2 Industrial, commercial, and transport units	1.2.1 Industrial or commercial units
		1.2.2 Road and rail networks and associated land
		1.2.3 Port areas
		1.2.4 Airports
	1.3 Mine, dump, and construction sites	1.3.1 Mineral extraction sites
		1.3.2 Dump sites
		1.3.3 Construction sites
	1.4 Artificial nonagricultural vegetated areas	1.4.1 Green urban areas
		1.4.2 Sport and leisure facilities
2 Agricultural areas	2.1 Arable land	2.1.1 Nonirrigated arable land
		2.1.2 Permanently irrigated land
		2.1.3 Rice fields
	2.2 Permanent crops	2.2.1 Vineyards
		2.2.2 Fruit trees and berry plantations
		2.2.3 Olive groves
	2.3 Pastures	2.3.1 Pastures
	2.4 Heterogeneous agricultural areas	2.4.1 Annual crops associated with permanent crops
		2.4.2 Complex cultivation patterns
		2.4.3 Land principally covered by agriculture, with significant areas of natural vegetation
		2.4.4 Agro-forestry areas
3 Forests and seminatural areas	3.1 Forests	3.1.1 Broad-leaved forest
		3.1.2 Coniferous forest
		3.1.3 Mixed forest
	3.2 Shrub and/or herbaceous vegetation associations	3.2.1 Natural grassland
		3.2.2 Moors and heathland
		3.2.3 Sclerophyllous vegetation
		3.2.4 Transitional woodland-shrub
	3.3 Open spaces with little or no vegetation	3.3.1 Beaches, dunes, and sand plains
		3.3.2 Bare rock
		3.3.3 Sparsely vegetated areas
		3.3.4 Burnt areas
		3.3.5 Glaciers and perpetual snow
4 Wetlands	4.1 Inland wetlands	4.1.1 Inland marshes
		4.1.2 Peatbogs
	4.2 Coastal wetlands	4.2.1 Salt marshes
		4.2.2 Salines
		4.2.3 Intertidal flats
5 Water bodies	5.1 Inland waters	5.1.1 Water courses
		5.1.2 Water bodies
	5.2 Marine waters	5.2.1 Coastal lagoons
		5.2.2 Estuaries
		5.2.3 Sea and Ocean

6.3.4 UMD LEGEND

A second legend based on the AVHRR dataset mentioned above was developed by the University of Maryland. The UMD legend essentially is a modified IGBP legend renouncing the IGBP classes 11 (*permanent wetlands*), 14 (*cropland/natural vegetation mosaics*), and 15 (*snow and ice*) (Table 6.4). Contrary to the IGBP classification based on unsupervised clustering of NDVI composites, UMD used a supervised classification tree algorithm considering 41 multitemporal metrics derived not only from NDVI values but from all five AVHRR bands (Hansen and Reed 2000; Hansen et al., 2000).

TABLE 6.3
IGBP Discover Nomenclature

Classification Code	IGBP Class
1	Evergreen needle-leaf forests
2	Evergreen broad-leaf forests
3	Deciduous needle-leaf forests
4	Deciduous broad-leaf forests
5	Mixed forests
6	Closed shrublands
7	Open shrublands
8	Woody savannas
9	Savannas
10	Grasslands
11	Permanent wetlands
12	Cropland
13	Urban and built-up
14	Cropland/natural vegetation mosaics
15	Snow and ice
16	Barren or sparsely vegetated
17	Water bodies

TABLE 6.4
UMD Nomenclature

Classification Code	UMD Class
0	Water bodies
1	Evergreen needle-leaf forests
2	Evergreen broad-leaf forests
3	Deciduous needle-leaf forests
4	Deciduous broad-leaf forests
5	Mixed forests
6	Woodlands
7	Wooded grasslands/shrublands
8	Closed bushlands or shrublands
9	Open shrublands
10	Grasslands
11	Croplands
12	Barren
13	Urban and built-up

Access for the dataset is provided through the University of Maryland's Global Land Cover Facility via the Web (http://glcf.umiacs.umd.edu/data/).

6.4   LEGEND TRANSLATION INTO THE LCCS

6.4.1   OBJECTIVES

The objectives of the LCCS translation process are as follows:

- Create a translation of the legends and their land, using LCCS classifiers.
- Show the feasibility, possibilities, and discrepancies of the translation.
- Evaluate known issues to overcome possible difficulties that may have been encountered (Section 6.6).

The initial background for this work was the intention to study the possibility of linking CORINE Land Cover to global land-cover activities and foster interaction and comparability between these land-cover mapping activities—an idea originating from the harmonization workshop held at FAO, Rome (Herold and Schmullius, 2004).

6.4.2   TRANSLATION PROCESS

Using the LCCS software, a translation of the legends was done for each single class. ACS and CLC translations were realized on the second and third level, respectively. All classes went through a first translation done by the GOFC-GOLD land-cover office and were then adjusted according to advice from GLCN-LCTC staff members. A translation form was prepared for every class. Problems that occurred during the translation were pointed out, with special attention being given to inconsistencies.

Legend properties and class descriptions of the ACS were found in its revision paper published by the U.S. Geological Survey (Anderson et al., 1976).

To produce the most suitable translation of CLC classes, they were studied in detail using the addendum to the CORINE technical guide (Bossard et al., 2000) and CEC (1994). Additional information was found on the Web portal of the European Topic Centre on Terrestrial Environment (http://www.eea.europa.eu/publications/COR0-landcover), which is a part of the EEA.

IGBP Discover and UMD classes were translated with the help of Hansen et al. (2000), Hansen and Reed (2000), and Loveland et al. (2000).

6.5   RESULTS

Translation is a way to assess the degree of consistency (or vagueness) of the processed legends. The process was not straightforward for all classes. Some problems occurred through all legends but differed in their extent and magnitude; others were legend specific. Although legend criteria can usually be translated with the LCCS, the criteria cannot often comply completely with the LCCS classifiers. Before taking a closer and more specific look at the individual legends, the most important general translation issues are discussed in detail:

- Threshold differences
- Occurrences of land-use and other non-land-cover terminology
- Difficulties due to mixed classes (cartographic standards)

Other particular issues are addressed in the legend-specific part of this chapter. These include translator judgments on the consistency of the class description and the quality of the LCCS translation. High consistency and high confidence point to a successful LCCS translation and, vice versa, low consistency and low quality refer to problems discussed in more detail hereafter.

6.5.1 Threshold Differences

Threshold differences for specific classifiers are of key importance for land-cover comparability, that is, vegetation/tree canopy cover in the case of vegetated areas, density thresholds for urban areas indicating the composition of impervious surfaces, or height thresholds for identifying trees. The difference should not exceed 5–10 points for being ignored. These differences, however, do not affect the evaluation of the class consistency because the values reported in the LCCS cannot be taken as a reference; therefore, they do not serve as an evaluation element for consistency.

For natural vegetation, a cover density threshold has to be defined when creating an LCCS class. In the legends analyzed, however, no vegetation-cover information is specified in some cases; that is, some provide only qualitative (i.e., "dense") or sometimes contradictory specifications. In such cases, the translator has to decide which values are most suitable. This choice was not made following a strict rule (e.g., defining the widest range from 100% to 15%), but following the conclusions drawn from other class descriptions.

6.5.2 Land-Use and Other Non-Land-Cover Terminology

There is a link between land cover and land use, and many applications often use both types of information. Hence, the need or desire to include this information in a multipurpose legend is obvious. However, this intention often results in a mix of land-cover and non-land-cover terminology and favors inconsistencies and a general vagueness of the meaning of classes. The LCCS, on the contrary, is designed primarily to describe land cover in a rather rigorous way. Thematic incompatibility or lack of suitable translations is found for some categories. In fact, LCCS does offer a range of possibilities to describe artificially covered surfaces—urban (built-up) as well as cultivated areas—but these capabilities are controlled and regulated by the attempt to describe these categories purely from a land-cover point of view. Part of the translated legends, especially CLC and ACS, are not restricted to "pure" land-cover and land-use terms.

Examples of affected classes often referred to are given below:

- Processes (CLC classes 133 *construction sites*, 324 *transitional woodland-shrub*; ACS class 76 *transitional areas*),
- Cultural practices (CLC classes 212 *permanently irrigated land*, 231 *pastures*; ACS category 3 *rangeland*).
- Environmental events (CLC class 334 *burnt areas*)
- An entire ecoregion (ACS category 8 *tundra*)

Other classes include very specific elements, for example, ACS class 22 *orchards, groves, vineyards, nurseries, and ornamental horticultural areas*, and ACS class 24 *other agricultural land*. Within this context for the CORINE legend, "nurseries of fruit trees and shrubs" are included in CLC class 211 *nonirrigated arable land* or "gravel accumulation along stream channels." Such specifics are not generally available in the LCCS but can be eventually accommodated by defining user-defined attributes. In all those cases, the actual land-cover characteristics often remain uncertain. This again implies imprecise class boundary definitions, leaving the possibilities of overlaps or gaps between classes, thus making interpretation susceptible to errors and increasing the time and resources required for mapping.

One main point of discussion, in the translation process, was on the definition of "pasture," especially regarding the translation of CLC class 231. As a consensus, the LCCS mode function was used to leave out the differentiation between "cultivated and managed terrestrial area(s)" and "natural and seminatural terrestrial vegetation." Certainly, this decision was a compromise. Pastures are covered with herbaceous vegetation used for grazing and are usually considered seminatural vegetation. Typically in the United States, "artificial" pasture, where nonnative domesticated forage plants have replaced the native herbaceous vegetation, is called rangeland—as is evident in the ACS. According to the ACS, the issue depends on how pasture is defined. Apparently, there is more than one definition, and the meaning of this term may differ from country to country or from technical terminology to common speech. Thus, the problem is merely semantic. Since CLC includes artificial pasture and the sowing of plants, the proper translation has to include this option. Furthermore, a specific thematic extension of CLC (e.g., up to 50% tree cover for specific pastures) has been neglected in the translations. Such issues are assumed to be rare; otherwise, they may lead to major inconsistencies among classes.

Non-land-cover distinction criteria cannot precisely define land-cover characteristics. Frequently, more than one land-cover type may be present within such a class. This becomes noticeable especially when observing the classes belonging to CLC 2.4 *heterogeneous agricultural areas*. These classes are so vague from a land-cover point of view that a perfect translation with the LCCS is a problem, and the result has to be seen as an approximation trying to represent the most relevant characteristics of the class. Similar observations exist for other classes, including CLC classes 212 *permanently irrigated land*, 322 *moors and heathland*, and 324 *transitional woodland-shrub*. Translation forces the creation of mixed classes because their definitions are not based on a land-cover perspective.

6.5.3 TRANSLATION OF MIXED UNIT CLASSES

The LCCS has a rigorous way of handling the mixed unit concept. In effect, the concept does not need to be addressed automatically in land-cover class ontology. It is more of a cartographic rule that is applied in particular cases when a particular type of geographic area (heterogeneous areas) needs to be represented in a map with the constraint of scale. Being scale sensitive, it cannot be considered in the classification system itself, which by definition should define the ontology of different land-cover features independently from the way they are represented in a specific map. Unfortunately, in the existing legends examined, mixed classes do not follow strict criteria and very often increase the vagueness and ambiguity of class definition.

One example is CLC class 243 *land principally occupied by agriculture, with significant areas of natural vegetation*. The class description defines the share of cultivated and natural/seminatural vegetation in the range of 25%–75% each. This share contradicts the class name where the term *principally* should indicate a prevalence of agriculture over natural vegetation. Even the high flexibility of the LCCS in handling cartographic mixed units cannot properly represent this contradiction.

For mixed forests, the LCCS offers the option "mixed" that can be selected when defining leaf phenology. However, the LCCS includes only broad-leaved deciduous and needle-leaved evergreen vegetation. The CLC, IGBP, and UMD class definitions do not have these restrictions, and not every mixed forest will follow this guideline either. Broad-leaved evergreen or needle-leaved deciduous species that possibly occur inside a population are excluded as per the definition. Nevertheless, this kind of translation was preferred to the creation of a spatial mixture of broad-leaved and needle-leaved trees for the reasons given in the previous example. The GLC2000 legend defines its mixed forest class as a thematic mixed unit. However, that is only where broad-leaved *or* needle-leaved species would occur (cf. Section 6.2.2)—which actually is not consistent.

Mixing of classes occurs not merely through explicit class descriptions; in some cases, it is a result of definition deficiencies. The ACS specifies a kind of "rest class" (ACS classes 17 *other urban or built-up land*, 24 *other agricultural land*); that is, classes collecting those area characteristics that do not match any of the characteristics described within the other, more specific thematic neighbor classes. Though in certain respects, gaps between classes are prevented, one type of inconsistency (definition gaps) is compensated by another (indistinct definition). A similar issue affects some "mixed" classes of the ACS (classes 16 *mixed urban or built-up land*, 77 *mixed barren land*, 85 *mixed tundra*), which limits the definition of mixed units to the particular hierarchical level.

6.5.4 LEGEND-SPECIFIC ISSUES

Consistency of class definitions is evaluated in four grades (*insufficient, fair, good*, and *very good*) and translation confidence in three grades (*fair, good*, and *very good*). To quantify both parameters, we assigned the following values to them.

Consistency
- *Insufficient* = 0
- *Fair* = 1
- *Good* = 2
- *Very good* = 3

Evaluation of a class definition's consistency follows some guidelines, which are decisive for the grade achieved by each class. A *very good* rating requires perfect class consistency without overlaps to any other class of the legend. Class boundaries should be clearly discernible, and class characteristics should use inherently concordant separation criteria. A *good* rating still assumes consistent core definition and separation criterion for the class, though possible definition uncertainties (e.g., due to land use or other terminology or lack of vegetation-cover specifications) may cause a blurred class boundary. To gain a *fair* rating, the core definition of the class has to allow a unique separation against its immediate neighbor classes, and/or the class has to provide legend-inherent consistency although overlaps in land cover cannot be excluded. A class's consistency is rated *insufficient* when it does not comply with any of the requirements mentioned. The class definition does not allow a clear separation from other classes of the legend (major overlaps) and/or is either ambiguous in the description of its land-cover/use features or does not sufficiently specify them.

In the case of asymmetric overlaps of classes, the more common or generic class is rated better, whereas the special class that introduces land use or other terms (and hence inconsistencies) is rated worse. Overlaps of classes can be asymmetric when, for instance, one class can be part of another class relating to its land-cover specifications but is defined further by non-land-cover characteristics. An example is apparent from the ACS *tundra* classes 8x, which specify a whole set of land-cover classes especially for this ecological zone. In this case, the more generic *rangeland* classes (representing natural/seminatural vegetation) or the basic class 74 *bare exposed rock* are not penalized for the overlap and achieve a higher consistency rating although they are affected just as much. Since the *tundra* classes cause these inconsistencies (non-land-cover terminology), their score will suffer from adequate penalties.

Confidence
- *Fair* = 0
- *Good* = 1
- *Very good* = 2

According to consistency, a *very good* rating can be attained only with absolute confidence in a translation that is complete and unambiguous. If another translation is conceivable, and yet the actual version is an appropriate choice to represent the class description, then the translation confidence is rated *good*. When the translation can reflect a class only with deviations to its definition and hence cannot fully agree with the class structure and all its details, it will achieve *fair* confidence. Whenever a translation is possible, the translator should have a *fair* confidence at least, or else a translation is actually impossible—thus making a rating below *fair* meaningless.

We present the evaluation scores for each legend in the following reviews of legend-specific issues and discuss them comparatively in the concluding Section 6.6.

6.5.4.1 ACS Issues

Insufficient consistency for most of *urban or built-up* (classes 1*x*), *agricultural land* (classes 2*x*), and the *tundra* category (classes 8*x*) is obvious from Figure 6.4. Simultaneously, these classes show a tendency toward a lower confidence rating. The *rangeland* and *forestland* categories are less problematic in both terms.

Primarily, the ACS is land-use/resource oriented. Thus, there may be discrepancies owing to a rather land-cover-oriented classification system. Furthermore, the Anderson system fulfills certain unfavorable conditions, which deteriorate the classification operations done with it:

1. Land-cover and land-use terms are used simultaneously, and they occur mixed with each other (examples: *rangeland* category or class 21 *cropland and pasture*).
2. Class definitions are unsystematic and inconsistent, and class boundaries appear barely understandable and arbitrary (examples: overlaps throughout the classification system, especially with the *tundra* category).
3. Important and commonly used characteristics are ignored (examples: cover density and leaf type).
4. Mixed classes are used inappropriately; they should not be part of a classification system but can be used within a legend. Obviously, the proper meanings of "classification" and "legend" were not considered sufficiently.

6.5.4.1.1 Urban or Built-Up

Class 1 *urban or built-up* is a pure land-use category. Most of the categories in level II cannot be accommodated by LCCS standard classifiers, since the LCCS is far less land-use oriented. Thus, it

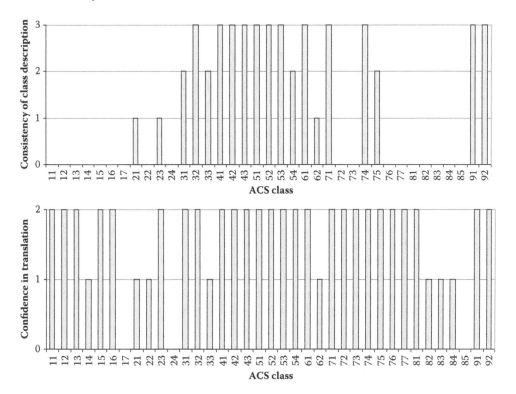

FIGURE 6.4 Evaluation of consistency of the original class description and the translator confidence in the quality of proposed translation to present the class concept within the LCCS for ACS level 2 classes (see Table 6.1 for class names).

is necessary to introduce user-defined attributes to describe and allow discrimination of the classes. The Anderson classes neglect cover density—certainly some of the most common classifiers in urban areas. Cover density should be considered at an additional level.

Furthermore, overlaps between the classes exist, originating from the two industrial classes 13 and 15 and from class 16 *mixed urban or built-up land*, which comprises a mixture of any of the level II urban classes. Since the mix can be complex and the LCCS cannot adopt this definition as it is, a user-defined attribute was added.

6.5.4.1.2 Agricultural Land

Again, we find a pure land-use category—making the description, with a primarily land-cover classification system, a bit uncomfortable. Noteworthy is a shared level II class for cropland and pasture, representing the American definition of pasture as being more intensely managed areas, including cultivation practices as seeding and fertilizing, which is opposite to rangeland with a native vegetation cover regulated only by grazing.

The emphasis of land use becomes obvious again in class 23 *confined feeding operations*: From a land-cover point of view, this class is rather a built-up object (and hence defined as such with the LCCS). Wetland agriculture is included as well, and it does not pertain to one of the wetland classes (6*x*). Note that the LCCS definitions, in favor of clarity, consider only terrestrial classifiers.

Class 24 *other agricultural land* summarizes land uses associated with any of the other level II classes of *agricultural land* and is meant to be negligible on smaller scales, but it hardly brings any benefit.

6.5.4.1.3 Rangeland

Rangeland refers to natural or seminatural vegetation grazed by herbivores. Rangeland areas are occupied by native herbaceous or shrubby vegetation and can be grazed by both domestic and wild herbivores.

In contrast to pastureland, generally, only native vegetation is present in rangeland areas, though Anderson et al. (1976) mention that some rangelands may present seeded or domesticated plant species. More intensive techniques (seeding, irrigation, fertilizing) are typical for pastureland, whereas rangelands are managed principally based on the stocking of grazing animals according to the duration and season of grazing. Thus, range management aims at sustaining, improving, or protecting natural resources comprising plant and animal life as well as soil and water and simultaneously using these resources for forage production and other purposes (e.g., recreation).

From the definition, it can be deduced that *rangeland* again is a land-use term. Vegetation cover can be very different, including prairies/steppes, shrub-/woodlands, savannas, and tundra. Tundra forms its own category in the ACS. Even forests used for grazing can be considered rangelands.

The Anderson system distinguishes between *herbaceous rangeland* and *shrub and brush rangeland* but does not specify any vegetation cover or other thresholds. Class 33 *mixed rangeland* defines the fraction of either herbaceous or shrubby rangeland as a more than one-third intermixture, which cannot be translated properly with the LCCS. Hence, a cartographic mixture according to LCCS rules had to be created, defining the large-sized (shrubby) vegetation as dominating to prevent a splitting into two parts. Alternatively, only two subclasses could accommodate the Anderson definition, with the other subclass specifying herbaceous species as dominant vegetation.

6.5.4.1.4 Forestland

Anderson et al. (1976) specify a minimum tree-crown cover of 10% for the *forestland* category, which is a rather low threshold (GLC2000 > 15%, CLC ≥ 30%, IGBP > 60%). Even areas with little or no forest growth (<10% crown cover) are accounted for when no other land use is obvious. Thus, clear-cuts are included in this category. Areas meeting the requirements for both *forestland* and *urban or built-up* land are assigned to the urban category. Analogously, areas that simultaneously comply with the condition for the *wetland* category are included in the *wetland* category, since the wetland character is supposed to be more important. As indicated above, grazed forestland is not assigned to the *rangeland* category but rather forms a part of *forestland*.

The Anderson classification system first distinguishes its *forestland* category into deciduous and evergreen species. That is undoubtedly exceptional, since no distinction into broad-leaved or needle-leaved vegetation accompanies or precedes those second-level classes. Of course, a third or following level can consider leaf type, but the primary criterion of the classification system is the shedding of leaves. Consequently, class 43 *mixed forestland* is not a forest species mixture in the common sense of broad-leaved and needle-leaved trees but a mixture of deciduous and evergreen plants. Therefore, a mixed forest in the Anderson system can be a pure broad-leaved (or needle-leaved) forest.

The LCCS does not capture mixed forestland composed of deciduous and evergreen species, nor does the LCCS allow the user to define leaf phenology independent of leaf type. More specifically, the user must specify either broad-leaved or needle-leaved to release the evergreen/deciduous option. On the one hand, this is a constraint of LCCS 2 software; on the other hand, the primary distinction according to leaf type is a common practice. That leads us to some inconveniences in the translation: for the classes 41 *deciduous forestland* and 42 *evergreen forestland*, a thematic mixture was created, each containing the broad-leaved and the needle-leaved part. A similar solution is unavailable for class 43 *mixed forestland*, so only a user-defined attribute can accommodate Anderson's class definition.

6.5.4.1.5   Water

Oceans are not considered in the ACS, since only inland waters are taken into account. That is valid for class 54 *bays and estuaries* as well. Those water areas are included only when considered to be inland water and hence are included within the total area of the United States.

LCCS translation can be carried out without problem; only a user-defined attribute has to be added to class 54.

6.5.4.1.6   Wetland

Anderson et al. (1976) divide wetlands into *forested wetland* and *nonforested wetland* on their level II categories. The evident overlap to forestland classes was mentioned above. Class 62 *nonforested wetland* comprises a part of herbaceous vegetation as well as nonvegetated wetlands (alluvial and tidal flats). Cultivated wetlands are classified as *agricultural land*, whereas grazed wetlands are retained here. Overlaps to the corresponding categories (*barren land, agricultural land, and rangeland*) are unavoidable.

6.5.4.1.7   Barren Land

Barren land is defined to show less than one-third vegetation or other cover. Wet, nonvegetated barren land is considered in class 62 *nonforested wetland*. Barren areas found in the tundra region are accounted for in the tundra category (class 83 *bare ground tundra*). Not included are those areas where it is evident from the data source that they will be returned to their former use (e.g., clearcuts). However, the *barren land* category covers the cases where neither the former nor the future land use is perceptible (class 76 *transitional areas*). Hence, overlaps occur again in the *barren land* category.

Following the LCCS, class 77 *mixed barren land* cannot be translated in the usual language. The possible land-use/land-cover features comprise any level II classes of *barren land* with none of them reaching the two-thirds threshold of the observed area. Only the usage of a user-defined attribute allows an LCCS translation.

6.5.4.1.8   Tundra

The tundra category is another peculiarity of the ACS. The term *tundra* describes an entire ecoregion rather than land cover. Although those regions certainly feature characteristic vegetation, tundra describes no specific life form but comprises a set of environmental factors (climate, soil, hydrology, etc.).

Class 81 *shrub and brush tundra* essentially is a clone of class 32 *shrub and brush rangeland*. Both classes show the same life forms, that is, the same land cover. Only the environmental attribute *Polar Arctic* was added to form a suitable equivalence to the Anderson class description. Cover density is described as "dense to open," yet no definition of such terms and their meanings are given.

As mentioned above, among *barren land* class 83 *bare ground tundra* actually results in a complete overlap with that category—a vegetation cover of less than one-third is specified. This threshold cannot be translated exactly with the LCCS, in which the maximum cover density was set to 40%.

Also, class 84 *wet tundra* can be part of another category and overlaps with *wetland*. Finally, the last tundra class offers the biggest trouble, in fact, in such a way that a translation with the LCCS becomes impossible. To classify a specific area as class 85 *mixed tundra*, a mixture of all level II tundra classes is imaginable as long as one type of tundra does not reach two-thirds of this area. Since that does not limit life form/vegetation cover, and since "tundra" is not even is a land-cover term that one can define within the LCCS, *mixed tundra* must remain without LCCS description.

6.5.4.1.9 Perennial Snow or Ice

Neither the definition nor the translation of Anderson's snow and ice category cause problems. The distinction between class 91 *perennial snowfields* and class 92 *glaciers* can be made by the presence or absence of (glacial) flow features.

6.5.4.2 CLC Issues

Comparable to the ACS, we can observe low consistency values in agricultural classes, especially among mixed agriculture classes (24*x*). However, in contrast to the ACS, we find higher consistency within urban classes but again lower values for natural/seminatural vegetation (Figure 6.5).

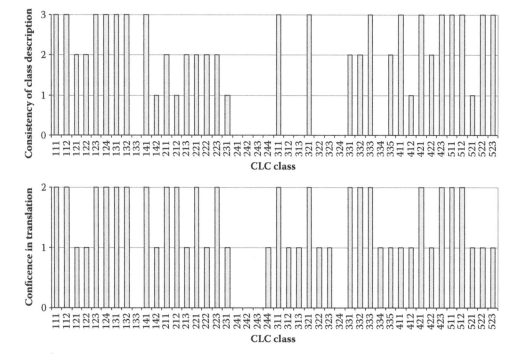

FIGURE 6.5 Evaluation of consistency of the original class description and the translator confidence in the quality of proposed translation to present the class concept within the LCCS for CORINE level 3 classes (see Table 6.2 for class names).

6.5.4.2.1 Artificial Surfaces

CLC definitions focus on land-use descriptions. Compared to the ACS, there is better consistency among the CLC classes. The main benefit is the inclusion of cover density. Yet, some issues are worth mentioning: First, CLC class 133 *construction sites* do not allow us to draw any conclusions about the actual (or past/future) land cover. Similar to CLC class 324 *transitional woodland-shrub*, which we discuss among the category *forests and seminatural areas* (see Section 6.5.4.2.3), the class definition refers to a process. Certainly, processes are of key importance for land-cover *change* mapping, making the purpose of considering them clearly comprehensible. However, they should not obstruct the very sense of a land-cover map that is to provide us with reliable land-cover information. Unfortunately, CLC class 133 (and likewise 324) withholds this information by not providing actual land-cover characteristics. Especially in areas that are most interesting—as they are affected by changing processes—it reveals the least information about the current land-cover status.

Classes 14*x artificial, nonagricultural vegetated areas* are another case where non-land-cover terminology causes ambiguities. Apart from definition uncertainties owing to the LCCS, both the classes 141 *green urban areas* and 142 *sport and leisure facilities* may represent an identical land-cover feature (e.g., parks), depending on their geographic occurrence (topology of urban fabric). Again, it becomes obvious that land-use criteria are generally unsuited to distinguish land cover in a consistent way, particularly when no significance is attached to class separation criteria.

6.5.4.2.2 Agricultural Areas

The CLC category *agricultural areas* contains (among others) level 2 subclasses *arable land* and *permanent crops*. Obviously, CLC again uses different criteria to define and separate these classes. This is confirmed at the third-level CLC classes, where we find permanent crops among CLC's *arable land* as well, namely inside class 212 *permanently irrigated land*. Thus, crops as defined in other agricultural classes can be part of CLC class 212 if irrigation infrastructure is used for water supply. On the other hand, CLC class 213 *rice fields* actually features the characteristics to identify this class as a subclass of CLC class 212.

CORINE Level 2 subclasses *pastures* and *heterogeneous agricultural areas*, which we have already discussed, are exemplary for translation difficulties as regards non-land-cover terminology and mixed classes. The translation of CLC class 231 *pastures* is roughly satisfying, whereas the translation of classes 241 *annual crops associated with permanent crops*, 242 *complex cultivation patterns*, 243 *land principally occupied by agriculture, with significant areas of natural vegetation*, and 244 *agro-forestry areas* is, for the most part, not even possible, at least not thoroughly, and is not comprehensively representative. The class design of these classes is heavily characterized by the use of land-use and topologic specifications and the lack of integrative class separation criteria. Thus, multiple sources of inconsistencies occur simultaneously, resulting in major difficulties in using those classes. This is not only valid for the translation presented here and the task of land-cover harmonization, but may also interfere with accuracy during the interpretation and classification process of CORINE itself. This is confirmed by the accuracy assessment of the EEA (2006), stating the highest subjectivity index percentages for classes 242, 243, and 324 (cf. Section 6.6).

6.5.4.2.3 Forests and Seminatural Areas

Forest classes are not defined properly with CLC classes 311 *broad-leaved forests* and 312 *coniferous forests*. The class names target different things: the first one reflects vegetation physiognomy, and the second describes floristics and refers to the cone-bearing conifers, which form a division named "pinophyta" in the recent taxonomic nomenclature. The classes are not consistently separated from each other. As a result, coniferous species with broad leaves can be part of both classes. In fact, the term "coniferous" usually may be applied in a similar manner as "needle-leaved"; however, technical terminology should be used correctly. The complementary term to "broad-leaved" is "needle-leaved."

CLC classes 32x (*shrub and/or herbaceous vegetation associations*) completely neglect physiognomic parameters. The classification does not take into consideration canopy cover, leaf type, or seasonality, but focuses on the definition of certain vegetation associations (CLC classes 322 *moors and heathland*, 323 *sclerophyllous vegetation*). Regarding land cover, this clearly results in class overlaps between the CLC shrub classes. Since mainly non-land-cover terminology is used to define the classes, no "neutral" shrub class exists within CLC; this causes a definition gap for shrubby land cover, which, for that reason, is assigned to CLC class 322 as per definition. CLC class 324 *transitional woodland-shrub* has contradictory definitions regarding (tree) canopy cover and sacrifices a clear land-cover description in favor of a debatable process definition. Indeed, the processes of forest degradation and regeneration can be an important factor for land-cover change (possibly driven by land-use change), but since both processes are contrary and not separated further, the usefulness of this class is rather limited. The descriptions of these classes by land-cover terms and hence the translation with the LCCS software cannot be definite. The moderate to unsatisfactory ratings regarding consistency of class description and confidence in the translation reflect this (see Figure 6.5).

Open spaces with little or no vegetation (classes 33x) show slight inconsistencies in the definition of classes 332 *bare rock* and 333 *sparsely vegetated areas*, which are caused by the share of vegetation cover; sparsely vegetated areas where 75% of the land surface is covered by rocks are included in class 332. This is contrary to the classification guidelines provided for class 333, which include areas with a vegetation cover from 15 up to 50 (or between 10% and 50%, both value ranges can be found within the guidelines). Another class is not in agreement with the requirements for a regular description of land cover: CLC class 334 *burnt areas* does not discriminate between any vegetation-cover type affected by fire. Hence, all life forms can or cannot be present in the concerned areas. By this definition, the class refers only to an environmental event; actual land cover remains unknown in any case, making translation with the LCCS arbitrary.

6.5.4.2.4   Wetlands

CORINE lacks the specification of vegetation cover for its wetland classes and includes both managed and natural wetlands. Majority of the classes of the *wetlands* category achieve moderate levels of consistency and translation confidence. The following points give the main reasons for the intermediate rating. CLC class 412 *peat bogs* does not refer to land cover; areas may be bare (and exploited) or vegetated; if vegetated, a separation to CLC class 411 *inland marshes* may be difficult. Furthermore, CORINE does not include all peat bogs because wooded peat bogs are assigned to the appropriate forest class (31x). Similar to *inland wetlands* (classes 41x), the classification of *coastal wetlands* (classes 42x) does not give priority to land cover: CLC classes 422 *salines* and 423 *intertidal flats* refer to land use and geographical (spatial) occurrence.

6.5.4.2.5   Water Bodies

Geographic terminology can be found again in CLC classes 521 *coastal lagoons* and 522 *estuaries*. Apart from this, translation of the category *water bodies* into the LCCS did not cause problems.

6.5.4.3   IGBP Discover/UMD Issues

As apparent from the evaluation scores in Figure 6.6, few problems occurred during the translation of the IGBP/UMD legend. For the most part, classes were outlined according to life forms and common land-cover classifiers. Thus, near-perfect translation into LCCS classifiers could be achieved for these classes. Difficulties appeared for some classes concerning only the IGBP legend, since all of the following classes were not (or not identically) present within the UMD variant.

6.5.4.3.1   Mixed Forests

A mixed forest is commonly defined as a mixture of broad-leaved and needle-leaved species. Within the IGBP legend, the four defined forest types (*evergreen needle-leaf forests, evergreen broad-leaf*

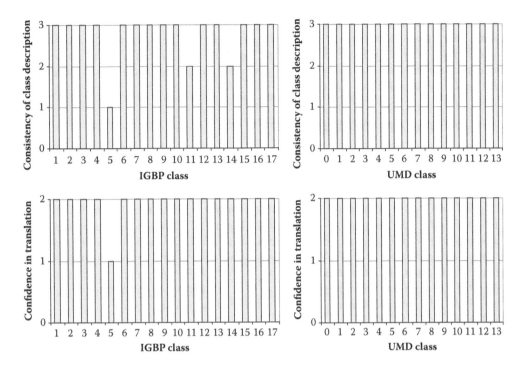

FIGURE 6.6 Evaluation of consistency of the original class descriptions and the translator confidence in the quality of proposed translations to present the class concept within the LCCS for IGBP (left) and UMD (right) classes (see Tables 6.3 and 6.4 for class names).

forests, deciduous needle-leaf forests, and deciduous broad-leaf forests) are supposed to build the mixture. A reasonable conclusion is that an area exclusively vegetated by needle-leaved (or broad-leaved) species, for example, an area with spruce and larch trees, will fall into this class as long as one part of the trees is evergreen and the other part deciduous.

This definition is contrary to the common meaning of the term *mixed forest* (specifying a mixture of broad-leaved and needle-leaved trees). In addition, during data interpretation, all possible combinations had to be considered, and as the correct (!) result, very different forest types had to be merged into one class. Furthermore, the 60% intermixture threshold leaves only a rather narrow range for valid IGBP *mixed forests* within a two-type intermixture (e.g., of needle-leaved evergreen and broad-leaved deciduous species)—one part may easily exceed this threshold value.

6.5.4.3.2 Permanent Wetlands

IGBP class *permanent wetlands* will inevitably produce inconsistencies in life forms. The class separation criterion used by other IGBP classes is life form. Considering that, introducing another separation criterion at the same classification level will not allow consistency among the classes. Consequently, some areas may meet the conditions of both classes, for example, a "wetland forest." On the other hand, a generic wetland class comprising all of its types does not permit identification and distinction of life forms—a clear deficiency of this approach.

6.5.4.3.3 Cropland/Natural Vegetation

Comparable to the *mixed forests* category, the 60% threshold value provides only a narrow definition for the intermixture. The LCCS defines a broader (perhaps more practical) range here, specifying between 50% and 80% for the first, and between 20% and 50% for the second, component of the mixed class. On the other hand, the limited capabilities of the LCCS in creating (spatial) mixed classes do not allow a proper translation according to the IGBP class definition. Natural vegetation is represented by its generic LCCS category only, and since one part of the mixed class had to be defined as dominating, *cropland* was chosen—according to the class name and its characterizing nature for the concerned areas.

6.6   CONCLUSIONS AND DISCUSSION

6.6.1   COMPARING LEGEND TRANSLATIONS

The occurrences and description of translation issues (i.e., concerning their quantity as well as their quality) help to compare the results obtained from the translation analyses. To assess the legend's overall performance in terms of its consistency and translation confidence, the evaluation results were summarized (Figure 6.7). Figure 6.7 shows the legend scores for both parameters in percent of the maximum score for full translation consistency and confidence. The range of these values indicates the differences faced across the legends during the translation process. Perhaps this evaluation, even though strictly oriented on the criteria introduced in Section 6.5.4, does not follow a metric system—a score twice as high does not make a legend twice as good. Nevertheless, quantification of the evaluation can provide an indicator of the legend translation.

It is apparent that both legends with higher scores (IGBP, UMD) have only about one-half to one-third of the class number compared with ACS and CLC. The more classes exist, the smaller are thematic class distances and the more likely are inconsistencies and overlaps between classes. Furthermore, ACS and CLC were not developed for global application. Thus, they cover a more narrow thematic range of land cover. In contrast, the IGBP/UMD legend consists of rather generic classes for coarse-resolution satellite data analysis with a clear focus on land cover. CLC and the ACS were developed for more detailed analysis and include much more specifications on land use, that is, more agricultural and urban classes. Hence, they are more susceptible to the resulting land-cover/non-land-cover terminology conflict. Thus, the lower score of CLC and the ACS is a consequence, especially since no consistent construction set like the LCCS was used for the legend creation, which could have helped prevent some inconsistencies.

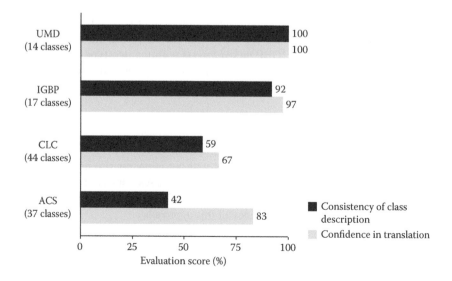

FIGURE 6.7   Evaluation scores for legend consistency and translation confidence.

CORINE shows reasonable efforts to ensure an intrinsic thematic consistency. This can be concluded from the enormous amount of guidelines that were provided supplementary to the CLC class descriptions. The initial CLC Technical Guide (CEC, 1994) offered a short definition of each class that was extensively extended by an addendum (Bossard et al., 2000), in order to limit the confusion. As shown, the potential of confusion can (at least partially) originate from inconsistencies in class definitions. Of course, the core definitions (and difficulties) of the CLC classes were not altered. A consistent land-cover description has to be valid for the total area of interest. CLC has to cover all particularities in its nomenclature, and sometimes consistency can be provided only by excluding or including specific particularities in the appropriate class description. That way a kind of "synthetic" (i.e., not class immanent) consistency is created. The vast number of guidelines given to the class descriptions are symptomatic.

Problematic in the context of "synthetic" consistency is not only that new uncertainty arises from every new individual case, but also that the user has to consider them altogether—during the whole chain of data classification, validation to interpretation, and analysis. There is an augmented susceptibility to errors and confusion resulting in the augmented effort to maintain inherent standardization. In the face of this and recalling the already mentioned mix of land-cover and non-land-cover terminology, automated classification may become challenging and impracticable.

6.6.2 CORINE—COMPARISON WITH VALIDATION DATA

To analyze the results derived from this work, we used a report published by the EEA providing information on the thematic accuracy of CORINE (EEA, 2006). They presented a comparison with Land use/cover area frame survey (LUCAS) *in situ* observations to derive accuracy statistics for the major CLC classes. The process of interpreting the LUCAS samples into CORINE categories revealed some interesting results worth discussing in the context of the LCCS translation results. Both the findings of EEA (2006) and the report presented here are plotted against each other in Figure 6.8.

The interpretation of the LUCAS reference data emphasized that subjectivity (hence different interpreters came up with different results) was noted for 18% of all samples. The most subjective CLC classes are shown in Table 6.5. The most prominent classes in this context are *land principally occupied by agriculture, with significant areas of natural vegetation* (243), *transitional woodland-shrub* (324), *complex cultivation patterns* (242), and *mixed forest* (313), where more than a third of the samples were labeled as subjective.

The analysis of CLC class definitions using the LCCS highlighted similar classes with problematic translation characteristics. This is emphasized in Figure 6.8. Obviously, classes with low translation confidence also exhibit larger amounts of subjectivity and thus inconsistencies in interpreting the LUCAS reference points. There also seems to be some relationship between the LCCS-assessed consistency of the class definition and overall agreement between the CORINE 2000-mapped classes and LUCAS reference information. The relationships are not deterministic, and this is not expected since a number of other factors influence mapping confidence and accuracy. Even though inconsistent land-cover definitions alone do not necessarily determine product quality, they eventually complicate the comparison and scaling of CORINE land-characterization features, particularly for complex and mixed unit classes.

The EEA (2006) report draws some general conclusions. In any future efforts, special attention should be paid to the less accurate classes, which means that there is a need to improve the definition of mapping rules and the use of multitemporal satellite data during interpretation. Of particular importance is the decomposition of CLC mixed classes (e.g., 242, 243) into pure land-cover classes based on LUCAS LC statistics. Both conclusions are encouraged by the results of this translation exercise.

With the observed difficulties in mind, it seems problematic to completely put CLC (level 3) on a common ground with a consistent land-cover description. The CORINE level 3 concept is intended

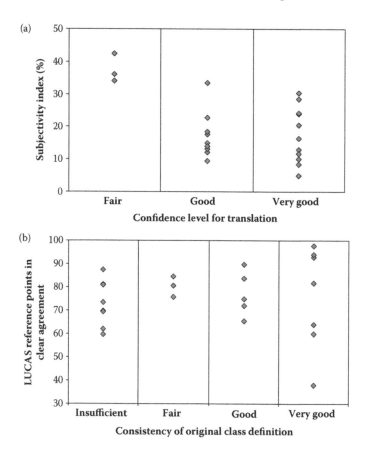

FIGURE 6.8 Comparing subjectivity index (top) and LUCAS reference point agreement (bottom) for representative classes derived in CLC2000 validation with results from the translation process. (From EEA, Thematic accuracy of CORINE land cover 2000. Assessment using LUCAS (land-use/cover area frame statistical survey)—Technical Report, 2006. Available at: http://reports.eea.europa.eu/technical_report_2006_7/en/technical_report_7_2006.pdf. With permission.)

TABLE 6.5
CLC2000 Classes with Largest Subjectivity Index in Interpreting LUCAS *In Situ* Observations

CLC2000 Class	Subjectivity Index (%)	Most Frequent Intermixing Classes[a]
243	42.3	242, 231, 211, 311, 323, 313, 324
324	36.1	312, 313, 311, 323
242	34.0	211, 243, 231
313	33.4	312, 311, 324

Note: The subjectivity index describes the percentage of all samples for that class with different corine class assignments from different interpreters.

[a] In order of importance.

not merely to account for "pure" land cover. Thus, CLC has much better potential of interoperability with global land-cover activities, for example, using the 2nd-level classes, aggregating several classes into a single one or/also splitting specific single classes, and for linkage with global land-cover activities. Further investigation in this direction will be necessary for CLC and should be carried out when the CORINE validation conclusions can be taken into consideration (cf. above). However, we will put forth some thoughts that represent a similar first step for ACS, since it shows comparable consistency issues.

6.6.3 USING LAND-COVER CLASSIFIERS

ACS also shows limitations in consistency performance and some peculiarities. On the basis of these experiences, we will use some issues to exemplify ways to address them. Obviously, inconsistencies of the classification system evolve to a great extent from using different separation criteria between classes within the same (1st-level) category, which result in cross-category overlaps and ambiguities. The proposed approach is to use the LCCS classifiers as independent means to characterize land cover in a nonhierarchical way.

For example, the most common classification criterion for vegetated areas is life form. Each vegetation category in the Anderson classification can be characterized by life form for the sake of consistency. Trees, shrubs, herbaceous vegetation, and nonvegetated areas occur multiple times and inside various categories across the whole ACS. Other independent LCCS classifiers may specify leaf type form or whether an area is terrestrial or aquatic/regularly flooded. For example, the category of a "forest wetland" is specified by the classifiers life form (trees) and the classifier aquatic and regularly flooded.

There is already some consensus on basic internationally used classifiers for land cover, which include the following:

- Vegetation life form (trees, shrubs, herbaceous vegetation, lichen and mosses, nonvegetated)
- Leaf type (needle-leaf, broad-leaf) and leaf longevity (deciduous, evergreen)
- Nonvegetated covers (bare soil/rock, built-up, snow, ice, water)
- Density of life form and leaf characteristics in percent cover
- Terrestrial versus aquatic/regularly flooded
- Artificiality of cover and land use

The majority of Anderson level 2 classes can be defined using a combination of these classifiers. Information about the climatic regime or eco region can be included as further classification details or as user-defined attributes. The translation exercise here provides the basis for such an effort. In the broader harmonization context, each land-cover map can be understood as different layers characterizing different land-cover classifiers. On this level, existing land-cover data can be much better compared and harmonized.

A translation in the LCCS language does not make an inconsistent legend design "better"; however, it provides a more consistent perspective describing known categories with standardized classifiers. Thus, this translation exercise takes the first step in defining avenues for land-cover harmonization in future efforts. For example, considerations of EEA, JRC, ESA, and GOFC-GOLD are currently underway to link the GlobCover product with the European CORINE mapping Program. The LCCS can help establish this link, and the first step has been taken with this work. However, an advanced solution will be arrived at by using LCCS classifiers in the development phase of land-cover products, that is, as done for GLC2000 and its successor GlobCover.

REFERENCES

Ahlqvist, O. 2008. In search for classification that support the dynamics of science—The FAO Land Cover Classification System and proposed modifications. *Environment and Planning B: Planning and Design*, 35(1), 169–186.

Anderson, J.R., Hardy, E.E., Roach, J.T., and Witmer, R.E. 1976. A land use and land cover classification system for use with remote sensor data. U.S. Geological Survey Professional Paper 964. Washington D.C.: USGS.

Arino, O., Leroy, M., Ranera, F., Gross, D., Bicheron, P., Nino, F., Brockman, C., et al. 2007. GLOBCOVER—a global land cover service with MERIS. In *Proceedings of Envisat Symposium 2007*, on CD Rom.

Bartholomé, E. and Belward, A.S. 2005. GLC2000: A new approach to global land cover mapping from Earth observation data. *International Journal of Remote Sensing*, 26, 1959–1977.

Belward, A. (Ed.). 1996. The IGBP-DIS global 1 km land cover data set (discover): Proposal and implementation plans. Report of the Land Cover Working Group of the IGBP-DIS. IGBP-DIS Working Paper 13.

Bossard, M., Feranec, J., and Otahel, J. 2000. CORINE land cover technical guide—Addendum 2000. Technical Report 40. EEA, Copenhagen. Available at: http://reports.eea.europa.eu/tech40add/en/tech40add.pdf

Büttner, G., Feranec, J., Jaffrain, G., Mari, L., Maucha, G., and Soukup, T. 2004. The European CORINE Land Cover 2000 Project. In *Paper Presented on 20th Congress of International Society of Photogrammetry and Remote Sensing*, July 12–23, 2004, Istanbul, Turkey.

CEC. 1994. CORINE Land cover—Technical guide. Available at: http://reports.eea.europa.eu/COR0-landcover/en.

Di Costanzo, M. and Ongaro, L. 2004. The Land Cover Classification System (LCCS) as a formal language: A proposal. *Journal of Agriculture and Environment for International Development*, 98(1), 117–164.

Di Gregorio, A. 2005. Land Cover Classification System—Classification concepts and user manual for software version 2, FAO Environment and Natural Resources Service Series, No. 8, Rome, 208 p. Available at: http://www.glcn-lccs.org

Di Gregorio, A. and Jansen, L.J.M. 1996a. Part I—Technical document on the Africover Land Cover Classification Scheme (pp. 4–33, 63–76), in FAO (1997). Africover Land Cover Classification.

Di Gregorio, A. and Jansen, L.J.M. 1996b. The Africover Land Cover Classification System: A dichotomous, modular-hierarchical approach. Working Paper with the Proposal for the International Working Group Meeting. Dakar, July 29–31, 1996. Rome: FAO.

EEA. 2006. The thematic accuracy of CORINE land cover 2000. Assessment using LUCAS (land use/cover area frame statistical survey). Technical Report 7/2006. Available at: http://reports.eea.europa.eu/technical_report_2006_7/en/technical_report_7_2006.pdf

GOFC-GOLD. 2009. Translating and evaluating land cover legends using the UN Land Cover Classification System (LCCS). GOFC-GOLD Report 43. Available at: http://www.fao.org/gtos/gofc-gold/series.html

Hansen, M.C. and Reed, B. 2000. A comparison of the IGBP Discover and University of Maryland 1 km global land cover products. *International Journal of Remote Sensing*, 21(6/7), 1365–1373.

Hansen, M.C., Defries, R.S., Townshend, J.R.G., and Sohlberg, R. 2000. Global land cover classification at 1 km spatial resolution using a classification tree approach. *International Journal of Remote Sensing*, 21(6/7), 1331–1364.

Herold, M. and Schmullius, C.C. 2004. Report on the harmonization of global and regional land cover products. Workshop Report at FAO, Rome, July 14–16, 2004. GOFC-GOLD Report 20.

Herold, M., Latham, J.S., Di Gregorio, A., and Schmullius, C.C. 2006a. Evolving standards in land cover characterization. *Journal of Land Use Science*, 1(2–4), 157–168.

Herold, M., Woodcock, C., Di Gregorio, A., Mayaux, P., Belward, A., Latham, J., and Schmullius, C.C. 2006b. A joint initiative for harmonization and validation of land cover datasets. *IEEE Transactions on Geoscience and Remote Sensing*, 44(7), 1719–1727.

IGBP. 1990. The International Geosphere-Biosphere Programme: A study of global change—The initial core projects. IGBP Global Change Report 12.

Jansen, L.J.M. 2004. Thematic harmonisation and analyses of Nordic data sets into Land Cover Classification System (LCCS) terminology. In G. Groom (Ed.), *Development in Image Application for Nordic Landscape Level Monitoring* (pp. 91–118). NMR Diverse Series. Copenhagen: Nordic Council of Ministers. Available at: http://www.norden.org/pub/miljo/miljo/sk/ANP2004705.pdf

JRC-IES. 2005. *CORINE Land Cover Updating for the Year 2000. Image2000 and CLC2000 Products and Methods.* Ispra, Italy.

Loveland, T.R., Reed, B.C., Brown, J. F., Ohlen, D.O., Zhu, Z., Yang, L., and Merchant, J.W. 2000. Development of a global land cover characteristics database and IGBP discover from 1 km AVHRR data. *International Journal of Remote Sensing*, 21(6/7), 1303–1330.

McConnell, W.J. and Moran, E.F. (Eds.). 2001. Meeting in the middle: The challenge of meso-level integration. An International Workshop on the Harmonization of Land Use and Land Cover Classification, Ispra, Italy, 17–20 October 2000. LUCC Report Series No. 5. Bloomington: LUCC Focus 1 Office, Indiana University.

Townsend, J.R. and Brady, M.A. 2006. A revised strategy for GOFC-GOLD. GOFC-GOLD Report 24.

Wyatt, B.K., Greatorex Davies, J.N., Hill, M.O., Parr, T.W., Bunce, R.G.H., and Fuller, R.M. 1994. Comparison of land cover definitions. Countryside 1990 Series, Department of the Environment, London.

7 Long-Term Satellite Data Records for Land-Cover Monitoring

Sangram Ganguly

CONTENTS

7.1   INTRODUCTION

The recent concerns about land-use/land-cover change have been highlighted by almost every other nation in the world in the wake of major changes in climate, frequent natural disasters, and human-induced changes, partly due to the differential demands for sustainability and functioning of life. Most of the large-scale changes in land cover in the last decade can be attributed to changes in vegetation and to urban expansion associated with continuing increase in food and fiber production, resource-use efficiency, and the wealth and well-being of a society. Although changes in land cover provide a positive stimulus for a nation's economic growth, these can significantly affect the functioning of the earth system.

Vegetation covers almost 75% of land surface. Its character, structure, and functional properties are critical for modeling the material and energy cycles in our climate system and for understanding the link between land-scale processes and climate variability. The large uncertainty in quantifying terrestrial carbon sinks/sources as characterized by vegetated land still poses a challenge for estimating net carbon fluxes in a multiparadigm modeling framework. However, with multiple satellite sensors onboard and robust physical algorithms in place, research has shown considerable promise in quantifying the changes and trends in large-scale terrestrial sink/source behavior vis-à-vis climate changes and human-induced changes. Over the past few decades, there has been a steep rise in generating research quality measurements by several international space missions (e.g., NOAA AVHRR, NASA TERRA/AQUA/AURA, Landsat, and SPOT), which have subsequently demonstrated their value for operational users and decision-making strategies.

Long-term monitoring of vegetated land cover is thus a topical issue in the light of the present concerns about climate change. Satellite remote sensing provides the ideal data for monitoring changes in land-surface characteristics at a range of scales, with sufficient spatial and temporal resolution. Advances in remote sensing, both in theory and instrumentation, have paved the way for better understanding of the partitioning of radiative energy between the earth's surface and the atmosphere (Diner et al., 1999; Justice et al., 1998; Tucker, 1986). As a result, studies on the retrieval of biophysical variables that act as a proxy to the amount of vegetation on the land surface and terrestrial productivity have gained momentum in recent decades.

7.2   LONG-TERM VEGETATION MONITORING WITH SATELLITE DATA

The advanced very high resolution radiometers (AVHRRs) onboard the NOAA series satellite platforms 7–16 provided the first long-term global time series of data suitable for vegetation sensing (Tucker et al., 2005). The NASA Moderate Resolution Imaging Spectroradiometer (MODIS) and Multi-angle Imaging SpectroRadiometer (MISR) onboard Terra and Aqua platforms started delivering high-quality spectral and angular measurement data from February 2000 (Justice et al., 2002). These data are expected to be improved by data from the planned Visible/Infrared Imager Radiometer Suite (VIIRS) instrument used in the NPOESS (National Polar-Orbiting Operational Environmental Satellite System) Preparatory Project (NPP) (Murphy et al., 2006). Other long-term sources of data for vegetation monitoring include the Sea-Viewing Wide Field-of-View (SeaWiFS), Systeme Pour l'Observation de la Terre (SPOT) VEGETATION, and Environmental Satellite (ENVISAT) Medium Resolution Imaging Spectrometers (MERIS).

Meaningful monitoring of vegetation requires a seamless and consistent long-term data record obtained from multiple instruments, but this is challenging because of sensor-related differences and methodological issues (Brown et al., 2006; Van Leeuwen et al., 2006; Vermote and Saleous, 2006). The challenges include modeling the highly variable radiative properties of global vegetation, scaling, and atmospheric correction of data. The sensor-related issues pertain to differences in sensors' spectral characteristics, spatial resolution, calibration, measurement geometry, and data information content (e.g., surface spectral reflectances). Therefore, the consistency among biophysical variables derived from different sensors has been a critical issue in establishing a proper consensus on vegetation monitoring over several decades.

Among the biophysical variables, leaf area index (LAI) and fraction of photosynthetically active radiation (FPAR) are recognized as the two most important variables representative of vegetation structure and functioning, which are commonly derived from satellite data (Running et al., 1986). Availability of data from multiple sensors in the recent decade allows for rich spectral and angular sampling of the radiation field reflected by vegetation canopies, thus enhancing the potential for obtaining accurate estimates of the biophysical variables. Long-term records of LAI and FPAR are required by various terrestrial biosphere models, such as the Terrestrial Ecosystem Model (TEM) (Melillo et al., 1993), Biome-BGC (Running and Gower, 1991), Simple Biospheric Model (SiB) (Sellers et al., 1986), Integrated Biosphere Simulated Model (IBIS) (Foley et al., 1996), Lund-Potsdam-Jena (LPJ) dynamic global vegetation model in Land Surface Model (LSM) (Bonan et al., 2003), and the Atmospheric-Vegetation Interactive Model (AVIM) (Jinjun et al., 1995), for investigating the response of ecosystems to changes in climate, carbon cycle, land cover, and land use.

7.3   EARTH SYSTEM DATA RECORDS OF VEGETATION LAI
FROM MULTIPLE SATELLITE-BORNE SENSORS

Long-term global vegetation monitoring requires temporally and spatially consistent datasets of vegetation biophysical variables, which are characteristic of vegetation structure and which function like LAI and FPAR. Such datasets are useful in many applications ranging from ecosystem monitoring to modeling of the exchange of energy, mass (e.g., water and CO_2), and momentum between the earth's surface and atmosphere (Demarty et al., 2007; Dickinson et al., 1986; Sellers et al., 1996; Tian et al., 2004). A crucial step in assembling these long-term datasets is establishing a link between data from earlier sensors (e.g., AVHRR) and present/future sensors (e.g., MODIS TERRA and NPOESS) such that the derived products are independent of sensor characteristics and represent the reality on the ground both in absolute value and variations in time and space (Van Leeuwen et al., 2006). Generating multidecadal globally validated datasets of LAI and FPAR with a physically based algorithm and of known accuracy is difficult, although several recent attempts have resulted in short-term research quality datasets from medium-resolution sensor data (Baret et al., 2007; Chen et al., 2002; Gobron et al., 1999; Knyazikhin et al., 1998; Plummer et al., 2006; Yang et al., 2006). Some recent studies (Ganguly et al., 2008b) have reported physically based approaches in deriving long-term LAI and FPAR products from AVHRR data, which are of quality comparable to that of the MODIS products. Sections 7.4 through 7.7 demonstrate the usefulness of such long-term data records in quantifying the large-scale changes in land cover owing to changes in climatic and anthropogenic factors.

7.4 VEGETATION VARIABILITY WITH SURFACE TEMPERATURE IN THE NORTHERN LATITUDES

The northern latitudes, 40°N–70°N, witnessed a persistent increase in growing-season vegetation greenness related to the unprecedented surface warming during 1981–1999 (Myneni et al., 1997; Slayback et al., 2003; Zhou et al., 2001). This greening was observed in Eurasia and less prominently in North America (Zhou et al., 2001). In fact, a decline in greenness was observed in parts of Alaska, boreal Canada, and northeastern Eurasia (Barber et al., 2000; Goetz et al., 2005). The multisensor consistent LAI dataset (Ganguly et al., 2008a) thus helped in reassessing these changes. The spatial trends (in %) in LAI for the growing-season, April to October, for the region 40°N–70°N were determined for the periods 1982–1999 and 1982–2006. The greening trend (Figure 7.1a) was evident in

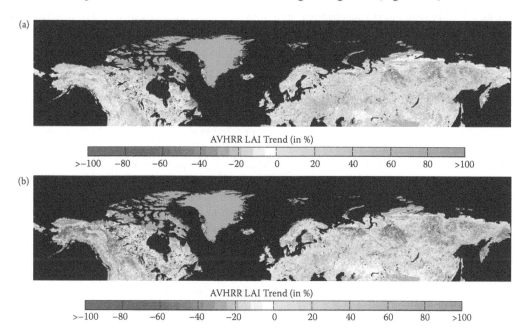

FIGURE 7.1 Trends in AVHRR LAI for the growing season, April to October, for the region 40°N–70°N, for the periods 1982–1999 (panel [a]) and 1982–2006 (panel [b]). For each 8-km AVHRR LAI pixel, the April-to-October mean LAI was regressed on time (years). The slope obtained from this regression, which if statistically significant based on the *t*-statistic at or lower than 10% level, was converted to a percent trend by multiplying by the number of years times 100 and dividing by the mean April-to-October AVHRR LAI of 1982.

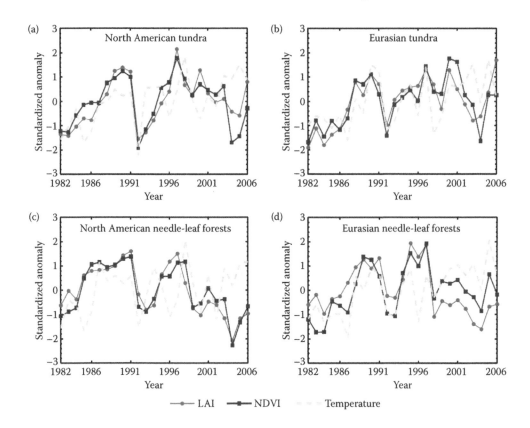

FIGURE 7.2 **(See color insert.)** Standardized April-to-October anomalies of AVHRR LAI (green), GIMMS AVHRR NDVI (blue), and GISS temperature (red dashed line) for Eurasian and North American needle-leaf forests (panels [c] and [d]) and tundra (panels [a] and [b]) from 1982 to 2006. (From Ganguly, S. et al., *Rem. Sens. Environ.*, 112, 4318–4332, 2008a.)

Eurasia, northern Alaska, Canada, and parts of North America, for 1982–1999. When this analysis was extended to 2006 (Figure 7.1b), it was found that large contiguous areas in North America, northern Eurasia, and southern Alaska showed a decreasing trend in growing-season LAI. This browning trend, especially in the boreal forests of southern Alaska and Canada and in the interior forests of Russia, has also been reported in recent studies (Angert et al., 2005; Goetz et al., 2005).

The spatial (40°N–70°N) and growing-season averages of standardized anomalies (anomalies normalized by their standard deviation) of LAI, normalized difference vegetation index (NDVI), and surface temperature (Hansen et al., 1999) are shown in Figure 7.2 for tundra and needle-leaf forests separately for North America and Eurasia. The anomaly of a given variable is defined as the difference between the growing-season mean in a given year and the growing-season mean over the 1982–2006 time interval. The results indicate that vegetation activity significantly correlates with the trends in surface temperature in the Eurasian and North American tundra over the entire period of the record (Table 3.2 in Ganguly et al., 2008a). This is consistent with the reports of persistent greening in the tundra and evidence of shrub expansion in northern Alaska and the pan-Arctic (Goetz et al., 2005; Tape et al., 2006).

A decreasing trend in vegetation greenness was observed after 1996–1997 despite a continuing warming trend in the North American needle-leaf forests. The regression model of LAI versus surface temperature and time was statistically significant at the 10% level for 1982–1999 but was statistically insignificant for 1982–2006 (Table 3.2 in Ganguly et al., 2008a). Similar patterns were observed in the Eurasian needle-leaf forests also. These results imply a decreasing trend in vegetation activity possibly due to warming-induced drought stress, as has been suggested previously (Barber et al., 2000; Bunn et al., 2006; Lapenis et al., 2005; Wilmking et al., 2004). There were also reports of declining growth and health of white spruce trees in Alaska, upsurge in insect disturbance in southern Alaska, and increase in fire frequency and severity in Alaska, Canada, and Siberia during the past 6–7 years of consistent warming (Soja et al., 2007). These changes justify the need for continued monitoring of vegetation activity in these northern regions in the face of unprecedented climatic changes.

7.5 VEGETATION VARIABILITY WITH PRECIPITATION IN THE SEMIARID TROPICS

The semiarid tropics are projected to be among the areas most affected by ongoing and future climate changes (Parry et al., 2007). In these regions, reduction in vegetation productivity and expansion of desertification are expected to take place owing to drier conditions due to continued warming trends accompanied by a reduction in precipitation (IPCC, 2007) and low adaptation capacity of the affected plant species (Parry et al., 2007). Over the past few decades, the tropical dry lands have experienced an increase in average air temperatures in the range of 0.2°C–2°C (IPCC, 2007) and modest but less homogeneous increases in precipitation (Gu et al., 2007; Zhang et al., 2007), which are more marked over ocean than over land.

In spite of these climate changes, which would suggest that tropical dry lands are already becoming drier, the satellite observations of vegetation greenness provide evidence that, similar to that in other parts of the globe and also over extensive portions of the semiarid tropics, primary productivity has been on the rise (Eklundh and Olsson, 2003; Herrmann et al., 2005; Pandya et al., 2004; Tucker and Nicholson, 1999). The availability of globally consistent climate datasets has led to a useful investigation that establishes quantitatively the correlation between climate and the global greening trends (Cao et al., 2004; Kawabata et al., 2001; Myneni et al., 1997; Nemani et al., 2003). Other driving factors, such as the changes in land cover and land use (Xiao and Moody, 2005) and fertilization effects due to atmospheric increases in carbon and nitrogen (Ichii et al., 2002), have been cited as reasons for the remaining portion of the trend, but a quantitative analysis of the effects of these factors on vegetation dynamics is still lacking. However, to help project the effects of climate change on ecosystems and societies, it is crucial to understand properly the changes in the drivers of ecosystem dynamics.

With the goal of identifying the relative contributions and spatial distribution of climate, socioeconomic, and land-use change in promoting the greening of the tropical dry lands, the changes in LAI in conjunction with the changes in climatic and land-use data for the period 1981–2006 are analyzed. The case study focuses on the semiarid tropics of the eastern hemisphere, where the largest contiguous dry lands are inhabited by nearly 1.7 billion people and are spread across 120 countries, most of which are among the poorest countries in the world and have the lowest human development index. Section 7.6 presents an analysis in which the greening of the semiarid tropics is compared with changes in precipitation across all the countries of the eastern hemisphere semiarid tropics. Section 7.7 presents the changes in vegetation greenness in the context of changes in socioeconomic and land-use change data, with particular focus on India, where high-resolution data are available at the national scale.

Availability of water critically limits plant growth in semiarid tropical regions, especially in grasslands where precipitation in the wet months is the primary driver of plant growth (Hickler et al., 2005; Nemani et al., 2003; Prince et al., 2007). This relationship provides a basis for evaluating the LAI product by examining the correlation between LAI and precipitation (Huffman et al., 2007).

For the purpose of analysis, the semiarid regions in the tropics and subtropics are defined as those with peak annual NDVI values in the range 0.12–0.55 (Figure 7.3). These regions approximately correspond to areas with annual total rainfall less than 700 mm. Using ancillary datasets such as the MODIS VCF (vegetation continuous fields) data (Hansen et al., 2003), MODIS Land Cover data (Friedl et al., 2010), and Tropical Rainfall Measuring Mission (TRMM) and Climatic

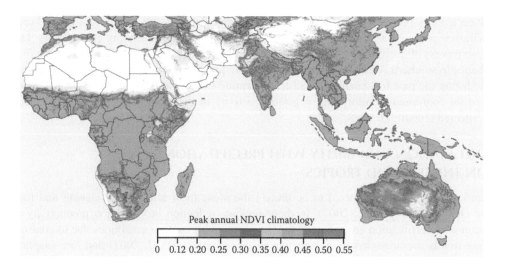

FIGURE 7.3 (See color insert.) Color map of peak annual NDVI climatology. Peak annual NDVI climatology was calculated by first estimating the 26-year (1981–2006) mean of monthly NDVI (monthly NDVI climatology) and then selecting the maximum value (per pixel, from 12-monthly climatological NDVI values). A spatial mask was applied on the color map based on peak annual NDVI climatology values in the range of 0.12–0.55. The NDVI data used is the AVHRR GIMMS NDVI product.

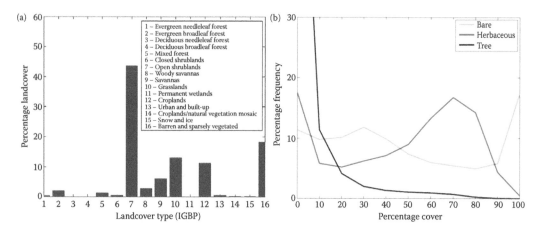

FIGURE 7.4 (See color insert.) Percentage distribution of IGBP land-cover classes (panel [a]) and frequency distribution of bare (red), herbaceous (blue), and tree (black) cover from MODIS VCF map, expressed as percentage of total number of pixels (panel [b]) for the peak annual NDVI climatology range of 0.12–0.55.

Research Unit (CRU) precipitation data (Huffman et al., 2007), a complete analysis of vegetation changes for particular land-cover types can be performed. On a global scale, the most prevalent land-cover types in the semiarid tropics are shrublands, grasslands, croplands, and, to a lesser extent, savannas (Figure 7.4a). Prevalence of herbaceous vegetation cover is also dominant in these areas (Figure 7.4b).

The LAI of semiarid vegetation fluctuates during the year depending on the vagaries of rainfall. The long-term average LAI values may be expected to be more stable, unless major shifts in precipitation, land-use practices, or a combination of both affect an ecosystem. To characterize the spatial distribution of such persistent changes in dry-land vegetation greenness and precipitation over the period of record, the percentage change in decadal means of annual maximum LAI and precipitation was calculated. Here, the annual maximum precipitation is defined as the total precipitation of the three wettest months during a year. For representative countries in the study area, the anomalies of annual maximum LAI and precipitation were correlated and compared with the countrywide decadal changes in total food production, irrigation area, fertilizer use, and macroeconomic indicators. The countries included in this study did not have border changes during the 26-year period, had a considerable portion (at least 40%) of their surface area within the tropical dry lands as defined above (Figure 7.3), and comprised at least 50 half-degree pixels to perform meaningful comparison with the precipitation dataset. In addition, the areas with increases in net irrigated area were used to test the hypothesis that changes in land use due to expansion in irrigated areas have been a major driver of increased vegetation greenness (NDVI) in India.

7.6   CLIMATE-DRIVEN INCREASES IN VEGETATION

Notable increases in annual maximum LAI were observed between 1981–1990 and 1995–2006 in over 70% of the tropical dry lands of the eastern hemisphere (Figure 7.5a), encompassing Turkey, large portions of the Middle Eastern countries, the Sahel, Horn of Africa, southern African countries, most of tropical Asia, and portions of Australia. About 29% of the area, principally distributed in eastern and southern Australia, southwest China, along the Namibian desert, and other portions of the coast of western Africa up to the Iberian peninsula, report decline in photosynthetic activity.

In general, the areas that have greened up (20%–60% from Figure 7.5a) within the semiarid tropics show increase in precipitation over two decades (Figure 7.5b). The increase in decadal precipitation is particularly marked along the Sudano-Sahelian semiarid tropics, the Horn of Africa, the Middle East, and Western Australia. More modest increases of greenness are found in most other regions, which are consistent with the findings of increases in tropical land precipitation (Gu et al., 2007; Zhang et al., 2007), which could be a consequence of the recent warming trends (Wentz et al., 2007). Reduction in precipitation in the range 20%–40% during the last decade occurred in Egypt, southern Ethiopia, and northern Kenya, especially in Pakistan, Afghanistan, and eastern Australia.

To investigate whether the observed increases in the photosynthetic capacity of the tropical dry lands of the eastern hemisphere are related to local changes in precipitation, the detrended anomalies of annual maximum LAI were correlated with the detrended anomalies of precipitation for the three wettest months in each of the four major regions in the study area (Figure 7.6). The Sahelian region consisted of Senegal, Mauritania, Mali, Burkina Faso, Niger, Nigeria, Chad, and Sudan; the southern African region consisted of Botswana, South Africa, and Namibia; and the South Asian region consisted of Afghanistan, Pakistan, and India. For the regions comprising Sahel, southern Africa, and Australia, significant ($p < .05$) positive linear correlations between the two variables are observed, supporting the hypothesis that changes in climate that brought increased rainfall especially since the early 1990s over most of the subtropical semiarid countries have promoted plant growth in these dry-land regions. The trends in the Palmer Drought Severity Index (Dai et al., 2004), which integrates atmospheric moisture with the evaporative demand of the vegetation, also point to increased moisture in the tropical dry lands and, therefore, enhanced vegetation growth. A recovery of total annual precipitation to the pre-1960 levels and consequent greening trends over the Sahel have been described by Tucker and Nicholson (1999) and Eklundh and Olsson (2003). The increase in greenness in South Asia (especially India) is not supported by enhanced precipitation and may, therefore, be due to other land-use factors such as irrigation and fertilizer use.

7.7   LAND-USE CHANGE-DRIVEN INCREASES IN VEGETATION

All the major countries in the study area, except Somalia, reported an increase in total food production over two decades of a systematic study period (FAOStat, 2007). The study indicates that land management and land-use changes may have contributed to the greening, especially where greening is not supported by changes in precipitation. Along with increased precipitation, the changes in land

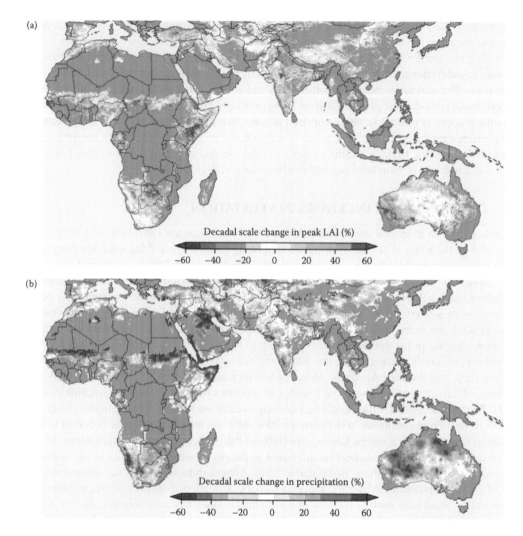

FIGURE 7.5 (See color insert.) (a) Percentage change in mean peak annual LAI between decade 1 (1981–1990) and decade 2 (1995–2006). For each year in a decade, the peak LAI was selected (per pixel from 12 LAI values). The mean peak LAI was calculated for each decade. Finally, the percentage change was calculated as [100 × (mean peak LAI decade2 − mean peak LAI decade1)/(mean peak LAI decade1)]. A spatial mask was applied on the color map based on peak annual NDVI climatology values in the range of 0.12–0.55 (all values outside this range appear in gray—masked out). (b) Percentage change in mean peak annual precipitation (mm/year) between decade 1 (1981–1990) and decade 2 (1995–2006). Peak precipitation for each year was calculated by summing the precipitation in the three wettest months. The mean peak annual precipitation for each decade and percentage change were calculated as in (a).

use such as transition from rain-fed to irrigated agriculture and increased use of mineral fertilizers, and probably other factors less documented, such as the improvements in agricultural practices and natural resource management (Niemeijer and Mazzucato, 2002; Reij et al., 2005; Tappan and McGahuey, 2007), are also considered to be strong factors likely to have increased the photosynthetic activity recorded by satellite data.

The role of land-use changes, helped by even modest changes in climate, in promoting large-scale increases in plant growth is particularly evident in India, where 52% of the country's land area is devoted to croplands (FAOStat, 2007). Although monthly average temperatures have been on the rise in India, the monthly precipitation trends point to a modest redistribution of the

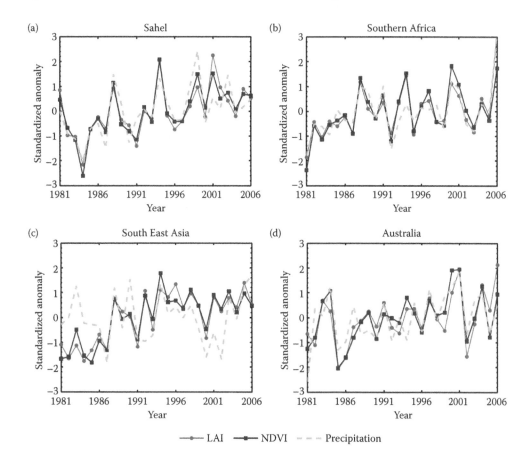

FIGURE 7.6 (See color insert.) Standardized anomalies of annual peak AVHRR LAI (green line), annual peak AVHRR NDVI (blue line), and annual peak (three wettest month CRU + TRMM) precipitation (red dashed line) for the semiarid regions (panels [a]–[d]) from 1981 to 2006.

monsoonal precipitation, and no significant increasing trend in total precipitation has been detected (Goswami et al., 2006). Yet, 80% of the semiarid dry lands of India display significant increases in decadal LAI. Analysis of the 1981–2006 trend in monthly LAI (Figure 7.7) shows that the largest increases in vegetation growth occurred during January and February, which correspond to the peak of the rabi (spring harvest in India) cropping season.

The rabi cropping season starts at the end of the summer monsoon (November) and extends through the following spring (February–May). Water for the rabi crops is supplied by the less abundant northeast (winter) monsoon, by the moisture accumulated from the southwest (summer monsoon) during the kharif (autumn harvest in India) season, or, increasingly, by irrigation. Irrigation, beyond making possible the cultivation of non-rainfed crops during the rabi season (i.e., a second rice crop), also supplements cropping-water requirements during the kharif season, when monsoon rains are delayed. It is, therefore, suggested that land-use changes have been the principal driver of enhanced plant growth detected from satellite in this predominantly water-scarce country.

Noteworthy are the states of Madhya Pradesh and Rajasthan, where decadal scale changes in LAI and changes in net irrigated area have been significantly higher than in the other states (Figure 7.8). Large-scale increases in decadal LAI are seen in Mandsaur, Jhalawar, Ujjain, Shajapur, Ratlam, and Kota districts (Figure 7.8). In particular, the semiarid region of Mandsaur district is spread over an area of 5554 km² with approximately 1600 inhabited villages, and water for irrigation is sustained through several macrolevel watersheds spread over around 15,500 ha across the Sitamau

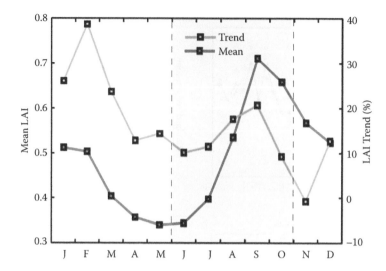

FIGURE 7.7 Long-term mean monthly LAI (red line) and percent trend in monthly LAI (green line) for India over the period 1982–2006. The long-term mean monthly LAI was calculated by averaging the maximum monthly LAI of each pixel over the period 1982–2006. The spatially averaged mean monthly LAI was then plotted for each month. The monthly trend of LAI was calculated as the slope of a linear regression fitted through the spatially averaged maximum LAI of each month as a function of the period 1982–2006. The percent trend is then calculated as follows: LAI trend (%) = [slope × 25/1982 monthly maximum LAI] × 100. Only pixels falling within the peak annual NDVI climatology mask of 0.12–0.55 were considered in the calculation.

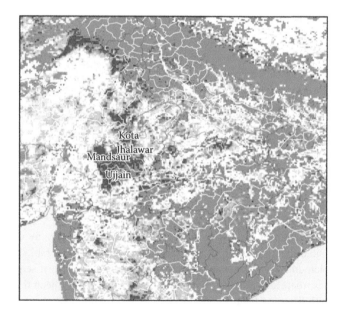

FIGURE 7.8 Percentage change in mean peak annual LAI as in Figure 7.5a for the semiarid districts of Mandsaur (state: Madhya Pradesh), Kota (state: Rajasthan), Jhalawar (state: Rajasthan), and Ujjain (state: Madhya Pradesh) in India. Decadal-scale change in LAI shows a percent increase of more than 50% in these districts. The gray boundaries are state boundaries, and the white boundaries depict district-level partition.

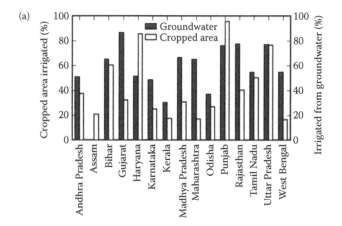

(a)

(b)

State	Fertilizer Consumption (kg/Ha)	Fraction Irrigated Sown Area (%)
Andhra Pradesh	128.4	37.58
Karnataka	90.9	24.90
Kerala	68.1	17.31
Tamil Nadu	114	50.32
Gujarat	77.7	32.13
Madhya Pradesh	36.4	30.96
Maharashtra	73.8	16.90
Rajasthan	28.5	40.45
Goa	32.3	16.73
Dadar & Nagar Haveli	35.5	29.82
Haryana	152.7	85.76
Punjab	174.9	95.36
Uttar Pradesh	126.5	76.30
Himachal Pradesh	41.5	22.75
Jammu & Kashmir	60.1	40.90
Bihar	87.1	60.47
Odisha	34	26.83
West Bengal	122.3	16.48
Nagaland	1.8	19.94
Arunachal Pradesh	2.8	25.66
Assam	42.7	20.62

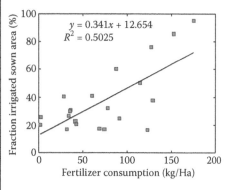

FIGURE 7.9 (See color insert.) (a) Percentage of cropped area that is irrigated (blue bar) and percentage of irrigated land utilizing groundwater (green bar) for each of the major Indian states. (b) Tabulates the fertilizer consumption (kg/ha) and fraction irrigated sown area (%) for different states (2002–2003) and shows the corresponding regression relation (orange squares represent states).

and Mandsaur blocks (http://fes.org.in/includeAll.php?pId=Mi0yNi0z). These districts are covered by the Chambal Valley Project, which facilitates large-scale building of dams for providing hydro-electric power and water for irrigation and agriculture.

Further rise in vegetation productivity can be explained by the increase in fertilizer use for the resource-demanding high-yield crop varieties, which replaced traditional cultivars once the supply of water is ensured through irrigation (Figure 7.9b) (Bhattaray and Narayanamoorthy, 2003; http://dacnet.nic.in/). As shown in Figure 7.9a, currently in most Indian states, more than half of the cropped area is irrigated, and the water used for irrigation is derived increasingly from groundwater sources, representing up to 80% of all water sources used for irrigation in some states, such as Punjab and Uttar Pradesh (Narayanamoorthy, 2002). Access to microcredit and to heavily subsidized electricity has led to the expansion of private wells to irrigate fields distant from major irrigation infrastructures (Shah, 2005), particularly benefiting Indian agriculture by increasing crop yields and total food production, thus alleviating rural poverty.

These greening trends, however, are not expected to continue strongly over these regions. The increase in annual LAI has already slowed down in India, and the slowdown has been reflected in a flattening of growth in total food production (FAOStat, 2007). The reasons for this slowdown are complex. Since the mid-1990s, a number of basins have been increasingly suffering from groundwater overexploitation and are at the risk of salinization (http://cgwb.gov.in/gw_profiles/gwprofiles.html), and commodity prices have been declining owing to globalization (Narayanamoorthy, 2007), reducing farmers' potential investments in production. While these factors can be reversed through better irrigation and proper policies, they can be further dampened by the current trends in climate, if here to stay.

7.8   CANONICAL CORRELATION ANALYSIS

The correlations observed between LAI and temperature in the northern regions and between LAI and precipitation in the semiarid areas raise a question about the mechanistic basis for these relations. It has been reported previously that large-scale circulation anomalies, such as the El Niño-Southern Oscillation (ENSO) and Arctic Oscillation (AO), explain similar correlations but at the hemispheric scale (Buermann et al., 2003). The canonical correlation analysis (CCA) is ideally suited for analyzing spatiotemporal data as it seeks to estimate dominant and independent modes of covariability between two sets of spatiotemporal variables (Barnett and Preisendorfer, 1987; Bjornsson and Venegas, 1997). The variables are linearly transformed into two new sets of uncorrelated variables called canonical variates, which explain the covariability between the two original variables, in a descending order. Thus, most of the covariability is captured by the first 2–3 canonical variates.

For the CCA in the North, each year is denoted as a variable (1982–2006, that is, 25 variables in total) and each pixel as an observation (the total number of observations is the number of vegetated pixels in the latitudinal zone 45°N and 65°N). The two sets of variables for CCA are the springtime (March–May) LAI and surface temperature anomalies at 1° resolution (Buermann et al., 2003). The anomalies were normalized by their respective standard deviations. Each of the set of 25 (time) variables was transformed to principal components (PCs) using singular value decomposition. In each case, only the first six PCs were retained as they explain a large fraction of the variance in the input set of variables. In CCA, each canonical variate is a time series, which accounts for a certain fraction of the covariability between the variables (PCs). In this analysis, the first two canonical variates derived from each set of six PCs explained about 50% of the covariability between the two sets of variables.

The September to November (SON) NINO3 index (http://www.cpc.ncep.noaa.gov/data/indices/wksst.for) is used to represent ENSO because the sea surface temperature anomalies then approach peak values during an ENSO cycle (Dai et al., 1997). Figure 7.10a shows that the correlation between SON NINO3 index and the first canonical variate related to LAI is very low ($r = 0.1$). The same is true for the correlation between SON NINO3 index and the first canonical variate related to temperature anomalies. This is in contrast to a strong correlation reported by Buermann et al. (2003) for 1982–1998. This decline in correlation may be due to weak ENSO activity and/or changes in teleconnection patterns since 1998–2000 (http://www.cpc.ncep.noaa.gov/products/CDB/Tropics/figt5.shtml). The correlation between the AO index and the second canonical variates of both LAI and temperature is reasonably strong (0.45 and 0.61, respectively; Figure 7.10b), consistent with the strong correlations reported by Buermann et al. (2003) for 1982–1998. Thus, the AO seems to continue to be a prominent driver of surface temperature (Thompson and Wallace, 1998) and plant growth variability in the northern latitudes.

CCA was also performed on standardized anomalies of annual maximum LAI and precipitation for the semiarid regions of 40°N–40°S latitudinal zone (cf. Section 7.6). The first two canonical variates explained about 50% of the covariability between annual peak LAI and precipitation anomalies. A reasonable correlation is seen between the September–November NINO3 index and

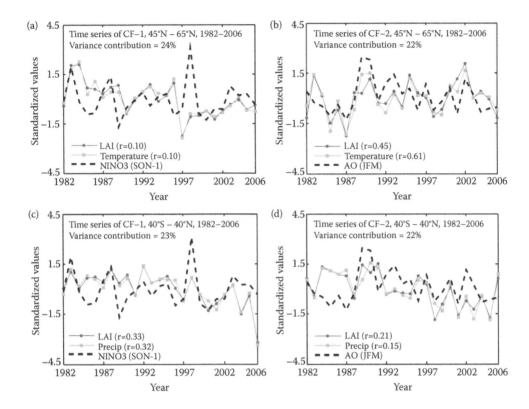

FIGURE 7.10 Correlation between standardized time series of the first canonical factor (CF-1, panels [a] and [c]) and second canonical factor (CF-2, panels [b] and [d]) with NINO3 and AO indices in the northern and tropical/subtropical regions.

the first canonical variates of LAI and precipitation (0.33 and 0.32, respectively; Figure 7.10c), consistent with several previous reports of the effects of ENSO on interannual variability in tropical and subtropical precipitation (Dai and Wigley, 2000; Ropelewski and Halpert, 1987). The correlation between the second canonical variates and the AO index is weak (Figure 7.10d), which is not surprising as the AO is not known to be a driver of precipitation and thus plant growth variability in these regions.

In summary, the strong ENSO-driven linked variations between northern vegetation greenness and surface temperature observed during the 1980s and 1990s have weakened since 2000. The effects of AO, however, continue to be strong. In the tropical and subtropical regions, the effect of ENSO on linked variations between semiarid vegetation greenness and precipitation continues to be apparent. These results further instill confidence in these long-term datasets.

7.9 LAND-SURFACE PHENOLOGY FROM MODIS: CHARACTERIZATION OF LAND-COVER DYNAMICS

Investigations focused on monitoring and modeling biospheric processes require accurate information about spatiotemporal dynamics in ecosystem properties. Because vegetation phenology affects terrestrial carbon cycling across a wide range of ecosystem and climate regimes (Baldocchi et al., 2001; Churkina et al., 2005; Richarson et al., 2009), accurate information on phenology is important to studies of regional-to-global carbon budgets that indirectly quantify the state of change in a particular land cover. The presence of leaves also affects land-surface albedo (Moore et al.,

1996; Ollinger et al., 2008) and exerts strong control on surface radiation budgets and the partitioning of net radiation between latent and sensible heat fluxes (Chen and Dudhia, 2001; Yang et al., 2001). Thus, the phenological dynamics of vegetated ecosystems affect a host of eco-physiological processes that affect hydrologic processes (Hogg et al., 2000), nutrient cycling (Cooke and Weih, 2005), and land–atmosphere interactions (Heimann et al., 1998).

In recent years, growing-season dynamics, including shifts in the timing of bud burst, leaf development, senescence, and changes in growing-season length, have been widely studied in the context of ecosystem responses to climate change (Cleland et al., 2007). Sections 7.4 and 7.5 show the trends in vegetation greenness, using AVHRR LAI data in the northern hemisphere and semiarid tropics. Complex phenological responses have also been observed in controlled experiments, where warming accelerated greening of plant canopies but elevated CO_2 and nitrogen fertilization delayed flowering (Cleland et al., 2007). Both biophysical and biochemical processes affect, and are diagnostic of, ecosystem–climate interactions. Therefore, there is a substantial need to accurately characterize the phenology of ecosystems and, by extension, the response of ecosystems to changes in climate (Morisette et al., 2009).

Moderate-resolution satellite remote sensing provides global high-temporal frequency measurements of land-surface properties and is, therefore, well suited for monitoring seasonal-to-decadal patterns and trends in regional-to-global phenology (de Beurs and Henebry, 2005; Reed et al., 1994; White et al., 1997; Zhang et al., 2003). Landsat MSS was the first space-borne sensor used to characterize the seasonality of vegetation at landscape and regional scales (Thompson and Wehmanen, 1979). However, detecting phenological transition dates requires higher temporal resolution than is afforded by Landsat-class instruments, and coarse-to-moderate spatial resolution sensors such as AVHRR (Goward et al., 1985), MODIS (Zhang et al., 2003), and SPOT-VEGETATION (Delbart et al., 2006) are more commonly used for this purpose. Indeed, the utility of such sensors for studies of land-surface phenology has been established over the last two decades (Justice et al., 1985) during which a number of different methods have been developed for detecting phenological transition dates. The most well-known methods include threshold-based techniques (Jönsson and Eklundh, 2002; White et al., 1997), methods based on spectral analysis (Jakubauskas et al., 2001; Moody and Johnson, 2001), and inflection point estimation in time series of vegetation indices (Moulin et al., 1997; Zhang et al., 2003). All these methods use time series of vegetation indices to identify the timing of phenological transition dates such as the start and end of the growing season.

Since 2000, MODIS has provided an excellent basis for regional-to-global scale studies of land-surface phenology (Ahl et al., 2006; Fisher et al., 2007; Zhang et al., 2003, 2006). Ganguly et al. (2010) present an overview and characterization of the new Collection 5 (C5) MODIS Global Land Cover Dynamics (MLCD) product, which is produced globally at a spatial resolution of 500 m and has been available from 2001 till now. The cardinal parameters produced as a part of the product include onset of greenness, maturity, senescence, and dormancy for every 500-m pixel. Based on these parameters, useful metrics like the growing season length can be calculated, and this has important implications in estimating the "net primary productivity" of a specific ecoregion. To illustrate the nature and scale of geographic patterns in interannual variability captured by the MLCD product, Figure 7.11 shows a map of anomalies in the timing of greenness onset and growing-season length for 2002 relative to 2001–2006 averages (computed as 2002 minus the multiyear average). This figure suggests that the onset of greening occurred later over much of North America relative to the 2001–2006 average, especially at mid-to-high latitudes and in the south-central United States. With the exception of the South Asian region, growing-season anomalies follow the same general pattern and are positive (i.e., shorter growing season) throughout much of the continent. The climatic force behind this pattern is unclear, but it is likely that the widespread drought in the northern hemisphere that prevailed till 2002 provides a partial explanation (Lotsch et al., 2005).

The MODIS Land Cover Dynamics Product is one of a number of remote-sensing-based products being used to generate regional-to-global scale maps of vegetation phenology (Ganguly et al.,

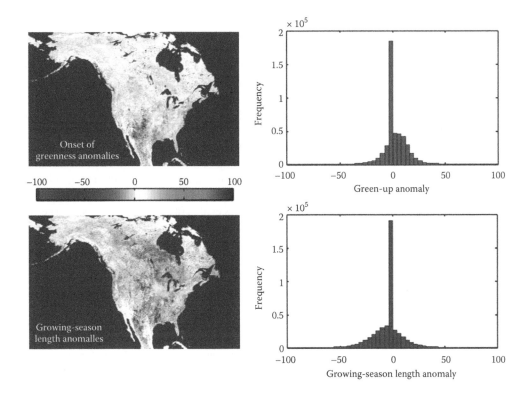

FIGURE 7.11 **(See color insert.)** Anomalies in the timing of green-up onset and growing-season length for 2002 relative to the 2001–2006 mean. Histograms show the frequency of green-up and growing-season length anomalies. Details about processing the MODIS data and deriving the phenological parameters have been described in depth by Ganguly et al. (2010).

2010). The results of several studies show that remote sensing of vegetation phenology can provide good qualitative estimates over large regions (e.g., temperate deciduous vegetation and agriculture). However, a number of important issues naturally remain to be resolved in order to address the uncertainties in the input satellite data as well as reassure the scientists who wish to use these products confidently. Besides providing better characterization of the error and uncertainty associated with the metrics like those from the MLCD product, ongoing efforts are focused on developing improved methods. These include preprocessing of the input data (including screening for snow, clouds, and aerosols) as well as creating better understanding of the nature and utility of the retrieved phenological values in environments that present challenges for remote sensing, including high latitude, arid, and tropical ecosystems.

7.10 STATE OF KNOWLEDGE AND FUTURE RESEARCH

This chapter presents an overview of analyses and techniques that can be routinely applied to study long-term changes in land-cover dynamics using coarse-resolution sensors and sensors with moderately higher resolution. The availability of long-term consistent datasets from sensors such as AVHRR and MODIS is the backbone for documenting the observed changes for large-scale regional-to-global studies. Analyzing the long-term dominant trends and changes in land cover instill confidence in utilizing this seamless, consistent product for large-scale terrestrial-biosphere models and for monitoring the global-scale vegetation dynamics in response to changes in climate and human activity. Despite the robustness of the methodological approach and the expected accuracy of the derived products, there are inevitably certain limitations. First, the data measurement uncertainties from different sensors can significantly affect the retrieval of a biophysical product. This requires better calibration and atmospheric correction algorithms, along with solar and view angle corrections for surface reflectance. Second, the global retrieval of biophysical products utilizes land-cover classification maps, which set the basis for identifying the spatial heterogeneity of distribution of biomes. Classification inaccuracies are critical factors, especially for regions that show dramatic changes in land-cover dynamics (e.g., changes from herbaceous to woody biomes). The validity of these long-term products during the 1980s and 1990s represents a more challenging problem in land-cover dynamics, as the present algorithms rely on a single land-cover map for the entire period. Finally, a direct validation of coarse-resolution products with ground measurements is a complicated task due to scaling of the plot-level measurements to sensor resolution, geo-location uncertainties, limited temporal and spatial sampling of ground data, field instrument calibration, sampling errors, and so on. (Buermann et al., 2002; Weiss et al., 2007; Yang et al., 2006). The accuracy of the direct validation exercise is a function of the area homogeneity, as the comparison of the field-level measurements with larger pixels of a satellite product is a valid exercise if performed over spatially homogeneous pixels.

The scientific community has a pressing need for these long-term datasets, and further research can continue along the following lines:

1. Scale dependency is a critical issue in retrieving the biophysical parameters such as LAI across multiple sensors. The scaling methodology described by Ganguly et al. (2008b) can be seen as a benchmark for retrieving LAI fields at any given resolution for any given sensor. The theory of canopy spectral invariants will provide a framework by which structural information can be maintained in a self-consistent manner across multiple scales (Ganguly et al., 2008b). This algorithm can thus be applied to retrieve LAI at finer resolutions (e.g., Landsat), thus allowing a better capture of the spatial heterogeneity of leaf dynamics. In future, to ensure data continuity of LAI, surface reflectances from VIIRS onboard NPOESS should be analyzed to maintain product consistency with the AVHRR and MODIS data. Open access to the Landsat archive now enables the scientific community to exploit these theoretical approaches in deriving a high-resolution, long-term parameter suite of biophysical variables, albeit the cost in processing and storage.

2. Discrepancies between field measurements and satellite observations also arise owing to the scaling problem. The understanding of scale dependency in the development of an algorithm will facilitate an improved validation scheme to better compare coarse-resolution retrievals with field measurements, as well as proper explanation of the physics behind intercomparing data of different resolutions and from multiple sensors.

3. The consistent long-term data record of LAI and FPAR can be used to produce a long-term GPP/NPP time series based on the MODIS NPP logic (Nemani et al., 2003). NPP is the source of most food, fiber, and fuel; changes in NPP integrate climatic, ecological, geochemical, and human effects on the biosphere (Nemani et al., 2003). The NPP algorithm inputs vegetation parameters (land-cover type, LAI, and FPAR) and daily climate data (incident solar radiation [IPAR], minimum and average air temperatures and humidity), and so estimates in productivity are sensitive to uncertainties in input LAI/FPAR (e.g., differences in LAI from multiple sensors). The availability of long-term products will thus improve future NPP estimates, which can also be used in deriving total biomass.

4. The multiyear global LAI dataset will be a significant input to different climate models for investigating the response of ecosystems to changes in climate, carbon cycle, land cover, and land use. An improvement over the long-term dataset will be to create a consistent dynamic vegetation layer or an improved phenology record covering the AVHRR, MODIS, and NPOESS eras. For example, the algorithm as developed by Ganguly et al. (2008b) also accounts for generation of consistent surface reflectances across multiple sensors,

thus extending the scope of research to create consistent vegetation indices such as the enhanced vegetation index (EVI). EVI has improved sensitivity in high biomass regions and improved vegetation monitoring through decoupling of the canopy background signal and reduction in atmospheric influences (Huete et al., 2006). Overall, long-term global datasets of LAI and phenology with a monthly temporal resolution will be an indispensable input to integrated climate–vegetation–land-surface models to quantify global land-cover change and terrestrial productivity in the context of climate change, land-use change, and anthropogenic influences.

5. Finally, following the case study in Section 7.7, research can be extended to develop a deterministic model for anticipating changes in crop productivity and/or vegetation greenness due to continued warming over the semiarid tropical regions (as projected in the IPCC, 2007), especially in countries such as India, China, and the Sahel. A convincing stride will be to explore the further sustainability of the greening trend as observed in a developing and highly populated country like India, where the greening due to land-use change is dominant, and in countries in the Sahel, where precipitation-induced greening is significant. Owing to overexploitation of groundwater for irrigation, changes in policies subsidizing the crop inputs, and subsequent projections in future warming trends, there would be a challenging food security scenario for a large number of developing countries in the semiarid tropics, with a rapidly increasing population.

REFERENCES

Ahl, D.E., Gower, S.T., Burrows, S.N., Shabanov, N.V., Myneni, R.B., and Knyazikhin, Y. 2006. Monitoring spring canopy phenology of a deciduous broadleaf forest using MODIS. *Remote Sensing of Environment*, 104, 88–95.

Angert, A., Biraud, S., Bonfils, C., Henning, C.C., Buermann, W., Pinzon, J., Tucker, C.J., and Fung, I. 2005. Drier summers cancel out the CO_2 uptake enhancement induced by warmer springs. *Proceedings of the National Academy of Sciences USA*, 102, 10823–10827.

Baldocchi, D., Falgle, E., and Wilson, K. 2001. A spectral analysis of biosphere-atmosphere trace fas flux densities and micrometeorological variables across hour to multi-year time scales. *Agricultural and Forest Meteorology*, 107(1), 1–276.

Barber, V.A., Juday, G.P., and Finney, B.P. 2000. Reduced growth of Alaskan white spruce in the twentieth century from temperature-induced drought stress. *Nature*, 405, 668–673.

Baret, F., Hagolle, O., Geiger, B., Bicheron, P., Miras, B., Huc, M., Berthelot, B., et al. 2007. LAI, fAPAR, and fCover CYCLOPES global products derived from VEGETATION Part 1: Principles of the algorithm. *Remote Sensing of Environment*, 110, 275–286.

Barnett, T.P. and Preisendorfer, R. 1987. Origins and levels of monthly and seasonal forecast skill for United States surface air temperatures determined by canonical correlation analysis. *Monthly Weather Review*, 115, 1825–1850.

Bhattaray, M. and Naraynamoorthy, A. 2003. Impact of irrigation on rural poverty in India: An aggregate panel-data analysis. *Water Policy*, 5, 443–458.

Bjornsson, H. and Venegas, S.A. 1997. A manual for EOF and SVD analysis of climate data. McGill University, CCGCR Report No. 97–1, Montreal, Quebec, pp. 52.

Bonan, G.B., Levis, S., Sitch, S., Vertenstein, M., and Oleson, K.W. 2003. A dynamic global vegetation model for use with climate models: Concepts and description of simulated vegetation dynamics. *Global Change Biology*, 9, 1543–1566.

Brown, M.E., Pinzon, J.E., Morisette, J.T., Didan, K., and Tucker, C.J. 2006. Evaluation of the consistency of long term NDVI time series derived from AVHRR, SPOT-Vegetation, SeaWIFS, MODIS, and Landsat ETM+. *IEEE Transactions on Geoscience and Remote Sensing*, 44(7), 1787–1793.

Buermann, W., Anderson, B., Tucker, C.J., Dickinson, R.E., Lucht, W., Potter, C.S., and Myneni, R.B. 2003. Interannual covariability in northern hemisphere air temperatures and greenness associated with El Niño-Southern oscillation and the Arctic oscillation. *Journal of Geophysical Research*, 108, 1–16.

Buermann, W., Wang, Y., Dong, J., Zhou, L., Zeng, X., Dickinson, R.E., Potter, C.S., and Myneni, R.B. 2002. Analysis of a multiyear global vegetation leaf area index dataset. *Journal of Geophysical Research*, 107, 1–15.

Bunn, A.G. and Goetz, S.J. 2006. Trends in satellite-observed circumpolar photosynthetic activity from 1982 to 2003: The influence of seasonality, cover type, and vegetation density. *Earth Interactions*, 10(12), 1–19.

Cao, M., Prince, S.D., Small, J., and Goetz, S.J. 2004. Remotely sensed interannual variations and trends in terrestrial net primary productivity 1981–2000. *Ecosystems*, 7, 233–242.

Chen, F. and Dudhia, J. 2001. Coupling an advanced land surface-hydrology model with the Penn State-NCAR MM5 modeling system. Part I: Model implementation and sensitivity. *Monthly Weather Review*, 129, 569–585.

Chen, J.M., Pavlic, G., Brown, L., Cihlar, J., Leblanc S.G., White, H.P., Hall, R.J., et al. 2002. Derivation and validation of Canada-wide coarse resolution leaf area index maps using high-resolution satellite imagery and ground measurements. *Remote Sensing of Environment*, 80, 165–184.

Churkina, G., Schimel, D., Braswell, B.H., and Xiao, X. 2005. Spatial analysis of growing season length control over net ecosystem exchange. *Global Change Biology*, 11, 1777–1787.

Cleland, E.E., Chuine, I., Menzel, A., Mooney, H.A., and Schwartz, M.D. 2007. Shifting plant phenology in response to global change. *Trends in Ecology & Evolution*, 22, 357–365.

Cooke, J.E.K. and Weih, M. 2005. Nitrogen storage and seasonal nitrogen cycling in *Populus*: Bridging molecular physiology and ecophysiology. *New Phytologist*, 167, 19–30.

Dai, A. and Wigley, T.M.L. 2000. Global patterns of ENSO-induced precipitation. *Geophysical Research Letters*, 27, 1283–1286.

Dai, A., Fung, I.Y., and Del Genio, A.D. 1997. Surface observed global land precipitation variations during 1900–88. *Journal of Climate*, 10, 2943–2962.

Dai, A., Trenberth, K.E., and Qian, T. 2004. A global data set of Palmer Drought Severity Index for 1870–2002: Relationship with soil moisture and effects of surface warming. *Journal of Hydrometeorology*, 5, 1117–1130.

De Beurs, K.M. and Henebry, G.M. 2005. Land surface phenology and temperature variation in the International Geosphere-Biosphere Program high latitude transects. *Global Change Biology*, 11(5), 779–790.

Delbart, N., Le Toan, T., Kergoat, L., and Fedotova, V. 2006. Remote sensing of spring phenology in boreal regions: A free of snow-effect method using NOAA-AVHRR and SPOT-VGT data 1982–2004. *Remote Sensing of Environment*, 101, 52–62.

Demarty, J., Chevallier, F., Friend, A.D., Viovy, N., Piao, S., and Ciais, P. 2007. Assimilation of global MODIS leaf area index retrievals within a terrestrial biosphere model. *Geophysical Research Letters*, 34, L15402, doi:10.1029/2007GL030014.

Dickinson, R.E. 1983. Land surface processes and climate surface albedos and energy-balance. *Advances in Geophysics*, 25, 305–353.

Dickinson, R.E., Hendersen-Sellers, A., Kennedy, P.J., and Wilson, M.F. 1986. Biosphere-Atmosphere Transfer Scheme (BATS) for the NCAR CCM, *NCAR Res.*, Boulder, CO, NCAR/TN-275-STR.

Diner, D.J., Asner, G.P., Davies, R., Knyazikhin, Y., Muller, J.P., Nolin, A.W., Pinty, B., Schaaf, C.B., and Stroeve, J. 1999. New directions in earth observing: Scientific applications of multiangle remote sensing. *Bulletin of the American Meteorological Society*, 80(11), 2209–2228.

Douville, H. and Royer, J.F. 1996. Influence of the temperate and boreal forests on the Northern Hemisphere climate in the Meteo-France climate model. *Climate Dynamics*, 13, 57–74.

Eklundh, L. and Olsson, L. 2003. Vegetation index trends for the African Sahel 1982–1999. *Geophysical Research Letters*, 30(1430), doi:10.1029/2002GL016772.

FAOStat. 2007. Food and Agriculture Organization 2007. FAO Statistical Databases: Agriculture, Fisheries, Forestry, Nutrition Food and Agriculture Organization, Rome.

Fisher, J.I. and Mustard, J.F. 2007. Cross-scalar satellite phenology from ground, Landsat, and MODIS data. *Remote Sensing of Environment*, 109, 261–273.

Foley, J.A., Prentice, I.C., Ramankutty, N., Levis, S., Pollard, D., Sitch, S., and Haxeltine, A. 1996. An integrated biosphere model of land surface processes, terrestrial carbon balance, and vegetation dynamics. *Global Biogeochemical Cycles*, 10, 603–628.

Friedl, M.A., Sulla-Menashe, D., Tan, B., Schneider, A., Ramankutty, N., Sibley, A., and Huang, X. 2010. MODIS collection 5 global land cover: Algorithm refinements and characterization of new datasets. *Remote Sensing of Environment*, 114(1), 168–182.

Ganguly, S., Friedl, M.A., Tan, B., Zhang, X., and Verma, M. 2010. Land surface phenology from MODIS: Characterization of the collection 5 global land cover dynamics product. *Remote Sensing of Environment*, 114(8), 1805–1816.

Ganguly, S., Samanta, A., Schull, M.A., Shabanov, N.V., Milesi, C., Nemani, R.R., Knyazikhin, Y., and Myneni, R.B. 2008a. Generating vegetation leaf area index Earth system data record from multiple sensors. Part 2: Implementation, analysis and validation. *Remote Sensing of Environment*, 112, 4318–4332.

Ganguly, S., Schull, M.A., Samanta, A., Shabanov, N.V., Milesi, C., Nemani, R.R., Knyazikhin, Y., and Myneni, R.B. 2008b. Generating vegetation leaf area index Earth system data record from multiple sensors. Part 1: Theory. *Remote Sensing of Environment*, 112, 4333–4343.

Gobron, N., Pinty, B., Verstraete, M., and Govaerts, Y. 1999. MERIS Global Vegetation Index (MGVI): Description and preliminary application. *International Journal of Remote Sensing*, 20, 1917–1927.

Goetz, S.J., Bunn, A.G., Fiske, G.J., and Houghton, R.A. 2005. Satellite-observed photosynthetic trends across boreal North America associated with climate and fire disturbance. *Proceedings of the National Academy of Sciences USA*, 102, 13521–13525.

Goswami, B.N., Venugopal, V., Sengupta, D., Madhusoodanan, M.S., and Xavier, P.K. 2006. Increasing trend of extreme rain events over India in a warming environment. *Science*, 314, 1442–1445.

Goward, S.N., Tucker, C.J., and Dye, D.G. 1985. North-American vegetation patterns observed with the NOAA-7 advanced very high-resolution radiometer. *Vegetation*, 64, 3–14.

Gu, G., Adler, R., Huffman, G., and Curtis, S. 2007. Tropical rainfall variability on interannual-to-interdecadal/longer-time scales derived from the GPCP monthly product. *Journal of Climate*, 20, 4033–4046.

Hansen, J., Ruedy, R., Glascoe, J., and Sato, M. 1999. GISS analysis of surface temperature change. *Journal of Geophysical Research*, 104, 30997–31022.

Hansen, M., DeFries, R.S., Townshend, J.R.G., Carroll, M., Dimiceli, C., and Sohlberg, R.A. 2003. Global percent tree cover at a spatial resolution of 500 meters: First results of the MODIS vegetation continuous fields algorithm. *Earth Interactions*, 7(10), 1–15.

Heimann, M., Esser, G., Haxeltine, A., Kaduk, J., Kicklighter, D.W., Knorr, W., Kohlmaier, G.H., et al. 1998. Evaluation of terrestrial carbon cycle models through simulations of the seasonal cycle of atmospheric CO_2: First results of a model intercomparison study. *Global Biogeochemical Cycles*, 12(1), 1–24.

Herrmann, S.M., Anyamba, A., and Tucker, C.J. 2005. Recent trends in vegetation dynamics in the African Sahel and their relationship to climate. *Global Environmental Change*, 15, 394–404.

Hickler, T., Eklundh, L., Seaquist, J.W., Smith, B., Ardö, J., Olsson, L., Sykes, M.T., and Sjöström, M. 2005. Precipitation controls Sahel greening trend. *Geophysical Research Letters*, 32, 1–4.

Hogg, E.H., Price, D.T., and Black, T.A. 2000. Postulated feedbacks of deciduous forest phenology on seasonal climate patterns in the western Canadian interior. *Journal of Climate*, 13, 4229–4243.

Huete, A.R., Didan, K., Shimabukuro, Y.E., Ratana, P., Saleska, S.R., Hutyra, L.R., Yang, W., Nemani, R.R., and Myneni, R. 2006. Amazon rainforests green-up with sunlight in dry season. *Geophysical Research Letters*, 33, doi: 10.1029/2005gl025583.

Huffman, G.J., Adler, R.F., Bolvin, D.T., Gu, G., Nelkin, E.J., Bowman, K.P., Hong, Y., Stocker, E.F., and Wolff, D.B. 2007. The TRMM multi-satellite precipitation analysis: Quasi-global, multi-year, combined-sensor precipitation estimates at fine scale. *Journal of Hydrometeorology*, 8(1), 38–55.

Ichii, K., Kawabata, A., and Yamaguchi, Y. 2002. Global correlation analysis for NDVI and climatic variables and NDVI trends: 1982–1990. *International Journal of Remote Sensing*, 23, 3873–3878.

IPCC. 2007. Climate change 2007: The physical science basis. In S. Solomon, D. Qin, M. Manning, Z. Chen, M. Marquis, K.B. Averyt, M. Tignor, and H.L. Miller (Eds.), *Contribution of Working Group I to the Fourth Assessment Report of the Intergovernmental Panel on Climate Change* (p. 996). Cambridge; New York: Cambridge University Press.

Jakubauskas, M.E., Legates, D.R., and Kastens, J.H. 2001. Harmonic analysis of time-series AVHRR NDVI data. *Photogrammetric Engineering and Remote Sensing*, 67, 461–470.

Jinjun, J. 1995. A climate-vegetation interaction model: Simulating physical and biological processes at the surface. *Journal of Biogeography*, 22, 445–451.

Jonsson, P. and Eklundh, L. 2002. Seasonality extraction by function fitting to time-series of satellite sensor data. *IEEE Transactions on Geoscience and Remote Sensing*, 40, 1824–1832.

Justice, C.O., Townshend, J.R.G., Holben, B.N., and Tucker, C.J. 1985. Analysis of the phenology of global vegetation using meteorological satellite data. *International Journal of Remote Sensing*, 6, 1271–1318.

Justice, C.O., Townshend, J.R.G., Vermote, E.F., Masuoka, E., Wolfe, R.E., Saleous, N., Roy, D.P., and Morisette, J.T. 2002. An overview of MODIS Land data processing and product status. *Remote Sensing of Environment.*, 83, 3–15.

Justice, C.O., Vermote, E., Townshend, J.R.G., DeFries, R., Roy, D.P., Hall, D.K., Salomonson, V.V., et al. 1998. The Moderate Resolution Imaging Spectroradiometer (MODIS): Land remote sensing for global change research. *IEEE Transactions on Geoscience and Remote Sensing*, 36, 1228–1249.

Kawabata, A., Ichii, K., and Yamaguchi, Y. 2001. Global monitoring of international changes in vegetation activities using NDVI and its relationship to temperature and precipitation. *International Journal of Remote Sensing*, 22, 1377–1382.

Knyazikhin, Y., Martonchik, J.V., Myneni, R.B., Diner, D.J., and Running, S.W. 1998. Synergistic algorithm for estimating vegetation canopy leaf area index and fraction of absorbed photosynthetically active radiation from MODIS and MISR data. *Journal of Geophysical Research*, 103, 32257–32274.

Lapenis, A., Shvidenko, A., Shepaschenko, D., Nilsson, S., and Aiyyer, A. 2005. Acclimation of Russian forests to recent changes in climate. *Global Change Biology*, 11, 2090–2102.

Lotsch, A., Friedl, M.A., Anderson, B.T., and Tucker, C.J. 2005. Response of terrestrial ecosystems to recent Northern Hemispheric drought. *Geophysical Research Letters*, 32, L06705, doi:10.1029/2004GL022043.

Melillo, J.M., McGuire, A.D., Kicklighter, D.W., Moore, B., Vorosmarty, C.J., and Schloss, A.L. 1993. Global climate-change and terrestrial net primary production. *Nature*, 363, 234–240.

Moody, A. and Johnson, D.M. 2001. Land-surface phenologies using the discrete Fourier transform. *Remote Sensing of Environment*, 75(3), 305–323.

Moore, K.E., Fitzjarrald, D.R., Sakai, R.K., Goulden, M.L., Munger, J.W., and Wofsy, S.C. 1996. Seasonal variation in radiative and turbulent exchange at a deciduous forest in Central Massachusetts. *Journal of Applied Meteorology*, 35, 122–134.

Morisette, J.T., Richardson, A.D., Knapp, A.K., Fisher, J.I., Graham, E.A., Abatzoglou, J., Wilson, B.E., et al. 2009. Tracking the rhythm of the seasons in the face of global change: Phenological research in the 21st Century. *Frontiers in Ecology and the Environment*, 7, 253–260.

Moulin, S., Kergoat, L., Viovy, N., and Dedieu, G. 1997. Global-scale assessment of vegetation phenology using NOAA/AVHRR satellite measurements. *Journal of Climate*, 10, 1154–1170.

Murphy, R.E. 2006. The NPOESS preparatory project. *Earth Science Satellite Remote Sensing. Heidelberg:* Springer Berlin, 182–198, doi:10.1007/978-3-540-37293-6.

Myneni, R.B., Keeling, C.D., Tucker, C.J., Asrar, G., and Nemani, R.R. 1997. Increased plant growth in the northern high latitudes from 1981–1991. *Nature*, 386, 698–701.

Narayanamoorthy, A. 2002. Indian irrigation: Five decades of development. *Water Resources Journal*, 212, 1–29.

Narayanamoorthy, A. 2007. Deceleration in agricultural growth. *Economic and Political Weekly*, 42(25), 2375–2379.

Nemani, R.R., Keeling, C.D., Hashimoto, H., Jolly, W.M., Piper, S.C., Tucker, C.J., Myneni, R.B., and Running, S.W. 2003. Climate-driven increases in global terrestrial net primary production from 1982 to 1999. *Science*, 300, 1560–1563.

Niemeijer, D. and Mazzucato, V. 2002. Soil degradation in the west African Sahel: How serious is it? *Environment*, 44(2), 20–31.

Ollinger, S.V., Richardson, A.D., Martin, M.E., Hollinger, D.Y., Frolking, S.E., Reich, P.B., Plourde, L.C., et al. 2008. Canopy nitrogen, carbon assimilation, and albedo in temperate and boreal forests: Functional relations and potential climate feedbacks. *Proceedings of the National Academy of Sciences USA*, 105(49), 19335–19340.

Pandya, M.R., Singh, R.P., and Dadhwal, V.K. 2004. A signal of increased vegetation activity of India from 1981 to 2001 observed using satellite-derived fraction of absorbed photosynthetically active radiation. *Current Science*, 87, 1122–1126.

Parry, M.L., Canziani, O.F., Palutikof, J.P., van der Linden, P.J., and Hanson, C.E. (Eds.). 2007. Climate change 2007: Impacts, adaptation and vulnerability. In *Contribution of Working Group II to the Fourth Assessment Report of the Intergovernmental Panel on Climate Change* (p. 1000). Cambridge: Cambridge University Press.

Plummer, S., Arino, O., Simon, W., and Steffen, W. 2006. Establishing an Earth observation product service for the terrestrial carbon community: The GLOBCARBON initiative. *Mitigation and Adaptation Strategies for Global Change*, 11, 97–111.

Prince, S.D., Wessels, K.J., Tucker, C.J., and Nicholson, S.E. 2007. Desertification in the Sahel: A reinterpretation of a reinterpretation. *Global Change Biology*, 13, 1308–1313.

Reed, B.C., Brown, J.F., VanderZee, D., Loveland, T.R., Merchant, J.W., & Ohlen, D.O. 1994. Measuring phenological variability from satellite imagery. *Journal of Vegetation Science*, 5, 703–714.

Reij, C., Tappan, G., and Belemvire, A. 2005. Changing land management practices and vegetation on the Central Plateau of Burkina Faso (1968–2002). *Journal of Arid Environments*, 63, 642–659.

Richardson, A.D., Braswell, B.H., Hollinger, D., Jenkins, J.P., and Ollinger, S.V. 2009. Near-surface remote sensing of spatial and temporal variation in canopy phenology. *Ecological Applications*, 19(6), 1417–1428.

Ropelewski, C.F. and Halpert, M. S. 1987. Global and regional scale precipitation pattern associated with El Nino/Southern Oscillation. *Monthly Weather Review*, 115, 1606–1626.

Running, S.W. and Gower, S.T. 1991. Forest-BGC, a general-model of forest ecosystem processes for regional applications. II. dynamic carbon allocation and nitrogen budgets. *Tree Physiology*, 9, 147–160.

Running, S.W., Peterson, D.L., Spanner, M.A., and Teuber, K.B. 1986. Remote-sensing of coniferous forest leaf-area. *Ecology*, 67, 273–276.

Scholze, M., Knorr, W., Arnell, N.W., and Prentice, I.C. 2006. A climate-change risk analysis for world ecosystems. *Proceedings of the National Academy of Sciences USA*, 103, 13116–13120.

Sellers, P.J., Dickinson, R.E., Randall, D.A., Betts, A.K., Hall, F.G., Berry, J.A., Collatz, G.J., et al. 1997. Modeling the exchanges of energy, water, and carbon between continents and the atmosphere. *Science*, 275, 502–509.

Sellers, P.J., Mintz, Y., Sud, Y.C., and Dalcher, A. 1986. A simple biosphere model (sib) for use within general-circulation models. *Journal of the Atmospheric Sciences*, 43, 505–531.

Sellers, P.J., Randall, D.A., Collatz, G.J., Berry, J.A., Field, C.B., Dazlich, D.A., Zhang, C., Collelo, G.D., and Bounoua, L. 1996. A revised land surface parameterization (SiB2) for atmosphere GCMs. Part II : The generation of global fields of terrestrial biophysical parameters from satellite data. *Journal of Climate*, 9, 706–737.

Shah, T. 2005. Groundwater and human development: Challenges and opportunities in livelihoods and environment. *Water Science and Technology*, 51, 27–37.

Slayback, D.A., Pinzon, J.E., Los, S.O., and Tucker, C.J. 2003. Northern hemisphere photosynthetic trends 1982–1999. *Global Change Biology*, 9, 1–15.

Soja, A.J., Tchebakova, N.M., French, N.H.F., Flannigan, M.D., Shugart, H.H., Stocks, B.J., Sukhinin, A.I., Parfenova, E.I., Chapin III, F.S., and Stackhouse Jr., P.W. 2007. Climate-induced boreal forest change: Predictions versus current observations. *Global and Planetary Change*, 56, 274–296.

Sud, Y.C., Shukla, J., and Mintz, Y. 1988. Influence of land surface-roughness on atmospheric circulation and precipitation—A sensitivity study with a general-circulation model. *Journal of Applied Meteorology*, 27, 1036–1054.

Tappan, G. and McGahuey, M. 2007. Tracking environmental dynamics and agricultural intensification in southern Mali. *Agricultural Systems*, 94, 38–51.

Tape, K., Sturm, M., and Racine, C. 2006. The evidence for shrub expansion in northern Alaska and the pan-Arctic. *Global Change Biology*, 12, 686–702.

Thompson, D.R. and Wehmanen, O.A. 1979. Using Landsat digital data to detect moisture stress. *Photogrammetric Engineering and Remote Sensing*, 45, 201–207.

Thompson, D.W.J. and Wallace, J.M. 1998. The Arctic oscillation signature in the winter time geo-potential height and temperature fields. *Geophysical Research Letters*, 25, 1297–1300.

Tian, Y., Dickinson, R.E., Zhou, L., Zeng, X., Dai, Y., Myneni, R.B., Knyazikhin, Y., et al. 2004. Comparison of seasonal and spatial variations of LAI/FPAR from MODIS and Common Land Model. *Journal of Geophysical Research*, 109, D01103, doi:10.1029/2003JD003777.

Tucker, C.J. and Nicholson, S.E. 1999. Variations in the size of the Sahara Desert from 1980 to 1997. *Ambio*, 28, 587–591.

Tucker, C.J., Fung, I.Y., Keeling, C.D., and Gammon, R.H. 1986. Relationship between atmospheric CO_2 variations and a satellite-derived vegetation index. *Nature*, 319, 195–199.

Tucker, C.J., Pinzon, J.E., Brown, M.E., Slayback, D.A., Pak, E.W., Mahoney, R., Vermote, E.F., and El Saleous, N. 2005. An extended AVHRR 8-km NDVI dataset compatible with MODIS and SPOT vegetation NDVI data. *International Journal of Remote Sensing*, 26, 4485–4498.

Van Leeuwen, W.J.D., Orr, B.J., Marsh, S.E., and Herrmann, S.M. 2006. Multi-sensor NDVI data continuity: Uncertainties and implications for vegetation monitoring applications. *Remote Sensing of Environment*, 100, 67–81.

Vermote, E.F. and Saleous, N.Z. 2006. Calibration of NOAA16 AVHRR over a desert site using MODIS data. *Remote Sensing of Environment*, 105, 214–220.

Weiss, M., Baret, F., Garrigues, S., and Lacaze, R. 2007. LAI and fAPAR CYCLOPES global products derived from VEGETATION. Part 2: Validation and comparison with MODIS collection 4 products. *Remote Sensing of Environment*, 110, 317–333.

Wentz, F.J., Ricciardulli, L., Hilburn, K., and Mears, C. 2007. How much more rain will global warming bring? *Science*, 317, 233–235.

White, M.A., Thornton, P.E., and Running, S.W. 1997. A continental phenology model for monitoring vegetation responses to interannual climatic variability. *Global Biogeochemical Cycles*, 11, 217–234.

Wilmking, M., Juday, G.P., Barber, V.A., and Zald, H.S.J. 2004. Recent climate warming forces contrasting growth responses of white spruce at tree line in Alaska through temperature thresholds. *Global Change Biology*, 10, 1724–1736.

Xiao, J. and Moody, A. 2005. Geographical distribution of global greening trends and their climatic correlates: 1982–1998. *International Journal of Remote Sensing*, 11, 2371–2390.

Yang, R., Friedl, M.A., and Ni, W. 2001. Parameterization of shortwave radiation fluxes for nonuniform vegetation canopies in land surface models. *Journal of Geophysical Research*, 106(D13), 14275–14286.

Yang, W., Shabanov, N.V., Huang, D., Wang, W., Dickinson, R.E., Nemani, R.R., Knyazikhin, Y., and Myneni, R.B. 2006. Analysis of leaf area index product from combination of MODIS and Aqua data. *Remote Sensing of Environment*, 104, 297–312.

Zhang, X., Friedl, M.A., Schaaf, C.B., Strahler, A.H., Hodges, J.C.F., Gao, F., Reed, B.C., and Huete, A. 2003. Monitoring vegetation phenology using MODIS. *Remote Sensing of Environment*, 84, 471–475.

Zhang, X., Friedl, M.A., and Schaaf, C.B. 2006. Global vegetation phenology from Moderate Resolution Imaging Spectroradiometer (MODIS): Evaluation of global patterns and comparison with in situ measurements. *Journal of Geophysical Research*, 111, G04017.

Zhang, X., Zwiers, F.W., Hegerl, G.C., Lambert, F.H., Gillett, N.P., Solomon, S., Stott, P.A., and Nozawa, T. 2007. Detection of human influence on twentieth-century precipitation trends. *Nature*, 448, 461–465.

Zhou, L., Tucker, C.J., Kaufmann, R.K., Slayback, D., Shabanov, N.V., and Myneni, R.B. 2001. Variations in northern vegetation activity inferred from satellite data of vegetation index during 1981 to 1999. *Journal of Geophysical Research*, 106, 20069–20083.

8 Preprocessing
Need for Sensor Calibration

Gyanesh Chander

CONTENTS

8.1 INTRODUCTION

The ability to detect and quantify land-cover and land-use changes in the earth's environment using remote sensing depends on sensors that can provide accurate and consistent measurements of the earth's surface features over time. A critical step in providing these measurements is having a process to standardize image data from different sensors onto a common scale. To take full advantage of remote sensing, the data must be inherently sound. This implies an ongoing need for calibration, validation, stability monitoring, and quality assurance. To use remotely sensed data and ensure science observations of high quality, scientists need to know the following:

- What part of the electromagnetic (EM) spectrum they are looking at (spectral)
- How much energy the instrument is receiving (radiometric)
- Where the energy is coming from:
  - Center of pixel location (geometric)
  - Bounds of the area from which the energy is coming (spatial)

The earth-observing (EO) sensors' calibration accuracy and consistency over time are critical performance parameters and have a direct effect on the quality of the land-cover data products derived from on-orbit observations. As more satellite observations become available to the science and user communities, the number of science data products and the applications derived for these products continue to increase. Long-term land-cover and land-use data are often constructed based on observations made by multiple EO sensors over a broad range of spectra and on a large scale in both time and space. These sensors, either of the same type or of different types, can be operated on the same platform or different ones. Even sensors of the same type can be developed and built with different technologies by different instrument vendors and operated over different time spans. Some sensors may have been built without adequate onboard calibration and may not have gone through a comprehensive system-level prelaunch characterization; therefore, they cannot firmly establish their calibration traceability or consistently maintain calibration stability.

The Global Earth Observation System of Systems (GEOSS) aims to deliver comprehensive "knowledge information products" in a timely manner to meet the needs of its nine "societal benefit areas." Accomplishment of this vision, starting from a system of disparate systems built for a wide range of applications, requires the creation of an internationally coordinated operational framework to facilitate interoperability and harmonization. The Committee on Earth Observation Satellites (CEOS), the world space agency committee, has taken up responsibility for the space segment of GEOSS. It is recognized that the success of GEOSS critically depends on the interoperability of a diverse system of systems, with data access and data-quality assurance being the two key aspects of interoperability. Specifically, the CEOS Working Group on Calibration and Validation (WGCV) has been given the task of developing a data-quality assurance strategy for the GEOSS with key guidelines. Several tasks and actions have been initiated to establish calibration consistency and standards across systems, including the establishment of CEOS reference standard test sites (http://calval.cr.usgs.gov/satellite/sites_catalog/) and a traceability chain for primary site data and "best practices" guidance on site characterization and applications. The recent development of Quality Assurance Framework for Earth Observation (QA4EO) (http://qa4eo.org/) is an example.

Land cover is one of the key terrestrial essential climate variables (ECVs) currently feasible for global implementation. Global Climate Observing System (GCOS) leads the international community in defining ECVs to meet the needs of the Intergovernmental Panel on Climate Change (IPCC) and the United Nations Framework Convention on Climate Change (UNFCCC). In an era in which the number of EO satellites is rapidly growing and measurements from satellite sensors are used to address urgent global issues, often through synergistic and operational combinations of data from multiple sources, it is imperative that scientists and decision makers be able to rely on the accuracy of earth observation data products. Thus, characterization and calibration of these sensors, particularly their relative biases, are vital to the success of developing reliable ECVs and an integrated GEOSS for coordinated and sustained observations of the earth. This chapter briefly summarizes the need for sensor calibration and reviews the various aspects of radiometric calibration. It also discusses the importance of cross-calibration between sensors.

8.2 NEED FOR SENSOR CALIBRATION

Remote sensing is the field of study associated with extracting information about an object without coming into physical contact with it (Schott, 2007). With several Internet-based mapping services, television, weather channels, and other day-to-day uses, satellite imagery has clearly become a part of mainstream information society. Nevertheless, for most operational remote-sensing applications, critical issues remain regarding the "consistency of quality" in remotely sensed data. Consistent data quality implies the adherence of data to appropriate standards in the underlying physical quantities that are measured. These well-calibrated data then ensure accuracy and enhance interoperability, which enables the development of advanced EO technologies beneficial to user communities. Calibration and validation (Cal/Val) can play an essential role in bringing remote sensing to mainstream consumers in an information society, provided it is an integral part of a quality assurance strategy.

The CEOS WGCV defines calibration as "the process of quantitatively defining the system responses to known, controlled signal inputs." However, this definition is too broad. In practice, calibration is the process of measuring and evaluating system parameters required to correct image products to create an accurate and consistent data product with physical units. Most users want access to ready-to-use data from stable and well-characterized sensor systems in such a manner that sensor characterization and calibration are essentially transparent to them. The radiometric, geometric, and spectral characteristics of sensors should be well understood to generate similar geophysical and biophysical products from dissimilar measurement systems. Thus, there is a strong need from the user community to have the data calibrated and artifacts removed before using the data for their applications. Here are a few key reasons why the user community depends on well-calibrated data to ensure sound and useful results from their applications.

8.2.1 Applications Based on Temporal Analysis

The data acquired by sensors are affected by the sun zenith angle, the earth–sun distance, the view zenith angle, the atmospheric conditions, topography, and the temporal evolution of the target characteristics. The application scientist is interested in studying the temporal characteristics of the "target" and is not interested in the other factors degrading the imagery. These effects should be isolated as much as possible to make the best use of the remote-sensing data. Temporal studies require an understanding of the changes in the target characteristics over time. If a sensor's response is not monitored and corrected, then changes in the sensor's response are likely to be incorrectly attributed to changes in the observed image (Allen, 1990; Allen and Walsh, 1993; Anderson et al., 2005; Andrade and Oliveira, 2004; De Colstoun et al., 2003; Cohen and Goward, 2004; Cohen et al., 2010; Gao et al., 2006b; Goetz et al., 2000; Huang et al., 2007, 2008, 2009, 2010; Roy et al., 1999, 2008; Senay and Elliott, 1997; Wulder et al., 2008a, 2008b, 2009).

8.2.2 Applications Based on Absolute Calibration

Studies have shown that the discrepancies between satellite at-sensor spectral radiance measurements within the same class of instruments in the reflective solar bands can be up to 20% (Helder MSS Calibration, Landsat Science Team Meeting). This is a far cry from meeting the climate-change detection requirements, which stipulates a 1% per decade stability in albedo (http://www. wmo.int/pages/prog/gcos/Publications/gcos-107.pdf). In general, the absolute radiometric calibration for sensors is specified to an uncertainty of less than 10%. For bright targets (playas, snow, clouds, etc.) where the signal level is high, 10% accuracy is acceptable. However, for low-reflectance targets (water, vegetation, grass, etc.) where the signal is very low, a 10% difference in spectral radiance can have adverse effects on science applications. For example, a difference in 10% spectral radiance in vegetation is likely to be the difference between living and dead plants. The user needs a complete product that includes not only image data but also product quality, reliability, and standardization. For example, to use the data from the Landsat series of sensors, Multispectral Scanner (MSS), Thematic Mapper (TM), Enhanced Thematic Mapper Plus (ETM+), Advanced Land Imager (ALI), and Operational Land Imager (OLI), for long-term climate-change studies and for generation of geophysical/biophysical variables from these datasets, it is imperative that the sensors are cross-calibrated to each other and brought to a common radiometric scale.

8.2.3 Applications Based on Mosaics

Individual images have limited sizes for many applications. Often, multiple images are required to create a mosaic at the local, regional, national, and global levels. These images can differ greatly in atmospheric condition, illumination geometry, and vegetation phenology. Minimizing differences among adjacent images is necessary to achieve efficiency in image analysis and to ensure consistency among products derived by using different images (Chander et al., 2009a; Eidenshink, 1992; Eidenshink and Faundeen, 1994; Gutman and Rukhovetz, 1996; Gutman et al., 1996, 1998, 2008; Hansen and Reed, 2000; Hansen et al., 2000, 2005, 2008; Justice and Townshend, 1994; Justice et al., 1998; Loveland and Belward, 1997; Loveland et al., 1991, 1995, 1999, 2000; Masek et al., 2008; Roy et al., 2010; Sohl et al., 2000; Wulder et al., 2010).

8.2.4 APPLICATIONS REQUIRING SURFACE REFLECTANCE CORRECTION

The use of satellite imagery over land for deriving quantities such as vegetation indices, leaf area index (LAI), and fraction of photosynthetically active radiation (FPAR) requires that the signal measured at the top of the atmosphere be corrected for atmospheric effects and converted to surface reflectance. The effects of atmospheric correction are dramatic and are undoubtedly the most important correction that can be made for a sensor that is looking through a significantly hazy atmosphere. Without such corrections, errors in land-cover mapping and other derived products can reach 20% or greater (Baret et al., 2007; Gao and Masek, 2006; Gao et al. 2006a; Markham et al., 1992; Masek et al., 2008; Moran et al., 1992, 2003; Santer et al., 2005; Singh, 1985; Teillet, 1989; Teillet et al., 1994; Vermote and Kotchenova 2008; Vermote et al., 1995, 1996, 2002, 2007, 2009).

8.3 TYPICAL PREPROCESSING CHAIN

The remote-sensing data has to go through significant preprocessing steps before the user community can use the data for scientific applications. Typical preprocessing steps (Figure 8.1) include artifact (http://landsat.usgs.gov/science_an_anomalies.php) correction, radiometric and geometric calibration, and atmospheric correction. The next few subsections provide a brief overview of radiometric calibration and discuss the various steps in it.

8.4 RADIOMETRIC CALIBRATION

Radiometry is the science of characterizing or measuring how much EM energy is present at, or associated with, some location or direction in space (Schott, 2007). In practice, the term is usually limited

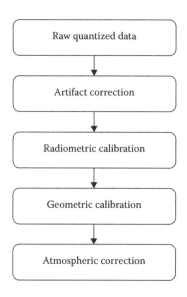

FIGURE 8.1 Typical preprocessing chain for remote-sensing data.

to the measurement of infrared (IR), visible, and ultraviolet (UV) light using optical instruments. In remote sensing, radiometry is the study and correction of degradation of imagery caused by instrumentation and atmospheric effects. Radiometric characterization and calibration are a prerequisite for creating high-quality science image data and, consequently, high-level downstream products. Radiometric calibration of these sensors helps characterize the operation of the instrument, but more importantly, calibration allows the remote-sensing data to be used in a quantitative sense.

An important part of the radiometric calibration process is identifying and quantifying factors that distort image information owing to instrument characteristics, atmospheric conditions, and noise, so that the gain factor and bias factor are accurately known. Then the conversion of digital number (DN) to at-sensor spectral radiance can be obtained accurately. Radiometric characterization is an integral part of instrument build, test, and on-orbit operations. It occurs late in the development process, so it is always at risk due to cost and schedule. Design and characterization are major contributors to utility of the data.

Research articles, special journal issues, reports, and books have occasionally provided overviews or reviews of satellite sensor radiometric calibration with some mention of vicarious calibration (Ahern et al., 1988, 1996; Bruegge and Butler, 1996; Butler et al., 2005; Chen, 1996; Dinguirard and Slater, 1999; Markham and Baker, 1985; Markham and Budge, 2004; Markham et al., 2004a; Nithianandam et al., 1993; Slater, 1980, 1984, 1985; Slater and Biggar, 1996; Slater et al., 1996, 2001; Teillet, 1997a, 1997b). The two levels of radiometric calibration are absolute radiometric calibration and relative radiometric calibration. Relative radiometric calibration typically attempts to correct distortions due to individual detector behavior, whereas absolute radiometric calibration attempts to correct distortions caused by overall system behavior and atmospheric effects. This section briefly discusses both methods.

8.4.1 Relative Radiometric Calibration

"Relative" radiometric calibration of a satellite image involves characterizing and correcting the response of individual detectors. Ideally, detectors constructed from the same material should respond identically to the same incident energy. Typically, however, detectors do not respond identically, resulting in differences in detector gain and bias levels that cause "striping" of the image data. This striping can be corrected by picking a reference detector, then shifting and scaling the responses of other detectors to the reference detector's gain and bias. This process is called relative radiometric calibration. Before relative calibration can be performed effectively, instrument artifacts have to be removed or reduced (Barker, 1983, 1984; Helder and Micijevic, 2004; Helder and Ruggles, 2004; Helder et al., 1992, 1997). Relative calibration is vital, so all of the detectors that detect the same radiance level report the same DN value. For applications that use image classification derived from statistical analysis of the DN in a given image, the relative calibration to remove striping is highly important.

8.4.2 Absolute Radiometric Calibration

"Absolute" radiometric calibration enables the conversion of image DNs to values with physical units of at-sensor spectral radiance (W m^{-2} sr^{-1} μm^{-1}). DNs from one sensor have no relation to a DN from a different sensor. Conversion to at-sensor spectral radiance and top-of-atmosphere (TOA) reflectance are the fundamental steps to compare products from different sensors. Both absolute and relative radiometric calibration can be performed before instrument launch ("prelaunch" calibration) and/or throughout the instrument's operating lifetime with the use of an internal calibration source ("onboard" calibration) and/or radiance measurements acquired from the earth's surface ("vicarious" calibration). To produce a good absolute calibration estimate, it is necessary to use image data that have been relatively calibrated. Extensive calibration activities, both prelaunch and postlaunch, are needed to derive the radiometric gain and characterize the sensor's performance.

8.4.2.1 Prelaunch Calibration

Prelaunch calibration is the work commonly done in a laboratory before instrument launch. Several reasons exist for doing prelaunch calibration. It allows the system to be tested to ensure that it operates properly before being integrated into the launch vehicle. Laboratory calibrations are easier to control and perform better than the methods used after launch. The calibration factors for the sensor are usually determined preflight, using controlled radiation sources and reflectance panels. However, it must be noted that prelaunch calibration ensures only accurate instrument performance before launch; additional calibration is required after launch to ensure adequate long-term performance. Two major kinds of preflight measurements are to characterize the instrument spectral response and the absolute calibration coefficients. The spectral response needs to be accurately characterized for out-of-band response, whereas the absolute calibration coefficients need to be checked against official standards provided by national standard laboratories.

The primary tool for prelaunch radiometric characterization and calibration is a spherical integrating source (SIS) illuminated by tungsten-halogen lamps. It can provide a "uniform," "stable," full-aperture source, but it is not inherently radiometrically calibrated or vacuum qualified, and the color temperature does not match that of the sun. SIS is normally used for absolute and relative calibration, and for testing linearity, dynamic range, and signal-to-noise ratio (SNR). Prelaunch measurements have to be performed in vacuum under thermal balance conditions. It is also necessary to operate the instrument at flight-representative thermal conditions because the instrument is sensitive to thermal infrared (TIR) emissions; hence, the thermal environment needs to be controlled and monitored. Calibration measurements are performed under steady state conditions to ensure that all instrument and thermal environment temperatures are stable. The instrument should be in its main operating mode, with IR detectors controlled and switched on, along with flight blackbodies at nominal operating conditions, full scan cycle operating, and continuous acquisition of instrument science packets (all bands). Note that calibration activities have priority over other tests, and if the configuration is changed during a measurement sequence then the test is stopped and repeated.

8.4.2.2 Postlaunch, Onboard Calibration

Onboard calibration systems usually use lamps and/or solar diffusers to calibrate reflective bands and use blackbody sources to calibrate thermal bands. The onboard calibrators should be used at system level before launch to demonstrate performance and provide transfer to orbit test. Multiple, well-designed systems should be used (full system, full aperture) to perform the prelaunch test, because lamps may not be stable through launch (particularly gas filled) and diffusers may degrade (materials, special handling procedures). The onboard calibrators usually consist of one or more lamps (usually available with every acquisition), a diffuser, detectors (used in conjunction with sun or lamp), and/or light emitting diodes (LEDs). A rigorous system would have multiple postlaunch onboard calibration methods, such as, lamp-based, diffuser-based, and lunar-based, along with ground-based vicarious calibration.

For lamp-based calibration, the processing system uses the detector's response to internal calibrator (IC) lamps on an image-by-image basis for radiometric calibration (Helder et al., 1998; Markham et al., 2004b). Before launch, the effective radiance of each lamp state for each reflective band detector is determined such that each detector's response to the internal lamp is compared to its response to an external calibrated source. The reflective band calibration algorithm for in-flight data uses a regression of the detector responses against the prelaunch radiances of the various lamp states. The slope of the regression represents the gain, and the intercept represents the bias. This methodology is required to assume that irradiance of the calibration lamps remains constant over time. Since there is no way to validate lamp radiance once on orbit, independent calibration is needed to verify the stability of onboard calibration devices.

Many aspects of radiometric response can be expected to degrade over time and with changes in environmental conditions; therefore, radiometric response must be continually recharacterized throughout the life of the system. The characterization frequency is dependent on the stability of the instrument. As the instrument ages, detector responses change and the instrument requires regular recalibration, so postlaunch calibration is vital to ensure the maintenance of high data quality. Onboard calibration sources give excellent temporal sampling with a high-precision view of the sensor's behavior as a function of time over periods of hours to months to allow trending of the system responses. Beyond this time period, it becomes necessary to verify the status of the lamps and the diffuser through independent means. The vicarious methods provide these independent data and give calibration information over periods of months to years.

8.4.2.3   Postlaunch, Vicarious Calibration

Note that the term "vicarious calibration" refers to all methods that do not rely on onboard systems. Vicarious calibration is an approach that attempts to estimate the at-sensor spectral radiance over a selected test site on the earth's surface, using surface measurements and radiative transfer code computations. In the radiance-based approach, measurements of the upwelling radiance from the test site are made with a well-calibrated radiometer. Downwelling radiance at select wavelengths is also measured to provide basis points for modeling atmospheric transmittance. These radiances are then used to further constrain the radiative transfer code calculations to predict the at-sensor spectral radiances at the TOA, as seen by the sensor.

Vicarious calibration techniques provide full-aperture calibrations with relatively high accuracy (but lower accuracy compared to laboratory methods). The biggest advantage of these vicarious calibrations is that the calibration is performed with the system operating in the mode in which the system collects its remote-sensing data. However, vicarious calibration techniques that involve field campaigns to obtain radiometric gains are expensive and labor-intensive, which limit the number of such calibrations possible for high-quality evaluation of sensor performance. Another factor limiting ground reference approaches is that the calibrations can be performed only when the system collects data over the test site.

For a satellite with a 16-day repeat cycle, the maximum number of calibrations possible during a given year over a given test site is 22. The actual number will certainly be smaller owing to local weather conditions and cloud cover obscuring the test site. Finally, the vicarious calibration approach depends on finding a good instrumented site with minimal cloud cover. It is clear that both the vicarious and onboard calibration systems are necessary for an accurate picture of the calibration status of earth-imaging sensors. There is a strong need for persistent calibrations of an instrument over its lifetime and for a variety of calibration methods to assess the true radiometric response of an instrument as accurately as possible.

8.4.3   Cross-Calibration

Sensor cross-calibration uses a well-calibrated sensor as a transfer radiometer to achieve characterization of other sensors using near-simultaneous observations of the earth. Regular cross-calibration is needed for several reasons:

- Data from multiple sensors are increasingly used to gain a more complete understanding of land-surface processes at a variety of scales. However, it is difficult and costly for any one nation to put sensors on an absolute radiometric scale.
- Data continuity requires consistency in quality and interpretation of image data acquired by different imaging sensors. Cross-calibration is the only viable solution to tie similar sensors [e.g., Landsat TM and ETM+; Terra and Aqua Moderate Resolution Imaging Spectroradiometer (MODIS)] and differing sensors (e.g., MODIS and ETM+) onto a common radiometric scale, thus playing an important role in mission continuity, interoperability, and data fusion.
- Cross-calibration is useful in situations where onboard references are not available [e.g., advanced very high resolution radiometer (AVHRR)] or where vicarious calibrations are limited.

- Cross-calibration between sensors is critical to coordinate observations from different sensors, exploiting their individual spatial resolutions, temporal sampling, and information contents to monitor surface processes over broad scales in both time and space.

As mentioned earlier, vicarious calibration can be labor-intensive and limit the number of calibrations performed. To overcome these limitations, there has been a significant increase in the use of cross-calibration techniques from near-simultaneous surface collections and the use of pseudo-invariant sites to monitor the long-term TOA reflectance trends from different sensors. These techniques, when coupled with "ground truth" information, can facilitate a better approach to validate the absolute calibration accuracies of the sensors involved and, most importantly, to evaluate their radiometric calibration stability and help address global earth observation concerns within the GEOSS.

8.5 RADIOMETRIC CALIBRATION VARIABLES

The beginning of wisdom is calling things by their correct names (Antisthenes, fifth century BC, Greece). It is important to spell out all the variables and units used in radiometry. Radiometric terms are consistent with those established by the Commission Internationale de l'Eclairage (CIE) and adopted by most international societies. The following is a list of variables used in the radiometric calibration procedure:

Q = Raw quantized pixel value [DN]
G = Detector gain or responsivity [DN/(W/(m^2·sr·μm))]
B = Detector bias or background response [DN]
L_λ = Spectral radiance at the sensor's aperture [W/(m^2·sr·μm)]
Q_{cal} = Quantized calibrated pixel value [DN]
Q_{calmin} = Minimum quantized calibrated pixel value (DN = 0/1) corresponding to LMIN$_\lambda$
Q_{calmax} = Maximum quantized calibrated pixel value (DN = 255) corresponding to LMAX$_\lambda$
LMIN$_\lambda$ = Spectral at-sensor radiance scaled to Q_{calmin} [W/(m^2·sr·μm)]
LMAX$_\lambda$ = Spectral at-sensor radiance scaled to Q_{calmax} [W/(m^2·sr·μm)]
$G_{rescale}$ = Band-specific rescaling gain factor [(W/(m^2·sr·μm))/DN]
$B_{rescale}$ = Band-specific rescaling bias factor [W/(m^2·sr·μm)]
α = Processing gain used to convert Q to Q_{cal} [unitless]
β = Processing bias used to convert Q to Q_{cal} [DN]

8.5.1 AT-SENSOR SPECTRAL RADIANCE FOR L0Rp PRODUCTS (Q-TO-L_λ)

Pixel values in the raw data products (L0Rp) are represented as Q. The detectors exhibit linear response to the earth's surface radiance. The response is quantized into 8-bit numbers that represent brightness values between 0 and 255 in the L0Rp. Band average detector gains (G) and biases (B) are used to convert the raw data (Q) to at-sensor spectral radiance (L_λ). This process is given by the relationships:

$$Q = G \times L_\lambda + B. \tag{8.1}$$

$$L_\lambda = \frac{(Q - B)}{G}. \tag{8.2}$$

G represents the sensor gain, and B represents the line-by-line biases based on the dark shutter responses acquired from each scan line. During processing, the absolute gains are combined with the detector relative gains and the band rescaling gains to obtain a detector-specific processing gain. Note that G and B are used for conversion to at-sensor spectral radiance from the L0Rp products. The remote-sensing user community receives the radiometrically and geometrically corrected Level 1 products.

8.5.2 AT-SENSOR SPECTRAL RADIANCE FOR L1 PRODUCTS (Q_{CAL}-TO-L_λ)

The users can receive the data only as Level 1 (L1) products. The pixel values in the L1 data are represented as Q_{cal}. These are the DNs that users receive with Level 1 Landsat products. During radiometric calibration, Q from L0Rp image data is converted to units of absolute radiance using 32-bit floating-point calculations. The absolute radiance values are then scaled to 8-bit values representing Q_{cal} before being output to the distribution media. Conversion from Q_{cal} in L1 products back to L_λ requires knowledge of the original rescaling factors (Chander et al., 2009b). This process is given by the relationship:

$$L_\lambda = \left(\frac{\text{LMAX}_\lambda - \text{LMIN}_\lambda}{Q_{cal\,max} - Q_{cal\,min}}\right)\left(Q_{cal} - Q_{cal\,min}\right) + \text{LMIN}_\lambda \tag{8.3}$$

or

$$L_\lambda = G_{rescale} \times Q_{cal} + B_{rescale},$$

where

$$G_{rescale} = \frac{\text{LMAX}_\lambda - \text{LMIN}_\lambda}{Q_{cal\,max} - Q_{cal\,min}} \quad \text{and}$$

$$B_{rescale} = \text{LMIN}_\lambda - \left(\frac{\text{LMAX}_\lambda - \text{LMIN}_\lambda}{Q_{cal\,max} - Q_{cal\,min}}\right)Q_{cal\,min}. \tag{8.4}$$

The $Q_{calmax} = 255$ and $Q_{calmin} = 0$ are typical for 8-bit radiometric resolution data. There may be other systems that use(d) different values. The absolute gains (G) are used for converting the Q in the L0Rp to spectral radiance, and the rescaling gains ($G_{rescale}$) are used to convert the Q_{cal} in the L1 data to spectral radiance. The conversion from Q to Q_{cal} is performed during the L1 product generation; accordingly, users with L1 data do not apply the absolute gains for conversion to at-sensor spectral radiance.

8.5.3 CONVERSION TO TOA REFLECTANCE (L_λ-TO-ρ_P)

A reduction in scene-to-scene variability can be achieved by converting the at-sensor spectral radiance to exoatmospheric TOA reflectance, also known as in-band planetary albedo. When comparing images from different sensors, there are three advantages in using TOA reflectance instead of at-sensor spectral radiance. First, it removes the cosine effect of different solar zenith angles due to the time difference between data acquisitions. Second, TOA reflectance compensates for different values of the exoatmospheric solar irradiance arising from spectral band differences. Third, the TOA reflectance corrects for the variation in the earth–sun distance between different data acquisition dates. These variations can be significant, geographically and temporally. The TOA reflectance of the earth is computed according to the equation:

$$\rho_\lambda = \frac{\pi \cdot L_\lambda \cdot d^2}{\text{ESUN}_\lambda \cdot \cos\theta_s}, \tag{8.5}$$

where ρ_λ is the planetary TOA reflectance [unitless], π is the mathematical constant approximately equal to 3.14159 [unitless], L_λ is the spectral radiance at the sensor's aperture [W/(m^2·sr·µm)], d is the earth–sun distance [astronomical units], $ESUN_\lambda$ is the mean exoatmospheric solar spectral irradiance [W/(m²·μm)], and θ_s is the solar zenith angle [degrees].

8.6 READY-TO-USE IMAGES

Remote-sensing data are not provided to the user community in a form such that they can focus on the scientific analysis of the data and not on geometric and radiometric issues. Instead, most standard products require substantial processing efforts by users. Typical processing efforts include detector normalization, bidirectional reflectance distribution function (BRDF), and atmospheric correction to improve radiometric consistency and masking of pixels contaminated by nonland features such as cloud and shadow. Depending on the scope of application, such processing effort often accounts for a significant portion of the total effort and can result in substantially reduced amount of time available for conducting the analysis the data are really intended for. The users should take responsibility to ensure that their datasets are artifact-corrected and well calibrated so that their specific application results become more reliable and traceable. The user community needs ready-to-use images.

8.7 SUMMARY

With Google Maps mapping service, television, weather channels, and other day-to-day uses, satellite imagery has clearly become a part of mainstream information society. Nevertheless, for most operational remote-sensing applications, critical issues remain regarding the consistency of quality in remotely sensed data. Consistent data quality implies the adherence of data to appropriate standards to the underlying physical quantities they measure. To take full advantage of remote sensing, the data must be inherently sound. This implies an ongoing need for calibration, validation, stability monitoring, and quality assurance.

REFERENCES

Ahern, F.J., Brown, R.J., Cihlar, J., Gauthier, R., Murphy, J., Neville, R.A., and Teillet, P.M. 1988. Radiometric correction of visible and infrared remote sensing data at the Canada Centre for Remote Sensing. In A.P. Cracknell and L. Hayes (Eds.), *Remote Sensing Yearbook* (pp. 101–127). Philadelphia, PA: Taylor and Francis.

Allen, J.D. 1990. Remote sensor comparison for crop area estimation using multitemporal data. In R. Mills (Ed.), *Proceedings of the 1990 IEEE International Geoscience and Remote Sensing Symposium* (pp. 609–612). Piscataway, NJ: IEEE.

Allen, T.R. and Walsh, S.J. 1993. Characterizing multitemporal alpine snowmelt patterns for ecological inferences. *Photogrammetric Engineering and Remote Sensing*, 59, 1521–1529.

Anderson, L.O., Shimabukuro, Y.E., Defries, R.S., and Morton, D. 2005. Assessment of deforestation in near real time over the Brazilian Amazon using multitemporal fraction images derived from Terra MODIS. *IEEE Geoscience and Remote Sensing Letters*, 2, 315–318.

Andrade, J.B. and Oliveira, T.S. 2004. Spatial and temporal-time analysis of land use in part of the semi-arid region of Ceará State, Brazil. *Revista Brasileira de Ciencia do Solo*, 28, 393–401.

Baret, F., Hagolle, O., Geiger, B., Bicheron, P., Miras, B., Huc, M., Berthelot, B., et al. 2007. LAI, fAPAR and fCover CYCLOPES global products derived from VEGETATION—Part 1: Principles of the algorithm. *Remote Sensing of Environment*, 110, 275–286.

Barker, J.L. 1983. Relative radiometric calibration of Landsat TM reflective bands. In *Landsat-4 Science Characterization Early Results, Proceedings of the Landsat-4 Science Characterization Early Results Symposium*, February 22–24, 1983, NASA Conference Publication 2355, Vol. III—Thematic Mapper (TM), Pt. 2, pp. 1–219. Greenbelt, MD: NASA.

Barker, J.L. 1984. Relative radiometric calibration of Landsat TM reflective bands. In *Landsat-4 Science Investigations Summary, Including December 1983 Workshop Results, Proceedings of the Landsat-4 Early Results Symposium*, February 22–24, 1983, and the Landsat Science Characterization Workshop, December 6, 1983, NASA Conference Publication 2326, Vol. 1, pp. 140–180. Greenbelt, MD: NASA.

Bruegge, C. and Butler, J. (Eds.) 1996. *Journal of Atmospheric and Oceanographic Technology, Special Issue on Earth Observing System Calibration.* Boston, MA: American Meteorological Society.

Butler, J.J., Johnson, B.C., and Barnes, R.A. 2005. The calibration and characterization of Earth remote sensing and environmental monitoring instruments. *Optical Radiometry, Experimental Methods in the Physical Sciences*, 41, 453–534.

Chander, G., Huang, C., Yang, L., Homer, C., and Larson, C. 2009a. Developing consistent Landsat data sets for large area applications: The MRLC 2001 protocol. *IEEE Geoscience and Remote Sensing Letters*, 6, 777–781.

Chander, G., Markham, B.L., and Helder, D.L. 2009b. Summary of current radiometric calibration coefficients for Landsat MSS, TM, ETM+, and EO-1 ALI sensors. *Remote Sensing of Environment*, 113, 893–903.

Chen, H.S. 1996. *Remote Sensing Calibration Systems: An Introduction.* Hampton, VA: Deepak Publishing.

Cohen, W.B. and Goward, S.N. 2004. Landsat's role in ecological applications of remote sensing. *BioScience*, 54, 535–545.

Cohen, W.B., Yang, Z., and Kennedy, R. 2010. Detecting trends in forest disturbance and recovery using yearly Landsat time series: II. TimeSync—Tools for calibration and validation. *Remote Sensing of Environment*, 114, 2911–2924.

De Colstoun, B., E.C., Story, M.H., Thompson, C., Commisso, K., Smith, T.G., and Irons, J.R. 2003. National Park vegetation mapping using multitemporal Landsat 7 data and a decision tree classifier. *Remote Sensing of Environment*, 85, 316–327.

Dinguirard, M. and Slater, P.N. 1999. Calibration of space-multispectral imaging sensors: A review. *Remote Sensing of Environment*, 68, 194–205.

Eidenshink, J.C. 1992. The 1990 conterminous US AVHRR data set. *Photogrammetric Engineering and Remote Sensing*, 58, 809–813.

Eidenshink, J.C. and Faundeen, J.L. 1994. The 1 km AVHRR global land data set: First stages in implementation. *International Journal of Remote Sensing*, 15, 3443–3462.

Gao, F. and Masek, J.G. 2006. Mapping wildland fire scar using fused Landsat and MODIS surface reflectance. In W. Emery and G. Wick (Eds.), *Proceedings of the 2006 IEEE International Geoscience and Remote Sensing Symposium* (pp. 4172–4175). Piscataway, NJ: IEEE.

Gao, F., Masek, J., Schwaller, M., and Hall, F. 2006a. On the blending of the Landsat and MODIS surface reflectance: Predicting daily Landsat surface reflectance. *IEEE Transactions on Geoscience and Remote Sensing*, 44, 2207–2218.

Gao, J., Liu, Y., and Chen, Y. 2006b. Land cover changes during agrarian restructuring in Northeast China. *Applied Geography*, 26, 312–322.

Goetz, S.J., Prince, S.D., Thawley, M.M., Smith, A.J., Wright, R., and Weiner, M. 2000. Applications of multi-temporal land cover information in the mid-Atlantic region: A RESAC initiative. In *Proceedings of the 2000 IEEE International Geoscience and Remote Sensing Symposium* (pp. 357–359). Piscataway, NJ: IEEE.

Gutman, G., Ignatov, A., and Olson, S. 1996. Global land monitoring using AVHRR time series. *Advances in Space Research*, 17, 51–54.

Gutman, G. and Rukhovetz, L. 1996. Towards satellite-derived global estimation of monthly evapotranspiration over land surfaces. *Advances in Space Research*, 18, 67–71.

Gutman, G., Tarpley, D., Ignatov, A., and Olson, S. 1998. Global AVHRR products for land climate studies. *Advances in Space Research*, 22, 1591–1594.

Gutman, G.G., Byrnes, R., Masek, J., Covington, S., Justice, C., Franks, S., and Headley, R. 2008. Towards monitoring land-cover and land-use changes at a global scale: The global land survey 2005. *Photogrammetric Engineering and Remote Sensing*, 74, 6–10.

Hansen, M.C., Defries, R.S., Townshend, J.R.G., and Sohlberg, R. 2000. Global land cover classification at 1 km spatial resolution using a classification tree approach. *International Journal of Remote Sensing*, 21, 1331–1364.

Hansen, M.C. and Reed, B. 2000. A comparison of the IGBP DISCover and University of Maryland 1 km global land cover products. *International Journal of Remote Sensing*, 21, 1365–1373.

Hansen, M.C., Roy, D.P., Lindquist, E., Adusei, B., Justice, C.O., and Altstatt, A. 2008. A method for integrating MODIS and Landsat data for systematic monitoring of forest cover and change in the Congo Basin. *Remote Sensing of Environment*, 112, 2495–2513.

Hansen, M.C., Townshend, J.R.G., DeFries, R.S., and Carroll, M. 2005. Estimation of tree cover using MODIS data at global, continental and regional/local scales. *International Journal of Remote Sensing*, 26, 4359–4380.

Helder, D., Boncyk, W., and Morfitt, R. 1997. Landsat TM memory effect characterization and correction. *Canadian Journal of Remote Sensing*, 23, 299–308.

Helder, D., Boncyk, W., and Morfitt, R. 1998. Absolute calibration of the Landsat Thematic Mapper using the internal calibrator. In T.I. Stein (Ed.), *Proceedings of the 1998 IEEE International Geoscience and Remote Sensing Symposium* (pp. 2716–2718). Piscataway, NJ: IEEE.

Helder, D.L. and Micijevic, E. 2004. Landsat-5 Thematic Mapper outgassing effects. *IEEE Transactions on Geoscience and Remote Sensing*, 42, 2717–2729.

Helder, D.L., Quirk, B.K., and Hood, J.J. 1992. A technique for the reduction of banding in Landsat Thematic Mapper images. *Photogrammetric Engineering and Remote Sensing*, 58, 1425–1431.

Helder, D.L. and Ruggles, T.A. 2004. Landsat Thematic Mapper reflective-band radiometric artifacts. *IEEE Transactions on Geoscience and Remote Sensing*, 42, 2704–2716.

Huang, C., Goward, S.N., Masek, J.G., Thomas, N., Zhu, Z., and Vogelmann, J.E. 2010. An automated approach for reconstructing recent forest disturbance history using dense Landsat time series stacks. *Remote Sensing of Environment*, 114, 183–198.

Huang, C., Goward, S.N., Schleeweis, K., Thomas, N., Masek, J.G., and Zhu, Z. 2009. Dynamics of national forests assessed using the Landsat record: Case studies in eastern United States. *Remote Sensing of Environment*, 113, 1430–1442.

Huang, C., Shao, Y., Li, J., Chen, J., and Liu, J. 2008. Temporal analysis of land surface temperature in Beijing utilizing remote sensing imagery. In *Proceedings of the 2008 IEEE International Geoscience and Remote Sensing Symposium* (pp. 1304–1307). Piscataway, NJ: IEEE.

Huang, C., Shao, Y., Liu, J., and Chen, J. 2007. Temporal analysis of urban forest in Beijing using Landsat imagery. *Journal of Applied Remote Sensing*, 1, 013534.

Justice, C.O. and Townshend, J.R. 1994. Data sets for global remote sensing: Lessons learnt. *International Journal of Remote Sensing*, 15, 3621–3639.

Justice, C.O., Vermote, E., Townshend, J.R.G., Defries, R., Roy, D.P., Hall, D.K., Salomonson, V.V., et al. 1998. The Moderate Resolution Imaging Spectroradiometer (MODIS): Land remote sensing for global change research. *IEEE Transactions on Geoscience and Remote Sensing*, 36, 1228–1249.

Loveland, T.R. and Belward, A.S. 1997. IGBP-DIS global 1 km land cover data set, DISCover: First results. *International Journal of Remote Sensing*, 18, 3289–3295.

Loveland, T.R., Merchant, J.W., Brown, J.F., Ohlen, D.O., Reed, B.C., Olson, P., and Hutchinson, J. 1995. Seasonal land-cover regions of the U.S. *Annals of the Association of American Geographers*, 85, 339–355.

Loveland, T.R., Merchant, J.W., Ohlen, D.O., and Brown, J.F. 1991. Development of a land-cover characteristics database for the conterminous US. *Photogrammetric Engineering and Remote Sensing*, 57, 1453–1463.

Loveland, T.R., Reed, B.C., Brown, J.F., Ohlen, D.O., Zhu, Z., Yang, L., and Merchant, J.W. 2000. Development of a global land cover characteristics database and IGBP DISCover from 1 km AVHRR data. *International Journal of Remote Sensing*, 21, 1303–1330.

Loveland, T.R., Zhu, Z., Ohlen, D.O., Brown, J.F., Reed, B.C., and Yang, L. 1999. An analysis of the IGBP global land-cover characterization process. *Photogrammetric Engineering and Remote Sensing*, 65, 1021–1032.

Markham, B.L. and Barker, J.L. (Eds.). 1985. *Photogrammetric Engineering and Remote Sensing, special issue on Landsat Image Data Quality Analysis (LIDQA)*. Bethesda, MD: ASPRS.

Markham, B.L., Halthore, R.N., and Goetz, S.J. 1992. Surface reflectance retrieval from satellite and aircraft sensors: Results of sensor and algorithm comparisons during FIFE. *Journal of Geophysical Research D: Atmospheres*, 97, 18785–18795.

Markham, B.L., Storey, J.C., Crawford, M.M., Goodenough, D.G., and Irons, J.R. (Eds.). 2004a. *IEEE Transactions on Geoscience and Remote Sensing, Special Issue on Landsat Sensor Performance Characterization*. Piscataway, NJ: IEEE.

Markham, B.L., Thome, K.J., Barsi, J.A., Kaita, E., Helder, D.L., Barker, J.L., and Scaramuzza, P.L. 2004b. Landsat-7 ETM+ on-orbit reflective-band radiometric stability and absolute calibration. *IEEE Transactions on Geoscience and Remote Sensing*, 42, 2810–2820.

Masek, J.G., Huang, C., Wolfe, R., Cohen, W., Hall, F., Kutler, J., and Nelson, P. 2008. North American forest disturbance mapped from a decadal Landsat record. *Remote Sensing of Environment*, 112, 2914–2926.

Morain, S.A. and Budge, A.M. (Eds.). 2004. *Postlaunch Calibration of Satellite Sensors, Proceedings of the International Workshop on Radiometric and Geometric Calibration*. New York: A.A. Balkema Publishers.

Moran, M.S., Jackson, R.D., Slater, P.N., and Teillet, P.M. 1992. Evaluation of simplified procedures for retrieval of land surface reflectance factors from satellite sensor output. *Remote Sensing of Environment*, 41, 169–184.

Moran, M.S., Bryant, R., Holifield, C.D., and McElroy, S. 2003. Refined empirical line approach for retrieving surface reflectance from EO-1 ALI images. *IEEE Transactions on Geoscience and Remote Sensing*, 41, 1411–1414.

Nithianandam, J., Guenther, B.W., and Allison, L.J. 1993. An anecdotal review of NASA Earth observing satellite remote sensors and radiometric calibration methods. *Metrologia*, 30, 207–212.

Roy, D.P., Giglio, L., Kendall, J.D., and Justice, C.O. 1999. Multi-temporal active-fire based burn scar detection algorithm. *International Journal of Remote Sensing*, 20, 1031–1038.

Roy, D.P., Ju, J., Lewis, P., Schaaf, C., Gao, F., Hansen, M., and Lindquist, E. 2008. Multi-temporal MODIS-Landsat data fusion for relative radiometric normalization, gap filling, and prediction of Landsat data. *Remote Sensing of Environment*, 112, 3112–3130.

Roy, D.P., Ju, J., Kline, K., Scaramuzza, P.L., Kovalskyy, V., Hansen, M., Loveland, T.R., Vermote, E., and Zhang, C. 2010. Web-enabled Landsat Data (WELD): Landsat ETM+ composited mosaics of the conterminous United States. *Remote Sensing of Environment*, 114, 35–49.

Santer, R., Ramon, D., Vidot, J., and Dilligeard, E. 2005. A surface reflectance model for aerosol remote sensing over land. In *ESA Special Publication, Vol. 572*, In H. Sawaya-Lacoste and L. Ouwehand (Eds.), *Proceedings of the 2004 Envisat & ERS Symposium* (pp. 2045–2054). Noordwijk, The Netherlands: ESA Publications Division.

Schott, J.R. 2007. *Remote Sensing: The Image Chain Approach*. New York: Oxford University Press.

Senay, G.B. and Elliott, R.L. 1997. NDVI as a means of characterizing temporal variability in land cover for use in ET modeling. In A. Ward and B.G. Wilson (Eds.), *Proceedings of the ASAE Annual International Meeting, Vol. 2* (pp. 1–5). St. Joseph, MI: American Society of Agricultural Engineers.

Singh, S.M. 1985. Earth's surface reflectance from the AVHRR channel 1 data. In *Advanced Technology for Monitoring and Processing Global Environmental Data, Proceedings RSS/CERMA Conference* (pp. 81–90). Reading, England: Remote Sensing Society.

Slater, P.N. 1980 . *Remote Sensing, Optics and Optical Systems*. Reading, MA: Addison-Wesley Publishing Company.

Slater, P.N. 1984. The importance and attainment of accurate absolute radiometric calibration. In P.N. Slater (Ed.), *Proceedings of SPIE, Critical Reviews of Technology* (pp. 34–40). Bellingham, WA: SPIE.

Slater, P.N. 1985. Radiometric considerations in remote-sensing. In D.A. Landgrebe (Ed.), *Proceedings of the IEEE, Special Issue on Perceiving Earth's Resources from Space* (pp. 997–1011). Piscatatway, NJ: IEEE.

Slater, P.N. and Biggar, S.F. 1996. Suggestions for radiometric calibration coefficient generation. *Journal of Atmospheric and Oceanic Technology*, 13, 376–382.

Slater, P.N., Biggar, S.F., Palmer, J.M., and Thome, K.J. 2001. Unified approach to absolute radiometric calibration in the solar-reflective range. *Remote Sensing of Environment*, 77, 293–303.

Slater, P.N., Biggar, S.F., Thome, K.J., Gellman, D.I., and Spyak, P.R. 1996. Vicarious radiometric calibrations of EOS sensors. *Journal of Atmospheric and Oceanic Technology*, 13, 349–359.

Sohl, T.L., Loveland, T.R., Sayler, K.L., Gallant, A.L., Auch, R., and Napton, D. 2000. Land cover trends project: A strategy for monitoring land cover change at a national scale. In T.I. Stein (Ed.), *Proceedings of the 2000 IEEE International Geoscience and Remote Sensing Symposium* (pp. 2002–2004). Piscataway, NJ: IEEE.

Teillet, P.M. 1989. Surface reflectance retrieval using atmospheric correction algorithms. In J. Gower, J. Cihlar, and D. Goodenough (Eds.), *Proceedings of the 1989 IEEE International Geoscience and Remote Sensing Symposium* (pp. 864–867). Piscataway, NJ: IEEE.

Teillet, P.M. 1997a. A status overview of Earth observation calibration/validation for terrestrial applications. *Canadian Journal of Remote Sensing*, 23, 291–298.

Teillet, P.M. (Ed.). 1997b. *Canadian Journal of Remote Sensing, Special Issue on Calibration/Validation*. Kanata, Ontario, Canada: Canadian Aeronautics and Space Institute.

Teillet, P.M., Fedosejevs, G., Ahern, F.J., and Gauthier, R.P. 1994. Sensitivity of surface reflectance retrieval to uncertainties in aerosol optical properties. *Applied Optics*, 33, 3933–3940.

Vermote, E.F., El Saleous, N.Z., and Holben, B.N. 1996. Aerosol retrieval and atmospheric correction. In G. D'Souza, A.S. Belward, and J.-P. Malingreau (Eds.), *Advances in the Use of NOAA AVHRR Data for Land Applications* (pp. 93–124). Boston, MA: Kluwer.

Vermote, E.F., El Saleous, N.Z., and Justice, C.O. 2002. Atmospheric correction of MODIS data in the visible to middle infrared: First results. *Remote Sensing of Environment*, 83, 97–111.

Vermote, E.F., El Saleous, N.Z., and Roger, J.-C. 1995. Operational atmospheric correction of AVHRR visible and near-infrared data. In *Proceedings of SPIE, Atmospheric Sensing and Modelling* (pp. 141–149). Bellingham, WA: SPIE.

Vermote, E.F., Justice, C.O., and Bréon, F.M. 2009. Towards a generalized approach for correction of the BRDF effect in MODIS directional reflectances. *IEEE Transactions on Geoscience and Remote Sensing*, 47, 898–908.

Vermote, E.F., and Kotchenova, S. 2008. Atmospheric correction for the monitoring of land surfaces. *Journal of Geophysical Research D: Atmospheres*, 113, D23S90.

Vermote, E.F., Roger, J.C., Sinyuk, A., Saleous, N., and Dubovik, O. 2007. Fusion of MODIS-MISR aerosol inversion for estimation of aerosol absorption. *Remote Sensing of Environment*, 107, 81–89.

Wulder, M.A., Ortlepp, S.M., White, J.C., and Coops, N.C. 2008a. Impact of sun-surface-sensor geometry upon multitemporal high spatial resolution satellite imagery. *Canadian Journal of Remote Sensing*, 34, 455–461.

Wulder, M.A., White, J.C., Alvarez, F., Han, T., Rogan, J., and Hawkes, B. 2009. Characterizing boreal forest wildfire with multi-temporal Landsat and LIDAR data. *Remote Sensing of Environment*, 113, 1540–1555.

Wulder, M.A., White, J.C., Coops, N.C., and Butson, C.R. 2008b. Multi-temporal analysis of high spatial resolution imagery for disturbance monitoring. *Remote Sensing of Environment*, 112, 2729–2740.

Wulder, M.A., White, J.C., Gillis, M.D., Walsworth, N., Hansen, M.C., and Potapov, P. 2010. Multiscale satellite and spatial information and analysis framework in support of a large-area forest monitoring and inventory update. *Environmental Monitoring and Assessment*, 170, 417–433.

9 Classification Trees and Mixed Pixel Training Data

Matthew C. Hansen

CONTENTS

9.1 INTRODUCTION

Research in the last decade on supervised land-cover classification has emphasized new distribution-free algorithms as high-performance alternatives to traditional classifiers. Such classifiers include decision trees, neural networks, nearest neighbor, and support vector machine algorithms. Distribution-free algorithms work on the spectral frontiers between land-cover classes, a marked improvement over conventional parametric classifiers reliant on the statistics of central tendency. A number of comparisons between distribution-free methods have been made, which have historically favored parametric techniques. Ince (1987) and Hardin and Thomson (1992) showed that nearest-neighbor classifiers were superior to parametric classifiers. Hansen et al. (1996) and Friedl and Brodley (1997) found comparable performance between a classification tree approach and a maximum likelihood one. Key et al. (1989), Bischof et al. (1992), and Gopal et al. (1999) tested the maximum likelihood classifier versus neural network classifiers and found that the neural network classifiers provide accuracies similar to or superior than that provided by the maximum likelihood classifier. Likewise, support vector machines have been compared to the maximum likelihood classifier and have been found to yield higher accuracies (Huang et al., 2002). Support vector machines, in turn, have been found to outperform decision trees and neural nets (Huang et al., 2002). However, variables such as the number of features, model parameter selection, and the number of training samples can affect the relative performance of distribution-free classifiers (Pal and Mather, 2003).

Critical to any supervised learning algorithm is training data. Distribution-free algorithms target interclass spectral frontiers in delineating decision boundaries with implications for appropriate training datasets. Although many studies correlate higher accuracies with increased training data (Foody et al., 1995), the performance of support vector machines has been shown to improve when specifically targeting populations of mixed pixels (Foody and Mathur, 2004a, 2004b, 2006). The same has been demonstrated for neural nets (Bernard et al., 1997; Foody, 1999).

Supervised land-cover characterizations typically rely on core exemplar training sites for model calibration. The use of a fuzzy classifier or a soft classifier permits the identification of spectrally ambiguous pixels by labeling them with an intermediate confidence value. For example, standard maximum likelihood classifiers can produce layers of per-pixel class membership probability. Although such algorithms may do a reasonable job in identifying mixed pixels, mixed pixels themselves are rarely used directly for model calibration. This study builds on the work of Foody and Mathur (2006), who used mixed training datasets along interclass spectral frontiers as an efficient approach to training a support vector machine algorithm. In this study, a classification tree algorithm is applied using targeted mixed pixel training sites, the results of which are compared to a classification derived using core area training sites of the kind recommended in traditional remote-sensing textbooks (Landgrebe, 2003; Lillesand and Kiefer, 2008; Verbyla, 1995). Comparisons with heritage algorithms are also included for reference. Results illustrate the value of including mixed pixel training in the derivation and interpretation of classification tree models.

9.2 TRAINING DATA

Supervised land-cover characterization approaches require samples of the cover types of interest. These samples are referred to as training data and are used to relate the labels to the independent variables, namely multispectral imagery and/or ancillary datasets. Training data are a critical component of the process. Sometimes training data already exist for use as with the USDA National Agricultural Statistics Service Cropland Data Layer, which uses labeled polygons from the Farm Service Agency (NASS, 2011). More typically, training data need to be derived by the analyst. A few principles should be followed in deriving a robust training dataset. First, training data must represent the major biogeographic variation found within the study area, with all land-cover themes of interest sampled within identified subregions. For example, one study area may include a montane zone, easily separated using a thermal brightness or land-surface temperature input. If this zone has only training data for one class, for example, forest, then the entire montane zone can easily be discriminated using a thermal input, resulting in the entire area being assigned to the forest class. For such a region, all classes of interest that exist in the montane zone need to be assigned training sites within it. Ancillary datasets, such as elevation data or ecoregion, may assist in identifying biogeographic subregions and may even be included as input variables. In addition to biogeographic variation, other factors such as illumination geometry, land use, and soil moisture may cause spectral variation within given cover types. This intraclass variability must also be accounted for in training-set derivation.

For any classification scheme, there are implied spectral boundaries, even if these are not explicitly stated, which when crossed represent the migration from one class to another. Accurate delineation of these boundaries is sought. For example, if a classification has the goal of mapping tree-cover categories, there is a range of canopy values associated with each class, for example, a 10%–30% canopy cover for woodland class. Given that the classes are typically defined within a range of physiognomic-structural attributes, training data should target the physical boundaries between classes. Exploration of the spectral space can be undertaken via preliminary analyses, including spectral scatter plots, unsupervised clustering, or principal component analysis. Any such method can be used to reveal the spectral variation within the data and assist with training-site derivation. A more straightforward approach to covering the spectral class frontiers with training is by targeting class boundaries in the spatial domain. This runs counter to classical instructional texts on land-cover mapping that emphasize the use of core, homogeneous sites for training-site delineation. Figure 9.1 captures this idea, which is based on the traditional notion that core exemplar sites are needed for training; attempting to derive sites on mixed pixels will only introduce errors to the training dataset. However, the core spectral regions for land-cover classes are the easy part of identification. Delineating only these regions leaves the characterization of more heterogeneous, mixed pixels to the algorithm. The algorithm, it must be stated, knows nothing about the biophysical nature of the spectral signatures. Leaving the decision making to the vagaries of an algorithm is not necessary, and this problem is largely remedied by developing training data within heterogeneous pixels.

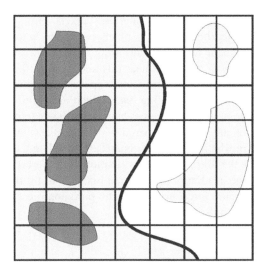

FIGURE 9.1 Example of training-data delineation based on heritage methods; core areas are selected for training, and mixed pixels are avoided. Two classes are shown, with delineated training sites overlain in a darker shade of gray.

Collecting mixed pixel training data typically involves employing digital image interpretation techniques. The first method is to directly allocate mixed pixels using the source imagery to be used in the classification. This requires the categorical labeling of the mixed pixel boundary as shown in Figure 9.1, based on expert interpretation. A more labor-intensive and costly approach is to use fine-scale data to quantify subpixel mixtures in allocating training labels. Very high spatial resolution data, such as IKONOS or other data, are suitable for direct quantification of crown cover, impervious surface, or other traits used in defining the classes of interest. When interpreting satellite data for training-set derivation, a physiognomic-structural vegetation characterization scheme is preferred. Land cover, defined as the observed biophysical state of the earth's surface, lends itself most unambiguously to physiognomic-structural definitions (DiGregorio and Jansen, 2000). In addition, the signal being mapped in multispectral and multitemporal space is correlated with vegetation structure and phenology in terms of life form and cover. Finally, physiognomic-structural definition sets based on measurable traits facilitate training-set derivation and product validation, especially if fine-scale data are available for interpretation (Hansen and Goetz, 2005).

9.3 STUDY AREA

The study area is a region of west-central Illinois, which is shown in Figure 9.2a as a subset of a Landsat image dated July 1, 2010. The area is dissected by gallery forests with drainage divides dominated by agricultural land uses. For the analysis, a RapidEye image dated August 14, 2010, was coregistered to the Landsat image and mapped at 5-m spatial resolution into forest and nonforest categories. The percentage of forest-mapped RapidEye pixels per Landsat pixel (36 RapidEye pixels per Landsat pixel) was used to label each Landsat pixel as forest or nonforest. In this study, forest was defined as tree assemblages having complete crown closure (100%) for trees ≥ 5 m in height at the RapidEye pixel scale. The 5-m RapidEye forest/nonforest product was aggregated to Landsat pixel scale, and all Landsat pixels in this subset were labeled as either forest or nonforest, using a 50% forest extent threshold at the 30-m pixel Landsat scale. Each pixel was labeled as mixed or pure based on a spatial buffer between the Landsat forest and nonforest pixels. The forest/nonforest boundary was buffered to contain two adjacent forest and two nonforest pixels. This 120-m buffer became a population of mixed forest and nonforest pixels, and the rest of the image a population of pure forest and nonforest pixels. This spatially defined population of pure/homogeneous and mixed/heterogeneous pixels is used to better reflect the actions of an on-screen interpreter, which is able to direct effort away from or toward mixed pixel populations. Only visible red band 3 (0.63–0.69 μm) and near-infrared band 4 (0.78–0.90 μm) were used as spectral inputs to better visualize the spectral feature space in graphic form. Issues of the curse of dimensionality (Hughes, 1968) and training data, which are significant in more complicated feature spaces, were not addressed. Decision trees have been shown to perform less well in feature spaces with higher dimensionality when compared with other classifiers (Pal and Mather, 2003). For a single Landsat image in which many of the bands are correlated, the curse of dimensionality is of limited concern. Figure 9.3 illustrates the distribution of the forest and nonforest populations for (1) the entire subset image, (2) the pure core area pixels, and (3) the mixed pixels. The total population consisted of 8708 pure forest pixels, 321,766 pure nonforest pixels, 110,252 mixed forest pixels, and 239,011 mixed nonforest pixels.

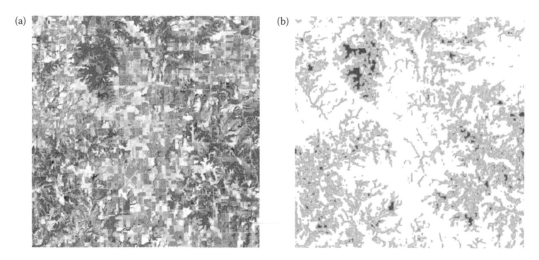

FIGURE 9.2   (See color insert.) (a) Landsat 5 image, WRS2 path/row 024/032, centered on 91 10 21.5W, 39 59 8.7N with dimensions 26.3 km by 26.3 km. Near-infrared band 4 is shown in red, and visible red band 3 is shown in cyan. (b) Reference labels derived from a RapidEye forest/nonforest classification. Dark and light green are ≥50% forest cover. Yellow and orange are <50% forest cover. Dark green and yellow represent spatially homogeneous forest and nonforest labels, respectively. Light green and orange represent spatially heterogeneous forest and nonforest labels, respectively. These mixed pixels constitute a 120-m buffer along forest/nonforest interfaces. Forest accounts for 82.5% of the image and nonforest 17.5%.

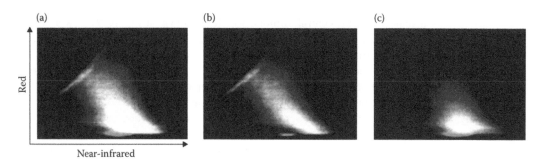

FIGURE 9.3   (See color insert.) (a) All forest and nonforest data from Figure 9.2, (b) forest and nonforest pixels greater than 60 m from forest/nonforest interfaces (pure population), and (c) forest and nonforest pixels within a 120-m buffer along forest/nonforest interfaces (mixed population).

9.4 HERITAGE ALGORITHMS AND CLASSIFICATION TREES

An initial test was done to emphasize the advantages of distribution-free classifiers compared with that of conventional methods. An isodata clustering, a maximum likelihood classifier, and a decision tree classifier were run on all of the data (Jensen, 2004). For the isodata method, 25 clusters were produced and compared to the RapidEye-derived Landsat-scale map depicted in Figure 9.2b. Each cluster was assigned the majority class based on the reference map. Figure 9.4 shows the results and the decision boundaries made per method, zoomed in on the spectral region where the pertinent boundaries exist. In this image, forest is shown to exist in a fairly uniform distribution. However, the unsupervised method, being unguided, creates arbitrary clusters regarding the classes of interest, only one of which is forest-dominated (Figure 9.4a). When this cluster is labeled as forest and the others as nonforest, the highest achievable classification accuracy is 88.3%. Although clustering is an interesting and valuable tool for analyzing data distributions, it does nothing in targeting the spectral frontiers between classes of interest. Figure 9.4b shows the decision boundary for the maximum likelihood algorithm. The mean/variance/covariance statistics are largely insensitive to the actual forest and nonforest class boundaries. Although the core areas are captured, the actual boundary between the forest and nonforest classes is largely missed. As a result, the accuracy of this test run is 91.0%. The classification tree algorithm splits the red/near-infrared spectral space using orthogonal splits until a predetermined threshold prevents further splitting (0.01 of the root deviance). The result is a 52-node tree with the terminal nodes displayed in Figure 9.4c. The tree algorithm works on the interclass boundary exclusively. The result is a fine-scale partitioning of the feature space, almost per "signature." The result is a set of rules that can vary by a single digital number in regions of confusion and that yield an accuracy of 95.0%. It is worth noting that the idea of an optimum hyperplane fitting between training labels is largely absent. Decision rules are made per quantization in the red and near-infrared bands as training labels overlap in spectral space.

The decision tree classifier works on the spectral boundaries between classes, as do other distribution-free models such as support vector machines, k-nearest neighbor, and neural networks. Being most familiar with decision trees (Hansen, 1996), I do not compare the robustness of the various choices of distribution-free algorithms. Conceptually, it is clear that an orthogonally splitting decision tree is less appropriate for creating a decision boundary than a support vector machine in a sparsely labeled feature space (Huang et al., 2002). However, it is posited here that training data should well populate the class boundaries and force a distribution-free classifier to create the optimum decision boundary as defined not by the classifier but by the labeled distribution. Figure 9.3c is an example. Here, the decision tree is delineating fine spectral features—in this case, as small as two "signatures" in size. Overfitting is a challenge common to distribution-free algorithms, and

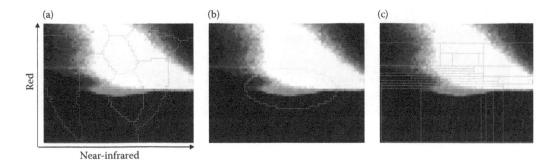

FIGURE 9.4 **(See color insert.)** Results of (a) unsupervised clustering, (b) maximum likelihood, and (c) classification tree algorithms on partitioning the red/near-infrared feature space for forest (shown in red) and nonforest (shown in cyan). Green boundaries indicate forest, orange nonforest. For this test, all data were used as inputs.

methods exist to best generalize a classifier. However, with purposely targeted training on mixed pixels across the feature space, overfitting concerns are reduced along the most important interclass spectral frontiers.

9.5 TRAINING DATA—PURE VERSUS MIXED SITE SELECTION

As just shown, distribution-free algorithms are highly appropriate for working along interclass spectral boundaries where mixed pixels are located. To best exploit these classifiers, one should develop training data that targets mixed pixels. The spectral plots of Figure 9.3a through 9.3c illustrate this idea. Figure 9.3a shows the entire population, Figure 9.3b shows the core homogeneous regions of Figure 9.2b, and Figure 9.3c shows the spatially buffered heterogeneous regions of Figure 9.2b. The core area spectral plot is biased toward the spectral regions that are relatively unambiguous and easily identified. The mixed pixel training emphasizes the spectral frontiers and forces the appropriate supervised algorithm to expend effort in delineating the optimal decision boundary. In this example, training data are derived from a data source of finer spatial resolution (a classified RapidEye image). Although this is preferred, it typically has high costs, both in terms of effort and data. However, even without a subpixel dataset for training-set derivation, an analyst can reliably label mixed pixels using photointerpretation skills or freely available ancillary information, such as GoogleEarth. In this landscape, 97.6% of mixed pixels (defined as 25%–75% forest cover) are found in the 120-m buffer and account for 9.6% of the study area. The intermediate forest cover areas are, by definition, found in the mixed pixel zone, and it is this area that requires robust training labels.

Figure 9.5 illustrates the results when running a classification tree algorithm on pure training data versus mixed training data. A single sample was taken from the pure and mixed zones shown in Figure 9.2b, in proportion to their presence in the overall landscape: 82.5% nonforest and 17.5% forest. A 7% sample of each population was selected, as anything greater would be larger than the total number of pure forest pixels. A single tree was built for both pure and mixed training data inputs. The resulting decision boundaries are shown in Figure 9.5. When using pure pixels as training data, as suggested by many remote-sensing textbooks, the model results in a very simple 5-node tree, as shown in Figure 9.5a. Given the lack of mixed pixel information, the algorithm is able to create a parsimonious set of rules that result in few, nearly pure terminal nodes. When using mixed

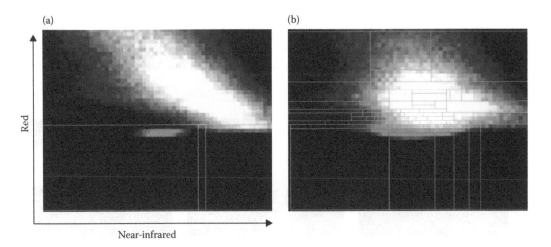

FIGURE 9.5 (See color insert.) Example decision boundaries made using a classification tree for (a) core site training dataset and (b) mixed pixel training dataset. For each model, a 7% sample of forest and nonforest were drawn for model generation from the populations shown in Figure 9.2b. Cyan represents nonforest and red represents forest, based on Figure 9.2a.

training data, the model becomes much more complex, consisting of 72 terminal nodes, and the resulting set of decision boundaries operate at fine scales within the feature space (Figure 9.5b). The accuracy of the core training model is 94.3% and of the mixed training model is 94.8%. However, the confidence of the core training data model is overstated. For classification trees, each node has a class membership probability that can be used on a per-pixel basis as a fuzzy confidence measure (Bankanza et al., 2009). For the set of pure training pixels, 4 of the 5 nodes reported nearly pure class membership. When applied to the entire image, 0.2% of pixels from this model have intermediate probability values (>10% and <90%). For the mixed training model, 23.8% of the pixels in the scene have node probabilities in this range. Fuzzy classifiers are one way of accommodating change analyses, and the increased ambiguity in the mixed training model offers a way to compare consecutive classifications well. As the ultimate goal of remote-sensing applications is monitoring, the advantages of a parsimonious classification tree model based on pure pixel training are lost, given the overstated confidence.

A series of model runs were performed using samples from all training, core training, and mixed training. Twenty-five models per sampling rates of 1%, 2%, 3%, 4%, 5%, 6%, and 7% were made for each training type. Figure 9.6 illustrates the results. For core sampling, the average accuracy was 94.26%, with little change as the sampling rate increased. For core sampling, a small training dataset will provide an accurate result from a few simple rules derived using the classification tree algorithm. This is an established advantage of tree-based classifiers (Hansen et al., 1996). However, the mixed pixel training performs consistently better, and the performance improves as sampling rate increases. Average accuracy for the mixed pixel training samples was 94.78%. The best average accuracy for the sample from the entire population of pure and mixed pixels was 95.01%. For pixels ranging from 25% to 75% forest cover, as defined by the RapidEye product, mean accuracies for the three training scenarios were 64.68% when using pure training, 66.13% when using mixed pixel training, and 66.66% when sampling the entire population. An overall improvement of 0.5% for all pixels and 1.4% for mixed pixels, in particular, was found when comparing pure versus mixed training models. Considering that rates of land-cover change are often in the range of 1% per year or even decade, such an improvement can be critical to monitoring objectives. Results indicate that mixed pixels are

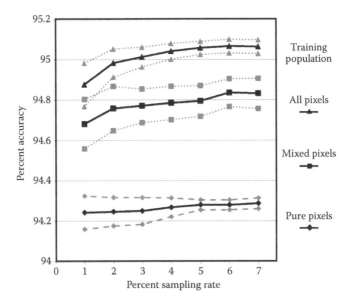

FIGURE 9.6 Accuracy of 25 tree models built per sampling rate (1%–7%) for three training dataset populations. Black represents the mean accuracy for the 25 model runs, and gray represents the ± standard deviation of the 25 models.

important in maximizing accuracies for classification tree models and that core sites will lead to overstated confidence and lower accuracies than training sets that include mixed pixel training.

9.6 DISCUSSION

Targeting mixed pixels as training data has advantages in pushing distribution-free algorithms toward delineation of optimal decision boundary. Two basic approaches to labeling mixed pixels for classification purposes are available. The first is the direct labeling of mixed pixels without the use of subpixel information. This is a viable approach and relies principally on the talents of the interpreter. A good interpreter will make accurate labels more often than not, and distribution-free classifiers are robust enough to tolerate errors in training labels, ensuring a good map product. Subpixel training using a higher spatial resolution reference is much more expensive in terms of efforts of analysts and the cost of data. An obvious application when deriving mixed pixels is their use in the estimation of fractional cover. Instead of a per-class confidence measure per pixel, a biophysical estimate of fractional cover is made. The various approaches to subpixel cover estimation can be divided into two general categories: those that rely on exemplar categorical reference data to model the intercategory variation and those that require calibration data along the entire range of mixtures. Examples of the former type include linear mixture models, fuzzy classifiers, and logistic regression approaches. Such methods do not rely on mixed pixels that define partial cover conditions. Methods that rely on calibration data at the subpixel level over the continuum of cover mixtures include distribution-free methods such as regression tree models. Models that exploit subpixel information directly in the calibration process should be able to perform better than those that only model subpixel cover variation, much in the same way that mixed pixel classification training outperforms core pixel classification training.

9.7 CONCLUSION

Mixed pixel training with distribution-free classifiers targets the spectral frontiers of interclass feature space. Although pure homogeneous training sites are easily delineated, they provide no information for assigning mixed pixels that can occupy a considerable portion of any scene. Given that human disturbance typically fragments a landscape, the accurate mapping of intermediate mixed pixels is important for monitoring applications. Robust delineation of mixed pixels improves the value of per-pixel probability or confidence measures, with implications for the comparison of consecutive characterizations for change assessment. Although deriving mixed pixel training data is costly, the results indicate improved land-cover characterizations compared to heritage methods.

Future work would intercompare various algorithms and feature spaces. For example, decision tree algorithms have been shown to perform less well in higher dimensional feature spaces when compared to even maximum likelihood classifiers (Pal and Mather, 2003). Improvement in our understanding of the effects of dimensionality and training-set delineation on performance of the algorithm is an area to be further researched. Concerning the core frontiers of mixed pixel landcover transitions, it is posited here that decision boundaries are defined more by the relative density of the training labels than by the chosen distribution-free algorithm. The idea of a hyperplane fitting a void is not the case when mixed pixels are sufficiently targeted for training. This implies a potential functional equivalency of algorithms as the data distributions dictate decision space frontiers, which is another topic for future research.

REFERENCES

Bankanza, J-R.B., Hansen, M.C., Roy, D.P., DeGrandi, G., and Justice, C.O. 2009. Wetland mapping in the Congo Basin using optical and radar remotely sensed data and derived topographical indices. *Remote Sensing of Environment*, 114, 73–86.

Bernard, A.C., Wilkinson, G.G., and Kanellopoulos, I. 1997. Training strategies for neural network soft classification of remotely sensed imagery. *International Journal of Remote Sensing*, 18, 1851–1856.

Bischof, H., Schneider, W., and Pinz, A.J. 1992. Multispectral classification of Landsat images using neural networks. *IEEE Transactions on Geoscience and Remote Sensing*, 30, 482–490.

DiGregorio, A. and Jansen, L. 2000. *Land Cover Classification System (LCCS): Classification Concepts and User Manual*. Rome: Food and Agricultural Organization of the United Nations, pp. 179.

Foody, G.M. 1999. The significance of border training patterns in classification by a feedforward neural network using back propagation learning. *International Journal of Remote Sensing*, 20, 3549–3562.

Foody, G.M. and Mathur, A. 2004a. A relative evaluation of multiclass image classification by support vector machines. *IEEE Transactions on Geoscience and Remote Sensing*, 42, 1335–1343.

Foody, G.M. and Mathur, A. 2004b. Toward intelligent training of supervised image classifications: Directing training data acquisition for SVM classification. *Remote Sensing of Environment*, 93, 107–117.

Foody, G.M. and Mathur, A. 2006. The use of small training sets containing mixed pixels for accurate hard image classification: Training on mixed spectral responses for classification by a SVM. *Remote Sensing of Environment*, 103, 179–189.

Foody, G.M., McCulloch, M.B., and Yates, W.B. 1995. The effect of training set size and composition on artificial neural network classification. *International Journal of Remote Sensing*, 16, 1707–1723.

Friedl, M.A. and Brodley, C.E. 1997, Decision tree classification of land cover from remotely sensed data, *Remote Sensing of Environment*, 61, 399–409.

Gopal, S., Woodcock, C., and Strahler, A. 1999. Fuzzy ARTMAP classification of global land cover from the 1 degree AVHRR data set. *Remote Sensing of Environment*, 676, 230–243.

Hansen, M.C. and Goetz, S.J. 2005. Land cover classification and change detection. In M.G. Anderson (ed.), *Encyclopedia of Hydrological Sciences* (pp. 853–874).New York: John Wiley and Sons.

Hansen, M., Dubayah, R., and DeFries, R. 1996. Classification trees: An alternative to traditional land cover classifiers. *International Journal of Remote Sensing*, 17, 1075–1081.

Hardin, P.J. and Thomson, C.N. 1992. Fast nearest neighbor classification models for multispectral imagery. *Professional Geographer*, 44, 191–201.

Huang, C., Davis, L.S., and Townshend, J.R.G. 2002. An assessment of support vector machines for land cover classification. *International Journal of Remote Sensing*, 23, 725–749.

Hughes, G.F. 1968. On the mean accuracy of statistical pattern recognizers. *IEEE Transactions on Information Theory*, IT-14, 55–63.

Ince, F. 1987. Maximum likelihood classification, optimal or problematic? A comparison with the nearest neighbor classification. *International Journal of Remote Sensing*, 8, 1829–1838.

Jensen, J. 2004. *Introductory Digital Image Processing*. Upper Saddle River, NJ: Prentice Hall, pp. 544.

Key, J., Maslanik, J.A., and Schweiger, A.J. 1989. Classification of merged AVHRR and SMMR Arctic data with neural networks. *Photogrammetric Engineering and Remote Sensing*, 55, 1331–1338.

Landgrebe, D.A. 2003. *Signal Theory Methods in Multispectral Remote Sensing* New York: John Wiley and Sons, pp. 508.

Lillesand, T.M. and Kiefer, R.W. 2008. *Remote Sensing and Image Interpretation*. New York: John Wiley and Sons, pp. 736.

National Agricultural Statistics Service. 2011. Cropland data layer. Available at: www.nass.usda.gov/research/Cropland/SARS1a.htm. USDA-NASS, Washington, D.C.

Pal, M. and Mather, P.M. 2003. An assessment of the effectiveness of decision tree methods for land cover classification. *Remote Sensing of Environment*, 86, 554–565.

Verbyla, D.L. 1995. *Satellite Remote Sensing of Natural Resources*. Boca Raton, FL: CRC Press, pp. 199.

10 Comparison between New Digital Image Classification Methods and Traditional Methods for Land-Cover Mapping

Alberto Jesús Perea Moreno and José Emilio Meroño De Larriva

CONTENTS

10.1 INTRODUCTION

Coarse-resolution satellite data have been used extensively for land-cover characterization and mapping. Recently available very high spatial resolution (VHR) (<5 m) remote-sensing data have opened the door for detailed investigation of land-cover characterization and mapping (Basso et al., 2001; Horie et al., 1992; Lobell et al., 2003; Pinter et al. 2003).

Traditional land-cover mapping is based on pixel-based classification. New digital image analysis algorithm, such as that used in object-oriented classification, is based on semantic information to interpret an image. This information is not represented by single pixels but by meaningful image objects and the mutual relationship between them (Abbas et al., 2007). The main difference between object-oriented classification and pixel-based classification is that the algorithm does not classify each single pixel but classifies image objects extracted through an image segmentation step. Image objects provide a more appropriate scale for mapping environmental features at multiple spatial scales and more relevant information than individual pixels (Gamanya et al., 2007).

In this chapter, we analyze and compare pixel-based classification, object-oriented classification, and Hierarchical Temporal Memory (HTM) algorithm and then implement these three methods using QuickBird data on a small area of Cordoba (Spain).

10.2 OBJECT-ORIENTED CLASSIFICATION

Object-oriented classification starts by segmenting an image into meaningful objects. A segmentation algorithm is a bottom-up region-merging technique. Each pixel is considered to be a separate object. Adjacent pairs of image objects are merged to form bigger segments based on a local homogeneity criterion that describes the similarity between adjacent image objects. As Xiaoxia et al. (2005) points out, a pair of image objects with the smallest increase in the defined criterion is merged. The process ends when the smallest increase in homogeneity exceeds a user-defined threshold, producing bigger objects when a higher threshold is used. This homogeneity criterion is a combination of color (spectral values) and shape properties (a combination of smoothness and compactness) that users can select. The procedure is controlled by the user who specifies the conditions as scale (size) or resolution of the objects (Xiaoxia et al., 2005). The result obtained is an image object that can be used in the next step during classification.

The next step is classification after image segmentation. The main schemes for object-based classification are supervised fuzzy logic nearest neighbor (NN) and fuzzy membership functions (Walker and Blaschke, 2008). The NN classifier uses representative training samples for each class and then the algorithm searches for the closest sample object in the feature space for each image object. The fuzzy NN classifier assigns a membership value between 0 and 1 based on the object's distance to its NN. The fuzzy membership function classification is based on fuzzy logic principles, that is, fuzzy rules are formed for the description of classes. As Hussain and Shan (2010) point out, in the case of VHR imagery with high spectral variability, a common problem is that an object may belong to one or more classes at the same time (Benz et al., 2004). To overcome the problem, fuzzy classification is used, which requires a selection of appropriate features to develop a rule set and define membership functions for every class of interest. The classification results depend on these input features, and a membership value is assigned to every class. The membership value varies between 0 and 1, and the value closer to 1 with no, or less, alternative assignment is regarded as the best result for a particular class.

Object-based approaches have been successful for land-use and land-cover classification (Frohn et al., 2005; Jensen et al. 2006). Gong and Howarth (1990) postulate that it is important to realize that conventional classifiers (maximum likelihood classifier and minimum distance classifier) do not recognize spatial patterns in the same way the human user does. To solve this problem, new algorithms were developed, and their main mission was to incorporate data different from the spectral features in order to improve the outcome of the purely spectral classification.

In analyzing very high resolution satellite data, segmentation of image pixels (object-oriented classification) into homogeneous objects has been explored in several studies through clustering routines and region-growing algorithms (Haralick and Shapiro, 1985; Ryherd and Woodcock, 1996). Woodcock and Strahler (1987) developed the concept of segmentation based on the theory of scale in remote sensing, which showed that the local variance of digital image data in relation to the spatial resolution can be used for selecting the appropriate image scale for mapping individual land-cover features (Johansen et al., 2009). An alternative to pixel-based classification may be to operate at the spatial scale of the objects of interest themselves, rather than to rely on the extent of image pixels (Flanders et al., 2003; Perea et al., 2009a; Platt and Rapoza 2008).

10.3 ALGORITHMS BASED ON THE HUMAN NEOCORTEX

A human brain is a continuous target for an enormous number of spatial and sequential patterns. These patterns constantly change and fleet through different divisions of the "old" brain, until they finally reach the neocortex (Hawkins and Blakeslee, 2005).

The difference between a computer and the human brain is that a computer tries to compute responses to predicaments, but sometimes this is not possible with complicated predicaments; the brain, in contrast, does not compute responses to predicaments but rather returns responses from memory, passing through different neurons. These responses are stored in the memory, which is represented by the neurons. The memory of an action is not programmed in the neurons; it is added to the neurons as the result of a learning process involving monotonous preparation. Another aspect of the brain's memory is that it creates associations routinely, which is why the term "autoassociative memory" is used. This autoassociative nature of human memory enables it to bring comprehensive patterns to the mind, regardless of whether the patterns are spatial or temporal, even if there is significant missing information about the patterns. At any time, memory can be stimulated by a very small bit of information, resulting in the remembering of entire bits at once. This continuous parade of memories makes up "thoughts" (Hawkins and Blakeslee, 2005).

Computer memory is intended to recall data precisely as it was stored at the beginning, whereas brain memory retains information only to the level of value, independent of the details. This attribute of brain memory is called *invariant representation*, and it gives stability to the recognition process by managing variations almost perfectly.

Hawkins and Blakeslee (2005) point out that memories are stored in the neocortex, and subsequently the brain recalls memories autoassociatively. The brain's memory system differs from that of computers because computers do not use invariant representations. The neocortex is also able to create predictions by linking invariant representations and recent information. This means that to predict the future with the help of past memories, it is essential to have a memory system that uses serial storage, autoassociative memory, and invariant representation. Scientists do not yet know how the cortex shapes invariant representations.

The storage of memories during learning process and subsequent application is more efficient than the use of mathematical equations applied by computers. Also, the fact that the procedure of building predictions, which concerns the fundamental nature of intelligence, requires a powerful memory system gives us good reason to believe that memory has an important role in intelligence (Hawkins and Blakeslee, 2005). Parts of this theory, known as the Memory-Prediction Theory (MPT), are modeled in the HTM technology developed by a company called Numenta (Hawkins and George, 2007a); this model simulates the structural and algorithmic properties of the neocortex, where spatial and temporal relations between features of the sensory signals are formed in a hierarchical memory architecture during a learning process. When a new pattern arrives, the recognition process can be viewed as choosing the stored representation that best predicts the pattern (Hawkins and George, 2007a). HTMs have been successfully applied to the recognition of relatively simple images, showing invariance across several transformations and robustness of noisy patterns (Hawkins and George, 2007b). This new algorithm is not a neural network. Classic neural networks, for example, multilayer perceptrons, are supervised learning models that are typically trained with an algorithm known as back-propagation. (We use "classic" to differentiate it from its newer forms, e.g., the Boltzmann Machine, that have stronger generative semantics.) Classic neural networks are generally not thought of as generative models. Although some instantiations of neural networks use space and time, they do not exploit temporal coherence as HTMs do. Neural networks generally require a large amount of data to train, and they often struggle with "over-fitting" (Hawkins and George, 2007b). Perea et al. (2009b) carried out a land-use classification of digital aerial photographs using a network based on HTM. Better results were attained, but this network was limited because the classification used only one pattern in an image.

10.4   DATA BASIS AND STUDY AREA

10.4.1   STUDY AREA

The study was performed in Cordoba Province, Spain, in Pedroches Valley, and includes the municipality of Peñarroya-Pueblonuevo. This is a rectangular area of 16 × 20 km and covers 32,000 ha (Figure 10.1). It is typical of the Andalusian region with dry crops and continental Mediterranean climate, characterized by long dry summers and mild winters.

To evaluate the QuickBird multispectral images for classification purposes, information from field visits was used. An area of 900 ha distributed over the study area was georeferenced using the submeter differential GPS TRIMBLE PRO-XRS equipped with a TDC-1 unit. Five hundred hectares of this area were visited to collect the spectral signature, and a total of 1100 independent and distant samples for every land use were georeferenced. Field data collected were used during the training stage of the classifications. Finally, 750 independent and distant samples, collected along 400 ha, were used to check the accuracy of the classifications. Also, reflectance data were analyzed to determine the land uses and their spectral similarities.

Following land-cover classes were prevalent in the study area: bare soil, cereals (corn [*Zea mays* L.], oats [*Avena sativa* L.], rye [*Secale cereale* L.], wheat [*Triticum aestivum* L.], and barley [*Hordeum vulgare* L.]), burnt crop stubble, alfalfa (*Medicago sativa* L.), other high-protein crops with

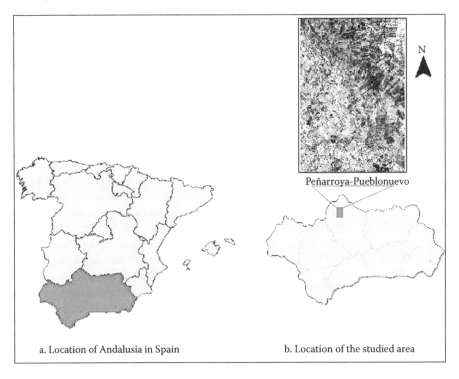

a. Location of Andalusia in Spain   b. Location of the studied area

FIGURE 10.1   Map of the study area.

early growth state (peas [*Pisum sativum* L.], beans [*Vicia faba* L.]), woodlands/scrublands (holly oak [*Quercus ilex* L.] and common retama [*Retama sphaerocarpa* [L.] Boiss.]), and urban soil.

10.4.2 DATA

Six multispectral images (QuickBird, Ortho Ready Standard Imagery, Digital Globe, Longmont, Colorado, USA), identified in UTM coordinates (Universal Transverse Mercator) and georeferenced in the WGS84 system, were used. These images were orthorectified and referenced to the European Datum 1950 of the International Ellipsoid. The images were codified in 16 bits, with a resolution of 2.4 m, and were composed of four bands (blue, green, red, and near infrared).

To determine the optimum time for the acquisition of imagery, we monitored the phenology of the study area's forest vegetation for 12 months. The images were acquired on April 27, 2007, when the crop canopy of all the species was full, in order to minimize phenological differences due to the variability of the topography among areas occupied by the same species. These images were taken with an incident angle of 1.07°, beginning at 11:22 AM, with a solar elevation angle of 62.9°.

10.5 METHODS

This section describes the steps followed in the methodological approach. The main classification steps include image and data preprocessing, supervised classification, object-oriented classification, HTM networks, and evaluation.

10.5.1 PREPROCESSING

Radiometric and geometric corrections were previously carried out by the distributor. No atmospheric corrections were needed. Also, an orthorectification process was carried out. To supply the statistical analysis with a redundant dataset, the image was subjected to two different spectral transformations: the principal component analysis (PCA) transformation and the normalized difference vegetation index (NDVI) thematic image generation.

This PCA statistical technique converts intercorrelated multispectral bands into a new set of uncorrelated components, the so-called principal components (PCs) (Zhang, 2004). The first principal component PC1 accounts for maximum variance. The high-resolution image replaces PC1 since it contains information common to all bands, whereas spectral information is unique to each band (Pohl, 1999). It is assumed that Pan data are very similar to the first PC image (Chavez et al., 1991). All four QuickBird bands are used as input to the PCA.

NDVI, known to be positively correlated with plant biomass (Mather, 1999), is defined as follows:

$$NDVI = \frac{(R_{NIR} - R_{RED})}{R_{NIR} + R_{RED}},$$

where R_{NIR} and R_{RED} are reflectances in the near-infrared band (R800 nm) and the red band (R690 nm), respectively.

10.5.2 SUPERVISED CLASSIFICATION

Maximum likelihood classification is one of the most popular methods of classification in remote sensing (Benedictsson et al., 1990; Foody et al., 1992). The maximum likelihood decision rule is based on the probability that a pixel belongs to a particular class. The basic equation assumes that these probabilities are equal for all classes and that the input bands have normal distributions.

Pixel-based supervised maximum likelihood image classification was performed in ERDAS Imagine 9.2® using the image formed by the PCs and NDVI. It is important that training samples be representative of the class sought to be identified. With the help of fieldwork investigation, knowledge of the data and of the classes desired was acquired before classification. Training samples (a set of pixels) of representative patterns and land-cover features recognized can be selected more determinately. Samples are selected elaborately, and the Seed Properties dialog and Area of Interest (AOI) tools can be used. Seed pixel is used as a model pixel, against which the pixels contiguous to it are compared, based on parameters (neighbourhood, geographic constraints, spectral euclidean distance) specified by the user.

Signature separability is a statistical measure of the distance between two signatures. Separability can be calculated for any combination of bands used in the classification. For distance (Euclidean) evaluation, the spectral distance between the mean vectors of each pair of signatures is computed. If the spectral distance between two samples is not significant for any pair of bands, then they may not be distinct enough to produce a successful classification. The spectral distance is also the basis of the minimum distance classification. Therefore, computing the distances between signatures can help predict the results of a minimum distance classification.

In the classification, signature separability functions were used to examine the quality of training site and class signature before performing the classification. Ismail and Jusoff (2008) postulated that signature separability contains all the available information about signature and class information for each class. The importance of using this is to determine how well each class is separated from each of the other classes. This function allows the operator to use statistical analysis to enhance the accuracy of the very subjective process of classification.

10.5.3 Object-Oriented Image Classification

Image objects were created using the image segmentation tool offered in eCognition® Developer 7.0. The segmentation process in this software is a bottom-up region-merging approach, where the smallest objects contain single pixels (Baatz et al., 2004). In the process, smaller objects were merged into larger objects based on three parameters: scale, color (spectral properties), and shape (smoothness and compactness). "The segmentation process was stopped when the smallest growth of an object exceeded a user-defined threshold, which is an arbitrary value (i.e., a scale parameter) that determines the maximum possible change in heterogeneity when several objects are merged" (Benz et al., 2004). The larger the scale parameter, the larger the size of the resultant objects. A scale parameter of 125 was selected, based on visual interpretation of the image segmentation results using different scale parameters. The value of 125 was considered appropriate to maximize both local homogeneity and global heterogeneity, as well as to produce a reasonable number of objects to process.

The homogeneity of segments was controlled by both spectral and shape percentages, as well as by weight for the relative contribution of each input band. Spectral (color) homogeneity was given an overall spectral factor percentage of 90%. Shape-homogeneity criteria included an overall shape-factor percentage of 10%, which was subdivided into smoothness (8%) and compactness (2%). A higher compactness value helps separate objects with different shapes but without much color contrast (e.g., rooftops vs. roads), whereas a higher smoothness weight helps identify objects that have a greater variability between features (Baatz et al., 2004).

eCognition Developer 7.0 offers two different classifiers: NN and membership functions. This experiment uses the NN classification, which assigns classes to image objects based on minimum distance measurements. The NN classifier can potentially use a variety of features; moreover, feature space can be defined for each single class independently In Figure 10.2, the process of an oriented based classification using the NN classifier is explained.

10.5.4 HTM Networks

An HTM network is a collection of linked nodes organized in a tree-shaped hierarchy. See Figure 10.3 for an example of an HTM network. HTM networks consist of several layers or levels

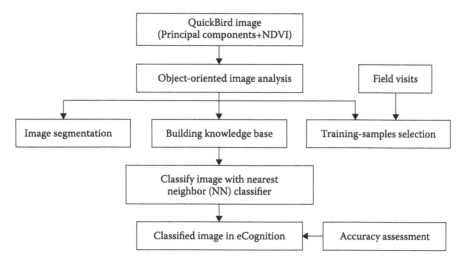

FIGURE 10.2 The methodology flowchart of object-oriented image analysis.

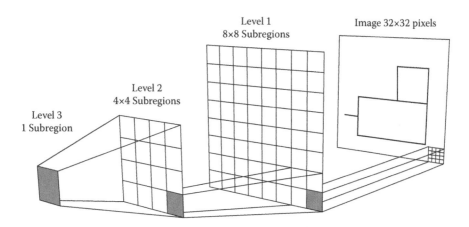

FIGURE 10.3 Example of an HTM network. (Adapted from Garalevicius, S.J., Memory-prediction framework for pattern recognition: Performance and suitability of the Bayesian model of visual cortex. In Wilson, D. and Sutcliffe, G. (Eds.), *FLAIRS Conference* (pp. 92–97). Florida, May 7–9, 2007.)

of nodes, with one node at the top level. HTM networks operate in two stages: the learning stage and the inference stage. During the learning stage, the network is exposed to training patterns, and it then builds a model of these data. During the inference stage, the network recognizes the new, usually unseen, test patterns. More concretely, during a (supervised) learning stage, the network learns which pattern belongs to which category, whereas during the inference stage the network will generate a belief distribution over these categories for every new pattern it recognizes. Belief distributions (represented by belief vectors) are a measure of belief that the input pattern belongs to one of the categories.

All of the nodes (except the top node used in supervised learning) process information in the same way, so we will now explain the operation of such a node.

10.5.4.1 Operation of Nodes during Learning

During the learning mode, the node is receiving inputs and measuring their statistics. The spatial pooler learns mapping from a potentially infinite number of input patterns to a finite number of quantization centers. The output of the spatial pooler, which is considered an input to the temporal pooler, is expressed in terms of its quantization centers. This stage can be seen as a preprocessing step for the temporal pooler, simplifying its input. The temporal pooler learns temporal groups, which are groups of quantization centers that frequently occur close together in time. The output of the temporal pooler is in terms of the temporal groups it has learned (George and Jaros, 2007).

10.5.4.2   Operation of Spatial Pooler during Learning

The spatial pooler has two stages of operation:

- During the learning stage it quantizes the input patterns and memorizes the quantization centers.
- Once these quantization centers are learned, it produces outputs in terms of these quantization centers during the inference stage (George and Jaros, 2007).

The spatial poolers from nodes at the first level receive raw data from the sensor, whereas the spatial poolers from nodes higher in the hierarchy receive the outputs from child nodes. The inputs to the spatial poolers of nodes higher in the hierarchy are the concatenations of the outputs of their child nodes. The input to the spatial pooler is represented by a row vector, and the role of the spatial pooler is to quantize this vector and build a matrix from these quantization centers.

This matrix is empty before training. The vectors in this matrix (the quantization centers) are called *coincidences*, and hence the matrix is called a coincidence matrix.

There are three spatial pooler algorithms: Gaussian, dot, and product. During learning, the dot and product algorithms work the same way. The Gaussian spatial pooler algorithm is used for nodes at the first level, whereas the dot/product learning algorithm is applied at level >1. The input of the spatial pooler at level $n + 1$ is a probability distribution over the temporal groups of the nodes at level n. A spatial pooler algorithm parameter specifies which algorithm to use, although it is common to use the same algorithm for every node up the hierarchy.

10.5.4.3   Operation of Temporal Pooler during Learning

The objective of *the temporal pooler* is to create temporal coherent groups from a sequence of spatial patterns. This mechanism pools patterns using their temporal proximity. If pattern A is frequently followed by pattern B, the temporal pooler can assign them to the same group. To this end, it builds a first-order time-adjacency matrix; after learning, this can be used to derive how likely a certain transition between each of the coincidences is. When a new input vector is presented during training, the spatial pooler represents it as one of its learned coincidences i. The temporal pooler then looks back in history a certain number of steps, which is represented by the parameter *transitionMemory*. After the learning stage and before inference, when the time-adjacency matrix is formed, the temporal pooler uses this matrix to create temporal groups.

10.5.4.4   Training the Network

To make classifications, we use a supervised mapper that replaces the temporal pooler at the highest level of an HTM network. For every training input pattern, the supervised mapper receives two inputs during learning: the coincidence from the spatial pooler and the category of the input vector from the category sensor. It has a mapping matrix that stores how many times a coincidence i belongs to a category c by incrementing element (c, i) every time it receives these inputs together.

10.5.4.5   Operation of Nodes during Inference

After training a node, it can be switched to inference mode. During inference, the level already has a model of the world (stored in the spatial and temporal pooler nodes). When the level receives an input from its children, it uses its internal model of the world to create an output to send to its parent(s).

10.5.4.6 Spatial Pooler during Inference

The three spatial pooler algorithms, Gaussian, dot, and product, work differently during the inference stage, but they all convert an input vector into a belief vector over coincidences. As stated before, the Gaussian spatial pooler algorithm is used in first level nodes, and the dot or product algorithms are used in the nodes higher in the hierarchy.

10.5.4.7 Operation of Temporal Pooler during Inference

During inference, the temporal pooler receives a belief vector over coincidences from the spatial pooler. It will then calculate a belief distribution over groups. In this mode, two different algorithms exist for the temporal pooler, *maxProp* and *sumProp*, governed by the parameter *temporalPoolerAlgorithm*. In maxProp inference mode, the maximum value per temporal group is set as output. In sumProp inference mode, smoother score is computed for the group based on the current input only.

10.5.4.8 Operation of Top Node during Inference

During inference *of the top node*, the spatial pooler works as described above. The supervised mapper receives a belief vector over coincidences from the spatial pooler and a category from the category sensor. It calculates a belief distribution over these categories. At this stage, it is necessary to choose between the two different temporal pooler algorithms, *maxProp* and *sumProp*, during inference, controlled by the parameter *mapperAlgorithm*.

10.5.4.9 Training Parameters

10.5.4.9.1 MaxDistance

This specifies the distance by which an input pattern has to differ from a stored pattern in order to be regarded as a different pattern for storage. This parameter is used in first-level nodes by the Gaussian spatial pooler algorithm. During learning, a new pattern is compared to existing coincidences. The maximum Euclidean distance at which two input vectors are considered the same during learning is established; that is, when the squared Euclidean distance between them is smaller than maxDistance, the new pattern is stored as a coincidence (Numenta Inc., 2008).

10.5.4.9.2 Sigma

During a node's inference stage, each input pattern is compared with the stored patterns, assuming that the stored patterns are centers of radial basis functions with Gaussian tuning. The sigma parameter specifies the standard deviation of this Gaussian. Select a parameter value based on the noise in the environment. Keep sigma high for noisy situations and low for nonnoisy situations (Numenta Inc., 2008).

During the inference state of the node, the spatial pooler generates a belief vector over learned coincidences for a given input pattern. The belief in a coincidence is represented as an unnormalized multidimensional Gaussian with the coincidence vector as its mean and a variance of sigma. A platform called NUPIC, developed by Numenta®, was used to implement our HTM network. The HTM network consists of three levels. The input level consists of 16 nodes, each receiving a feature and the corresponding delta. Level 2 consists of four nodes, each receiving the output of four input-level child nodes. Level 3 consists of one top-level node.

10.5.4.10 Evaluation

To evaluate the quality of classifications, a total of 750 verification points were taken to compare real cover (true terrain) and that obtained by classification.

The overall accuracy, Kappa statistic, and the producer's and user's accuracy were calculated for each of the classifications. The overall accuracy was calculated through the plot ratio, correctly classified, and divided by the total number included in the evaluation process. The Kappa statistic is an alternative measure of classification accuracy that subtracts the effect from random accuracy;

it quantifies how much better a particular classification is in comparison with a random classification. Some authors have suggested the use of a subjective scale where Kappa values <40% are poor, 40%–55% fair, 55%–70% good, 70%–85% very good, and >85% excellent (Monserud and Leemans, 1992).

A statistical study was made with the spectral response of all the pixels of the image included in the validating sites to assess the accuracy of the pixel-based classification.

On the other hand, a Kappa analysis and pairwise Z-test were calculated to determine if the two classifications were significantly different ($\alpha = 0.05$) (Congalton and Green, 1999; Dwivedi et al., 2004; Zar, 2007):

$$\hat{K} = \frac{|p_0 - p_c|}{1 - p_c}$$

and

$$Z = \frac{\left|\hat{K}_1 - \hat{K}_2\right|}{\sqrt{\widehat{\text{var}}\left(\hat{K}_1\right) + \widehat{\text{var}}\left(\hat{K}_2\right)}},$$

where p_0 represents actual agreement, p_c represents "chance agreement," and $\hat{K}_1, \hat{K}_2$ represent the Kappa coefficients for the two classifications, respectively. The Kappa coefficient is a measure of the agreement between observed and predicted values and of whether that agreement is by chance (Congalton and Green, 1999). Pairwise Z-scores and probabilities (p-values) were calculated for every combination of the two classifications. Using a two-tailed Z-test ($\alpha = 0.05$ and $Z_{\alpha/2} = 1.96$), if the p-value was ≥0.025, then the classifications were not considered as having significant statistical difference (Zar, 2007).

10.6   RESULTS

10.6.1   Supervised Classification

The maximum likelihood classification method, especially the cereal and alfalfa classes had the worst classification accuracy (Table 10.1). This is consistent with other studies, because the cereal

TABLE 10.1

Producer's and User's Accuracy, Overall Accuracy, and Kappa Statistic for Supervised and Object-Oriented Classifications

Category	Supervised Classification Image Principal Components and NDVI		Object-Oriented Classification	
	Pa (%)	Ua (%)	Pa (%)	Ua (%)
Bare soil	93.50	92.30	99.40	94.40
Cereal	78.40	87.80	86.60	94.50
Burnt crop stubble	33.30	99.00	94.40	85.00
Other high-protein crops	99.50	42.80	100.00	71.40
Alfalfa	97.60	47.30	88.30	64.20
Woodlands and scrublands	100.00	83.30	100.00	84.47
Urban soil	71.00	86.80	80.00	100.0
Overall accuracy (%)	85.26		92.00	
Kappa statistic (%)	77.36		89.60	

and alfalfa classes are very similar in the spectrum, and therefore it is quite difficult to avoid the spectral overlapping effect.

In heterogeneous areas such as urban areas, conventional pixel-based classification approaches have very limited applications because of the very similar spectral characteristics among different land-cover types (e.g., bare soil) and high spectral variation within the same land-cover class.

10.6.2 Results of the Object-Oriented Classification

The result of segmentation is a new image that divides the original image into regions so that each contains similar pixels. After the process of segmentation, a new image was obtained, divided into 13,811 regions that were later classified (Figure 10.4).

The classification accuracy was assessed using randomly selected points for which land cover was determined using the information from field visits. The accuracy assessment of this classifier can be found in Table 10.1.

The object-oriented classification approach yields a higher accuracy than the supervised classification, with an overall accuracy of 92% and a Kappa coefficient of 89.60%.

The supervised classification had problems with several classes. The cereal class, for example, was often misclassified as alfalfa. Small patches of alfalfa were often contained within larger plots of cereal. The class burnt crop stubble was also confused with alfalfa. As with the cereal class, this misclassification can be attributed to small patches not being identified.

Under the object-based approach, for most of the cases both user and producer accuracies for the individual classes were higher than those obtained using the pixel-based method. For some classes, producer or user accuracy reached a value of 100%, for example, for urban soil and burnt crop stubble. For other classes, for example, woodlands and scrublands, producer and user accuracies increased but remained low.

The highest producer accuracies were for high-protein crops and woodlands and scrublands categories, all with the value of 100%. In contrast, the lowest value was for urban soil (80%). As for user accuracy, the best results were achieved for the urban soil category (100%); as with producer accuracy, the lowest value was for the alfalfa category (64.20%), owing to misclassification of high-protein crops during image classification.

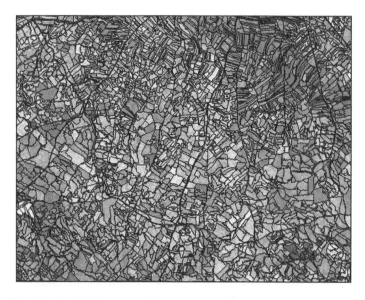

FIGURE 10.4 (See color insert.) Segmented image using a scale parameter of 125.

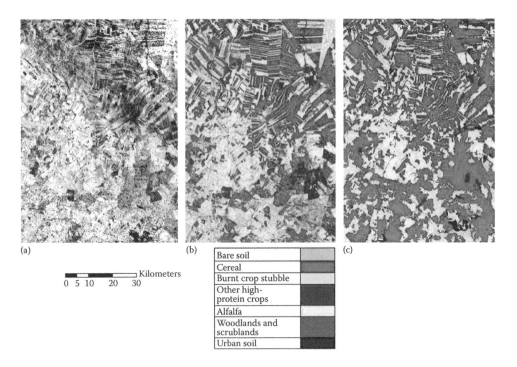

(a)   (b)   (c)

Bare soil	
Cereal	
Burnt crop stubble	
Other high-protein crops	
Alfalfa	
Woodlands and scrublands	
Urban soil	

Kilometers
0 5 10   20   30

FIGURE 10.5 (See color insert.) Example of comparison between QuickBird image (a), supervised classification of the image formed by the principal component and the NDVI index (b), and the oriented-based classification (c).

The classification result of the object-oriented method was consistently high relative to many other methods, and there was statistically significant difference between this type of classification and the supervised classification (p-value < .0001) (Chen et al., 2004; Enderle and Weih, 2005). When high-resolution imagery is used in heterogeneous landscapes, conventional pixel-based classification approaches that utilize only spectral information have very limited usefulness. Figure 10.5 shows an example of comparison between the QuickBird image, supervised classification of the image formed by the principal component and the NDVI index, and the object-oriented classification.

10.6.3 Results of HTM Classification

We investigated the effect of the parameters *maxDistance* and *Sigma* on overall accuracy, Kappa coefficient, and the average number of coincidences and temporal groups learned in the bottom-level nodes. The other parameters (*transitionMemory* and *topNeighbors*) were set to 5 and 1, respectively. These are default values, and different values had a negative effect on the performance of the system. We varied the values for *Maxdistance* and set *Sigma* to the square root of *Maxdistance*. This is a reasonable starting value for *Sigma*, because distances between coincidences are calculated as the squared Euclidean distance instead of the standard Euclidean distance.

The highest overall accuracy was obtained with an intermediate value for *Maxdistance*: 3. This might indicate that with a lower value for *Maxdistance*, the HTM would see variations in input patterns owing to noise as different coincidences. On the other hand, when *Maxdistance* is higher than the optimal value, the spatial pooler will pool together patterns that have different causes.

It was found that the accuracy of HTM classification is higher than that of the pixel-based classification. The overall accuracy and Kappa coefficient are significantly higher, achieving the values 94.43% and 92.32%, respectively. Producer and user accuracies were better in this type of

TABLE 10.2

Producer's and User's Accuracy, Overall Accuracy, and Kappa Statistic for Supervised and HTM Classifications

Category	Supervised Classification Image Principal Components and NDVI		HTM Classification	
	Pa (%)	Ua (%)	Pa (%)	Ua (%)
Bare soil	93.50	92.30	100.00	90.63
Cereal	78.40	87.80	88.46	96.46
Burnt crop stubble	33.30	99.00	95.00	100.00
Other high-protein crops	99.50	42.80	100.00	78.70
Alfalfa	97.60	47.30	96.89	74.32
Woodlands and scrublands	100.00	83.30	100.00	96.22
Urban soil	71.00	86.80	100.00	100.00
Overall accuracy (%)	85.26		94.43	
Kappa statistic (%)	77.36		92.32	

classification, and these were statistically much higher than that in the supervised classification (p-value < .0001).

However, a comparison of the Z-scores and p-values for the object-based classification and HTM classification indicates that there was no statistically significant difference between them (Z-score = 1.437 and p-value = .0769).

The results of HTM classification (Table 10.2) show a marked improvement in both producer and user accuracies in most categories, compared with purely spectral classifications. Furthermore, this algorithm achieves accuracy rates and Kappa coefficients both above 90% in some cases. Producer accuracy increases in all cases except for alfalfa—which does not increase in value but is nevertheless above 90%. The alfalfa category is confused with the category of high-protein crops for the aforementioned reasons. User accuracy increases in all categories except for bare soil (90.63%), but it is a reasonable value.

Figure 10.6 presents a comparison between the QuickBird image, the supervised classification of the image formed by the principal component and the NDVI index, and the HTM classification.

10.7 CONCLUSIONS

The object-oriented classifier outruns the pixel-based method overwhelmingly. It yields an overall accuracy of 89.33%, whereas the overall accuracy for the supervised classification method is only 70.89%. The variation between accuracies of different classes is significantly narrowed down in the object-oriented classification. In particular, the object-oriented approach shows superior performance in classifying built-up areas. The object concept enables the use of various features, making full use of high-resolution image information. Beyond purely spectral information, image objects contain additional attributes that can be used for classification. With different parameters, the multiscale approach offers the possibility of easily adapting image object resolution to specific requirements, data, and tasks. In addition, HTM classification considers spatial and temporal relations between features of the sensory signals, which are formed in a hierarchical memory architecture during a learning process, thereby improving the results obtained by the supervised classification.

In this research, new digital image classification methods have been evaluated for classification, and the results are satisfactory for land-cover mapping. The proposed techniques were successfully

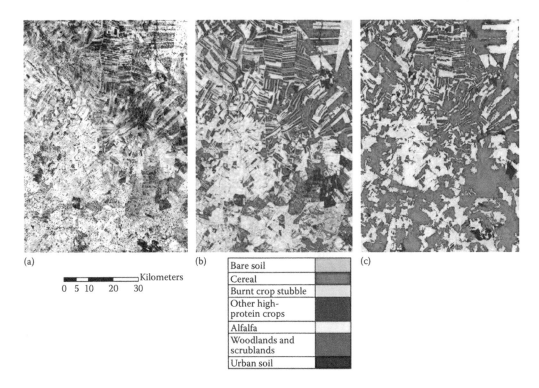

(a)   (b)   (c)

Bare soil	
Cereal	
Burnt crop stubble	
Other high-protein crops	
Alfalfa	
Woodlands and scrublands	
Urban soil	

Kilometers
0 5 10   20   30

FIGURE 10.6 **(See color insert.)** Example of comparison between QuickBird image (a), supervised classification of the image formed by the principal component and the NDVI index (b), and the HTM classification (c).

tested with QuickBird images. The results presented in this chapter show the efficiency and higher accuracy of polygon-based classification and HTM networks. It is recommended that these techniques be tested on VHR data, such as QuickBird images or digital aerial photos, especially in areas where more specific classes can be generated. In contrast, traditional classification techniques, especially pixel-based approaches, are limited in that they typically produce a characteristic "salt and pepper" effect and are unable to extract objects of interest.

REFERENCES

Abbas, T.A., Ali, S.H., and Ali, I.H. 2007. Object-oriented classification of forest images using soft computing approach. In *4th International Conference: Sciences of Electronic, Technologies of Information and Telecommunications*, Tunisia, March 25–29.

Baatz, M., Benz, U., Dehghani, S., and Heynen, M., 2004. *eCognition User Guide 4*. Munich: Definiens Imagine GmbH.

Basso, B., Ritchie, J.T., Pierce, F.J., Braga, R.P., and Jones, J.W. 2001. Spatial validation of crop models for precision agriculture. *Agricultural Systems,* 68, 97–112.

Benedictsson, J.A., Swain, P.H., and Ersoy, O.K. 1990. Neural network approaches versus statistical methods in classification of multisource remote-sensing data. *IEEE Transactions on Geoscience and Remote Sensing,* 28(4), 540–551.

Benz, U.C., Hofmann, P., Willhauck, G., Lingenfelder, I., and Heynen, M. 2004. Multiresolution, object-oriented fuzzy analysis of remote-sensing data for GIS ready information. *ISPRS Journal of Photogrammetry & Remote Sensing,* 58, 239–258.

Chavez, P.S., Sides, S.C., and Anderson, J.A. 1991. Comparison of three different methods to merge multiresolution and multispectral data: Landsat TM and SPOT Panchromatic. *Photogrammetric Engineering and Remote Sensing,* 57(3), 295–303.

Chen, D., Stow, D.A., and Gong, P. 2004. Examining the effect of spatial resolution and texture window size on classification accuracy: An urban environment case. *International Journal of Remote Sensing*, 25(11), 2177–2192.

Congalton, R. and Green, K. 1999. *Assessing the Accuracy of Remotely Sensed Data: Principles and Practices*. Boca Raton, FL: CRC/Lewis Publishers.

Dwivedi, R.S., Kandrika, S., and Ramana, K.V. 2004. Comparison of classifiers of remote-sensing data for land-use/land-cover mapping. *Current Science*, 86(2), 328–335.

Enderle, D. and Weih, R.C. 2005. Integrating supervised and unsupervised classification methods to develop a more accurate land-cover classification. *Journal of the Arkansas Academy of Science*, 59, 65–73.

Flanders, D., Hall-Beyer, M., and Pereverzoff, J. 2003. Preliminary evaluation of eCognition object-based software for cut block delineation and feature extraction. *Canadian Journal of Remote Sensing*, 29, 441–452.

Foody, G.M., Campbell, N.A., Trood, N.M., and Wood, T.F. 1992. Derivation and application of probabilistic measures of class membership from the maximum likelihood classification. *Photogrammetric Engineering & Remote Sensing*, 58(9), 1335–1341.

Frohn, R.C., Hinkel, K.M., and Eisner, W.R. 2005. Satellite remote-sensing classification of thaw lakes and drained thaw lake basins on the North Slope of Alaska. *Remote Sensing of Environment*, 97, 116–126.

Gamanya, R., Maeyer, P.D., and Dapper, M.D. 2007. An automated satellite image classification design using object-oriented segmentation algorithms: A move towards standardization. *Expert Systems with Applications*, 32, 616–624.

Garalevicius, S.J. 2007. Memory-prediction framework for pattern recognition: Performance and suitability of the Bayesian model of visual cortex. In Wilson, D. and Sutcliffe, G. (Eds.), *FLAIRS Conference* (pp. 92–97). Florida, May 7–9.

George, D. and Jaros, B. 2007. The HTM learning algorithms. Available at: http://www.numenta.com/htm-overview/education/Numenta_HTM_Learning_Algos.pdf

Gong, H. and Howarth, P.J. 1990. An assessment of some factors influencing multispectral land-cover classification. *Photogrammetric Engineering and Remote Sensing*, 56, 597–603.

Haralick, R.M. and Shapiro, L.G. 1985. Image segmentation techniques. *Computer Vision, Graphics, and Image Processing*, 29, 100–132.

Hawkins, J. and Blakeslee, S. 2005. *On Intelligence*. New York: Owl Books Henry Holt and Company.

Hawkins, J. and George, D. 2007a. Hierarchical temporal memory, concepts, theory, and terminology. Available at: http://www.numenta.com/htm-overview/education/Numenta_HTM_Concepts.pdf

Hawkins, J. and George, D. 2007b. Hierarchical temporal memory, comparison with existing models. Available at: http://www.numenta.com/htm-overview/education/HTM_Comparison.pdf

Horie, T., Yajima, M., and Nakagawa, H. 1992. Yield forecasting. *Agricultural Systems*, 40, 211–236.

Hussain, E. and Shan, J. 2010. Rule inheritance in object-based classification for urban land-cover mapping. In American Society for Photogrammetry & Remote Sensing (Ed.), *ASPRS 2010 Annual Conference*, San Diego, California, April 26–30.

Ismail, M.S. and Jusoff, K. 2008. Satellite data classification accuracy assessment based on reference dataset. *International Journal of Computer and Information Science and Engineering*, 2(1), 96–102.

Jensen, J.R., García-Quijano, M., Hadley, B., Im, J., Wang, Z., Nel, A.L., Teixeira, E., and Davis, B.A. 2006. Remote sensing agricultural crop type for sustainable development in South Africa. *Geocarto International*, 21, 5–18.

Johansen, K., Phinn S., Witte, C., Philip, S., and Newton, L. 2009. Mapping banana plantations from object-oriented classification of Spot-5 imagery. *Photogrammetric Engineering and Remote Sensing*, 75(9), 1069–1081.

Lobell, D.B., Asner, G.P., Ortiz-Monasterio, I.J., and Benning, T.L., 2003. Remote sensing of regional crop production in the Yaqui Valley, Mexico: Estimates and uncertainties. *Agriculture, Ecosystems and Environment*, 94, 205–220.

Mather, P.M. 1999. *Computer Processing of Remotely Sensed Images—An Introduction*, 2nd ed. Chichester: John Wiley & Sons.

Monserud, R.A. and Leemans, R. 1992. Comparing global vegetation maps with the Kappa statistic. *Ecological Modelling*, 62, 275–293.

Numenta Inc. 2008. Advanced NuPIC programming. Available at: http://www.numenta.com/archives/education/nupic_prog_guide.pdf

Perea, A.J., Meroño, J.E., and Aguilera, M.J. 2009a. Oriented-based classification in aerial digital photography for land-use discrimination. *Interciencia*, 34, 612–616.

Perea, A.J., Meroño, J.E., and Aguilera, M.J. 2009b. Application of Numenta® Hierarchical Temporal Memory for land-use classification. *South African Journal of Science*, 105(9–10), 370–375.

Pinter, P.J., Hatfield, J.L., Schepers, J.S., Barnes, E.M., Moran, M.S., Daughtry, C.S.T., and Upchurch, D.R. 2003. Remote sensing for crop management. *Photogrammetric Engineering and Remote Sensing,* 69(6), 647–664.

Platt, R.V. and Rapoza, L. 2008. An evaluation of an object-oriented paradigm for land-use/land-cover classification. *Professional Geographer,* 60, 87–100.

Pohl, C. 1999. Tools and methods for fusion of images of different spatial resolution. *International Archives of Photogrammetry and Remote Sensing,* 32 Valladolid, 3–4 June, part 7-4-3 W6.

Ryherd, S. and Woodcock, C. 1996. Combining spectral and texture data in the segmentation of remotely sensed images. *Photogrammetric Engineering and Remote Sensing,* 62(2), 181–194.

Walker, J.S. and Blaschke, T. 2008. Object-based land-cover classification for the Phoenix metropolitan area: Optimization vs. transportability. *International Journal of Remote Sensing,* 29(7), 2021–2040.

Woodcock, C.E. and Strahler, A.H. 1987. The factor of scale in remote sensing. *Remote Sensing of Environment,* 21, 311–332.

Xiaoxia, S., Jixian, Z., and Zhengjun, L. 2005. A comparison of object-oriented and pixel-based classification approaches using QuickBird imagery. In *3rd International Symposium on Remote Sensing and Data Fusion Over Urban Areas (URBAN 2005),* China, August 27–29.

Zar, J.H. 2007. *Biostatistical Analysis,* 4th ed. Upper Saddle River, NJ: Prentice Hall.

Zhang, Y. 2004. Understanding image fusion. *Photogrammetric Engineering and Remote Sensing,* 70(6), 657–661.

11 Land-Cover Change Detection

Xuexia (Sherry) Chen, Chandra P. Giri, and James E. Vogelmann

CONTENTS

11.1 INTRODUCTION

Land cover is the biophysical material on the surface of the earth. Land-cover types include grass, shrubs, trees, barren, water, and man-made features. Land cover changes continuously. The rate of change can either be dramatic and abrupt, such as the changes caused by logging, hurricanes, and fire, or subtle and gradual, such as regeneration of forests and damage caused by insects (Verbesselt et al., 2010). Previous studies have shown that land cover has changed dramatically during the past several centuries and that these changes have severely affected our ecosystems (Foody, 2010; Lambin et al., 2001). Lambin and Strahlers (1994b) summarized five types of causes for land-cover changes: (1) long-term natural changes in climate conditions, (2) geomorphological and ecological processes, (3) human-induced alterations of vegetation cover and landscapes, (4) interannual climate variability, and (5) human-induced greenhouse effect. Tools and techniques are needed to detect, describe, and predict these changes to facilitate sustainable management of natural resources.

Accurate and up-to-date information on land-cover change is needed for many applications. Carbon pools and fluxes are receiving more attention owing to the impact of the global carbon cycle on climate (Houghton et al., 1999; Jenkins et al., 2001). Land-cover change is a source of increased atmospheric CO_2, which can affect global climate and may cause further changes in land cover (Foody, 2010). Forest disturbances such as fire, disease, insect outbreaks, drought, hurricanes, and harvesting, which result in land-cover change, disturb the carbon accumulated in woody biomass and soils, with different effects on the global carbon budget. Forest disturbances such as fires and harvesting release carbon into the atmosphere through oxidation and decomposition, and the postevent vegetation recovery processes sequester carbon from the atmosphere (Masek et al., 2008). Detailed quantification of carbon pools and fluxes depends on our understanding of the spatial distribution of the biomass in various land-cover types and their changes during disturbances (Masek and Collatz, 2006; Wulder et al., 2004).

Detection of land-cover change aims at describing the status of land cover at different times to identify the actual differences in the variables of interest (Green et al., 1994; Singh, 1989). One of the objectives of land-cover change detection is to understand better the relationships and interactions between humans and the environment in order to manage and use resources in a better way for sustainable development (Lu et al., 2004b). It is particularly important to differentiate natural changes from human-induced changes and, when warranted, to make interventions to control or mitigate some of the more negative effects of change (Lambin and Strahlers, 1994b).

Stand replacement changes, such as those caused by urbanization, logging, and forest fires, are prominent and are characterized by large changes in the structure and function of vegetation, which can be easily detected by remotely sensed data (Chambers et al., 2007; Skole and Tucker, 1993; Spanner et al., 1994). Subtle changes arising from slight disturbances, such as those caused by drought, insect attack, and forest thinning, are more difficult to detect and map than the stand replacement changes (Chen and Cihlar, 1996; Myneni et al., 1997; Spanner et al., 1994; Vogelmann et al., 2009). Some long-term ecosystem changes, such as those caused by global warming, can best be detected using a broad detection scale and time period. These changes are best measured by statistical analysis because the changes may not be seen by simply comparing two images (Cohen et al., 2010; Kennedy et al., 2010; Westerling et al., 2006).

The number of satellite-based remotely sensed datasets available for analysis has increased markedly ever since Landsat data first became available in 1972, and now there is much information that can be used for land-cover monitoring investigations (Cohen and Goward, 2004). Recently, the archive of Landsat data became available to the public at no cost, making it more feasible to acquire and use large volumes of multitemporal imagery for monitoring land-cover and land-use change (Huang et al., 2010; Woodcock et al., 2008). These accumulated remotely sensed datasets are especially useful for monitoring long-term ecosystem effects.

During the past four decades, many change detection techniques have been developed and applied to assess land-cover changes. Several review papers and books have summarized and compared the various detection techniques (Canty, 2009; Coppin and Bauer, 1996; Coppin et al., 2004; Gao, 2008; Green et al., 1994; Kennedy et al., 2009; Lu et al., 2004b; Singh, 1989; Wulder and Franklin, 2007). The goals of this chapter are to summarize several important aspects of change detection, including data selection, data preprocessing, detection methods, national land-cover change projects, and future development. We emphasize the various objectives and advantages of change detection techniques found in the literature, especially the new and advanced ones frequently used and reported as being successful in recent publications.

11.2 DATA SELECTION

The success of using remotely sensed data for land-cover change detection depends on careful selection of the data source. The important attributes of remotely sensed data sources are spatial, temporal, spectral, and radiometric resolution (Lu et al., 2004b; Weber, 2001). Spatial resolution is an indication of the scale of observation (Woodcock and Strahler, 1987). It represents the size of the area on the ground from which the measurements were recorded in an image pixel. The smaller is the pixel, the higher is the likelihood of a sensor recording spatially fine details. Temporal resolution is related to how frequently a sensor can revisit the same location on the earth's surface. Spectral resolution refers to the spectral differences at wavelength intervals that a sensor is capable of detecting (Lillesand and Kiefer, 1994). The finer is the spectral resolution, the narrower is the wavelength range for a particular channel or band. Radiometric resolution is the smallest difference of the electromagnetic energy that can be detected by a sensor (Lillesand and Kiefer, 1994). The finer is the radiometric resolution of a sensor, the more sensitive is the sensor to detect small differences in reflected or emitted energy from the targets. The number of bits influences radiometric resolution properties. For example, Landsat Thematic Mapper (TM) and Enhanced Thematic Mapper Plus (ETM+) data are recorded as 8 bits and IKONOS data as 11 bits. Accordingly, the 11 bits of IKONOS can theoretically divide the spectral space into more bits than the 8 bits of Landsat.

Selection of remotely sensed data also depends on the targets of land-cover change analysis and needs to match the methods of land-cover change analysis. There is a trade-off between spatial resolution and temporal resolution. High spatial resolution images generally have the advantage of better geometric details of land cover at local scales. However, it is also generally more difficult to acquire good cloud-free images with these sensors for multitemporal analysis owing to low repeat coverage (i.e., low temporal resolution). Meanwhile, low spatial resolution images that have high-frequency revisit coverage are usually best for characterizing broad-scale phenomena that cover large areas. Conversely, high spatial resolution data are often used for detecting land-cover change information requiring high levels of spatial details. The low spatial resolution data over large areas generally cost less than the high spatial resolution data that cover small areas.

High (<10 m) and moderate (10–100 m) spatial resolution remotely sensed data, such as IKONOS, Landsat TM and ETM+, and Satellite Pour l'Observation de la Terre (SPOT), are usually used for local or regional assessments, but there are many cases where Landsat data have been used for regional to global assessments (Chen et al., 2002, 2004; Giri et al., 2007, 2011; Millward et al., 2006; Morisette et al., 2003; Rollins, 2009; Roy et al., 2010). The advanced very high resolution radiometer (AVHRR) and Moderate Resolution Imaging Spectroradiometer (MODIS) sensors on TERRA and AQUA have been widely used for routine monitoring of continental- to global-scale vegetation changes (Giri et al., 2005; Myneni et al., 1997; Zhan et al., 2002; Zhou et al., 2001). The advantages of available, daily, and low-cost imagery, albeit at low spatial resolution, have made it possible to routinely develop information on large-area land-cover change.

Comparison of the National Oceanic and Atmospheric Administration (NOAA) AVHRR and Landsat TM/ETM+ time series data demonstrated that both sensors can enable derivation of enhanced vegetation-related variables, such as trends in time and the shift in phenological cycles (Stellmes et al., 2010). The high temporal resolution of AVHRR data with coarse spatial resolution is particularly suitable for enhanced time series methods, whereas the Landsat data are better for revealing ecosystem changes occurring at a fine spatial scale (Stellmes et al., 2010). Important fine-scale land-cover changes cannot be captured by coarse-scale time series (Stellmes et al., 2010).

Remotely sensed data can be characterized as being either two-dimensional or three-dimensional on the basis of spatial dimensions. The two-dimensional data refer to "normal" images, such as Landsat and AVHRR data, which capture the characteristics of the land surface in X and Y directions. The three-dimensional data contain extra information in the vertical direction and are exemplified by lidar (light detection and ranging) data. Lidar data can provide information about the vertical structure and volume of the surface or vegetation canopy. Three-dimensional data can provide the vertical attributes of surface features, which can help detect various forest types and other land-cover types quickly based on their different vertical structures (Antonarakis et al., 2008; Lefsky et al., 1999; Zimble et al., 2003) as well as characterize land-cover changes (Rosso et al., 2006). Lidar data are collected as single points or profiles, so the land surface features collected are sampled with noncontiguous data rather than fully imaged data (Wulder and Franklin, 2007). Most lidar data contain a vertical resolution between 1 and 5 m depending on the instrument, flying status, and user needs. Several studies have demonstrated the great potential of using lidar data for change detection (Rosso et al., 2006; Vepakomma et al., 2008; Wulder et al., 2009; Zimble et al., 2003).

Interferometric Synthetic Aperture Radar (InSAR) images are sensitive to surface roughness, shape, and structure, so they can provide additional vertical information. Several studies have demonstrated that InSAR is useful for quantifying urban impervious surfaces and monitoring water-level changes in wetlands (Hong et al., 2010; Kim et al., 2009; Yang et al., 2009).

Many analysts use the same sensor, with the same radiometric and spatial resolution properties, with anniversary or very near anniversary acquisition dates for change detection; this is to ensure data consistency and eliminate the effects of unwanted external sources of variability such as sun angle variation and phenological differences (Lu et al., 2004b). However, data on seasonal difference, such as both leaf-on and leaf-off images, contain important information on the understory vegetation, and these can also be used to differentiate deciduous canopy and evergreen canopy (Yang et al., 2001). Thus, change analyses done using data at different times of the year can provide different, yet useful, information on change.

11.3 DATA PREPROCESSING

Although a digital change between two images can be related to a real change in land cover, it is also important to recognize that the change may be related to a range of other parameters, including image misregistration, differing atmospheric conditions, sensor differences, and different viewing conditions. Once the data have been selected, preprocessing is applied to minimize the effects of bias arising from various changes attributed to "noise" and instrument "artifacts." Among the various steps of data preprocessing for change detection, multitemporal image geometric correction and radiometric correction are the most important.

11.3.1 GEOMETRIC CORRECTION

Raw images usually contain certain geometric distortions relative to the platform, the sensor, the total field of view, the atmosphere, and the earth (Lillesand and Kiefer, 1994; Toutin, 2004). Geometric correction can remove or reduce the distortions caused by these factors so that the images can be correctly registered in a geographic information system (GIS). Inaccurate geospatial registration of either images or field inventory data can cause nonlinear effects of false detection owing to comparison of different land-cover features (Dai and Khorram, 1998; Le Hégarat-Mascle et al., 2005; Stow and Chen, 2002; Verbyla and Boles, 2000; Weber et al., 2008). If multiple images are used for change detection, then precise image geometric correction is essential.

Absolute geometric correction usually includes the use of collected ground control points to compensate for the spatial distortion of the uncorrected images. Sometimes, relative geometric correction methods are used to correct one image to a reference image, which is assumed to have a precise reference geometric system. High-resolution imagery usually needs to have high-accuracy ground control points for geometric correction, whereas nonparametric methods are suitable for low-resolution imagery (Wulder and Franklin, 2007). Toutin (2004) summarized several frequently used geometric processing methods, including both the nonparametric and the parametric methods (Toutin, 2004; Wulder and Franklin, 2007).

The geometric approaches currently available are not "perfect" because coregistration of multitemporal images always has associated residual errors in rectification models. The effect of misregistration on change detection was evaluated by Dai and Khorram (1998), and they concluded that a georegistration accuracy of less than one-fifth of a pixel is required to control the total change detection error of less than 10%. It was also found that among the seven Landsat TM bands, the near-infrared band was the most sensitive to misregistration when change detection is concerned.

11.3.2 RADIOMETRIC CORRECTION

Radiometric consistency among different remotely sensed datasets is difficult to attain because of differences in sensor characteristics, atmospheric condition, solar angle, sensor observation angle, and phenological characteristics (Chen et al., 2005a; Du et al., 2002; Song et al., 2001; Teillet et al., 2007; Weber, 2001). To make different datasets more comparable, it is important to apply radiometric corrections to the data (Chander et al., 2009b; Chen et al., 2005a, 2005b; Teillet et al., 2007).

Two types of radiometric corrections, the absolute correction and the relative correction, are commonly used to enable comparisons among remotely sensed images across sensors and across time (Dinguirard and Slater, 1999; Du et al., 2002; Schott, 1997; Schott et al., 1988; Song et al., 2001; Vicente-Serrano et al., 2008).

Absolute radiometric correction is aimed at extracting the true surface reflectance of scene targets on the surface of the earth. Generally, absolute radiometric correction is a two-step process. The first step is to convert the digital number (DN) of the sensor measurements to spectral radiance measured by satellite sensors (Chander et al., 2009a; Lillesand and Kiefer, 1994; Schott, 1997). The second step is to convert the sensor-detected radiance into ground surface reflectance using an atmospheric transmission model (Lillesand and Kiefer, 1994; Schott, 1997). This approach requires input of simultaneous atmospheric properties and sensor calibration parameters or the reasonable estimations of these parameters, which are difficult to obtain in many cases, especially for historical data (Chavez, 1996; Du et al., 2002; Masek et al., 2006; Song et al., 2001). A variety of methods have been developed to derive the atmospheric coefficients for absolute radiometric correction processes, such as the dark object subtraction (DOS) (Chavez, 1996; Song et al., 2001), the modified dense dark vegetation (MDDV) (Liang et al., 1997), and the second simulation of the satellite signal in the solar spectrum (6 S) (Masek et al., 2006; Vermote et al., 1997). Even after accurate absolute radiometric corrections and atmospheric corrections, multisensor images are not necessarily comparable because of variation in spectral and spatial resolution (Schroeder et al., 2006; Teillet et al. 1997, 2007).

Relative radiometric correction aims at reducing unexpected variation among multiple images by adjusting the radiometric properties of target images to match the radiometric properties of a reference image (Hall et al., 1991; Kennedy et al., 2010; Schroeder et al., 2006). In this approach, reflectance of invariant targets (e.g., urban, barren, and dense forests) within multiple scenes is used to facilitate interscene comparisons and generate normalization regression functions. If the correction works out as expected, images will appear to have been acquired from the same sensor, with the same calibration, and under the same atmospheric conditions. For relative radiometric correction, it is not imperative that images be corrected for surface reflectance. Some relative methods interpret the radiometric relationships between the target image and the reference image by linear regression (Chen et al., 2005a; Du et al., 2002; Elvidge et al., 1995; Schott et al., 1988; Song et al., 2001; Vicente-Serrano et al., 2008), whereas some others use orthogonal regression (Canty et al., 2004; Kennedy et al., 2010). Reference images are usually either the most recent scenes or the ones least affected by atmospheric effects and instrument artifacts (Vicente-Serrano et al., 2008). A variety of relative radiometric methods have been developed, including the use of pseudo-invariant features (PIF) (Salvaggio, 1993; Schott et al., 1988), automatic scattergram-controlled regression (ASCR; Elvidge et al., 1995), principal component analysis (PCA; Du et al., 2002), ridge method (Andréfouët et al., 2001; Song et al., 2001), multivariate alteration detection (MAD) method (Canty et al., 2004; Kennedy et al., 2010), and temporally invariant cluster (TIC) method (Chen et al., 2005a; Vicente-Serrano et al., 2008).

Besides absolute surface reflectance, the relative radiometric correction can be applied to raw DNs, radiance, and top-of-atmosphere (TOA) reflectance (Vicente-Serrano et al., 2008). Relative correction can correct the artifacts originating from atmosphere, sensor, and other sources in one process and is therefore widely used. These methods have some shortcomings. For example, the moisture changes in PIF can affect the accuracy of the approach, and the accuracy of isolating the pseudo-invariant features depends on the user's ability and knowledge (Salvaggio, 1993; Schott et al., 1988). With the ridge method, the identification of a regression function is based on the visual observation of the density ridge. If most of the collocated pixels contain subtle and systemic changes owing to factors such as phenological responses to different growth seasons, then the density ridge may contain biased distortions and the regression function may be difficult to identify or will contain bias errors. Subtle landscape changes can be "normalized away" in the process, if most pixels exhibit similar changes between observations.

Several studies have compared both absolute and relative correction methods (Schroeder et al., 2006; Song et al., 2001; Vicente-Serrano et al., 2008). Use of absolute correction alone decreased the consistency of the common scale of a nearly continuous 20-year Landsat TM/ETM+ image dataset, but relative normalization performed better (Schroeder et al., 2006). Teillet et al. (1997) found that normalized difference vegetation index (NDVI) was significantly affected by differences in spectral bandwidth and spatial resolution when they compared sensor-specific spectral band data derived from Airborne Visible InfraRed Imaging Spectrometer (AVIRIS) and spatial resolutions from Satellite Pour l'Observation de la Terre (SPOT) High Resolution Visible (HRV), Landsat TM, NOAA AVHRR, EOS (Earth Observation Satellite) MODIS, and Envisat Medium Resolution Image Spectrometer (MERIS). This demonstrated that after radiometric corrections, multisensor images were not necessarily comparable because of variation in spectral and spatial resolution (Schroeder et al., 2006; Teillet et al., 1997, 2007).

Generally, relative radiometric correction methods are simpler than absolute radiometric correction methods, and in several studies relative methods have provided satisfactory, consistent time series data for detecting land-cover changes (Andréfouët et al., 2001; Chen et al., 2005a, 2005b; Kennedy et al., 2010; Schroeder et al., 2006; Song et al., 2001). The choice of methods has not been settled and remains quite application-dependent.

11.3.3 OTHER CORRECTIONS

If the study area of interest is a mountainous region, a topographic correction is needed to reduce topographic effects. For example, slope and altitude variations induce significant changes in irradiance and upwelling radiance. Several methods can be used to reduce the slope-aspect effects: the Spectral reflectance Image Extraction from Radiance with Relief and Atmospheric correction (SIERRA) method (Lenot et al., 2009); the classical cosine correction method; the statistical, Minnaert, and C-correction approaches (Meyer et al., 1993); and the sun-terrain-sensor (SCS) model (Gu and Gillespie, 1998).

The bidirectional reflectance distribution function (BRDF) describes the differences in surface reflectance when the measurement is under different view zenith angles, solar zenith angles, and relative azimuth angles (Los et al., 2005; Susaki et al., 2004; Vierling et al., 1997). If the BRDF is known, the reflectance observation from different viewing and illumination angles can be corrected to a standard view and illumination geometry, which can exclude false land-cover changes (Los et al., 2005). Various BRDF models, including the SIERRA model and the Ross-Thick/Li-Sparse-Reciprocal (RTLSR) kernel-driven model, have been applied to AVHRR, MODIS, and the airborne hyperspectral imagery to generate nadir BRDF-adjusted reflectance (Lenot et al., 2009; Los et al., 2005; Privette et al., 1997; Román et al., 2009; Shepherd and Dymond, 2000). Implementing a BRDF model is problematic because different land-cover types, such as bare soil, open canopy vegetation, urban, and agriculture, have different BRDF characteristics, and the model parameters need to be evaluated before any application. The use of an inappropriate BRDF model, especially with strong angular effects, can introduce large reflectance level errors to the imagery (Lenot et al., 2009). Once the appropriate geocorrection, radiometric correction, and other necessary corrections are applied, the image data are ready for change detection analysis.

11.4 LAND-COVER CHANGE DETECTION METHODS

There are two types of remote-sensing change detection: map-to-map comparison and image-to-image comparison (Coppin et al., 2004; Green et al., 1994; Singh, 1989). In map-to-map comparison, individual land-cover maps are generated independently using different dates of imagery, and then the results are compared. The overall effectiveness of this approach depends on the classification accuracy of the images on two different dates. The actual differences in land cover can be influenced by many factors, including different classification systems and different mapping techniques

(Mas, 1999; Muchoney and Haack, 1994). The image-to-image comparisons involve analyzing the spectral characteristics of two or more images and identifying the actual spectral differences caused by the variables of interest (Coppin et al., 2004). Many different image-to-image comparisons have been successfully employed, and some of these are further described by Coppin et al. (2004). Some methods can provide only change or no-change detection results, whereas others can provide a complete matrix of change directions (Giri et al., 2007; Masek et al., 2008; Xian et al., 2009; Yuan et al., 2005). In this chapter, we focus our discussion on image-to-image comparison since it is more widely used.

The change detection methodologies are not independent of the data sources, so investigation of the data sources is important before selecting any usable detection approaches. In addition, different methods can also generate different change maps even using the same data. In the following sections, we describe several important categories for change detection. This does not imply any ranking or qualitative judgment. Some detection methods are not introduced in this chapter owing to the limitation of the chapter length and the currency of the methods. We recommend that interested readers refer to other previously published review articles (Canty, 2009; Coppin et al., 2004; Lu et al., 2004b; Singh, 1989; Wulder and Franklin, 2007).

11.4.1 SPECTRAL INDICES

Spectral indices derived from satellite data are widely used for land-cover change studies. They can reduce the data volume for processing and analysis and provide combined information that is more strongly related to changes in the scene than any single band (Coppin et al., 2004).

NDVI is a widely used vegetation index, which can reduce atmospheric and illumination effects by using the difference and the ratio of red and near-infrared bands (Rouse et al., 1974; Schott, 1997). NDVI values strongly correlate with green vegetation, and changes in NDVI indicate changes in biological activities (Chen et al., 2005a; Verbesselt et al., 2010; Yang et al., 1997; Zhou et al., 2001). NDVI decreases significantly after green biomass is removed, so it is widely used for mapping and monitoring fire disturbance, forest clear-cut activity, urbanization, and other land-cover changes (Chen et al., 2005b, 2006; Díaz-Delgado et al., 2003; Escuin et al., 2008; Hayes and Sader, 2001; Lunetta et al., 2006; Masek et al., 2008; Verbesselt et al., 2010; White et al., 1996).

The enhanced vegetation index (EVI) is calculated by using the reflectance of blue, red, and near-infrared bands (Huete et al., 2002; Miura et al., 2001). It was developed to contain the correction of canopy background and atmospheric scattering effects (Gao et al., 2000; Miura et al., 2001; Xiao et al., 2003). EVI is more sensitive in high biomass regions than NDVI and is strongly responsive to canopy structure characteristics (Chen et al., 2004, 2005a, 2011a; Huete et al., 2002; Pocewicz et al., 2007). EVI has been used for postfire forest regeneration and phenological analysis during change detection (Chen et al., 2005a, 2011b; Ganguly et al., 2010; Liang et al., 2011; Lupo et al., 2007).

The normalized burn ratio (NBR) is a spectral index that normalizes the reflectance of near-infrared (Landsat band 4) and mid-infrared (Landsat band 7) bands to monitor fire-affected areas (García and Caselles, 1991; Key and Benson, 2006). Since 2001, the change in NBR between two images (dNBR) has been used to map burned areas in the United States using pre- and postfire Landsat imageries for the Monitoring Trends in Burn Severity (MTBS) project (Eidenshink et al., 2007; Key and Benson, 2006). In addition to mapping burned areas, NBR has also been used to interpret burn severity and postfire vegetation regeneration, which reveals the magnitude of postfire ecological change (Chen et al., 2011a; Epting et al., 2005; Escuin et al., 2008; García and Caselles, 1991; Hall et al., 2008; Key and Benson, 2006; Soverel et al., 2010; Veraverbeke et al., 2010; Wimberly and Reilly, 2007; Wulder et al., 2009).

PCA is a linear transformation depending on the statistical relationships among pixel values rather than on the physical characteristics of the scene (Collins and Woodcock, 1994). The data axes are rotated into principal axes, or components, that represent the maximum data variance (Muchoney and Haack, 1994). When the PCA is used to detect changes between images, the proportion of change in an image must be relatively small so that the statistical analysis produces meaningful results (Collins and Woodcock, 1994). The advantage of PCA has been well illustrated in several land-cover change studies (Bateson and Curtiss, 1996; Cakir et al., 2006; Du et al., 2002; Mas, 1999; Millward et al., 2006).

The tasseled cap (TC) algorithm transforms the Landsat bands into three major characteristics: soil brightness, vegetation greenness, and soil vegetation wetness (Crist, 1985). The changes in these indices over time can be used to detect land-cover changes. The TC transformation parameters are independent of the image scenes, and this has been reported in previous studies (Crist, 1985; Huang et al., 2005). Fung's (1990) research indicated that most land-cover changes were reflected in terms of changes in brightness and greenness and thus were captured by the first two TC variables. The increase in greenness over time indicated an increase of vegetation, and the increase in brightness indicated an increase in bare soil or urbanization (Lunetta et al., 2004). The wetness is sensitive to surface moisture changes and can be used in detecting forest disturbances (Jin and Sader, 2005). The wetness index also has a strong and positive relationship to forest stand age, which can be used as a healthy forest growth indicator (Wulder et al., 2004). MODIS TC was also developed to align with the TM TC for maintaining continuity among sensors (Lobser and Cohen, 2007).

The disturbance index (DI) was derived from the Landsat TC data to record the normalized spectral distance of an investigated pixel from a nominal "mature forest" class to a "bare soil" class (Healey et al., 2005; Masek et al., 2008). It was originally designed to detect the unvegetated spectral signatures and the stand-replacing disturbance from all other forest changes (Healey et al., 2005). DI is derived from the statistics of forest reflectance from individual scenes, so it is relatively insensitive to the variation of solar geometry, BRDF effects, and vegetation phenology among multitemporal scenes (Masek et al., 2008). DI is adapted by the Landsat Ecosystem Disturbance Adaptive Processing System (LEDAPS) project to produce wall-to-wall maps of the stand-clearing forest disturbance and regrowth for the North American continent using early (~1990) and late (~2000) images (Masek et al., 2008). This detection method works best for dark closed-canopy forests and has certain limitations in areas with sparse tree cover owing to the difficulty in acquiring a "mature" forest signal from the scene (Masek et al., 2008).

The integrated forest index (IFI) is a recently developed spectral index representing the probability that a pixel is a forest-cover type based on the whole-scene statistical analysis (Huang et al., 2008, 2009; Masek et al., 2008). The IFI value for a pixel is calculated by its normalized distance to the center of forest training pixels in a multiple dimensional spectral space (Huang et al., 2008). The IFI is an inverse measure; the lower is the pixel's IFI value, the more likely is it a forest pixel (Huang et al., 2008). The forest training pixels are identified using local histogram spectral windows within the image. This index has been integrated into the automated vegetation change tracker (VCT) model for forest-cover change detection using Landsat time series stacks (LTSS) (Chen et al., 2011b; Huang et al., 2008, 2009, 2010; Masek et al., 2008; Thomas et al., 2011).

11.4.2 Spectral Mixture Analysis

In spectral mixture analysis (SMA), the signal recorded for a pixel is assumed to be a mixture of the radiances of the component end-members contained within that pixel. Knowing or deriving spectrally "pure" end-members of all the components within a pixel allows one to quantify the end-member fractions occurring within the pixel, using linear or nonlinear mixture approaches (Bateson and Curtiss, 1996; Byambakhuu et al., 2010; Chen et al., 2004; Foody and Cox, 1994; Holben and Shimabukuro, 1993; Ray and Murray, 1996). In the linear (i.e., first-order) approach, the mixed spectrum can be expressed by the linear combination of the spectra of the pure components, based on their fractional area. In linear spectral unmixing, the number of resolvable end-members in the inverted function is limited by the number of spectral bands (b), and the maximum number of end-members that can be derived is $b + 1$. The nonlinear mixture method considers second-order mixture effects such as photon scattering among components. It is based on the assumption that the reflected signal arises from nonlinear mixing among pixel end-member components. Although nonlinear mixture modeling can reduce the residual term and improve accuracy, estimating component fractions by using this approach is more complex and is difficult to simulate than when assuming linear mixtures (Ray and Murray, 1996).

Some previous land-cover change studies have aimed at detecting changes that are smaller than individual pixels. The SMA techniques were used to quantify cover fractions of the interested ground components such as forest canopy, pasture, second growth, impervious surface, and damaged vegetation (Adams et al., 1995; Chambers et al., 2007; Lu et al., 2004a; Yang et al., 2003). Image end-members were developed and used to unmix the multitemporal images into end-member fractions. Fraction image differencing results were then compared among multi-images to analyze land-cover change detection.

SMA is especially appropriate and practical for detecting image-element changes over time, using coarse-resolution images. The subtle natural ecosystem changes, such as vegetation regeneration and thinning, are relevant concerns. The spectral mixture analysis is scene-independent of the field training data because the end-members can be selected from the individual scene. Thus, multi-temporal fraction images can be effectively used for land-cover change detection without radiometric correction among scenes. Spectral end-members in one region that have been found to produce good estimates may not work well for another region. Therefore, careful examination and comparison of results from this method and other available methods are necessary before using the method across a wide range of areas, multiple surface conditions, and various datasets.

11.4.3 BITEMPORAL CHANGE DETECTION

Bitemporal change detection enables comparison of land cover of the same area, based on a two-point time-scale. The method requires careful selection of dates because the detected changes may reveal differences in phenology and not the feature differences of interest (Weber, 2001). Weber suggested the use of environmental criteria, especially growing degree days and accumulated precipitation, for performing match calculations so that appropriate remotely sensed images can be selected for land-cover change detection. For bitemporal change detection, images from the summer peak greenness period work best because they minimize the reflectance difference from the same cover type, caused by seasonal vegetation phenology, such as leaf-off conditions, autumn coloration, and sun angle difference (Coppin et al., 2004). In addition, different cover types tend to be the most spectrally stable and comparable during peak summer (Yang et al., 2001). Even if the image data are collected on anniversary dates or in seasons of peak summer, some factors may still affect the spectral signals and add "noise," including the variation in precipitation, temperature, and atmospheric conditions, during the change detection.

Bitemporal detection often uses one image to subtract another. This can be done either using the "original" image information (e.g., radiance or reflectance data) or derived imagery (e.g., spectral indices, unmixing fractions). Both images used need to be georegistered and radiometrically corrected (Coppin et al., 2004). The positive and negative values represent the change in two different directions, and the zero values represent no change. In reality, thresholds are often used to identify change and no-change areas. The thresholds can be selected using interactive and/or manual procedures or through statistical reports (Lu et al., 2004b). The threshold selection requires the skills of an analyst in order to exclude the external influences caused by atmospheric conditions, sun angle, soil moisture, and phenology dynamics. Bitemporal differences of NDVI, EVI, NBR, and DI have been widely used to detect vegetation change (Chen et al., 2011b; Escuin et al., 2008; García and Caselles, 1991; Hall et al., 2008; Hayes and Sader, 2001; Masek et al., 2008; Soverel et al., 2010; Wulder et al., 2004).

The detection time period varies for different detection targets. Long time periods are often best for describing long-term changes such as forest stand-clearing disturbances (e.g., logging, fire). After clear-cut harvest, the detectable forest recovery period is about 10–11 years for the remotely sensed spectral recovery (Cohen et al., 2010; Masek et al., 2008; Wulder et al., 2004). Landsat intervals of up to 5 years may be nearly as accurate for detecting forest change at 1- or 2-year inter-vals in some areas (Jin and Sader, 2005). In some events, there are no prominent differences in spec-tral signals to characterize the land-cover changes, such as partial harvest, insect damage, thinning, storm damage, and ground fire. These disturbance events are often difficult to discern, so detectable periods in these cases may need to be shorter, such as 1–2 years, to minimize the detection errors (Jin and Sader, 2005; Masek et al., 2008). Lunetta et al. (2004) compared detection results with dif-ferent time intervals using near-anniversary Landsat 5 TM data. The study results demonstrated that a minimum of a 3- to 4-year temporal data acquisition frequency was required to detect land-cover change events in north central North Carolina.

One of the major limitations of the bitemporal detection approach is that it uses only two dates of imagery in the process. Thus, neither is there a way to separate the older disturbances from the more recent ones, nor are there clues about when the disturbances occurred during the detec-tion period. Many applications demand temporally more detailed information on landscape trends, which requires analysis of more datasets acquired at regular time intervals.

11.4.4 Multitemporal Change Detection

Multitemporal change detection, also called time-trajectory analysis, compares the land cover of the same area over long time intervals with multiple imagery (Coppin et al., 2004). Multitemporal change detection typically needs to have sufficiently long records of data to capture the variability or trends due to land-cover changes. Two long-term archives of satellite data, Landsat and AVHRR, meet the requirements for time trajectory analysis at local and regional scales (Stellmes et al., 2010). The Landsat program has provided invaluable global data with 30 m × 30 m resolution since the launch of the first Landsat satellite on July 23, 1972. The Landsat program has provided the longest-running time series of systematically consistent remotely sensed data at medium resolution—a great benefit for monitoring the earth's surface characteristics (Cohen and Goward, 2004). The archive of Landsat data has been made available to the public at no cost, making it possible to acquire a large volume of multitemporal images for monitoring land-cover and land-use change (Woodcock et al., 2008). The Web-Enabled Landsat Data (WELD) project provides 30-m composites of Landsat ETM+ mosaics at weekly, monthly, seasonal, and annual periods for the conterminous United States (CONUS) and Alaska (Roy et al., 2010). In addition, NOAA's AVHRR has continued to acquire data at 1-km resolution since 1978, and these data have been widely used for multitemporal change detec-tion (Pouliot et al., 2011; Reed, 2006). Some other sensors, such as ASTER, MODIS, MERIS, and SPOT, cover a shorter time period, about 10 years, and can also be used for multitemporal change detection (Fensholt et al., 2009; Stellmes et al., 2010). The MODIS Global Land Cover Dynamics Products are recently available to users for investigating changing surface conditions relative to climate forcing, disturbance, and human management (Ganguly et al., 2010). All these products are ready for use in multitemporal change detection from regional to global scales.

The high temporal resolution data can better capture phenological characteristics and partially compensate for the coarse spatial resolution. Coarse-scale hypertemporal data will be suitable for monitoring land-cover change across large areas and identifying areas of interest for further investigation using fine-resolution data (Stellmes et al., 2010). The comparison between time-series AVHRR and TM/ETM+ data (Stellmes et al., 2010), as well as MODIS and Landsat data (Fisher and Mustard, 2007), indicates that time-series analysis derived from different sensor systems can yield comparable results regarding the direction of trends and their spatial patterns (Ganguly et al., 2010; Stellmes et al., 2010). Time-trajectory coarse-resolution data, such as AVHRR and MODIS data, have been shown to be very powerful for assessing inter- and intraseasonal phenological phenomena (Chen et al., 2001; Ganguly et al., 2010; Reed, 2006; Reed and Yang, 1997; Yang et al., 1997).

Change vector (CV) analysis is a detection method for identifying the nature and magnitude of land-cover change in a multitemporal feature space (Coppin et al., 2004; Lambin and Strahlers, 1994a). The change vector tool compares biophysical indicators, such as the NDVI, in the time trajectory. The vector difference between successive time trajectories is calculated as a vector in a multitemporal feature space. The length of the change vector represents the magnitude of the inter-annual change, and the direction represents the nature of the change (Lambin and Strahlers, 1994a). This approach can be easily extended to other biophysical indicators such as surface temperature and various spectral indices (Lambin and Strahlers, 1994a; Lu et al., 2004b; Xian et al., 2009). A suitable threshold is usually used to determine the change or no-change area.

VCT is a highly automated algorithm that can detect forest disturbance and postdisturbance recovery history using LTSS (Huang et al., 2010). The LTSS is an annual or biennial temporal sequence of Landsat images acquired during the peak growing season over a path/row tile of the World Reference System (WRS). The VCT approach contains two processing steps (Huang et al., 2008, 2009, 2010). The first step is to clip all images by common area, generate a cloud and shadow mask, and calculate spectral indices for individual images. The spectral indices include the NDVI, IFI, and NBR. In the second step, the indices and masks are analyzed on the basis of the spectral-temporal characteristics of land cover and are used to derive disturbance maps. Postdisturbance processes are also tracked using the spectral trajectory in the detected disturbance area. The VCT can detect most stand-clearing disturbances and some non-stand-clearing events. The most detectable changes include forest harvest, fire, and urban development, as well as some thinning and selective logging (Huang et al., 2010; Thomas et al., 2011). The VCT has been used at many locations across the United States, and the overall accuracies are about 80% for disturbances mapped at individual year level (Huang et al., 2010). It has also been used to assess forest change and fragmentation in Alabama and Mississippi (Li et al., 2009a, 2009b). LANDFIRE updating and analysis process also uses the VCT to provide land-cover disturbance history (Vogelmann et al., 2011).

Landsat-based Detection of Trends in Disturbance and Recovery (LandTrendr) is a recently developed change detection tool to process and analyze yearly LTSS to identify both abrupt disturbance and long-term change induced by human and natural processes (Kennedy et al., 2010). The LandTrendr uses straight-line segments to simplify the key features of the spectral trajectories, and a land-cover change map can be generated based on the starting and ending points of segments (Kennedy et al., 2010). The LandTrendr can detect multiple changes, including insect-induced mortality, insect-induced damage followed by fire, clear-cut harvest, stability followed by fire, and recovery from earlier fire damage (Kennedy et al., 2010). In addition, an image time series visualization and data collection tool, TimeSync, was developed for calibrating and validating LandTrendr performance using human interpretation of spectral trajectories (Cohen et al., 2010). This tool consists of four major components: an image chip window, a spectral trajectory window, Google Earth, and a Microsoft Access database (Cohen et al., 2010). The outputs of these two independent tools, LandTrendr and TimeSync, indicated that the overall accuracy interpreted by TimeSync was over 90% in 388 forested plots (Cohen et al., 2010). Detection of medium- and low-intensity disturbances is improved when compared with previously available methods using coarser time density image data (Cohen et al., 2010).

Phenology cycle analysis is another type of multitemporal change detection. The measurement of land-cover phenological characteristics can help separate the surface normal phenology conditions from the variation caused by land-cover change or climate change (de Beurs and Henebry, 2004; Reed and Yang, 1997; Stellmes et al., 2010). The variables of phenology, such as the start of season, growing-season length, and overall growing-season productivity, have a strong relationship with vegetation cover types. If the time series data are dense enough and cover a long time period, it is also possible to detect and separate the gradual and abrupt changes in vegetation cover (Stellmes et al., 2010). The seasonal variation of spectral indices, especially those strongly relative to vegetation performance, can be used to interpret the vegetation phenology.

11.4.5 INTEGRATION OF MULTIPLE SOURCE DATA AND MULTIPLE DETECTION METHODS

No single data source or detection method will be effective in all environments with respect to change detection. Before the start of a new investigation, we recommend conducting preliminary comparative analyses to understand better the limitations and strengths of the various datasets and methods for the study sites of interest. Several previous studies have evaluated numerical detection methods and provided some guidelines on selecting the detection approach. Epting et al. (2005) evaluated 13 spectral indices across four wildfire burn sites in Alaska. They found that the NBR had the highest correlation with the field-based composite burn index (CBI) estimates using both post-burn and pre-/postburn approaches. In an earlier study by Muchoney (1994), a number of change detection techniques were compared, including PCA, image differencing, spectral-temporal (layered temporal) change classification, and postclassification change differencing. The results indicated that image differencing and PCA were the best approaches to determine forest defoliation (Muchoney and Haack, 1994). Yuan and Elvidge (1998) systematically tested 75 change detection methods and concluded that the band-differencing techniques, based on automated scattergram-controlled regression (ASCR) normalization and NDVI, outperformed most other techniques. Many more comparative studies can be found in the literature (Cakir et al., 2006; Chen et al., 2011b; Fung, 1990; Lambin and Strahlers, 1994b; Lyon et al., 1998; Macleod and Congalton, 1998; Michener and Houhoulis, 1997; Millward et al., 2006; Yuan and Elvidge, 1998).

Multiple data sources and detection methods can be integrated and used for change detection. For example, Wulder et al. (2009) integrated lidar data and multitemporal Landsat data to provide improved opportunities for detecting postfire conditions. Millward et al. (2006) used remotely sensed data from three different sensors, TM, ETM+, and SPOT, to perform time series analysis to assess land-cover change over a 12-year period. Zhan et al. (2000) selected five change detection algorithms, including three spectral methods and two texture methods, to create a voting system for generating confidence in the change detection products.

Independent training and validation data need to be used for accuracy assessment. Field-collected data and high-resolution aerial photos are often used to assess accuracy or help set up thresholds for change and nonchange areas (Chen et al., 2011b; Epting et al., 2005; Hall et al., 2008; Lunetta et al., 2004). Sometimes, manual evaluation through visual comparisons and the analyst's knowledge of the region can be used if field information is lacking (Masek et al., 2008). Selection of the detection approach depends on the project goals and on whether the benefits of higher accuracy from integration of multiple methods outweigh the cost of the additional training data and computation time.

11.5   NATIONAL AND GLOBAL LAND-COVER CHANGE DATASETS

Several projects generate the National Land-Cover Database (NLCD) and the change databases. The NLCD provides land-cover data for the United States (http://www.mrlc.gov/nlcd.php). The currently available database includes the NLCD 1992, NLCD 2001, NLCD 2006, and land-cover change maps of 1992–2001 and 2001–2006 for the United States. The NLCD products indicated that 2.99% of the land cover was mapped as changed from 1992 to 2001 (Fry et al., 2009), and less than 2% was changed from 2001 to 2006 (Fry et al., 2011). Overall land-cover thematic accuracies at Anderson Level II and Level I were 58% and 80% for NLCD 1992, and 78.7% and 85.3% for NLCD 2001 (Wickham et al., 2010). In addition, NLCD 2001 was used as the baseline to generate NLCD 2006 by extracting and updating changed areas using pairs of Landsat scenes in the same season in 2001 and 2006 (Xian et al., 2009). Multi-Index Integrated Change (MIIC) was used for the change detection, which is an integration method using NBR, NDVI, CV, and a relative CV (Fry et al., 2011; Jin et al., 2010; Xian et al., 2009).

The Coastal Change Analysis Program (C-CAP) was initiated by NOAA to provide a national land-cover and land-change database for the coastal regions of the United States (Dobson et al., 1995; Portolese et al., 1998). The differences in bitemporal satellite imagery were used to detect upland and tidal land-cover change (Portolese et al., 1998). The thresholds were derived from aerial photos and field data to generate land-cover change/no-change masks (Portolese et al., 1998). The currently available dates for coastal land-cover maps are 1992, 1996, 2001, and 2005 (http://www.

csc.noaa.gov/digitalcoast/data/ccapregional/index.html). This program monitors habitats in coastal intertidal areas, wetlands, and adjacent uplands. Land-cover and land-change maps are provided every 1–5 years, and the monitoring cycle depends on the rate and magnitude of change in the study regions. The C-CAP program can improve our understanding of coastal ecosystems and provide a feedback to habitat managers on management policies and programs (Dobson et al., 1995).

Land Cover Trends (LCT) is a research project using satellite images and other data to assess the land-cover/land-use change rates, causes, and consequences between the early 1970s and 2000 in the United States (Loveland et al., 1999). In this project, a hybrid of available change detection approaches and a statistical sampling approach was used for change detection based on an ecoregion framework (Gallant et al., 2004; Stehman et al., 2003; Loveland et al., 1999, 2002). The selected specific methods depended on the characteristics of the specific ecoregion. Automated approaches were combined with manual interpretation to generate reliable products (Loveland et al., 2002; Sohl et al., 2004). The LCT focuses on the geographic understanding of regional and national land change across the United States and provides valuable information for managing environmental and natural resources (http://landcovertrends.usgs.gov/).

Monitoring Trends in Burn Severity (MTBS) is a fire-occurrence and burn-severity database (http://www.mtbs.gov/). This dataset provides burn severity data for perimeters of fires greater than 200 ha in the eastern United States and 400 ha in the western United States from 1984 to the present. Fire perimeters and burn severity products at 30-m resolution were generated from the comparison of pre- and postfire Landsat imagery (Eidenshink et al., 2007). The differences of pre- and postfire NBR were calculated and compared with field inventory data, the CBI, to identify the burn severity. These fire records also provide study sites to monitor fuel consumption and postfire landscape recovery over time (Chen et al., 2011a; Eidenshink et al., 2007). The MTBS database can be used to evaluate the environmental impacts due to large wildland fires and to improve land management in the United States (Chen et al., 2011a; Eidenshink et al., 2007).

Global land-cover maps are important for assessing global land-cover change. Currently, there are several global land-cover products, such as the International Geosphere-Biosphere Programme Data and Information System (IGBP-DIS), the MODIS global land-cover products, University of Maryland (UMD) global land-cover products, Global Land Cover 2000 (GLC2000), the GlobCover Land Cover, the global mangroves forest, and the gross forest-cover loss (GFCL) datasets. The IGBP-DIS used AVHRR data from 1992 to 1993 to generate the 1-km land-cover data for global terrestrial surfaces (Loveland and Belward, 1997; Loveland et al., 2000). The MODIS global land-cover products provide yearly land-cover type, land-cover dynamics, and vegetation continuous fields at 1-km resolution to study land-cover changes (Friedl et al., 2002; Ganguly et al., 2010). The 1-km resolution UMD global land-cover products were generated using AVHRR from 1992 to 1993 (Hansen and Reed, 2000). The GLC2000 products are at 1-km resolution and were generated on the basis of the images collected in 2000 by the VEGETATION sensor on-board SPOT 4 and a few other earth-observing sensors (Bartholome and Belward, 2005). The GlobCover Land Cover v2 product is a global land-cover map at 300-m resolution, which was derived from a time series of MERIS sensor image composites from 2004 to 2006 (GlobCover, 2011). The global mangrove forests were mapped at 30-m resolution by using Global Land Survey (GLS) data and the Landsat archive (Giri et al., 2011). The GFCL was estimated from 2000 to 2005 using MODIS and ETM+ satellite data (Hansen et al., 2010). Some comparative studies of these global land-cover and land-cover change datasets have been done to address their strengths and weaknesses (Giri et al., 2011; Hansen and Reed, 2000; Herold et al., 2008; Jung et al., 2006; Latifovic et al., 2004).

11.6 FUTURE DIRECTIONS

New change indicators or algorithms will continue to be developed, and the capacity for change detection will be enhanced in the future. Since the neighbor objects in nature tend to be correlated with each other, the spatial context and adjacent pixel information can be used to improve accuracy and reliability of change detection. Zhang et al. (2007) combined canonical correlation analysis and contextual Bayes decision for change detection using bitemporal images. There are also few recently developed methods such as temperature and spatial structure indicators (Lambin and Strahlers, 1994b) and MODIS tasseled cap indices (Lobser and Cohen, 2007). Fuzzy models, which consider uncertainty, may be a new direction in change detection. Fuzzy models allow the analyst to be specific about minimum, maximum, and average extents of land-cover types, to report the fuzzy area itself as a fuzzy number, and to justify descriptive qualifications of the results (Fisher, 2010). This can provide richer information and is especially useful when the ecosystem changes are operating at a scale finer than the spatial resolution of the sensor.

Continuity of data systems has been recognized as a major concern for future efforts in numerous applications including change detection analysis (Bailey et al., 2007). Both Landsat-5 and -7 would continue to collect data until December 2012 when the Landsat Data Continuity Mission (LDCM) is scheduled for launch (Wulder et al., 2011). Both Landsat sensors have experienced operating problems earlier; therefore, temporal and spatial discontinuities of Landsat data are likely if one or both of them fail before the launch of LDCM (Wulder et al., 2008, 2011). Multiple, international sources of data, such as the Indian Remote Sensing (IRS) Resourcesat-1 and CBERS (China–Brazil Earth Resources Satellite), provide Landsat-like data (Chander, 2007; Wulder et al., 2008). The data from these sensors can be potentially incorporated into existing analyses to help bridge a possible gap in Landsat data continuity. In addition, MODIS data have been evaluated and compared with NOAA AVHRR and have been used to generate multitemporal composite data for land-cover change mapping (Batra et al., 2006; Chuvieco et al., 2005; Gallo et al., 2005; Ressl et al., 2009; Stellmes et al., 2010). The Visible and Infrared Imaging Radiometer Suite (VIIRS), as part of the National Polar-Orbiting Operational Satellite System (NPOESS), can be considered the operational successor to AVHRR and MODIS (Townshend and Justice, 2002). These coarse-resolution sensors together will continuously support weather forecasting, long-term climate research, and global change detection.

As different types of satellite data become more accessible in the future, change detection using multisource data will become a key area of research and development. For example, multisource GIS data have been integrated into existing protocols for change detection applications and analyses, such as automatic change detection of road networks, areas, and terrain features (Li, 2010). Image analysis and display systems have been developed to integrate graphical user interfaces, database management systems, and spatial statistics (Castilla et al., 2009). The spatial patterns of changed areas can be directly converted to GIS shape-files for display and used for further statistical and management applications (Castilla et al., 2009). Remotely sensed data are routinely used as part of the GIS-based forest inventory. For instance, multidate Landsat data have been used to estimate stand age after forest harvest in a regenerating lodgepole pine (*Pinus contorta*) forest (Wulder et al., 2004). Change detection data can also be integrated into biogeochemical models for assessing forest net ecosystem productivity and ecosystem carbon flux (Goward et al., 2008; Masek and Collatz, 2006).

In addition, it is important to further develop and refine automated change detection methodology and algorithms. This becomes particularly relevant because more image datasets are being acquired, but current approaches can be quite time-consuming. Automation of the image change analysis process can save much time and effort and provide important information for further analysis (Castilla et al., 2009; Cohen et al., 2010; Dai and Khorram, 1997; Huang et al., 2010; Li, 2010; Yuhaniz and Vladimirova, 2009).

11.7 SUMMARY

This chapter summarizes recent literatures on data selection, data preprocessing, methods, and the future directions for image-based change detection investigations. There are no universal methods that can be applied for all data sources; different data sources and different change detection objectives may require different methods of analyses. In general, we consider that various approaches and data sources can be used together and that they will complement each other well.

Analysts should consider several important suggestions and observations when applying remotely sensed data to land-cover change detection:

1. The spatial, temporal, spectral, and radiometric resolution characteristics of data sources and the influence of phenology are all key variables that need careful consideration during data selection.
2. Geometric and radiometric corrections are required to ensure that the observed changes are "real" changes occurring on the land surface.
3. It is especially advantageous to comprehensively review and test several change detection methods and then select a few for further investigation based on empirical evidence.
4. The complexity of the approach does not necessarily guarantee improvement in the accuracy of the final change results. Depending on the goals of the investigation, very good results can be obtained using very simple methodology.
5. Preselect the appropriate change detection methods based on the desired outcome and the accuracy requirements. A comparison of the preselected change detection methods followed by an integration of the best ones is the most effective way to detect land-cover change. This can provide consistent and high-accuracy detection datasets for multiple applications.

ACKNOWLEDGMENTS

This study was made possible in part by ASRC Research and Technology Solutions (ARTS) under U.S. Geological Survey contracts 08HQCN0007 and G08PC91508. We thank Dr. Shengli Huang and Dr. Kevin Gallo, who reviewed an earlier draft of the chapter and offered suggestions for improving the manuscript. Any use of trade, product, or firm names is for descriptive purposes only and does not imply endorsement by the U.S. Government.

REFERENCES

Adams, J.B., Sabol, D.E., Kapos, V., Filho, R.A., Roberts, D.A., Smith, M.O., and Gillespie, A.R. 1995. Classification of multispectral images based on fractions of endmembers: Application to land-cover change in the Brazilian Amazon. *Remote Sensing of Environment*, 52, 137–154.

Andréfouët, S., Muller-Karger, F.E., Hochberg, E.J., Hu, C., and Carder, K.L. 2001. Change detection in shallow coral reef environments using Landsat 7 ETM+ data. *Remote Sensing of Environment*, 78, 150–162.

Antonarakis, A.S, Richards, K.S., and Brasington, J. 2008. Object-based land cover classification using airborne Lidar. *Remote Sensing of Environment*, 112, 2988–2998.

Bailey, G.B., Berger, M., Jeanjean, H., and Gallo, K.P. 2007. The CEOS constellation for land surface imaging. In *Sensors, Systems, and Next-Generation Satellites XI*, Florence, Italy, September 17–20, 2007, Proceedings of SPIE, Vol. 6744: Bellingham, Washington, Society of Photo-Optical Instrumentation Engineers (SPIE), article number 674425. Available at: http://dx.doi.org/10.1117/12.740854

Bartholome, E. and Belward, A.S. 2005. GLC2000: A new approach to global land cover mapping from Earth observation data. *International Journal of Remote Sensing*, 26, 1959–1977.

Bateson, A. and Curtiss, B. 1996. A method for manual end member selection and spectral unmixing. *Remote Sensing of Environment*, 55, 229–243.

Batra, N., Islam, S., Venturini, V., Bisht, G., and Jiang, L. 2006. Estimation and comparison of evapotranspiration from MODIS and AVHRR sensors for clear sky days over the southern Great Plains. *Remote Sensing of Environment*, 103, 1–15.

Byambakhuu, I., Sugita, M., and Matsushima, D. 2010. Spectral unmixing model to assess land cover fractions in Mongolian steppe regions. *Remote Sensing of Environment*, 114, 2361–2372.

Cakir, H.I., Khorram, S., and Nelson, S.A.C. 2006. Correspondence analysis for detecting land cover change. *Remote Sensing of Environment*, 102, 306–317.

Canty, M.J. 2009. *Image Analysis, Classification, and Change Detection in Remote Sensing: With Algorithms for ENVI/IDL*, 2nd edition. Boca Raton, FL: CRC Press.

Canty, M.J., Nielsen, A.A., and Schmidt, M. 2004. Automatic radiometric normalization of multitemporal satellite imagery. *Remote Sensing of Environment*, 91, 441–451.

Castilla, G., Guthrie, R.H., and Hay, G.J. 2009. The Land-cover Change Mapper (LCM) and its application to timber harvest monitoring in western Canada. *Photogrammetric Engineering & Remote Sensing*, 75, 941–950.

Chambers, J.Q., Fisher, J.I., Zeng, H., Chapman, E.L., Baker, D.B., and Hurtt, G.C. 2007. Hurricane Katrina's carbon footprint on U.S. Gulf Coast forest. *Science*, 318, 1107.

Chander, G. 2007. Initial data characterization, science utility and mission capability evaluation of candidate Landsat mission data gap sensors. In Landsat Data Gap Study, editor. Technical Report. Available at: http://calval.cr.usgs.gov/LDGST.php

Chander, G., Markham, B.L., and Helder, D.L. 2009a. Summary of current radiometric calibration coefficients for Landsat MSS, TM, ETM+, and EO-1 ALI sensors. *Remote Sensing of Environment*, 113, 893–903.

Chander, G., Xiong, X., Angal, A., Choi, T., and Malla, R. 2009b. Cross-comparison of the IRS-P6 AWiFS sensor with the L5 TM, L7 ETM+, and Terra MODIS sensors. *Proceedings of SPIE*, 7474, 74740Z.

Chavez, P.S. 1996. Image-based atmospheric corrections—Revisited and improved. *Photogrammetric Engineering and Remote Sensing*, 62, 1025–1036.

Chen, J.M. and Cihlar, J. 1996. Retrieving leaf area index of boreal conifer forest using Landsat TM images. *Remote Sensing of Environment*, 55, 153–162.

Chen, J.M., Pavlic, G., Brown, L., Cihlar, J., Leblanc, S.G., White, H.P., Hall, R.J., et al. 2002. Derivation and validation of Canada-wide coarse-resolution leaf area index maps using high-resolution satellite imagery and ground measurements. *Remote Sensing of Environment*, 80, 165–184.

Chen, X., Liu, S., Zhu, Z., Vogelmann, J., Li, Z., and Ohlen, D. 2011a. Estimating aboveground forest biomass carbon and fire consumption in the U.S. Utah High Plateaus using data from the Forest Inventory and Analysis Program, Landsat, and LANDFIRE. *Ecological Indicators*, 11, 140–148.

Chen, X., Vierling, L., and Deering, D. 2005a. A simple and effective radiometric correction method to improve landscape change detection across sensors and across time. *Remote Sensing of Environment*, 98, 63–79.

Chen, X., Vierling, L., Deering, D., and Conley, A. 2005b. Monitoring boreal forest leaf area index across a Siberian burn chronosequence: A MODIS validation study. *International Journal of Remote Sensing*, 26, 5433–5451.

Chen, X., Vierling, L., Rowell, E., and DeFelice, T. 2004. Using lidar and effective LAI data to evaluate IKONOS and Landsat 7 ETM+ vegetation cover estimates in a ponderosa pine forest. *Remote Sensing of Environment*, 91, 14–26.

Chen, X., Vogelmann, J.E., Rollins, M., Ohlen, D., Key, C.H., Yang, L., Huang, C., and Shi, H. 2011b. Detecting post-fire burn severity and vegetation recovery using multitemporal remote sensing spectral indices and field-collected Composite Burn Index data in a ponderosa pine forest. *International Journal of Remote Sensing*, 32(23), 7905–7927.

Chen, X., Xu, C., and Tan, Z. 2001. An analysis of relationships among plant community phenology and seasonal metrics of Normalized Difference Vegetation Index in the northern part of the monsoon region of China. *International Journal of Biometeorology*, 45, 170–177.

Chen, X.-L., Zhao, H.-M., Li, P.-X., and Yin, Z.-Y. 2006. Remote sensing image-based analysis of the relationship between urban heat island and land use/cover changes. *Remote Sensing of Environment*, 104, 133–146.

Chuvieco, E., Ventura, G., Martín, M.P., and Gómez, I. 2005. Assessment of multitemporal compositing techniques of MODIS and AVHRR images for burned land mapping. *Remote Sensing of Environment*, 94, 450–462.

Cohen, W. and Goward, S. 2004. Landsat's role in ecological applications of remote sensing. *BioScience*, 4, 535–545.

Cohen, W.B., Yang, Z., and Kennedy, R. 2010. Detecting trends in forest disturbance and recovery using yearly Landsat time series: 2. TimeSync—Tools for calibration and validation. *Remote Sensing of Environment*, 114, 2911–2924.

Collins, J.B. and Woodcock, C.E. 1994. Change detection using the Gramm-Schmidt transformation applied to mapping forest mortality. *Remote Sensing of Environment*, 50, 267–279.

Coppin, P., Jonckheere, I., Nackaerts, K., Muys, B., and Lambin, E. 2004. Digital change detection methods in ecosystem monitoring: A review. *International Journal of Remote Sensing*, 25, 1565–1596.

Coppin, P.R. and Bauer, M.E. 1996. Digital change detection in forest ecosystems with remote sensing imagery. *Remote Sensing Reviews*, 13, 207–234.

Crist, E.P. 1985. A TM tasseled cap equivalent transformation for reflectance factor data. *Remote Sensing of Environment*, 17, 301–306.

Dai, X. and Khorram, S. 1997. Development of a new automated land cover change detection system from remotely sensed imagery based on artificial neural networks. In *Geoscience and Remote Sensing* (pp. 1029–1031, vol. 1022). IGARSS '97. Remote Sensing—A Scientific Vision for Sustainable Development, 1997 IEEE International.

Dai, X. and Khorram, S. 1998. The effects of image misregistration on the accuracy of remotely sensed change detection. *IEEE Transactions on Geoscience and Remote Sensing,* 36, 1566–1577.

de Beurs, K.M. and Henebry, G.M. 2004. Land surface phenology, climatic variation, and institutional change: Analyzing agricultural land cover change in Kazakhstan. *Remote Sensing of Environment,* 89, 497–509.

Díaz-Delgado, R., Lloret, F., and Pons, X. 2003. Influence of fire severity on plant regeneration by means of remote sensing imagery. *International Journal of Remote Sensing,* 24, 1751–1763.

Dinguirard, M. and Slater, P.N. 1999. Calibration of space-multispectral imaging sensors: A review. *Remote Sensing of Environment,* 68, 194–205.

Dobson, J.E., Bright, E.A., Ferguson, R.L., Field, D.W., Wood, L.L., Haddad, K.D., Iredale III, H., Jensen, J.R., Klemas, V.V., Orth, R.J., and Thomas, J.P. 1995. NOAA Coastal Change Analysis Program (C-CAP): Guidance for Regional Implementation. NOAA Technical Report NMFS 123. Seattle, WA: U.S. Department of Commerce.

Du, Y., Teillet, P.M., and Cihlar, J. 2002. Radiometric normalization of multitemporal high-resolution satellite images with quality control for land cover change detection. *Remote Sensing of Environment,* 82, 123–134.

Eidenshink, J., Schwind, B., Brewer, K., Zhu, Z., Quayle, B., and Howard, S. 2007. A project for monitoring trends in burn severity. *Fire Ecology,* 3, 3–21.

Elvidge, C.D., Yuan, D., Weerackoon, R.D., and Lunetta, R.S. 1995. Relative radiometric normalization of Landsat Multispectral Scanner (MSS) data using an automatic scattergram-controlled regression. *Photogrammetric Engineering and Remote Sensing,* 61, 1255–1260.

Epting, J., Verbyla, D., and Sorbel, B. 2005. Evaluation of remotely sensed indices for assessing burn severity in interior Alaska using Landsat TM and ETM+. *Remote Sensing of Environment,* 96, 328–339.

Escuin, S., Navarro, R., and Fernández, P. 2008. Fire severity assessment by using NBR (Normalized Burn Ratio) and NDVI (Normalized Difference Vegetation Index) derived from LANDSAT TM/ETM images. *International Journal of Remote Sensing,* 29, 1053–1073.

Fensholt, R., Rasmussen, K., Nielsen, T.T., and Mbow, C. 2009. Evaluation of earth observation based long term vegetation trends—Intercomparing NDVI time series trend analysis consistency of Sahel from AVHRR GIMMS, Terra MODIS and SPOT VGT data. *Remote Sensing of Environment,* 113, 1886–1898.

Fisher, J.I. and Mustard, J.F. 2007. Cross-scalar satellite phenology from ground, Landsat, and MODIS data. *Remote Sensing of Environment,* 109, 261–273.

Fisher, P.F. 2010. Remote sensing of land cover classes as type 2 fuzzy sets. *Remote Sensing of Environment,* 114, 309–321.

Foody, G.M. 2010. Assessing the accuracy of land cover change with imperfect ground reference data. *Remote Sensing of Environment,* 114, 2271–2285.

Foody, G.M. and Cox, D.P. 1994. Sub-pixel land cover composition estimation using a linear mixture model and fuzzy membership functions. *International Journal of Remote Sensing,* 15, 619–631.

Friedl, M.A., McIver, D.K., Hodges, J.C.F., Zhang, X.Y., Muchoney, D., Strahler, A.H., Woodcock, C.E., et al. 2002. Global land cover mapping from MODIS: Algorithms and early results. *Remote Sensing of Environment,* 83, 287–302.

Fry, J.A., Coan, M.J., Homer, C.G., Meyer, D.K., and Wickham, J.D. 2009. Completion of the National Land Cover Database (NLCD) 1992–2001 land cover change retrofit product. U.S. Geological Survey Open-File Report 2008–1379, 18 pp.

Fry, J.A., Xian, G., Jin, S., Dewitz, J.A., Homer, C.G., Yang, L., Barnes, C.A., Herold, N.D., and Wickham, J.D. 2011. Completion of the 2006 National Land Cover Database for the conterminous United States. *Photogrammetric Engineering & Remote Sensing,* 77(9), 859–864.

Fung, T. 1990. An assessment of TM imagery for land-cover change detection. *IEEE Transactions on Geoscience and Remote Sensing,* 28, 681–684.

Gallant, A.L., Loveland, T.R., Sohl, T.L., and Napton, D.E. 2004. Using an ecoregion framework to analyze land-cover and land-use dynamics. *Environmental Management,* 34, S89–S110.

Gallo, K., Ji, L., Reed, B., Eidenshink, J., and Dwyer, J. 2005. Multi-platform comparisons of MODIS and AVHRR normalized difference vegetation index data. *Remote Sensing of Environment,* 99, 221–231.

Ganguly, S., Friedl, M.A., Tan, B., Zhang, X., and Verma, M. 2010. Land surface phenology from MODIS: Characterization of the Collection 5 global land cover dynamics product. *Remote Sensing of Environment,* 114, 1805–1816.

Gao, J. 2008. *Digital Analysis of Remotely Sensed Imagery,* 1st edition. Dubuque, IA: McGraw-Hill Professional.

Gao, X., Huete, A.R., Ni, W., and Miura, T. 2000. Optical-biophysical relationships of vegetation spectra without background contamination. *Remote Sensing of Environment*, 74, 609–620.

García, M.J.L. and Caselles, V. 1991. Mapping burns and natural reforestation using thematic mapper data. *Geocarto International*, 6, 31–37.

Giri, C., Ochieng, E., Tieszen, L.L., Zhu, Z., Singh, A., Loveland, T., Masek, J., and Duke, N. 2011. Status and distribution of mangrove forests of the world using earth observation satellite data. *Global Ecology and Biogeography*, 20, 154–159.

Giri, C., Pengra, B., Zhu, Z., Singh, A., and Tieszen, L.L. 2007. Monitoring mangrove forest dynamics of the Sundarbans in Bangladesh and India using multi-temporal satellite data from 1973 to 2000. *Estuarine, Coastal and Shelf Science*, 73, 91–100.

Giri, C., Zhu, Z., and Reed, B. 2005. A comparative analysis of the Global Land Cover 2000 and MODIS land cover data sets. *Remote Sensing of Environment*, 94, 123–132.

GlobCover. 2011. European Space Agency Ionia Globcover Portal. Available at: http://ionia1.esrin.esa.int/index.asp

Goward, S.N., Masek, J.G., Cohen, W., Moisen, G., Collatz, G.J., Healey, S., Houghton, R.A., et al. 2008. Forest disturbance and the North American carbon flux. *EOS, Transactions, American Geophysical Union*, 89, 105–106.

Green, K., Kempka, D., and Lackey, L. 1994. Using remote sensing to detect and monitor land-cover and land-use change. *Photogrammetric Engineering & Remote Sensing*, 60, 331–337.

Gu, D. and Gillespie, A. 1998. Topographic normalization of Landsat TM images of forest, based on subpixel Sun-Canopy-Sensor geometry. *Remote Sensing of Environment*, 64, 166–175.

Hall, F.G., Strebel, D.E., Nickeson, J.E., and Goetz, S.J. 1991. Radiometric rectification: Toward a common radiometric response among multidate, multisensor images. *Remote Sensing of Environment*, 35, 11–27.

Hall, R.J., Freeburn, J.T., de Groot, W.J., Pritchard, J.M., Lynham, T.J., and Landry, R. 2008. Remote sensing of burn severity: Experience from western Canada boreal fires. *International Journal of Wildland Fire*, 17, 476–489.

Hansen, M.C. and Reed, B. 2000. A comparison of the IGBP DISCover and University of Maryland 1 km global land cover products. *International Journal of Remote Sensing*, 21, 1365–1373.

Hansen, M.C., Stehman, S.V., and Potapov, P.V. 2010. Quantification of global gross forest cover loss. *Proceedings of the National Academy of Sciences of USA*, 107, 8650–8655.

Hayes, D.J. and Sader, S.A. 2001. Comparison of change-detection techniques for monitoring tropical forest clearing and vegetation regrowth in a time series. *Photogrammetric Engineering and Remote Sensing*, 67, 1067–1075.

Healey, S.P., Cohen, W.B., Zhiqiang, Y., and Krankina, O.N. 2005. Comparison of tasseled cap-based Landsat data structures for use in forest disturbance detection. *Remote Sensing of Environment*, 97, 301–310.

Herold, M., Mayaux, P., Woodcock, C.E., Baccini, A., and Schmullius, C. 2008. Some challenges in global land cover mapping: An assessment of agreement and accuracy in existing 1 km datasets. *Remote Sensing of Environment*, 112, 2538–2556.

Holben, B.N. and Shimabukuro, Y.E. 1993. Linear mixing model applied to coarse spatial resolution data from multispectral satellite sensors. *International Journal of Remote Sensing*, 14, 2231–2240.

Hong, S.-H., Wdowinski, S., Kim, S.-W., and Won, J.-S. 2010. Multi-temporal monitoring of wetland water levels in the Florida Everglades using interferometric synthetic aperture radar (InSAR). *Remote Sensing of Environment*, 114, 2436–2447.

Houghton, R.A., Hackler, J.L., and Lawrence, K.T. 1999. The U.S. carbon budget contributions from land-use change. *Science*, 285, 574–578.

Huang, C., Goward, S., Masek, J.G., Thomas, N., Zhu, Z., and Vogelmann, J.E. 2010. An automated approach for reconstructing recent forest disturbance history using dense Landsat time series stacks. *Remote Sensing of Environment*, 114, 183–198.

Huang, C., Goward, S.N., Schleeweis, K., Thomas, N., Masek, J.G., and Zhu, Z. 2009. Dynamics of national forests assessed using the Landsat record: Case studies in eastern United States. *Remote Sensing of Environment*, 113, 1430–1442.

Huang, C., Song, K., Kim, S., Townshend, J.R.G., Davis, P., Masek, J.G., and Goward, S.N. 2008. Use of a dark object concept and support vector machines to automate forest cover change analysis. *Remote Sensing of Environment*, 112, 970–985.

Huang, C., Wylie, B., Yang, L., Homer, C., and Zylstra, G. 2005. Derivation of a tasseled cap transformation based on Landsat 7 at-satellite reflectance (pp. 1–10). USGS. Available at: http://landcover.usgs.gov/pdf/tasseled.pdf

Huete, A., Didan, K., Miura, T., Rodriguez, E.P., Gao, X., and Ferreira, L.G. 2002. Overview of the radiometric and biophysical performance of the MODIS vegetation indices. *Remote Sensing of Environment*, 83, 195–213.

Jenkins, J.C., Birdsey, R., and Pan, Y. 2001. Biomass and NPP estimation for the mid-Atlantic region (USA) using plot-level forest inventory data. *Ecological Applications*, 11, 1174–1193.

Jin, S. and Sader, S.A. 2005. Comparison of time series tasseled cap wetness and the normalized difference moisture index in detecting forest disturbances. *Remote Sensing of Environment*, 94, 364–372.

Jin, S., Yang, L., Xian, G., Danielson, P., and Homer, C. 2010. A multi-index integrated change detection method for updating the National Land Cover Database, oral presentation. In *AGU 2010 Fall Meeting*, San Francisco, California.

Jung, M., Henkel, K., Herold, M., and Churkina, G. 2006. Exploiting synergies of global land cover products for carbon cycle modeling. *Remote Sensing of Environment*, 101, 534–553.

Kennedy, R.E., Townsend, P.A., Gross, J.E., Cohen, W.B., Bolstad, P., Wang, Y.Q., and Adams, P. 2009. Remote sensing change detection tools for natural resource managers: Understanding concepts and tradeoffs in the design of landscape monitoring projects. *Remote Sensing of Environment*, 113, 1382–1396.

Kennedy, R.E., Yang, Z., and Cohen, W.B. 2010. Detecting trends in forest disturbance and recovery using yearly Landsat time series: 1. LandTrendr—Temporal segmentation algorithms. *Remote Sensing of Environment*, 114, 2897–2910.

Key, C.H. and Benson, N.C. (Eds.). 2006. Landscape assessment (LA) sampling and analysis methods. USDA Forest Service, Rocky Mountain Research Station, General Technical Report, RMRS-GTR-164-CD.

Kim, J.-W., Lu, Z., Lee, H., Shum, C.K., Swarzenski, C.M., Doyle, T.W., and Baek, S.-H. 2009. Integrated analysis of PALSAR/Radarsat-1 InSAR and ENVISAT altimeter data for mapping of absolute water level changes in Louisiana wetlands. *Remote Sensing of Environment*, 113, 2356–2365.

Lambin, E.F. and Strahlers, A.H. 1994a. Change-vector analysis in multitemporal space: A tool to detect and categorize land-cover change processes using high temporal-resolution satellite data. *Remote Sensing of Environment*, 48, 231–244.

Lambin, E.F. and Strahlers, A.H. 1994b. Indicators of land-cover change for change-vector analysis in multi-temporal space at coarse spatial scales. *International Journal of Remote Sensing*, 15, 2099–2119.

Lambin, E.F., Turner, B.L., Geist, H.J., Agbola, S.B., Angelsen, A., Bruce, J.W., Coomes, O.T., et al. 2001. The causes of land-use and land-cover change: Moving beyond the myths. *Global Environmental Change*, 11, 261–269.

Latifovic, R., Zhu, Z.-L., Cihlar, J., Giri, C., and Olthof, I. 2004. Land cover mapping of North and Central America—Global Land Cover 2000. *Remote Sensing of Environment*, 89, 116–127.

Le Hégarat-Mascle, S., Ottlé, C., and Guérin, C. 2005. Land cover change detection at coarse spatial scales based on iterative estimation and previous state information. *Remote Sensing of Environment*, 95, 464–479.

Lefsky, M.A., Cohen, W.B., Acker, S.A., Parker, G.G., Spies, T.A., and Harding, D. 1999. Lidar remote sensing of the canopy structure and biophysical properties of Douglas-Fir Western Hemlock Forests. *Remote Sensing of Environment*, 70, 339–361.

Lenot, X., Achard, V., and Poutier, L. 2009. SIERRA: A new approach to atmospheric and topographic corrections for hyperspectral imagery. *Remote Sensing of Environment*, 113, 1664–1677.

Li, D. 2010. Remotely sensed images and GIS data fusion for automatic change detection. *International Journal of Image and Data Fusion*, 1, 99–108.

Li, M., Huang, C., Zhu, Z., Shi, H., Lu, H., and Peng, S. 2009a. Assessing rates of forest change and fragmentation in Alabama, USA, using the vegetation change tracker model. *Forest Ecology and Management*, 257, 1480–1488.

Li, M., Huang, C., Zhu, Z., Wen, W., Xu, D., and Liu, A. 2009b. Use of remote sensing coupled with a vegetation change tracker model to assess rates of forest change and fragmentation in Mississippi, USA. *International Journal of Remote Sensing*, 30, 6559–6574.

Liang, L., Schwartz, M.D., and Fei, S. 2011. Validating satellite phenology through intensive ground observation and landscape scaling in a mixed seasonal forest. *Remote Sensing of Environment*, 115, 143–157.

Liang, S., Fallah-Adl, H., Kalhrri, S., JaJa, J., Kaufman, Y.J., and Townshend, J.R.G. 1997. An operational atmospheric correction algorithm for Landsat Thematic Mapper imagery over the land. *Journal of Geophysical Research*, 102, 17173–17186.

Lillesand, T.M. and Kiefer, R.W. 1994. *Remote Sensing and Image Interpretation*. New York: John Wiley & Sons, Inc.

Lobser, S.E. and Cohen, W.B. 2007. MODIS tasselled cap: Land cover characteristics expressed through transformed MODIS data. *International Journal of Remote Sensing*, 28, 5079–5101.

Los, S.O., North, P.R.J., Grey, W.M.F., and Barnsley, M.J. 2005. A method to convert AVHRR Normalized Difference Vegetation Index time series to a standard viewing and illumination geometry. *Remote Sensing of Environment*, 99, 400–411.

Loveland, T.R. and Belward, A.S. 1997. The IGBP-DIS global 1km land cover data set, DISCover: First results. *International Journal of Remote Sensing*, 18, 3289–3295.

Loveland, T.R., Reed, B.C., Brown, J.F., Ohlen, D.O., Zhu, Z., Yang, L., and Merchant, J.W. 2000. Development of a global land cover characteristics database and IGBP DISCover from 1 km AVHRR data. *International Journal of Remote Sensing*, 21, 1303–1330.

Loveland, T.R., Sohl, T.L., Sayler, K., Gallant, A., Dwyer, J., Vogelmann, J.E., and Zylstra, G.J. 1999. Land cover trends: Rates, causes, and consequences of late-twentieth century U.S. land cover change. U.S. Environmental Protection Agency, EPA/600/R-99/105, pp. 52.

Loveland, T.R., Sohl, T.L., Stehman, S.V., Gallant, A.L., Sayler, K.L., and Napton, D.E. 2002. A strategy for estimating the rates of recent United States Land-Cover Changes. *Photogrammetric Engineering and Remote Sensing*, 68, 1091–1099.

Lu, D., Batistella, M., and Moran, E. 2004a. Multitemporal spectral mixture analysis for Amazonian land-cover change detection. *Canadian Journal of Remote Sensing*, 30, 87–100.

Lu, D., Mausel, P., Brondízio, E., and Moran, E. 2004b. Change detection techniques. *International Journal of Remote Sensing*, 25, 2365–2401.

Lunetta, R.S., Johnson, D.M., Lyon, J.G., and Crotwell, J. 2004. Impacts of imagery temporal frequency on land-cover change detection monitoring. *Remote Sensing of Environment*, 89, 444–454.

Lunetta, R.S., Knight, J.F., Ediriwickrema, J., Lyon, J.G., and Worthy, L.D. 2006. Land-cover change detection using multi-temporal MODIS NDVI data. *Remote Sensing of Environment*, 105, 142–154.

Lupo, F., Linderman, M., Vanacker, V., Bartholomé, E., and Lambin, E.F. 2007. Categorization of land-cover change processes based on phenological indicators extracted from time series of vegetation index data. *International Journal of Remote Sensing*, 28, 2469–2483.

Lyon, J.G., Yuan, D., Lunetta, R.S., and Elvidge, C.D. 1998. A change detection experiment using vegetation indices. *Photogrammetric Engineering and Remote Sensing*, 64, 143–150.

Macleod, R.D. and Congalton, R. 1998. A quantitative comparison of change-detection algorithms for monitoring eelgrass from remotely sensed data. *Photogrammetric Engineering and Remote Sensing*, 64, 207–216.

Mas, J.F. 1999. Monitoring land-cover changes: A comparison of change detection techniques. *International Journal of Remote Sensing*, 20, 139–152.

Masek, J.G. and Collatz, G.J. 2006. Estimating forest carbon fluxes in a disturbed southeastern landscape: Integration of remote sensing, forest inventory, and biogeochemical modeling. *Journal of Geophysical Research*, 111, G01006:doi:10.1029/2005JG000062.

Masek, J.G., Huang, C., Wolfe, R., Cohen, W., Hall, F., Kutler, J., and Nelson, P. 2008. North American forest disturbance mapped from a decadal Landsat record. *Remote Sensing of Environment*, 112, 2914–2926.

Masek, J.G., Vermote, E.F., Saleous, N.E., Wolfe, R., Hall, F.G., Huemmrich, K.F., Feng, G., Kutler, J., and Teng-Kui, L. 2006. A Landsat surface reflectance dataset for North America, 1990–2000. *Geoscience and Remote Sensing Letters, IEEE*, 3, 68–72.

Meyer, P., Itten, K.I., Kellenberger, T., Sandmeier, S., and Sandmeier, R. 1993. Radiometric corrections of topographically induced effects on Landsat TM data in an alpine environment. *ISPRS Journal of Photogrammetry and Remote Sensing*, 48, 17–28.

Michener, W.K. and Houhoulis, P.F. 1997. Detection of vegetation changes associated with extensive flooding in a forested ecosystem. *Photogrammetric Engineering and Remote Sensing*, 63, 1363–1374.

Millward, A.A., Piwowar, J.M., and Howarth, P.J. 2006. Time-series analysis of medium-resolution, multisensor satellite data for identifying landscape change. *Photogrammetric Engineering and Remote Sensing*, 72, 653–663.

Miura, T., Huete, A.R., Yoshioka, H., and Holben, B.N. 2001. An error and sensitivity analysis of atmospheric resistant vegetation indices derived from dark target-based atmospheric correction. *Remote Sensing of Environment*, 78, 284–298.

Morisette, J.T., Nickeson, J.E., Davis, P., Wang, Y., Tian, Y., Woodcock, C.E., Shabanov, N., et al. 2003. High spatial resolution satellite observations for validation of MODIS land products: IKONOS observations acquired under the NASA Scientific Data Purchase. *Remote Sensing of Environment*, 88, 100–110.

Muchoney, D.M. and Haack, B.N. 1994. Change detection for monitoring forest defoliation *Photogrammetric Engineering and Remote Sensing*, 60, 1243–1251.

Myneni, R.B., Keeling, C.D., Tucker, C.J., Asrar, G., and Nemani, R.R. 1997. Increased plant growth in the northern high latitudes from 1981 to 1991. *Nature*, 386, 698–702.

Pocewicz, A., Vierling, L.A., Lentile, L.B., and Smith, R. 2007. View angle effects on relationships between MISR vegetation indices and leaf area index in a recently burned ponderosa pine forest. *Remote Sensing of Environment*, 107, 322–333.

Portolese, J., Hart, T.F., Jr., and Henderson, F.M. 1998. TM-based coastal land cover change analysis and its application for state and local resource management needs. In *Geoscience and Remote Sensing Symposium Proceedings* (pp. 882–884, vol. 882). IGARSS '98. 1998 IEEE International.

Pouliot, D., Latifovic, R., Fernandes, R., and Olthof, I. 2011. Evaluation of compositing period and AVHRR and MERIS combination for improvement of spring phenology detection in deciduous forests. *Remote Sensing of Environment*, 115, 158–166.

Privette, J.L., Eck, T.F., and Deering, D.W. 1997. Estimating spectral albedo and nadir reflectance through inversion of simple BRDF models with AVHRR/MODIS-like data. *Journal of Geophysical Research*, 102, 29529–29542.

Ray, T.W. and Murray, B.C. 1996. Nonlinear spectral mixing in desert vegetation. *Remote Sensing of Environment*, 55, 59–64.

Reed, B.C. 2006. Trend analysis of time-series phenology of North America derived from satellite data. *GIScience & Remote Sensing*, 43, 24–38.

Reed, B.C. and Yang, L. 1997. Seasonal vegetation characteristics of the United States. *Geocarto International*, 12, 65–71.

Ressl, R., Lopez, G., Cruz, I., Colditz, R.R., Schmidt, M., Ressl, S., and Jiménez, R. 2009. Operational active fire mapping and burnt area identification applicable to Mexican Nature Protection Areas using MODIS and NOAA-AVHRR direct readout data. *Remote Sensing of Environment*, 113, 1113–1126.

Rollins, M.G. 2009. LANDFIRE: A nationally consistent vegetation, wildland fire, and fuel assessment. *International Journal of Wildland Fire*, 18, 235–249.

Román, M.O., Schaaf, C.B., Woodcock, C.E., Strahler, A.H., Yang, X., Braswell, R.H., Curtis, P.S., et al. 2009. The MODIS (Collection V005) BRDF/albedo product: Assessment of spatial representativeness over forested landscapes. *Remote Sensing of Environment*, 113, 2476–2498.

Rosso, P.H., Ustin, S.L., and Hastings, A. 2006. Use of lidar to study changes associated with Spartina invasion in San Francisco Bay marshes. *Remote Sensing of Environment*, 100, 295–306.

Rouse, J.W., Haas, Jr., R.H., Deering, D.W., Schell, J.A., and Harlan, J.C. 1974. Monitoring the vernal advancement and retrogradation (green wave effect) of natural vegetation. NASA/GSFC Type III Final Report, p. 371. Greenbelt, MD.

Roy, D.P., Ju, J., Kline, K., Scaramuzza, P.L., Kovalskyy, V., Hansen, M., Loveland, T.R., Vermote, E., and Zhang, C. 2010. Web-enabled Landsat Data (WELD): Landsat ETM+ composited mosaics of the conterminous United States. *Remote Sensing of Environment*, 114, 35–49.

Salvaggio, C. 1993. Radiometric scene normalization utilizing statistically invariant features. In *Proceedings of the Workshop on Atmospheric Correction of Landsat Imagery* (pp. 155–159), Los Angeles, CA.

Schott, J.R. 1997. *Remote Sensing—The Image Chain Approach*. New York: Oxford University Press.

Schott, J.R., Salvaggio, C., and Volchok, W.J. 1988. Radiometric scene normalization using pseudoinvariant features. *Remote Sensing of Environment*, 26, 1–14, IN11, 15–16.

Schroeder, T.A., Cohen, W.B., Song, C., Canty, M.J., and Yang, Z. 2006. Radiometric correction of multi-temporal Landsat data for characterization of early successional forest patterns in western Oregon. *Remote Sensing of Environment*, 103, 16–26.

Shepherd, J.D. and Dymond, J.R. 2000. BRDF correction of vegetation in AVHRR imagery. *Remote Sensing of Environment*, 74, 397–408.

Singh, A. 1989. Digital change detection techniques using remotely sensed data. *International Journal of Remote Sensing*, 10, 989–1003.

Skole, D. and Tucker, C. 1993. Tropical deforestation and habitat fragmentation in the Amazon—Satellite data from 1978 to 1988. *Science*, 260, 1905–1909.

Sohl, T.L., Gallant, A.L., and Loveland, T.R. 2004. The characteristics and interpretability of land surface change and implications for project design. *Photogrammetric Engineering and Remote Sensing*, 70(4), 439–448.

Song, C., Woodcock, C.E., Seto, K.C., Lenney, M.P., and Macomber, S.A. 2001. Classification and change detection using Landsat TM data: When and how to correct atmospheric effects? *Remote Sensing of Environment*, 75, 230–244.

Sophiayati Yuhaniz, S. and Vladimirova, T. 2009. An onboard automatic change detection system for disaster monitoring. *International Journal of Remote Sensing*, 30, 6121–6139.

Soverel, N.O., Perrakis, D.D.B., and Coops, N.C. 2010. Estimating burn severity from Landsat dNBR and RdNBR indices across western Canada. *Remote Sensing of Environment*, 114, 1896–1909.

Spanner, M., Johnson, L., Miller, J., McCreight, R., Freemantle, J., and Runyon, J. 1994. Remote sensing of seasonal leaf area index across the Oregon transect. *Ecological Applications*, 4, 258–271.

Stehman, S.V., Sohl, T.L., and Loveland, T.R. 2003. Statistical sampling to characterize recent United States land-cover change. *Remote Sensing of Environment*, 86, 517–529.

Stellmes, M., Udelhoven, T., Röder, A., Sonnenschein, R., and Hill, J. 2010. Dryland observation at local and regional scale—Comparison of Landsat TM/ETM+ and NOAA AVHRR time series. *Remote Sensing of Environment*, 114, 2111–2125.

Stow, D.A. and Chen, D.M. 2002. Sensitivity of multitemporal NOAA AVHRR data of an urbanizing region to land-use/land-cover changes and misregistration. *Remote Sensing of Environment*, 80, 297–307.

Susaki, J., Hara, K., Kajiwara, K., and Honda, Y. 2004. Robust estimation of BRDF model parameters. *Remote Sensing of Environment*, 89, 63–71.

Teillet, P.M., Fedosejevs, G., Thome, K.J., and Barker, J.L. 2007. Impacts of spectral band difference effects on radiometric cross-calibration between satellite sensors in the solar-reflective spectral domain. *Remote Sensing of Environment*, 110, 393–409.

Teillet, P.M., Staenz, K., and William, D.J. 1997. Effects of spectral, spatial, and radiometric characteristics on remote sensing vegetation indices of forested regions. *Remote Sensing of Environment*, 61, 139–149.

Thomas, N.E., Huang, C., Goward, S.N., Powell, S., Rishmawi, K., Schleeweis, K., and Hinds, A. 2011. Validation of North American forest disturbance dynamics derived from Landsat time series stacks. *Remote Sensing of Environment*, 115, 19–32.

Toutin, T. 2004. Review article: Geometric processing of remote sensing images: Models, algorithms and methods. *International Journal of Remote Sensing*, 25, 1893–1924.

Townshend, J.R.G. and Justice, C.O. 2002. Towards operational monitoring of terrestrial systems by moderate-resolution remote sensing. *Remote Sensing of Environment*, 83, 351–359.

Vepakomma, U., St-Onge, B., and Kneeshaw, D. 2008. Spatially explicit characterization of boreal forest gap dynamics using multi-temporal lidar data. *Remote Sensing of Environment*, 112, 2326–2340.

Veraverbeke, S., Lhermitte, S., Verstraeten, W.W., and Goossens, R. 2010. The temporal dimension of differenced Normalized Burn Ratio (dNBR) fire/burn severity studies: The case of the large 2007 Peloponnese wildfires in Greece. *Remote Sensing of Environment*, 114, 2548–2563.

Verbesselt, J., Hyndman, R., Zeileis, A., and Culvenor, D. 2010. Phenological change detection while accounting for abrupt and gradual trends in satellite image time series. *Remote Sensing of Environment*, 114, 2970–2980.

Verbyla, D.L. and Boles, S.H. 2000. Bias in land cover change estimates due to misregistration. *International Journal of Remote Sensing*, 21, 3553–3560.

Vermote, E.F., Tanre, D., Deuze, J.L., Herman, M., and Morcette, J.J. 1997. Second Simulation of the Satellite Signal in the Solar Spectrum, 6S: An overview. *IEEE Transactions on Geoscience and Remote Sensing*, 35, 675–686.

Vicente-Serrano, S.M., Pérez-Cabello, F., and Lasanta, T. 2008. Assessment of radiometric correction techniques in analyzing vegetation variability and change using time series of Landsat images. *Remote Sensing of Environment*, 112, 3916–3934.

Vierling, L.A., Deering, D.W., and Eck, T.F. 1997. Differences in arctic tundra vegetation type and phenology as seen using bidirectional radiometry in the early growing season. *Remote Sensing of Environment*, 60, 71–82.

Vogelmann, J.E., Kost, J.R., Tolk, B., Howard, S., Short, K., Chen, X., Huang, C., Pabst, K., and Rollins, M.G. 2011. Monitoring landscape change for LANDFIRE using multi-temporal satellite imagery and ancillary data. *IEEE Journal of Selected Topics in Applied Earth Observations and Remote Sensing*, 4, 252–2011.

Vogelmann, J.E., Tolk, B., and Zhu, Z. 2009. Monitoring forest changes in the southwestern United States using multitemporal Landsat data. *Remote Sensing of Environment*, 113, 1739–1748.

Weber, K.T. 2001. A method to incorporate phenology into land cover change analysis [Abstract]. *Journal of Range Management*, 54, 202.

Weber, K.T., Théau, J., and Serr, K. 2008. Effect of coregistration error on patchy target detection using high-resolution imagery. *Remote Sensing of Environment*, 112, 845–850.

Westerling, A.L., Hidalgo, H.G., Cayan, D.R., and Swetnam, T.W. 2006. Warming and earlier spring increase western US forest wildfire activity. *Science*, 313, 940–943.

White, J.D., Ryan, K.C., Key, C.C., and Running, S.W. 1996. Remote sensing of forest fire severity and vegetation recovery. *International Journal of Wildland Fire*, 6, 125–136.

Wickham, J.D., Stehman, S.V., Fry, J.A., Smith, J.H., and Homer, C.G. 2010. Thematic accuracy of the NLCD 2001 land cover for the conterminous United States. *Remote Sensing of Environment*, 114, 1286–1296.

Wimberly, M.C. and Reilly, M.J. 2007. Assessment of fire severity and species diversity in the southern Appalachians using Landsat TM and ETM+ imagery. *Remote Sensing of Environment*, 108, 189–197.

Woodcock, C.E., Allen, R., Anderson, M., Belward, A., Bindschadler, R., Cohen, W., Gao, F., et al. 2008. Free access to Landsat imagery. *Science*, 320, 1011.

Woodcock, C.E. and Strahler, A.H. 1987. The factor of scale in remote sensing. *Remote Sensing of Environment*, 21, 311–332.

Wulder, M.A. and Franklin, S.E. (Eds.). 2007. *Understanding Forest Disturbance and Spatial Pattern: Remote Sensing and GIS Approaches*. New York: Taylor & Francis Group.

Wulder, M.A., Skakun, R.S., Kurz, W.A., and White, J.C. 2004. Estimating time since forest harvest using segmented Landsat ETM+ imagery. *Remote Sensing of Environment*, 93, 179–187.

Wulder, M.A., White, J.C., Alvarez, F., Han, T., Rogan, J., and Hawkes, B. 2009. Characterizing boreal forest wildfire with multi-temporal Landsat and lidar data. *Remote Sensing of Environment*, 113, 1540–1555.

Wulder, M.A., White, J.C., Goward, S.N., Masek, J.G., Irons, J.R., Herold, M., Cohen, W.B., Loveland, T.R., and Woodcock, C.E. 2008. Landsat continuity: Issues and opportunities for land cover monitoring. *Remote Sensing of Environment*, 112, 955–969.

Wulder, M.A., White, J.C., Masek, J.G., Dwyer, J., and Roy, D.P. 2011. Continuity of Landsat observations: Short term considerations. *Remote Sensing of Environment*, 115, 747–751.

Xian, G., Homer, C., and Fry, J. 2009. Updating the 2001 National Land Cover Database land cover classification to 2006 by using Landsat imagery change detection methods. *Remote Sensing of Environment*, 113, 1133–1147.

Xiao, X., Braswell, B., Zhang, Q., Boles, S., Frolking, S., and Moore, B. 2003. Sensitivity of vegetation indices to atmospheric aerosols: Continental-scale observations in Northern Asia. *Remote Sensing of Environment*, 84, 385–392.

Yang, L., Homer, C., Hegge, K., Chengquan, H., Wylie, B., and Reed, B. 2001. A Landsat 7 scene selection strategy for a national land cover database. In *Geoscience and Remote Sensing Symposium* (pp. 1123–1125, vol. 1123). IGARSS '01. IEEE 2001 International.

Yang, L., Jiang, L., Lin, H., and Liao, M. 2009. Quantifying sub-pixel urban impervious surface through fusion of optical and InSAR imagery. *GIScience & Remote Sensing*, 46, 161–171.

Yang, L., Xian, G., Klaver, J.M., and Deal, B. 2003. Urban land-cover change detection through sub-pixel imperviousness mapping using remotely sensed data. *Photogrammetric Engineering and Remote Sensing*, 69, 1003–1010.

Yang, W., Yang, L., and Merchant, J.W. 1997. An assessment of AVHRR/NDVI-ecoclimatological relations in Nebraska, U.S.A. *International Journal of Remote Sensing*, 18, 2161–2180.

Yuan, D. and Elvidge, C. 1998. NALC land cover change detection pilot study: Washington D.C. area experiments. *Remote Sensing of Environment*, 66, 166–178.

Yuan, F., Sawaya, K.E., Loeffelholz, B.C., and Bauer, M.E. 2005. Land cover classification and change analysis of the Twin Cities (Minnesota) Metropolitan Area by multitemporal Landsat remote sensing. *Remote Sensing of Environment*, 98, 317–328.

Zhan, X., Defries, R., Townshend, J.R.G., Dimiceli, C., Hansen, M., Huang, C., and Sohlberg, R. 2000. The 250 m global land cover change product from the Moderate Resolution Imaging Spectroradiometer of NASA's Earth Observing System. *International Journal of Remote Sensing*, 21, 1433–1460.

Zhan, X., Sohlberg, R.A., Townshend, J.R.G., DiMiceli, C., Carroll, M.L., Eastman, J.C., Hansen, M.C., and DeFries, R.S. 2002. Detection of land cover changes using MODIS 250 m data. *Remote Sensing of Environment*, 83, 336–350.

Zhang, L., Liao, M., Yang, L., and Lin, H. 2007. Remote sensing change detection based on canonical correlation analysis and contextual bayes decision. *Photogrammetric Engineering and Remote Sensing*, 73, 311–318.

Zhou, L., Tucker, C.J., Kaufmann, R.K., Slayback, D., Shabanov, N.V., and Myneni, R.B. 2001. Variations in northern vegetation activity inferred from satellite data of vegetation index during 1981 to 1999. *Journal of Geophysical Research*, 106, 20069–20083.

Zimble, D.A., Evans, D.L., Carlson, G.C., Parker, R.C., Grado, S.C., and Gerard, P.D. 2003. Characterizing vertical forest structure using small-footprint airborne lidar. *Remote Sensing of Environment*, 87, 171–182.

12 Supervised Classification Approaches for the Development of Land-Cover Time Series

Darren Pouliot, Rasim Latifovic, Ian Olthof, and Robert Fraser

CONTENTS

12.1 INTRODUCTION

Land cover is a fundamental earth-surface attribute shaped by geologic, hydrologic, climatic, atmospheric, and land-use processes occurring at a range of space-time scales. Land cover, in turn, affects these processes through feedback mechanisms such as plant respiration, which both absorbs and releases carbon, water, oxygen, and other biochemical elements from or to the environment. Therefore, knowledge of land cover is essential to understand earth-surface processes relevant for managing land and preserving natural environments. Examples include climate and weather modeling (Bonan, 2004), carbon budget assessment (Turner et al., 2004), water supply and quality analysis (Chang, 2003), evaluation of terrestrial and aquatic ecosystem integrity (Eshleman, 2004; Fraser et al., 2009), investigation of effects of farming practices on erosion and on nutrient contamination in lakes and rivers (Potter, 2004), evaluation of suitability of land for infrastructure development or biofuel production (Fischer et al., 2010), wildlife population modeling (Kerr and Ostrovsky, 2003), and biodiversity assessment (Kerr, 2001).

The importance of land cover in earth-surface processes has prompted the development of methods to monitor land-cover status at local to global scales. Remote sensing is a practical approach for this because of its capacity to cover large areas with frequent revisits at low cost. Mapping land cover by remote sensing with moderate-resolution (30 m) data has been successful in achieving an accuracy target of 85% (Foody, 2002). At coarser spatial resolutions (0.25–1 km), accuracy is reduced by the effect of mixed pixels and depends on landscape homogeneity and thematic resolution (Latifovic and Olthof, 2004). Mapping accuracy is a crucial consideration for land-cover monitoring. A simple monitoring approach that compares land-cover maps from two dates has proven to be difficult for producing reliable change information. This method of postclassification comparison has been widely used for change detection analysis (Coppin et al., 2004; Lu et al., 2004; Singh, 1989). An accuracy estimate for this method can be made from the product of the accuracies of the two input maps, assuming that the errors between maps are independent (Stow et al., 1980). Thus, for maps with 85% accuracy, the accuracy of change derived from postclassification comparison is theoretically ~72%. For land-cover monitoring applications, this accuracy is too low, especially for regions with moderate to low rates of change.

An alternative to postclassification comparison is change detection derived directly from spectral data acquired at different times. Spectral change detection methods such as image differencing, image ratioing, change vector, and principal components analysis were among the first to be developed and evaluated (Singh, 1989). These methods achieved higher accuracy than postclassification comparison when classifications were derived using pattern recognition techniques (Coppin et al., 2004; Lu et al., 2004; Singh, 1989). However, spectral change detection methods typically provide information on the location and magnitude of change only. In some cases, detected changes can be attributed to their causes, such as fire, harvesting, or insect defoliation (Fraser et al. 2005). Potapov et al. (2008) classified changes as being caused by either fire or other agents. The incentive to use postclassification comparison for monitoring, despite its low accuracy, results from the rich information on the types of land-cover transitions it provides (e.g., from class x to class y). This offers a strong motive for pursuing additional research to improve accuracy of mapping methods used for producing land-cover time series. Therefore, classification techniques and their use for monitoring and developing land-cover time series are the subjects of this review. Specifically, supervised classification approaches are focused on because they are widely used and have the potential to build a more automated monitoring framework. Most of the attention is devoted to two aspects: (1) classification for developing land-cover time series and (2) postprocessing techniques to reduce the occurrence of false change. Postprocessing includes incorporation of expert rules, fuzzy information, class transition probabilities or limiting the changes between maps to those identified using a separate change detection method. The required degree of postprocessing is related to the quality of the initial mapping and thus to the factors affecting land-cover classification accuracy.

Mapping based on visual interpretation has been used to generate land-cover time series (Barson, 2008; Hurd et al., 2009; Kleeschulte and Büttner, 2008), but it is not addressed in this review. Approaches with significant potential for automation are considered more desirable because it allows time series to be generated rapidly and for short time steps. Further, the focus is on "hard" classification as opposed to the fuzzy or fractional approaches because of the detailed land-cover transition information that it can provide.

12.2   IMPLEMENTING CLASSIFICATION FOR LAND-COVER TIME SERIES

Land-cover time series developed from supervised classification typically employs classifier retraining or classifier extension. For the former, training data are collected or modified for each map in the time series. In the latter, the classifier is trained from one sample and used to generate time series without retraining. This is also referred to as signature extension (Minter, 1978; Olthof et al., 2005), within-scene generalization (Pax-Lenny et al., 2001; Woodcock et al., 2001), or as static training approach (Pouliot et al., 2009). The term "classifier extension" is used here instead of "signature extension" because it is descriptive of the process in an intuitive way, and many modern classifiers do not use signatures in the classical statistical sense. For time series applications, this is further specified as temporal classifier extension as opposed to spatial classifier extension where data from different areas are classified. For developing land-cover time series following a supervised classification approach, some form of either classifier retraining or extension is required to provide an initial estimate of a pixel's class.

12.2.1 CLASSIFIER RETRAINING

To generate a land-cover time series using classifier retraining, for each time step, the classifier is trained from a sample of reference and corresponding satellite data. The sample can be new or an updated version of the sample from a previous time step. The major advantage of this approach over classifier extension is that it is less sensitive to radiometric variability present in the satellite data record, which can result from environmental and atmospheric conditions (Coppin et al., 2004). The disadvantage is that developing and maintaining the training database can be costly and needs to be carefully completed to avoid introducing sample bias in the classification results. One way to reduce cost is to use the sample collected at time t and then update it with new training data for use at time $t + 1$. This can be difficult because it requires identifying sample points that have changed between t and $t + 1$. Change detection methods that achieve high accuracy, particularly with low omission error, can be used for this purpose. Latifovic and Pouliot (2005) and Xian et al. (2009) implemented classifier retraining approaches to generate $t + 1$ land-cover data by first detecting change using t and $t + 1$ spectral data and then sampling from the t land cover in areas not detected as change. This approach reduced the cost of acquiring additional samples and allowed the classifier to be trained directly from the spectral data used for the classification, thereby reducing concerns associated with radiometric variability between time periods. In both cases, it performed well as long as sufficient and well-distributed samples were acquired from the no-change areas.

12.2.2 CLASSIFIER EXTENSION

Training a classifier from one dataset and using it to classify data in other periods is an alternative to classifier retraining, which eliminates the need for collecting training data for each time step and thereby reduces cost. This allows automated classification for all time steps in the series, which is attractive for developing time series over long periods with frequent updates. It is particularly useful for generating historical time series where it may not be possible to acquire data needed for retraining. It also has advantages for near real-time monitoring since no additional training data are required, which simplifies processing and eliminates time lags between training data collection and satellite data acquisition. For simple two-class problems, high accuracies ($\geq 95\%$) have been obtained with this approach (Pax-Lenny et al., 2001; Pouliot et al., 2009; Woodcock et al., 2001). However, for higher levels of thematic detail, simple classifier extension has not been found sufficient. Figure 12.1 shows the results of various attempts at classifier extension for different sensors and classification problems. It shows that the agreement between classifications generated using classifier extension is very poor when a large number of classes are used. At approximately 20 classes, the agreement starts to increase rapidly as the number of classes decrease, but it still does not achieve high accuracy until only two classes are present. Early studies of classifier extension for spatial extension of crop classes found that the approach was unsatisfactory (Minter, 1978; Myers, 1983). In a more recent study, Olthof et al. (2005) also found the overall performance of classifier extension to be low. Thus, as with classifier retraining, this approach requires additional processing to ensure that sufficient consistency is maintained in the land-cover time series.

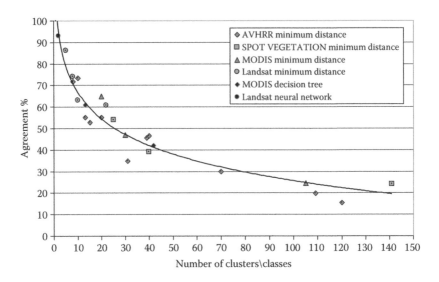

FIGURE 12.1   Consistency between various attempts at classifier extension using different sensors and baseline cluster\land-cover maps. Results for the Landsat neural network are taken from Pax-Lenny et al. (2001). (Adapted from Pax-Lenney et al., *Remote Sens. Environ.*, 77, 241–250, 2001.)

12.3   FACTORS AFFECTING LAND-COVER TIME SERIES ACCURACY AND TEMPORAL CONSISTENCY

Developing methods to achieve high land-cover mapping accuracy is an essential first step for using land-cover data to monitor change. In addition to achieving mapping accuracy, achieving high inter-map consistency is of particular importance. Consistency refers to the agreement between maps in areas that have not changed between dates. It is different from accuracy because land-cover labels can be incorrect relative to the true ground condition but can be the same between maps where change has not occurred. Thus, maps with somewhat lower accuracy can still be used to identify changes if they are consistent and differ only in actual land-cover changes. The following reviews the factors to be considered when developing accurate land-cover maps for developing land-cover time series with high consistency.

12.3.1   CLASSIFICATION METHODS

Research on classification methods has shown that techniques such as decision trees, neural networks, and support vector machines produce higher accuracies than statistical methods, particularly with nonnormal training data distributions (Arora and Foody, 1997; Huang et al., 2002; Meyer et al., 2003; Pal and Mather, 2003, 2005; Peddle, 1994). They are free of statistical assumptions, allowing classes to contain multimodal frequency distributions, thereby providing more freedom for class definition. They can also incorporate data from different measurement scales, including categorical variables. However, no single classifier has consistently been shown to strongly outperform the others in a range of classifier applications (Meyer et al., 2003). Decision trees are much faster to train than neural networks and support vector machines (Huang, 2002; Pal and Mather, 2003) but can be sensitive to overtraining (Jain et al., 2000). Overtraining is a condition where a classifier achieves high accuracy with the training data but performs poorly with samples outside the training set. Support vector machines are less sensitive to overtraining as they are designed specifically to avoid overspecifying class decision boundaries (Jain et al., 2000). For this reason, support vector machines may be best suited for classifier extension, but to our knowledge, no research has been done to test this hypothesis. Training data requirements may also be reduced for these classifiers as only samples defining class boundaries are required unlike statistical classifiers that need the complete class distribution. However, determining these class boundaries *a priori* for effective sampling is not trivial (Foody and Mathur, 2004, 2006).

12.3.2 TRAINING DATA

The size (i.e., the number of samples) of the training dataset and the sampling strategy used are two factors that influence the classifier's performance. It is important to ensure that a representative class distribution is achieved in the training set (Jensen, 1996). Several studies have shown that accuracy and sample size are related, where increases in sample size lead to higher accuracy until a saturation level is reached and additional samples make little improvement (Foody and Arora, 1997; Foody et al., 1995; Huang et al., 2002; Pal and Mather, 2003). Statistical sampling theory can be used to determine the required sample size, but studies show that training requirements are often specific to the classifier used and the complexity of the classification problem (Foody and Mathur, 2004; Foody et al., 2006). The number of predictive features used in the classification has been shown to increase the required sample size. Research suggests that the sample size for each class should be 10–30 times the number of input features used (Mather, 1999). For classifier extension, training data should also include samples from several years to capture class temporal variance.

Sampling design can strongly affect the representativeness of the sample and thus the resulting classification, which can lead to inconsistency in land-cover time series. For land-cover classification, stratified random sampling is often used as it ensures that rare classes are included. In random sampling, rare classes may be missed, but overall accuracy can still be high because the sample ensures that the most frequently occurring classes in the map are well characterized. Huang et al. (2002) evaluated the difference between sampling at a constant rate (percent of class area) and constant size (fixed number of samples per class) for stratified sampling and found that sampling at a constant rate improved results slightly. To illustrate the effects of sampling design on classification accuracy, we present a simple simulation. A Landsat scene was clustered into 30 spectral clusters using the k-means classifier with an initial systematic sample of every 100th pixel. This same cluster and image data were then sampled using a stratified random sample with a constant sample size for each cluster. This training sample was used to re-create the reference cluster map with the same minimum distance decision rule used in the initial k-means clustering. The agreement between the original cluster map generated with a systematic sample and the result of this stratified random sampling was compared for different sample sizes. Samples sizes varied from 1 to 4000 samples per cluster.

The interaction between sample size and sample design shows that for a small range of sample sizes, the two cluster images strongly agree (Figure 12.2). However, there is still an almost 2% difference in the highest agreement observed. For sample sizes greater than 500/cluster, the agreement diminishes because more local spatial clustering of samples occurs in some of the smaller clusters, biasing the spectral signatures to local regions of the image and essentially causing the signatures to drift from the original obtained with the k-means classifier. Owing to this sampling sensitivity, it is of particular importance to ensure that sampling is standardized for approaches using classifier retraining to avoid introducing sample-related bias in the classification results for different time steps.

12.3.3 THEMATIC RESOLUTION AND SEPARABILITY

Classification accuracy decreases nonlinearly with increasing thematic detail owing to a reduction in interclass separability (Fraser et al., 2009; Latifovic and Olthof, 2004; Latifovic and Pouliot, 2005). For generating land-cover time series, the developer needs to carefully consider the separability of

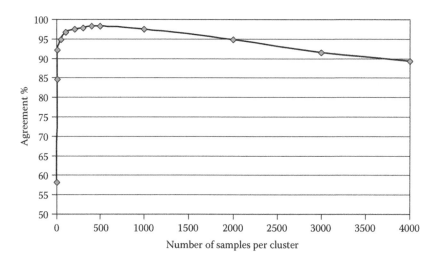

FIGURE 12.2 The effect of sampling on classification agreement for a reference cluster image generated using a systematic sample and then re-creating this cluster map using a stratified sample with a fixed number of samples per cluster.

the classes in both space and time and determine if some classes can be merged to enhance temporal consistency in the classified data (Colwell et al., 1980).

The selection of features used in the classification algorithm is a critical factor controlling class separability and accuracy. For land-cover mapping, features typically include spectral data or derivatives such as band transformations, vegetation indices, or image texture. Additional features may include topographic, climate, or other discriminatory spatial layers. For many classification problems, increasing the number of features relative to the training sample size does not necessarily increase accuracy. This is known as the "curse of dimensionality" or "peaking" phenomenon (Jain et al., 2000; Pal and Mather, 2003, 2005). The use of an unnecessarily large number of features is undesirable as it greatly enhances computational complexity and makes quality control more difficult. Feature selection is a complex problem as it is often not possible to evaluate exhaustively all feature combinations, and some optimization approach is required. Classic separability analysis approaches such as Bhattacharyya distance may not be appropriate as these are based on statistical assumptions that are not relevant to neural net, support vector, or decision tree classifiers, especially when nonratio data are used. There are various feature selection algorithms ranging from simple stepwise to more complex genetic search approaches. Jain et al. (2000) provide a good overview of feature selection algorithms. For a detailed description, see Liu and Motoda (2008). Pouliot et al. (2009) applied a simple forward stepwise selection method in which all features were evaluated in combination with the current set to determine the feature that offers the maximum improvement in classifier accuracy. The same analysis confirms that more features were needed for temporal classifier extension than for the nonextension cases. The study found that the additional features helped account for greater temporal variance encountered when extending the classifier in time.

Feature selection for developing land-cover time series also needs to consider the temporal consistency and quality of features. For classifier extension, the use of within-season features derived from anniversary-date spring or fall reflectance observations may be problematic as seasonal dynamics can have high interannual variance, particularly at high latitudes. This variation may be sufficient to cause substantial classification errors. Figure 12.3 shows an example of the seasonal normalized difference vegetation index (NDVI) observations for a deciduous forest, which are extracted from Moderate Resolution Imaging Spectroradiometer (MODIS) 250-m data. In this example, a feature derived from the period between day 140 and 200 will show an almost 15% increase in NDVI. Reflectance normalization may help alleviate the problem, but as different land covers develop or green up at different

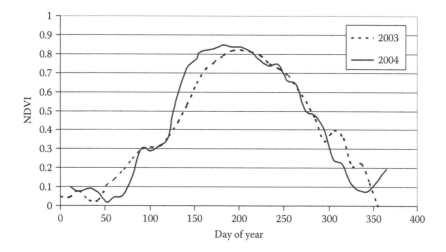

FIGURE 12.3 Comparison of MODIS NDVI for 2 years for a deciduous forest sample.

rates, simple normalization may be ineffective. If a classification problem strongly depends on seasonal temporal features, then a classifier retraining approach may be more robust than classifier extension.

12.3.4 DATA CONSISTENCY

Poor data consistency between years is a fundamental reason why simple approaches for developing consistent land-cover time series do not often perform well. Low consistency arises from known sources of error or undesirable variability in measuring conditions such as geolocation, atmospheric effects, sun-sensor geometry, topography, surface moisture, and seasonal phenology. The effects of these factors on reflectance variability have been widely studied and are not addressed here except for a few examples that directly evaluate the effect on change detection or classifier extension.

Several studies have evaluated the effect of geolocation accuracy on change detection (Dai and Khorram, 1998; Roy, 2000; Townshend et al., 1992; Wang and Ellis, 2005). These show that change detection accuracy is strongly affected by geolocation error relative to the image resolution and spatial heterogeneity. Dai and Khorram (1998) recommend an allowable error of 1/5 of a pixel for change detection studies.

The effect of atmosphere was evaluated by Pax-Lenny et al. (2001) who observed improvement in classifier extension accuracy with various approaches to atmospheric correction. However, the best result was obtained using a histogram-matching normalization technique that corrected for atmosphere, sensor calibration, phenology, and moisture differences between the image dates. Olthof et al. (2005) also found that image normalization improved classifier extension compared to an atmospheric correction approach based on dense dark vegetation.

12.4 ADDITIONAL PROCESSING TO IMPROVE TEMPORAL CONSISTENCY

Several methods can be used to suppress or correct errors between land-cover maps. Methods specific to developing land-cover time series are described here. The general idea is to use additional rules and weighting strategies to reduce the amount of false change detected.

12.4.1 EXPERT RULES

Expert rules are conditional "if ... then" checks derived from expert knowledge of the processes governing the system. For example, in land-cover mapping, forests typically do not grow at high

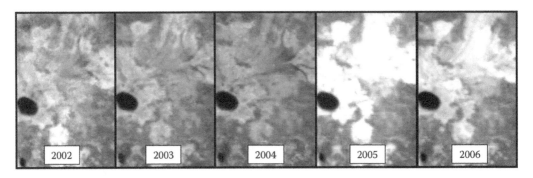

FIGURE 12.4 (See color insert.) Example of wetland interannual spectral variability as seen in MODIS at 250-m spatial resolution, with bands displayed as red = band 2, green = band 6, and blue = band 1.

elevations, and thus a simple rule can be used to constrain mapping forest pixels as a function of elevation. A similar result can possibly be achieved by including elevation data directly in the classification. Its effectiveness will depend on the classifier and training data used. The training data will need to include a range of vegetation and elevation combinations so that the classifier can develop a suitable decision boundary without creating confusion between other classes or overtraining. The main disadvantage is that training data requirements can be significantly increased. Expert rules are a powerful tool because they integrate knowledge and data sources not easily incorporated into supervised classification, including those related to environmental processes governing land-cover distribution. The basis of these rules in true and accepted knowledge of environmental processes allows them to be reliably extended over space and time.

For developing land-cover time series, the temporal dimension provides unique opportunities for rule definition. For example, wetlands are sensitive to moisture conditions, and this can result in high interannual spectral variability (Figure 12.4). Thus, in land-cover time series, wetland pixels often migrate between the wetland class when wet and the low vegetation cover class when dry. For consistently classifying a wetland that exhibits this temporal variation, a simple rule that functions as a temporal filter is one solution. Such a rule may state that if the temporal neighbors of a pixel classified as low vegetation for the previous year and the next year are both wetland, then the pixel should be reclassified as wetland (i.e., $Class_{time-1}$ = wetland, $Class_{time}$ = low vegetation, and $Class_{time+1}$ = wetland). Fraser et al. (2009) applied a similar rule for reducing the false change associated with cropland interannual variability.

12.4.2 FUZZY REASONING

Fuzzy reasoning is a framework that expands the simple "if … then" condition checks to include uncertainty. Uncertainty involves two aspects: (1) relevance, which refers to how much an information source contributes to the decision being made and (2) reliability, which refers to data quality (Liu and Mason, 2009). One of the main advantages of fuzzy reasoning is that it can be designed to incorporate several sources of information that contribute to a decision. It recognizes that no one piece of information is entirely correct and that a better result can be achieved by considering all information sources, their relevance to the problem, and uncertainty in their measurement. The biggest challenge with fuzzy approaches is the difficulty in determining a function that translates information to relevance, otherwise known as fuzzy membership. This is often a subjective process and can strongly affect the results.

Classification algorithms provide a hard output designating the class that each pixel belongs to, but many also provide a soft or fuzzy output that measures the strength of class membership or confidence in class assignment (Lillesand and Keifer, 2007). These soft classification outputs are useful in developing methods to improve land-cover time series based on fuzzy reasoning by combining them with other data and information sources. Combining fuzzy information can be based on ranking sources of fuzzy information and summing these ranks, various approaches to weighted linear combination, or combining probabilistic information using Bayes or Dempster–Shafer theory. Dempster–Shafer theory is an extension of Bayesian analysis to better handle uncertainty. In Bayesian analysis, information either supports (probability) or refutes a decision (1-probability), whereas in Dempster–Shafer theory, information is allowed to be uncertain as to which outcome it supports (Liu and Mason, 2009). The use of prior probabilities in classification is an example of a Bayesian approach to fuzzy reasoning, which has had a long history in land-cover classification. The general strategy is to derive class prior probabilities used for modifying the class membership function. A simple method to derive prior probabilities is to take the class proportions from an existing land-cover map. This can be derived for small subregions of the map area to better reflect the local class proportions (Strahler, 1980). Inclusion of class proportions for land-cover time series development has been carried out by Latifovic and Pouliot (2005), Caccetta et al. (2007), and Friedl et al. (2010). Dempster–Shafer theory has been evaluated for image classification by Peddle (1995a, 1995b). It has also been used by Comber et al. (2004) and Latifovic and Pouliot (2005) for developing land-cover time series by combing several sources of information thought to be relevant to the classification problem being addressed.

12.4.3 Transition Matrices

Transition matrices are essentially a collection of rules or transition probabilities that determine allowable or likely "from–to" class transitions for the set of classes in a map. Some transitions are not possible or may not occur over the evaluated time period, such as the conversion of barren land to forest in a single year. Thus, the use of a transition matrix integrates this knowledge and offers a powerful means to reduce errors between maps. The matrices can be developed based on expert opinion or by empirical analysis. Values in the matrix can be binary, allowing or disallowing certain transitions from occurring, or can be fuzzy representing the likelihood of a given transition.

In an expert approach, the transition matrix is based on an expert's knowledge of the landscape and the factors affecting change such as vegetation growth rates, historical probability of fire, planned future level of harvesting, or other human-related land-use change activities. The approach provides the means to include knowledge of the past and future. The disadvantage of this approach is that it is subjective, and so results can vary depending on the expert(s) employed to develop the matrix. Clark et al. (2010) used a binary matrix as part of an expert-based temporal filter approach to remove the occurrence of temporal anomalies in the time series. Several different approaches to incorporate fuzzy expert definitions have been developed. Melgani and Serpico (2003) defined transitions for within-season changes from April to May and used them as part of a Markov approach to simultaneously classify images based on their spatial-temporal dependencies. Latifovic and Pouliot (2005) developed transition matrices for a 5-year temporal interval for both forward and backward updating of land-cover maps. Caccetta et al. (2007) used the framework of Bayesian probability networks to develop transition matrices to include multiple time steps and additional information sources.

The empirical approach is based on examining transitions between existing maps or initial classification results as part of an iterative method (Bruzzone and Serpico, 1997; Bruzzone et al., 1999; Liu et al., 2008). This approach has typically been developed globally for entire images. Liu et al. (2008) showed that improvements could be made with local analysis. This method can be used to quickly generate a more objective transition matrix compared to expert definitions, but errors in the initial land covers used to determine transition probabilities can bias the results, as they are not easily factored out. In both expert and empirical approaches, a training set is typically required to define parameters used in combining the spatial and temporal information. In this case, the training set can be large because samples for each from–to class combination should be included.

12.4.4 Change Area Constraints

Detecting change directly from spectral data provides the most control over the changes detected as well as error rates through careful designation of decision bounds. In postclassification comparison, the thresholds used to detect change are defined by class boundaries, which can be narrow depending on the number of classes and features used. The thresholds are more likely to be sensitive to variability in the data, resulting in detection of false change. For example, mixed and deciduous forests have very similar spectral characteristics, and data variability can easily be sufficient to cause pixels to flip between these classes between years. Thus to avoid this occurrence, a common approach in developing land-cover time series is to update the land cover only in areas where changes have been confidently detected (Fraser et al., 2009; Hurd et al., 2009; Latifovic and Pouliot, 2005; McDermid et al., 2008; Xian et al., 2009). This complicates the updating procedure by including this additional step and requires that the change detection method be designed to capture all desired changes. This approach has been used mostly to capture abrupt changes resulting from fire or vegetation removal due to human activities such as forest harvesting. However, it is logical to extend this to include gradual changes that occur over longer periods.

12.5 EXAMPLES OF LAND-COVER TIME SERIES

Development of land-cover time series is a rapidly growing area of research. Efforts to date have focused on approaches to reduce false change between maps using methods of varying complexity. Time series consistency has been a central objective, but other considerations such as information content, cost, automation, implementation complexity, and historical legacies have also been important factors for different products. The following provides an overview of various global and national efforts for land-cover time series development that have been based on supervised classification.

12.5.1 Global

The MODIS land-cover time series was designed to support scientific investigations on the state of land cover at the global scale (Friedl et al., 2002, 2010). The version 5 product is generated annually at 500-m spatial resolution based on the 17-class International Geosphere-Biosphere legend. Data are classified using a decision tree with boosting, which improves accuracy and provides the class conditional probability. The method follows a classifier retraining approach in which each year the classifier is trained from a global database of samples collected by visual interpretation of Landsat data. The training database is carefully maintained to avoid including change samples and to expand it to cover more effectively the global distribution of land-cover types. Several postprocessing steps are used to improve accuracy and enhance temporal consistency by combining the conditional probabilities from the decision tree with prior probabilities designed to correct training sample bias and provide information on spatial class proportions. Simple rules are used to account for problematic classes such as wetlands, where specific thresholds are applied to the posterior probabilities (i.e., class conditional probabilities adjusted by sample bias and spatial proportions) to optimize the classification. The product is not meant to be used for simple postclassification analysis as too much change would likely be detected. It is more appropriately used in modeling where confusions between classes would be less problematic. For example, carbon sequestration between shrub and deciduous forest is more similar than that between shrub and grass (Turner et al., 2004). By not introducing too many constraints, the MODIS algorithm ensures that the product will effectively capture interannual land-cover variability, which is important in some applications. Results of cross-validation show the 2005 land-cover map as having an overall accuracy of ~75%.

12.5.2 AUSTRALIA

Australia produced a forest\nonforest time series from Landsat data at multiyear intervals starting in 1972, with recent time steps generated annually (Caccetta et al., 2007). The product was developed as part of Australia's National Carbon Accounting System. The initial classification was based on linear discriminant analysis applied using a classifier extension approach, but additional training data were used to recalibrate class thresholds to make them more specific to the mapping zone and image scene. Training data are typically collected by an experienced interpreter. The final step involves combining spatial-temporal transition probabilities based on a Bayesian conditional probability network approach. One of the key advantages is that transition probabilities for multiple dates can be incorporated. Recent work has sought to increase the number of forest-cover classes monitored (Furby et al., 2008).

12.5.3 CANADA

For monitoring in Canada, Latifovic and Pouliot (2005) generated a national land-cover time series from 1-km Advanced Very High Resolution Radiometer (AVHRR) data for 1985 to 2005 at 5-year time steps. The product was developed as a proof of concept where the main design objective was to maximize land-cover consistency. The method employed an updating strategy to modify an existing land-cover map to other dates in the time series. Updating was constrained to areas detected as change using a modified spectral change vector analysis considered to be more accurate than postclassification comparison. A minimum distance classifier was used for the initial classification that was retrained for each change object using unchanged samples from the change object's local neighborhood. The distance measure between the pixel and spectral signature for each class was used to define the class memberships. This information, along with expert-based transition matrices and class prior probabilities, was combined based on the Dempster–Shafer theory to provide adjusted pixel memberships from which the maximum value was taken as the updated class. The method was designed to update land cover both forward and backward in time, and thus it required separate transition matrices for each temporal direction. A problem was encountered with the classifier retraining method as the base map used in the analysis contained fire classes whose spectral signature migrated toward that of a mature forest and was not representative of the original fire condition. To address this concern, samples for this class were taken from the original image data and land-cover map and were applied using classifier extension. Accuracy for the time series at a thematic resolution of 12 classes was 62%. Using only samples where the dominate class occupied more than 60% of the pixel increased the accuracy to 74%, showing the effect of mixed pixels on accuracy at this spatial resolution.

12.5.4 UNITED STATES

In the United States, three national land-cover products were developed based on Landsat data for circa 1992, 2001, and 2006, with the next update planned for 2011. Accuracy assessed for the 1992 product in the Western United States showed a range from 38% to 70% for the Anderson Level II legend with 21 classes. For the Anderson Level I legend, the accuracy was much higher ranging from 82% to 85% for 7 classes (Wickham et al., 2004). Accuracy of the 2001 product ranged from 73% to 77% for 29 classes based on several sample regions distributed across the country (Homer et al., 2004). Comparison of these maps for change detection was not recommended because of differences in processing methodologies, legends, and data quality, which would lead to the detection of substantial false change. To address the need for change detection over this period, a change product was derived from these maps using a combination of legend reclassification, postclassification comparison, land-cover reclassification, and spectral change detection as described by Fry et al. (2009). For the 2006 land cover, a change updating approach was employed so that it could be compared to a revised version of the 2001 map (Xian et al., 2009; Fry et al., 2011). The approach used a Multi-Index Integrated Change Analysis method that combined several change features using a set of complex expert rules and class specific thresholds to detect change. Within each Landsat scenes, a decision tree was retrained by sampling the 2001 land cover in unchanged areas with the 2006 image data. This decision tree was then used to classify pixels in change areas. Overall accuracy for several sample areas across the United States ranged from 78% to 89 % (Xian et al., 2009).

REFERENCES

Arora, M.K. and Foody, G.M. 1997. Log-linear modeling for the evaluation of the variables affecting the accuracy of probabilistic, fuzzy and neural network classifications. *International Journal of Remote Sensing*, 18, 785–798.

Barson, M. 2008. Developing land cover and land use datasets for the Australian continent—A collaborative approach. In J.C. Campbell, K.B. Jones, J.H. Smith, and M.T. Koeppe (Eds.), *Proceedings of the North America Land Cover Summit 2006* (pp. 45–74).Washington, D.C., September 20–22.

Bonan, G.B. 2004. Biogeophysical feedbacks between land cover and climate. In R. Defires, G. Asner, and R. Houghton (Eds.), *Ecosystems and Land Use Change* (pp. 61–72). Washington, D.C.: American Geophysical Union.

Bruzzone, L. and Serpico, S.B. 1997. An iterative technique for the detection of land cover transitions in multitemporal remote-sensing images. *IEEE Transactions on Geoscience and Remote Sensing*, 35, 858–867.

Bruzzone, L., Prieto, D.F., and Serpico, S.B. 1999. A neural-statistical approach to multitemporal and multisource remote-sensing image classification. *IEEE Transactions on Geoscience and Remote Sensing*, 37, 1350–1359.

Caccetta, P.A., Furby, S.L., O'Connell, J., Wallace, J.F., and Wu, X. 2007. Continental monitoring: 34 years of land cover change using Landsat imagery. In *32nd International Symposium on Remote Sensing of Environment*, June 25–29, San José, Costa Rica.

Chang, M. 2003. *Forest Hydrology: An Introduction to Water and Forests*. Boca Raton, FL: CRC Press, 373 pp.

Clark, M.L., Aide, T.M., Grau, R.H., and Riner, G. 2010. A scalable approach to mapping annual land cover at 250 m using MODIS time series data: A case study in the Dry Chaco ecoregion of South America. *Remote Sensing of Environment*, 114, 2816–2832.

Colwell, J.E., Davis, G., and Thomson, F. 1980. Detection and measurement of changes in the production and quality of renewable resources. USDA Forest Service Final Report 145300–4-F, ERIM, Ann Arbor, MI, USA.

Comber, A.J., Law, A.N.R., and Lishman, J.R. 2004. A comparison of Bayes', Dempster-Shafer and Endorsement theories for managing knowledge uncertainty in the context of land cover monitoring. *Computers, Environment and Urban Systems*, 28, 311–327.

Coppin, P., Jonckheere, I., Nackaerts, K., and Muys, B. 2004. Digital change detection methods in ecosystem monitoring: A review. *International Journal of Remote Sensing*, 25, 1565–1596.

Dai, X., and Khorram, S. 1998. The effects of misregistration on the accuracy of remotely sensed change detection. *IEEE Transactions on Geoscience and Remote Sensing*, 36, 1567–1577.

Eshleman, K.N. 2004. Hydrological consequences of land use change: A review of the state-of-the-science. In R. Defires, G. Asner, and R. Houghton (Eds.), *Ecosystems and Land Use Change* (pp. 13–30). Washington, D.C.: American Geophysical Union.

Fischer, G., Prieler, S., van Velthuizen, H., Lensink, S.M., Londo, M., and de Wit, M. 2010. Biofuel production potentials in Europe: Sustainable use of cultivated land and pastures. Part I: Land productivity potentials. *Biomass and Bioenergy*, 34, 159–172.

Foody, G.M. 2002. Status of land cover classification accuracy assessment. *Remote Sensing of Environment*, 70, 627–633.

Foody, G.M. and Arora, M.K. 1997. An evaluation of some factors affecting the accuracy of classification by an artificial neural network. *International Journal of Remote Sensing*, 18, 799–810.

Foody, G.M. and Mathur, A. 2004. Toward intelligent training of supervised image classifications: Directing training data acquisition for SVM classification. *Remote Sensing of Environment*, 93, 107–117.

Foody, G.M. and Mathur, A. 2006. The use of small training sets containing mixed pixels for accurate hard image classification: training on mixed spectral responses for classification by SVM. *Remote Sensing of Environment*, 103, 179–189.

Foody, G.M., Mathur, A., Sanchez-Hernandez, C., and Boyd, D.S. 2006. Training set size requirements for the classification of a specific class. *Remote Sensing of Environment*, 104, 1, 1–14.

Foody, G.M., McCulloch, M. B., and Yates, W. B. 1995. The effect of training set size and composition on artificial neural network classification. *International Journal of Remote Sensing*, 16, 1707–1723.

Fraser, R.H., Abuelgasim, A., and Latifovic, R. 2005. A method for detecting large-scale forest cover change using coarse spatial resolution imagery. *Remote Sensing of Environment*, 95, 414–427.

Fraser, R.H., Olthof, I., and Pouliot, D. 2009. Monitoring land cover change and ecological integrity in Canada's national parks. *Remote Sensing of Environment*, 113, 1397–1409.

Friedl, M.A., McIver, D.K., Hodges, J.C.F., Zhang, X.Y., Muchoney, D., Strahler A.H., Woodcock, C.E., et al. 2002. Global land cover mapping from MODIS: Algorithms and early results. *Remote Sensing of Environment*, 83, 287–302.

Friedl, M.A., Sulla-Menashe, D., Tan, B., Schneider, A., Ramankutty, N., Sibley, A., and Huang, X. 2010. MODIS collection 5 global land cover: Algorithm refinements and characterization of new datasets. *Remote Sensing of Environment*, 114, 168–182.

Fry, J.A., Coan, M.J., Homer, C.G., Meyer, D.K., and Wickham, J.D. 2009. Completion of the National Land Cover Database (NLCD) 1992–2001 land cover change retrofit product, U.S. Geological Survey Open-File Report 2008–1379, 18 p.

Fry, J.A., Xian, G., Jin, S., Dewitz, J.A., Homer, C.G., Yang, L., Barnes, C.A., Herold, N.D., and Wickham, J.D. 2011. Completion of the 2006 National Land Cover Database for the Conterminous United States. *Photogrammetric Engineering and Remote Sensing*, 77(9), 858–864.

Furby, S.L., Caccetta, P.A., Wallace, J.F., Wu, X. O'Connell, J., Collings, S., Traylen, A., and Deveraux, D. 2008. Recent development in Landsat-based continental scale land cover change monitoring in Australia. *The International Archives of the Photogrammetry, Remote Sensing and Spatial Information Sciences*, XXXVII, 1491–1496.

Homer, C., Huang, C., Yang, L., Wylie B., and Coan, M. 2004. Development of a 2001 national land cover database for the United States. *Photogrammetric Engineering and Remote Sensing*, 70, 829–840.

Huang, C., Davis, L.S., and Townshend, J.R.G. 2002. An assessment of support vector machines for land cover classification. *International Journal of Remote Sensing*, 23, 725–749.

Hurd, J., Civco, D., Arnold, C., Wilson, E., and Rozum, J. 2009. Connecticut's changing landscape: Multi-temporal land cover data for Connecticut. The Fifth International Workshop on the Analysis of Multitemporal Remote Sensing Images, Connecticut, July 28–30, pp. 262–269.

Jain, A.K., Duin, R.P.W., and Mao, J. 2000. Statistical pattern recognition: A review. *IEEE Transactions on Pattern Analysis and Machine Intelligence*, 22, 4–37.

Jensen, J. 1996. *Introductory Digital Image Processing: A Remote Sensing Perspective*. Englewood Cliffs, NJ: Prentice-Hall.

Kerr, J. 2001. Global biodiversity patterns: From description to understanding. *Trends in Ecology and Evolution*, 16, 424–425.

Kerr, J. and Ostrovsky, M. 2003. From space to species: Ecological applications for remote sensing. *Trend in Ecology and Evolution*, 18, 299–305.

Kleeschulte, S. and Büttner, G. 2008. European land cover mapping—The CORINE experience. In J.C. Campbell, K.B. Jones, J.H. Smith, and M.T. Koeppe (Eds.), *Proceedings of the North America Land Cover Summit 2006* (pp. 31–44). Washington, DC, September 20–22.

Latifovic, R. and Olthof, I. 2004. Accuracy assessment using sub-pixel fractional error matrices of global land cover products derived from satellite data. *Remote Sensing of Environment*, 90, 153–165.

Latifovic, R. and Pouliot, D.A. 2005. Multi-temporal land cover mapping for Canada: Methodology and products. *Canadian Journal of Remote Sensing*, 31, 347–363.

Latifovic, R., Pouliot, D., and Nastev, M. 2010. Earth observation based land over for regional aquifer characterization. *Canadian Water Resources Journal*, 35, 1–18.

Latifovic, R., Zhu, Z., Cihlar, J., Giri, C., and Olthof, I. 2004. Land cover mapping of North and Central America—Global Land Cover 2000. *Remote Sensing of Environment*, 89, 116–127.

Liu, D., Song, K., Townshed, J.R.G., and Gong, P. 2008. Using local transition probability models in Markov random fields for forest change detection. *Remote Sensing of Environment*, 112, 2222–2231.

Liu, H. and Motoda, H. 2008. *Computational Methods of Feature Selection*. Boca Raton, FL: Chapman & Hall/CRC, 419 pp.

Liu, J.G. and Mason, P.J. 2009. *Essential Image Processing and GIS for Remote Sensing*. West Sussex, United Kingdom: John Wiley & Sons, 443 pp.

Lu, D., Mausel, P., Brondizio, E., and Moran, E. 2004. Change detection techniques. *International Journal of Remote Sensing*, 20, 2365–2407.

Lillesand, T.M. and Kiefer, R.W. 2007. *Remote Sensing and Image Interpretation*, 6th ed. New York: John Wiley and Sons, 804 pp.

Mather, P.M. 1999. *Computer Processing of Remotely-Sensed Images: An Introduction*. Chichester and New York: Wiley, 292 pp.

McDermid, G.J., Linke, J., Pape, A.D., Laskin, D.N., McLane, A.J., and Franklin, S.E. 2008. Object-based approaches to change analysis and thematic map update: Challenges and limitations. *Canadian Journal of Remote Sensing*, 34, 462–466.

Melgani, F. and Serpico, S.B. 2003. A Markov random field approach to spatio-temporal contextual image classification. *IEEE Transactions on Geoscience and Remote Sensing*, 41, 2478–2487.

Meyer, D., Leisch, F., and Hornik, K. 2003. The support vector machine under test. *Neurocomputing*, 55, 169–186.

Minter, T.C. 1978. Methods of extending crop signatures from one area to another. Proceedings, the LACIE symposium, a technical description of the large-area crop inventory experiment (LACIE), October 23–26, 1978, Houston, TX.

Myers, V.I. 1983. Remote-sensing applications in agriculture. In J.E. Colwell and R.N. Colwell (Eds.), *Manual of Remote Sensing* (pp. 2111–2228). Falls Church, VA: American Society of Photogrammetry. 2.

Olthof, I., Butson, C., and Fraser, R. 2005. Signature extension through space for northern land-cover classification: A comparison of radiometric correction methods. *Remote Sensing of Environment*, 95, 290–302.

Pal, M. and Mather, P.M. 2003. An assessment of the effectiveness of decision tree methods for land cover classification. *Remote Sensing of Environment*, 86, 554–565.

Pal, M. and Mather, P.M. 2005. Support vector machines for classification in remote sensing. *International Journal of Remote Sensing*, 26, 1007–1011.

Pax-Lenney, M., Woodcock, C.E., Macomber, S.A., Sucharita, G., and Song, C. 2001. Forest mapping with generalized classifier and Landsat TM data. *Remote Sensing of Environment*, 77, 241–250.

Peddle, D. 1995a. Knowledge formulation for supervised evidential classification. *Photogrammetric Engineering and Remote Sensing*, 61, 409–417.

Peddle, D. 1995b. MERCURY⊕: An evidential reasoning image classifier. *Computers and Geosciences*, 21, 1163–1176.

Peddle, D.R., Foody, G.M., Zhang, A., Franklin, S.E., and LeDrew, E.F. 1994. Multisource image classification II: An empirical comparison of evidential reasoning and neural network approaches. *Canadian Journal of Remote Sensing*, 20, 396–407.

Potapov, P., Hansen, M., Stehman, S., Loveland, T., and Pittman, K. 2008. Combining MODIS and Landsat imagery to estimate and map boreal forest cover loss. *Remote Sensing of Environment*, 112, 3708–3719.

Potter, K.W., Douglas, J.C., and Brick, E.M. 2004. Impacts of agriculture on aquatic ecosystems in the humid United States. In R. Defires, G. Asner, and R. Houghton (Eds.), *Ecosystems and Land Use Change* (pp. 61–72). Washington, DC: American Geophysical Union.

Pouliot, D., Latifovic, R., Fernandes, R., and Olthof, I. 2009. Evaluation of annual forest disturbance monitoring using decision trees and MODIS 250m data. *Remote Sensing of Environment*, 113, 1749–1759.

Roy, D.P. 2000. The impacts of misregistration upon composited wide field of view satellite data and implications for change detection. *IEEE Transactions on Geoscience and Remote Sensing*, 38, 2017–2032.

Singh, A. 1989. Digital change detection techniques using remotely-sensed data. *International Journal of Remote Sensing*, 10, 989–1003.

Stow, D.A., Tinney, L.R., and Estes, J.E. 1980. Deriving land use/land cover change statistics from Landsat: A study of prime agricultural land. *Proceedings of the 14th International Symposium on Remote Sensing of Environment* held in Ann Arbor in 1980 (pp. 1227–1237).

Strahler, A. 1980. The use of prior probabilities in maximum likelihood classification of remotely sensed data. *Remote Sensing of Environment*, 10, 135–163.

Townshend, J.R.G., Justice, C.O., Gurney, C., and McManus, J. 1992. The impact of misregistration on change detection. *IEEE Transactions on Geoscience and Remote Sensing*, 30, 1054–1060.

Turner, D.P., Ollinger, S.V., and Kimball, J.S. 2004. Integrating remote sensing and ecosystem process models for landscape- to regional-scale analysis of the carbon cycle. *Bioscience*, 54, 573–584.

Wang, H. and Ellis, E.C. 2005. Image misregistration error in change measurements. *Photogrammetric Engineering and Remote Sensing*, 71, 1037–1044.

Wickham, J.D., Stehman, S.V., Smith, J.H., and Yang, L. 2004. Thematic accuracy of the 1992 national land-cover data for the western United States. *Remote Sensing of Environment*, 91, 452–468.

Woodcock, C.E., Macomber, S.A., Pax-Lenney, M., and Cohen, W.B. 2001. Monitoring large areas for forest change using Landsat: Generalization across space, time and Landsat sensors. *Remote Sensing of Environment*, 78, 194–203.

Xian, G., Homer, C., and Fry, J. 2009. Updating the 2001 national land cover database land cover classification to 2006 using Landsat change detection methods. *Remote Sensing of Environment*, 113, 1133–1147.

13 Forest-Cover Change Detection Using Support Vector Machines

Chengquan Huang and Kuan Song

CONTENTS

13.1 INTRODUCTION

Forest is a critical component of the earth's surface, covering about 30% of the land area (e.g., FAO, 2001). Forest-cover changes, especially those of anthropogenic origin, have a wide impact on critical environmental processes including energy balance, water cycle, and biogeochemical processes. Understanding such changes, as well as their causes, requires that the changes be quantified. Reliable and up-to-date information on forest and forest change is required not only for resource management and ecological applications but also for addressing many pressing issues ranging from local to global scales, including carbon assessment, ecosystem dynamics, sustainability, and the vulnerability of natural and human systems (Band, 1993; Houghton, 1998; Lal, 1995; Pandey, 2002; Schimel, 1995). With its ability to obtain repeated observations of the earth's surface, satellite remote sensing is a primary data source for forest change monitoring.

Several key steps are required for mapping forest change using remotely sensed data, including defining the scope and objectives, selecting suitable satellite datasets, performing image geometric and radiometric correction, and detecting change. The key issues in each step are discussed in Section 13.2. Although a typical land-cover change study also includes accuracy assessment and change analysis, these are not discussed in this chapter because accuracy assessment methods have been detailed in many other publications (e.g., Congalton, 1991; Janssen and Wel, 1994; Stehman and Czaplewski, 1998), and analysis of a change product is often application driven. The main purpose of this chapter is to introduce an advanced machine learning algorithm called support vector machine (SVM) for mapping forest-cover change. The SVM is a statistical learning algorithm designed to achieve optimal classification accuracy through structural risk minimization (SRM) (Vapnik, 1995). Such a design allows the SVM to produce more accurate results than other machine learning algorithms commonly used in remote-sensing image classifications (e.g., Chan et al., 2001; Huang et al., 2002; Pal and Mather, 2005). The SVM was introduced to the remote-sensing community nearly a decade ago (Huang et al., 2002; Zhu and Blumberg, 2002). Since then it has seen increased use in remote-sensing-based studies of land cover and land-cover change (Huang et al., 2008; Knorn et al., 2009; Pal and Mather, 2005). Mountrakis et al. (2011) reviewed the use of the SVM in remote-sensing applications. This chapter offers a detailed description of the SVM algorithm and demonstrates its uses in forest change detection through a case study in eastern Paraguay.

13.2 MAJOR CONSIDERATIONS IN REMOTE-SENSING-BASED FOREST CHANGE DETECTION

13.2.1 DEFINING SCOPE AND OBJECTIVES

The first step in a study of forest change mapping is to define its scope and objectives. Among the issues to be considered are geographic coverage, spatial resolution and minimum mapping unit, temporal intervals, and change types. Broadly defined, land-cover change includes both modification within the same cover type and conversion from one type to another (Meyer and Turner, 1994). Conversion of forestland to agriculture, urban, and other nonforest uses is often referred to as deforestation, and the reverse is called afforestation or reforestation. In general, a forest harvest followed by immediate regrowth as part of a forest rotation process is not considered conversion. However, it may be difficult to distinguish between forest rotation and deforestation or afforestation if changes are mapped using images acquired at two time points only (see the discussion on bitemporal approach in Section 13.2.4). Examples of forest modification include thinning and various natural and human disturbances that result in partial removal of forest canopy. It should be noted that not all possible changes are detectable using available satellite images, nor do they have equal importance in different applications. Which change types should be mapped in a particular change mapping effort should be defined based on the intended uses of the derived change products and the ability to map those change types reliably using available satellite datasets.

13.2.2 SATELLITE DATA SELECTION

Once the scope of a change detection study is defined, the next step is to select suitable satellite images. Nowadays, users often have many satellite datasets to choose from. Whether a particular satellite dataset is suitable for forest change analysis is determined by its spatial and temporal characteristics. In general, Landsat images or images with Landsat-class spatial resolutions (i.e., hectare or subhectare resolutions) are suitable for analysis over large areas, because they are often available for very large areas, yet their pixel sizes are small enough for characterizing most logging, harvest, and many other human activities. In particular, the Landsat archive produced by a series of six Landsat systems provides one of the longest image records of the earth's land surface at subhectare spatial resolutions (Goward and Williams, 1997; Goward et al., 2006). An added benefit of using Landsat data is minimum or no data cost. The U.S. Geological Survey (USGS) adopted a no-cost data policy for all Landsat images in its archive in 2008, and no-cost-access policies are being planned for images to be acquired by future Landsat missions. Moderate Resolution Imaging Spectroradiometer (MODIS), advanced very high resolution radiometer (AVHRR), or other moderate-to-coarse spatial resolution datasets have been used to quantify large-scale clearing of forests in the tropical region (e.g., Hansen and DeFries, 2004; Morton et al., 2006). However, most changes of anthropogenic origin result in change patches that are smaller than the pixel size of those datasets, and therefore those changes cannot be detected reliably using those datasets (Justice and Townshend, 1988). Images with spatial resolutions finer than that of Landsat images will be needed to map selective logging and other fine-scale changes (Asner et al., 2005). Though such images of high spatial resolution are becoming increasingly available, obtaining large-area coverage using high-resolution images remains quite challenging, both technically and financially.

13.2.3 IMAGE CORRECTION REQUIREMENTS

Satellite images need to be corrected to achieve high levels of geometric integrity and radiometric consistency before they can be used to map land-cover change. Comparison of misaligned pixels often results in large quantities of spurious changes (Townshend et al., 1992). Inconsistent image radiometry can also result in false changes or make it difficult to derive accurate change products. Although accurate pixel alignment may be achievable using image-to-image registration techniques (e.g., Flusser and Suk, 1994; Kennedy and Cohen, 2003; Pratt, 1974), for most terrains, orthorectification or terrain correction is required to achieve satisfactory geolocation accuracy (Gao et al., 2009), which is also required for the images and the derived change products to be used together with ground measurements or other georeferenced datasets.

Radiometric inconsistency can arise from sensor degradation, other instrument errors, and changes in atmospheric conditions (Jensen, 1996). Such inconsistencies can be reduced or minimized by using the best available calibration methods (Chander and Markham, 2003; Chander et al., 2004, 2009; Markham and Barker, 1986) and effective atmospheric correction algorithms (e.g., Liang et al., 1997; Teillet and Fedosejevs, 1995; Vermote and Kotchenova, 2008). Because atmospheric correction was quite challenging owing to intensive computing requirements and lack of necessary *in situ* atmospheric measurements, radiometric normalization techniques were developed to achieve relative radiometric correction (Elvidge et al., 1995; Heo and FitzHugh, 2000; Vicente-Serrano et al., 2008). Recently, an automated atmospheric correction algorithm was developed and implemented in the Landsat Ecosystem Disturbance Adaptive Processing System (LEDAPS), which allows rapid processing of large quantities of Landsat images (Huang et al., 2009a; Masek et al., 2006). Because the radiometry of most surface types is also a function of vegetation phenology, use of images acquired during anniversary week, month, or season of different years is also necessary to minimize radiometric variations arising from differences in vegetation phenology (Lunetta et al., 2004).

13.2.4 CHANGE DETECTION APPROACHES

Satellite images were used to map land-cover change soon after the launch of the first Landsat in 1972 (Gordon, 1980; Todd, 1977). Since then, many change detection algorithms have been developed, tested, and used in land-cover change studies. Comprehensive reviews of these algorithms have been provided in several publications (Coppin et al., 2004; Lu et al., 2004; Singh, 1989). Most of these algorithms are bitemporal, that is, they use two time points data for change analysis, where each time point may be represented by images acquired in a single date or in multiple dates centered around that time point. One straightforward approach for bitemporal change detection is postclassification comparison (Figure 13.1a). In this approach, a land-cover classification is developed for each time point, and changes are mapped by comparing the two classifications. A main drawback of this approach is that owing to the compounding effect of errors in two separate classifications, the derived change map may have substantially more errors than the map derived using each individual classification (Stow et al., 1980).

Alternatively, changes can be detected using an image comparison approach or a bitemporal classification approach. In the image comparison approach (Figure 13.1b), changes are detected by comparing the images acquired at the two time points using simple techniques, such as differencing, ratioing, and regression; more complex techniques, such as change vector analysis, principal component analysis, or other forms of spectral transformation; or hybrid methods that combine some of these techniques. In general, these techniques rely on threshold values derived from local knowledge to separate change from no-change pixels. However, owing to spatial and temporal variations in vegetation composition and phenology and residual among-scene radiometric inconsistencies, such localized threshold values typically vary from one image to another. Therefore, they are generally not transferrable among images (Song et al., 2001). As a result, use of the image comparison approach in studies requiring a large number of Landsat images can be quite labor-intensive and time-consuming.

In the bitemporal classification approach (Figure 13.1c), images acquired at two time points are classified simultaneously, with change classes being included and mapped as part of the classification. Both unsupervised clustering algorithms and supervised machine learning algorithms can be used. Use of a clustering method in the bitemporal classification approach is often labor-intensive and time-consuming because human inputs are required to label the spectral clusters. However, given adequate expert knowledge and human inputs, this approach can yield highly reliable forest change products (Huang et al., 2007, 2009b; Steininger et al., 2001). Alternatively, one can use supervised machine learning algorithms (Chan, 1998; Chan et al., 2001; Huang et al., 2008), which are often more efficient than unsupervised methods when the required training data are available (Huang et al., 2003). Later in this chapter, we demonstrate the use of the SVM and several other machine learning algorithms for forest-cover change detection using the bitemporal classification approach.

In addition to the bitemporal change detection techniques described above, algorithms capable of analyzing three or more images at a time have been developed (e.g., Cohen et al., 2002; Coppin and Bauer, 1996; Lunetta et al., 2004). Recently, algorithms have also been developed for mapping forest change using Landsat time series stacks (LTSS), where each LTSS consists of one image every year

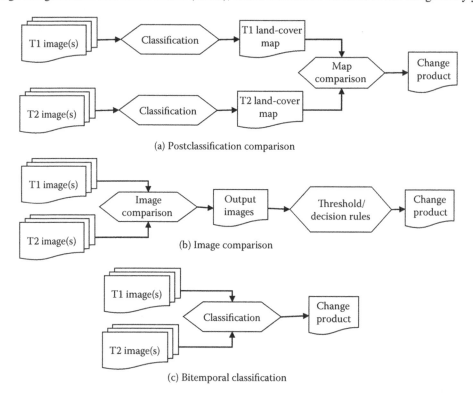

(a) Postclassification comparison

(b) Image comparison

(c) Bitemporal classification

FIGURE 13.1   Schematic diagrams of three bitemporal change detection approaches.

or every 2 years for two decades or longer (Huang et al., 2010; Kennedy et al., 2007, 2010). Although an LTSS can be divided into a sequence of image pairs and each pair can be analyzed using a bitemporal method, such an approach is far less efficient than a method designed for analyzing the entire LTSS simultaneously. By considering the rich temporal information provided by dense time series observations, the latter approach also allows detection of trends that may not be obvious and therefore may not be detectable using images acquired at two time points (Cohen et al., 2010; Huang et al., 2010; Kennedy et al., 2010).

13.3 THE SVM ALGORITHM

13.3.1 A Brief Overview

Vladimir N. Vapnik, a Russian mathematician and electrical engineer, is widely acknowledged as the inventor of the SVM. He attributed the development of the SVM to advances in mathematical reasoning and statistical learning over the last half century. The mathematical formulation of the SVM has been detailed in many publications (e.g., Burges, 1998; Vapnik, 1995, 1998). The description in this section follows the work of Vapnik (1995), Burges (1998), and Huang et al. (2002).

The inductive principle behind the SVM is SRM. This theory was designed to minimize overfitting, a problem common to classification models developed using neural networks and decision trees (Foody and Arora, 1997; Friedl et al., 1999; Paola and Schowengerdt, 1995). According to Vapnik (1995), the risk of a learning machine (R) is bounded by the sum of the empirical risk estimated from training samples (R_{emp}) and a confidence interval (Ψ):

$$R \leq R_{emp} + \Psi.$$

The SRM strategy is to keep the empirical risk (R_{emp}) fixed and minimize the confidence interval (Ψ), which is achieved by maximizing the margin between a separating hyperplane and the closest data points (Figure 13.2a). Here a separating hyperplane refers to a plane in a multidimensional

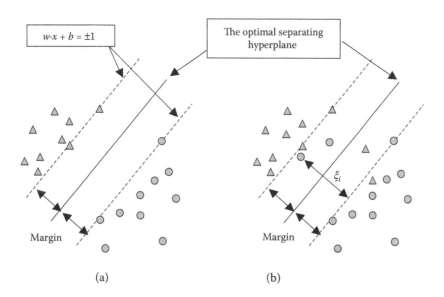

(a)   (b)

FIGURE 13.2 The SVM minimizes the risk of a classifier for (a) separable data samples by defining an optimal separating hyperplane as class boundary. This theory is extended to (b) nonseparable data samples by introducing a slack variable ξ_i and a penalty (C) (see Section 13.3.3).

space that separates the data samples of two classes. The optimal separating hyperplane is the separating hyperplane that maximizes the margin from closest data points to that hyperplane.

Apparently, the optimal separating hyperplane concept of the SVM requires a two-class problem. Two strategies can be employed to adapt a two-class method to an N-class classification problem (Gualtieri and Cromp, 1998). One is to construct a classifier for each pair of classes, which would result in $N(N - 1)/2$ classifiers. In this "one-against-one" strategy, a voting mechanism is used to determine the final class label for each data point. The other strategy is to break the N-class case into N two-class cases, in each of which a classification model is trained to classify one class against all others. In this "one-against-the-rest" strategy, a pixel is labeled with the class with which the pixel has the highest confidence value (Vapnik, 1995). Hsu and Lin (2002) demonstrated that the two approaches yielded similar accuracies, but the "one-against-one" strategy was faster.

Another key element of the SVM is the incorporation of regularization, a technique designed to achieve stable solutions in solving least square problems that consist of noisy data. In the SVM, this technique also makes it possible to define optimal classification boundaries for classes that are not 100% separable (see Section 13.3.3), which are common in land-cover and other real-world classification problems.

13.3.2 THE OPTIMAL SEPARATING HYPERPLANE

Let the training data of two separable classes with k samples be represented by

$$(x_1, y_1), \ldots, (x_k, y_k),$$

where $x \varepsilon R^n$ is an n-dimensional vector, and y is class label, having values of 1 or -1. Suppose the two classes can be separated by two hyperplanes parallel to the optimal hyperplane (Figure 13.2a):

$$w \cdot x_i + b \geq 1 \quad \text{for } y_i = 1, \quad i = 1, 2, \ldots, k, \tag{13.1}$$

$$w \cdot x_i + b \leq -1 \quad \text{for } y_i = -1, \tag{13.2}$$

where $w = (w_1, \ldots, w_n)$ is a vector of n elements. Inequalities (13.1) and (13.2) can be combined into a single inequality:

$$y_i [w \cdot x_i + b] \geq 1 \quad i = 1, \ldots, k. \tag{13.3}$$

As shown in Figure 13.2a, the optimal separating hyperplane is the one that separates the data with maximum margin. This hyperplane can be found by minimizing the norm of w, or the following function:

$$F(w) = \frac{1}{2}(w \cdot w) \tag{13.4}$$

under inequality constraint (13.3).

The saddle point of the following Lagrangean gives solutions to the above optimization problem:

$$L(w, b, \alpha) = \frac{1}{2}(w \cdot w) - \sum_{i=1}^{k} \alpha_i \{ y_i [w \cdot x_i + b] - 1 \}, \tag{13.5}$$

where $\alpha_i \geq 0$ are Lagrange multipliers (Sundaram, 1996). Solution to this optimization problem requires that the gradient of $L(w, b, \alpha)$ with respect to w and b vanishes, giving the following conditions:

$$w = \sum_{i=1}^{k} y_i \alpha_i x_i. \tag{13.6}$$

$$\sum_{i=1}^{k} \alpha_i y_i = 0. \tag{13.7}$$

Substitute (13.6) and (13.7) into (13.5), then the optimization problem becomes as follows:
Maximize

$$L(\alpha) = \sum_{i=1}^{k} \alpha_i - \frac{1}{2} \sum_{i=1}^{k} \sum_{j=1}^{k} \alpha_i \alpha_j y_i y_j (x_i \cdot x_j), \tag{13.8}$$

under constraints

$$\alpha_i \geq 0, \quad i = 1, \ldots, k.$$

Given an optimal solution $\alpha^0 = (\alpha_1^0, \ldots, \alpha_k^0)$ to (13.8), the solution w^0 to (13.5) is a linear combination of training samples:

$$w^0 = \sum_{i=1}^{k} y_i \alpha_i^0 x_i. \tag{13.9}$$

According to the Kuhn–Tucker theory (Sundaram, 1996), only points that satisfy the equalities in (13.1) and (13.2) can have nonzero coefficients α_i^0. These points lie on the two parallel hyperplanes and are called support vectors (Figure 13.2a). Let $x^0(1)$ be a support vector of one class and $x^0(-1)$ of the other, then the constant b^0 can be calculated as follows:

$$b^0 = \frac{1}{2} \left[w^0 \cdot x^0(1) + w^0 \cdot x^0(-1) \right]. \tag{13.10}$$

The decision rule that separates the two classes can be written as follows:

$$f(x) = \text{sign} \left(\sum_{\text{support vector}} y_i \alpha_i^0 (x_i \cdot x) - b^0 \right). \tag{13.11}$$

13.3.3 Dealing with Nonseparable Cases

As discussed earlier, an important assumption in deriving the above solution is that the data points are separable in the feature space. It is easy to see that there is no optimal solution if the data points cannot be separated without error. To resolve this problem, a penalty value C for misclassification errors and positive slack variables ξ_i are introduced (Figure 13.2b). These variables are incorporated into constraints (13.1) and (13.2) as follows:

$$w \cdot x_i + b \geq 1 - \xi_i \quad \text{for } y_i = 1, \tag{13.12}$$

$$w \cdot x_i + b \leq -1 + \xi_i \quad \text{for } y_i = -1, \tag{13.13}$$

where $\xi_i \geq 0$, and $i = 1, \ldots, k$.

The objective function (13.4) then becomes

$$F\left(w,\xi\right)=\frac{1}{2}\left(w\cdot w\right)+C\left(\sum_{i=1}^{k}\xi_{i}\right)^{l},\tag{13.14}$$

where C is a preset penalty value for misclassification errors. When $l=1$, the solution to this optimization problem is similar to that of the separable case.

13.3.4 SUPPORT VECTOR MACHINES

To generalize the above method to nonlinear decision functions, the SVM implements the following idea: it maps the input vector x into a high-dimensional feature space H and constructs the optimal separating hyperplane in that space. Suppose the data are mapped into a high-dimensional space H through a mapping function Φ:

$$\Phi:R^{n}\to H.\tag{13.15}$$

A vector x in the feature space can be represented as $\Phi(x)$ in the high-dimensional space H. Since the only way in which the inputs appear in the training problem (13.8) is in the form of dot products of two vectors, the training algorithm in the high-dimensional space H will depend only on data in this space through dot products, that is, on functions of the form $\Phi(x_i)\cdot\Phi(x_j)$. If there is a kernel function K such that

$$K(x_i,x_j)=\Phi(x_i)\cdot\Phi(x_j),\tag{13.16}$$

we will only need to use K in the training program without knowing the explicit form of Φ. The same trick can be applied to the decision function (13.11) because the only form in which the data appear is in the form of dot products. Thus, if a kernel function K can be found, a classifier can be trained and used in the high-dimensional space without knowing the explicit form of the mapping function. The optimization problem (13.8) can be rewritten as follows:

$$L\left(\alpha\right)=\sum_{i=1}^{k}\alpha_{i}-\frac{1}{2}\sum_{i=1}^{k}\sum_{j=1}^{k}\alpha_{i}\alpha_{j}y_{i}y_{j}K\left(x_{i},x_{j}\right).\tag{13.17}$$

And the decision rule expressed in equation (13.11) becomes as follows:

$$f\left(x\right)=\text{sign}\left(\sum_{\text{support vector}}y_{i}\alpha_{i}^{0}K\left(x_{i},x\right)-b^{0}\right).\tag{13.18}$$

A kernel that can be used to construct an SVM must meet the Mercer's condition (Courant and Hilbert, 1953). The following two types of kernels meet this condition (Vapnik, 1995): the polynomial kernels,

$$K\left(x_{1},x_{2}\right)=\left(x_{1}\cdot x_{2}+1\right)^{p},\tag{13.19}$$

and the radial basis functions (RBFs),

$$K\left(x_{1},x_{2}\right)=e^{-\gamma(x_{1}-x_{2})^{2}}.\tag{13.20}$$

13.4 ASSESSING SVM FOR FOREST-COVER CHANGE MAPPING IN PARAGUAY

Researchers at the Global Land Cover Facility (GLCF) conducted a forest change study in Paraguay for the period between 1990 and 2000, using an iterative clustering-supervised labeling method (Huang et al., 2009b). Although this method required intensive local expert knowledge and was time-consuming, it resulted in a highly reliable forest change product. For the 2000 epoch, forest and nonforest were classified with accuracies over 90% in many parts of the country (Huang et al., 2009b). Major classes in the developed forest change product included persisting forest, persisting nonforest, and forest loss, where "persisting" indicated that a pixel had the same class in 1990 and 2000. Forest gain was not included as a class in this product, because very little forest gain was observed in Paraguay. This map was used as reference data to evaluate several machine learning algorithms for forest change detection, including the SVM, the maximum likelihood classifier (MLC), Bayesian MLC, a contextual MLC, and decision trees. The SVM has been implemented in many computer software packages. A partial list of the packages is available at http://www.support-vector-machines.org/SVM_soft.html. The libsvm packages available at http://www.csie.ntu.edu.tw/~cjlin/libsvm/ were used in this case study (Chang and Lin, 2001). It implements the "one-against-one" strategy in dealing with more than two classes (see Section 13.3.1).

13.4.1 A Brief Overview of the Evaluated Algorithms

The MLC for a pixel X in n-band imagery is expressed as follows:

$$p_k(X) = \mathrm{Exp}\left(-\frac{1}{2}\ln|\Sigma_k| - \frac{1}{2}(X - \mu_k)' \Sigma_k^{-1}(X - \mu_k)\right), \tag{13.21}$$

where n is the number of bands, X is the image data of n bands, $p_k(X)$ is the probability of the pixel X belonging to class k, μ_k is the mean vector of class k, Σ_k is the variance-covariance matrix of class k, and $|\Sigma_k|$ is the determinant of Σ_k.

The Bayesian MLC brings the different occurrence probability of each class, that is, prior probability, into consideration. The philosophy of this classifier is "What is easier to find is what will be found." It is expressed as follows:

$$p_k(X) = \ln p(\varpi_k) - \frac{1}{2}\ln|\Sigma_k| - \frac{1}{2}(X - \mu_k)' \Sigma_k^{-1}(X - \mu_k), \tag{13.22}$$

where $p(\varpi_k)$ denotes the prior probability of class k. In our tests, this parameter was estimated from the previous MLC result as an approximation.

Besag (1986) derived an extended version of the Bayesian MLC by incorporating a heuristic measurement of contextual information as follows:

$$p_k(X) = \ln p(\varpi_k) - \frac{1}{2}\ln|\Sigma_k| - \frac{1}{2}(X - \mu_k)' \Sigma_k^{-1}(X - \mu_k) + \ln(\beta n), \tag{13.23}$$

where βn is a heuristic measure of the contextual information. The basic philosophy is that if the pixels surrounding the central pixel X are of the same class k as the center pixel X, then the probability of X being in k will be boosted. β is a heuristic parameter, which was set to 0.1 in our experiment, and n is the number of same-class pixels around X within a window of given size, which was 5×5 in our experiment. As this method is by nature a heuristic extension of the Bayesian classifier, the result was produced after several iterations until it reached a stable state where less than 1% of the pixels changed between iterations.

Decision tree defines classification rules by breaking a classification problem that is often very complex into multiple stages of simpler decision-making processes (Safavian and Landgrebe, 1991). Owing to demonstrated robustness in classifying remotely sensed data (Hansen et al., 1996), it has been used to produce land-cover classifications at regional (Homer et al., 2004; Huang et al., 2003) and global scales (DeFries et al., 1998; Friedl et al., 2002; Hansen et al., 2000). Depending on the algorithms used to grow a tree (i.e., to generate the classification rules), different flavors of decision trees have been developed (Breiman et al., 1984; Quinlan, 1993). The algorithm evaluated in this study is a univariate decision tree called C5.0 (Quinlan, 1993). Some details of this proprietary software are provided at http://rulequest.com/see5-info.html.

13.4.2 Experimental Design and Results

The selected algorithms were evaluated using training and test data selected across eastern Paraguay at two scales: single-scene and multiple-scene. In both cases, training samples were selected from the 10×10 km areas centered at the latitude–longitude degree intersections (Figure 13.3). In the single-scene test, only training samples selected within a scene were used to train a classifier, which was then used to classify that scene. This was replicated for all scenes in eastern Paraguay. In the multiscene test, training samples from different scenes were pooled together to train the selected algorithms. The developed classification models were than used to classify each scene separately.

In both tests, the pixels in a scene that were not selected as training samples were used to derive accuracy estimates using a standard accuracy assessment method as described by Congalton et al. (1983). The average accuracies or range of the accuracies for the forest loss class derived using each method in both the single-scene and multiscene tests are listed in Table 13.1.

Table 13.1 shows that in both the single-scene and multiscene tests, the SVM produced better accuracies than other algorithms. The decision tree had a slightly lower accuracy than the SVM in the single-scene test, but the difference increased to 5% in the multiscene test. The Bayesian MLC and contextual MLC performed marginally better than the original MLC in the single-scene test. All three MLC algorithms, however, produced substantially lower accuracies than the SVM and the decision tree, when tested at the multiscene level. For the three MLC methods, the accuracy differences between the single-scene test and multiscene test ranged from 15% to 28% in absolute value, demonstrating that these methods are less suitable for multiscene applications than the SVM and the decision tree.

13.5 DISCUSSION AND CONCLUSION

The SVM is an advanced machine learning algorithm that has several advantages over other machine learning algorithms commonly employed in land-cover studies. As a nonparametric method, it does not require *a priori* knowledge of the structure and statistical distribution of the data to be analyzed. Founded on the SRM theory, the SVM is designed to search for optimal class boundaries in formulating a classification model. Such a model should have less tendency of overfitting than those created using neural networks and decision trees, and therefore it may perform better when applied to unseen samples. As a result, the SVM consistently produces more accurate results than many other classification algorithms, as has been demonstrated in the case study described in this chapter and in many other studies (e.g., Chan et al., 2001; Huang et al., 2002; Pal and Mather, 2005). Thanks to its incorporation of regularization, a technique designed to achieve stable solutions in solving least square problems that consist of noisy data, the SVM can also tolerate considerable levels of measurement errors. In a series of experiments in eastern Paraguay, Song (2010) demonstrated that the SVM maintained relatively stable performances when the percentage of contaminated training data increased from 0% to as much as 30%. However, the accuracies of decision trees and neural networks degraded substantially as noises in the training data increased.

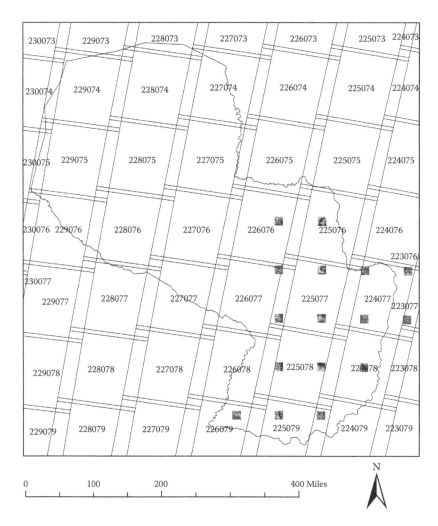

FIGURE 13.3 The SVM and the other classification algorithms evaluated in this study were trained using training samples selected from the 10 × 10 km areas centered at the 1° latitude–longitude intersections (small square images in eastern Paraguay). The pixels in each scene that were not used in training were used to calculate accuracy estimates. The slightly tilted rectangles show the nominal boundary of Landsat scenes, and the six-digit number in each rectangle gives the path (first three digits) and row (last three digits) number of each scene.

TABLE 13.1

Accuracies of the Forest Loss Class Derived Using Different Machine Learning Algorithms in Single-Scene and MultiScene Tests

Method	Single-Scene Test	Multscene Test
MLC	85%	60%–70%
Bayesian MLC	86%	60%–70%
Contextual MLC	88%	60%–70%
Decision tree	90%	80%
SVM	92%	85%

With these advantages, the SVM has seen increased use in remote-sensing applications (Mountrakis et al., 2011). Its value for bitemporal forest change analysis has been demonstrated in many studies (e.g., Knorn et al., 2009; Kuemmerle et al., 2009). As with other supervised methods, the SVM needs to be trained using training data. However, training data collection based on field-work, high-resolution images, or local expert knowledge is often expensive and time consuming, especially over very large areas. One advantage of the SVM is that, in general, it does not require as much training samples as other supervised methods do (Huang et al., 2002; Song, 2010), especially when training samples that are likely to be support vectors are targeted in training data collection (Foody and Mathur, 2004; Foody et al., 2006). Furthermore, by using a chain classification approach, Knorn et al. (2009) demonstrated that training data collected within a single scene may allow SVM classification of multiple neighboring scenes. The need for field data, high-resolution images, or local expert knowledge may be minimized or greatly reduced if training data can be derived automatically based on the images to be analyzed. An example of such an automatic training data delineation algorithm is the training data automation (TDA) method developed by Huang et al. (2008) for forest change analysis using Landsat images. Now that global Landsat datasets for four epochs over the last several decades have been assembled (Gutman et al., 2008; Tucker et al., 2004), a combination of the TDA or other automatic training data delineation methods with the SVM makes it feasible to map forest change at regional to global scales.

REFERENCES

Asner, G.P., Knapp, D.E., Broadbent, E.N., Oliveira, P.J.C., Keller, M., and Silva, J.N. 2005. Selective logging in the Brazilian Amazon. *Science*, 310, 480–482.

Band, L.E. 1993. Effect of land surface representation on forest water and carbon budgets. *Journal of Hydrology*, 150, 749–772.

Besag, J. 1986. On the statistical analysis of dirty pictures. *Journal of the Royal Statistical Society, Series B*, 48, 259–302.

Breiman, L., Friedman, J.H., Olshend, R.A., and Stone, C.J. 1984. *Classification and Regression Trees*. Belmont, CA: Wadsworth International Group, 358 pp.

Burges, C.J.C. 1998. A tutorial on support vector machines for pattern recognition. *Data Mining and Knowledge Discovery*, 2, 121–167.

Chan, J.C.-W. 1998. A neural network approach to land use/land cover change detection. Dissertation Thesis. Hong Kong: University of Hong Kong, 240 pp.

Chan, J.C.-W., Huang, C., and DeFries, R.S. 2001. Enhanced algorithm performance for land cover classification using bagging and boosting. *IEEE Transactions on Geoscience and Remote Sensing*, 39, 693–695.

Chander, G. and Markham, B. 2003. Revised Landsat-5 TM radiometric calibration procedures and postcalibration dynamic ranges. *IEEE Transactions on Geoscience and Remote Sensing*, 41, 2674–2677.

Chander, G., Helder, D.L., Markham, B.L., Dewald, J.D., Kaita, E., Thome, K.J., Micijevic, E., and Ruggles, T.A. 2004. Landsat-5 TM reflective-band absolute radiometric calibration. *IEEE Transactions on Geoscience and Remote Sensing*, 42, 2747–2760.

Chander, G., Markham, B.L., and Helder, D.L. 2009. Summary of current radiometric calibration coefficients for Landsat MSS, TM, ETM+, and EO-1 ALI sensors. *Remote Sensing of Environment*, 113, 893–903.

Chang, C.-C. and Lin, C.-J. 2001. *LIBSVM : A Library for Support Vector Machines*. Available at: http://www.csie.ntu.edu.tw/~cjlin/libsvm/ (last accessed January 2011).

Cohen, W.B., Maiersperger, T.K., Fiorella, M., Spies, T.A., Alig, R.J., and Oetter, D.R. 2002. Characterizing 23 years (1972–95) of stand replacement disturbance in western Oregon forests with Landsat imagery. *Ecosystems*, 5, 122–137.

Cohen, W.B., Yang, Z.G., and Kennedy, R. 2010. Detecting trends in forest disturbance and recovery using yearly Landsat time series: 2. TimeSync—Tools for calibration and validation. *Remote Sensing of Environment*, 114, 2911–2924.

Congalton, R. 1991. A review of assessing the accuracy of classifications of remotely sensed data. *Remote Sensing of Environment*, 37, 35–46.

Congalton, R.G., Oderwald, R.G., and Mead, R.A. 1983. Assessing Landsat classification accuracy using discrete multivariate analysis statistical techniques. *Photogrammetric Engineering and Remote Sensing*, 49, 1671–1678.

Coppin, P. and Bauer, M. 1996. Digital change detection in forested ecosystems with remote sensing imagery. *Remote Sensing Reviews*, 13, 234–237.

Coppin, P., Lambin, E., Jonckheere, I., Nackaerts, K., and Muys, B. 2004. Digital change detection methods in ecosystem monitoring: A review. *International Journal of Remote Sensing*, 25, 1565–1596.

Courant, R. and Hilbert, D. 1953. *Methods of Mathematical Physics,* New York: John Wiley.

DeFries, R.S., Hansen, M., Townshend, J.R.G., and Sohlberg, R. 1998. Global land cover classifications at 8km spatial resolution: The use of training data derived from Landsat imagery in decision tree classifiers. *International Journal of Remote Sensing*, 19, 3141–3168.

Elvidge, C.C., Yuan, D., Weerackoon, R.D., and Lunetta, R.S. 1995. Relative radiometric normalization of Landsat Multispectral Scanner (MSS) data using an automatic scattergram-controlled regression. *Photogrammetric Engineering and Remote Sensing*, 61, 1255–1260.

FAO. 2001. Global Forest Resources Assessment 2000—Main report. FAO Forestry Paper, Rome: Food and Agriculture Organization of the United Nations.

Flusser, J. and Suk, T. 1994. A moment-based approach to registration of images with affine geometric distortion. *IEEE Transaction on Geoscience and Remote Sensing*, 32, 382–387.

Foody, G.M. and Arora, M.K. 1997. An evaluation of some factors affecting the accuracy of classification by an artificial neural network. *International Journal of Remote Sensing*, 18, 799–810.

Foody, G.M. and Mathur, A. 2004. Toward intelligent training of supervised image classifications: Directing training data acquisition for SVM classification. *Remote Sensing of Environment*, 93, 107.

Foody, G.M., Mathur, A., Sanchez-Hernandez, C., and Boyd, D.S. 2006. Training set size requirements for the classification of a specific class. *Remote Sensing of Environment*, 104, 1–14.

Friedl, M.A., Brodley, C.E., and Strahler, A.H. 1999. Maximizing land cover classification accuracies produced by decision trees at continental to global scales. *IEEE Transactions on Geoscience and Remote Sensing*, 37, 969–977.

Friedl, M.A., Zhang, X.Y., Muchoney, D., Strahler, A.H., Woodcock, C.E., Gopal, S., Schneider, A., et al. 2002. Global land cover mapping from MODIS: Algorithms and early results. *Remote Sensing of Environment*, 83, 287–302.

Gao, F., Masek, J., and Wolfe, R. 2009. An automated registration and orthorectification package for Landsat and Landsat-like data processing. *Journal of Applied Remote Sensing*, 3, 033515, doi: 10.1117/1.3104620.

Gordon, S. 1980. Utilizing Landsat imagery to monitor land use change: A case study of Ohio. *Remote Sensing of Environment*, 9, 189–196.

Goward, S., Irons, J., Franks, S., Arvidson, T., Williams, D., and Faundeen, J. 2006. Historical record of landsat global coverage: Mission operations, NSLRSDA, and international cooperator stations. *Photogrammetric Engineering and Remote Sensing*, 72, 1155–1169.

Goward, S.N. and Williams, D.L. 1997. Landsat and earth systems science: Development of terrestrial monitoring. *Photogrammetric Engineering and Remote Sensing*, 63, 887–900.

Gualtieri, J.A. and Cromp, R.F. 1998. Support vector machines for hyperspectral remote sensing classification. In *The 27th AIPR Workshop: Advances in Computer Assisted Recognition*, October 27, 1998, Washington, D.C., 221–232.

Gutman, G., Byrnes, R., Masek, J., Covington, S., Justice, C., Franks, S., and Headley, R. 2008. Towards monitoring land-cover and land-use changes at a global scale: The Global Land Survey 2005. *Photogrammetric Engineering and Remote Sensing*, 74, 6–10.

Hansen, M., Dubayah, R., and DeFries, R. 1996. Classification trees: An alternative to traditional land cover classifiers. *International Journal of Remote Sensing*, 17, 1075–1081.

Hansen, M., DeFries, R.S., Townshend, J.R.G., and Sohlberg, R. 2000. Global land cover classification at 1 km spatial resolution using a classification tree approach. *International Journal of Remote Sensing*, 21, 1331–1364.

Hansen, M.C. and DeFries, R.S. 2004. Detecting long-term global forest change using continuous fields of tree-cover maps from 8-km advanced very high resolution radiometer (AVHRR) data for the years 1982–99. *Ecosystems*, 7, 695–716.

Heo, J. and FitzHugh, T.W. 2000. A standardized radiometric normalization method for change detection using remotely sensed imagery. *Photogrammetric Engineering and Remote Sensing*, 66, 173–181.

Homer, C., Huang, C., Yang, L., Wylie, B., and Coan, M. 2004. Development of a 2001 national land cover database for the United States. *Photogrammetric Engineering and Remote Sensing*, 70, 829–840.

Houghton, R.A. 1998. Historic role of forests in the global carbon cycle. In G.H. Kohlmaier, M. Weber, and R.A. Houghton (Eds.), *Carbon Dioxide Mitigation in Forestry and Wood Industry* (pp. 1–24). Berlin: Springer.

Hsu, C.-W. and Lin, C.-J. 2002. A comparison of methods for multiclass support vector machines. *IEEE Transactions on Neural Networks,* 13, 415–425.

Huang, C., Davis, L.S., and Townshend, J.R.G. 2002. An assessment of support vector machines for land cover classification. *International Journal of Remote Sensing*, 23, 725–749.

Huang, C., Homer, C., and Yang, L. 2003. Regional forest land cover characterization using medium spatial resolution satellite data. In M. Wulder and S. Franklin (Eds.), *Methods and Applications for Remote Sensing of Forests: Concepts and Case Studies* (pp. 389–410). Boston: Kluwer Academic Publishers.

Huang, C., Kim, S., Altstatt, A., Townshend, J.R.G., Davis, P., Song, K., Tucker, C.J., et al. 2007. Rapid loss of Paraguay's Atlantic forest and the status of protected areas—A Landsat assessment. *Remote Sensing of Environment*, 106, 460–466.

Huang, C., Song, K., Kim, S., Townshend, J.R.G., Davis, P., Masek, J., and Goward, S.N. 2008. Use of a dark object concept and support vector machines to automate forest cover change analysis. *Remote Sensing of Environment*, 112, 970–985.

Huang, C., Goward, S.N., Masek, J.G., Gao, F., Vermote, E.F., Thomas, N., Schleeweis, K., et al. 2009a. Development of time series stacks of Landsat images for reconstructing forest disturbance history. *International Journal of Digital Earth*, 2, 195–218.

Huang, C., Kim, S., Altstatt, A., Song, K., Townshend, J.R.G., Davis, P., Rodas, O., et al. 2009b. Assessment of Paraguay's forest cover change using Landsat observations. *Global and Planetary Change*, 67, 1–12.

Huang, C., Goward, S.N., Masek, J.G., Thomas, N., Zhu, Z., and Vogelmann, J.E. 2010. An automated approach for reconstructing recent forest disturbance history using dense Landsat time series stacks. *Remote Sensing of Environment*, 114, 183–198.

Janssen, L.L.F. and Wel, F. 1994. Accuracy assessment of satellite derived land cover data: A review. *IEEE Photogrammetric Engineering and Remote Sensing*, 60, 419–426.

Jensen, J.R. 1996. *Introductory Digital Image Processing: A Remote Sensing Perspective*. Englewood Cliffs, NJ: Prentice-Hall, 379 pp.

Justice, C.O. and Townshend, J.R.G. 1988. Selecting the spatial resolution of satellite sensors required for global monitoring of land transformations. *International Journal of Remote Sensing*, 9, 187–236.

Kennedy, R.E. and Cohen, W.B. 2003. Automated designation of tie-points for image-to-image coregistration. *International Journal of Remote Sensing*, 24, 3467–3490.

Kennedy, R.E., Cohen, W.B., and Schroeder, T.A. 2007. Trajectory-based change detection for automated characterization of forest disturbance dynamics. *Remote Sensing of Environment*, 110, 370–386.

Kennedy, R.E., Yang, Z.G., and Cohen, W.B. 2010. Detecting trends in forest disturbance and recovery using yearly Landsat time series: 1. LandTrendr—Temporal segmentation algorithms. *Remote Sensing of Environment*, 114, 2897–2910.

Knorn, J., Rabe, A., Radeloff, V.C., Kuemmerle, T., Kozak, J., and Hostert, P. 2009. Land cover mapping of large areas using chain classification of neighboring Landsat satellite images. *Remote Sensing of Environment*, 113, 957–964.

Kuemmerle, T., Chaskovskyy, O., Knorn, J., Radeloff, V.C., Kruhlov, I., Keeton, W.S., and Hostert, P. 2009. Forest cover change and illegal logging in the Ukrainian Carpathians in the transition period from 1988 to 2007. *Remote Sensing of Environment*, 113, 1194–1207.

Lal, R. 1995. *Sustainable Management of Soil Resources in the Humid Tropics*. New York: United Nations University Press, 146 pp.

Liang, S., Vallah-Adl, H., Kalluri, S., Jaja, J., Kaufman, Y.J., and Townshend, J.R.G. 1997. An operational atmospheric correction algorithm for Landsat Thematic Mapper imagery over the land. *Journal of Geophysical Research*, 102, 17173–17186.

Lu, D., Mausel, P., Brondízio, E., and Moran, E. 2004. Change detection techniques. *International Journal of Remote Sensing*, 25, 2365–2407.

Lunetta, R.S., Johnson, D.M., Lyon, J.G., and Crotwell, J. 2004. Impacts of imagery temporal frequency on land-cover change detection monitoring. *Remote Sensing of Environment*, 89, 444–454.

Markham, B.L. and Barker, J.L. 1986. Landsat MSS and TM post-calibration dynamic ranges, exoatmospheric reflectances and at-satellite temperatures. *EOSAT Landsat Technical Notes*, 1, 3–8.

Masek, J.G., Vermote, E.F., Saleous, N.E., Wolfe, R., Hall, F.G., Huemmrich, K.F., Feng, G., Kutler, J., and Teng-Kui, L. 2006. A Landsat surface reflectance dataset for North America, 1990–2000. *IEEE Geoscience and Remote Sensing Letters*, 3, 68–72.

Meyer, W.B. and Turner II, B.L. (Eds.). 1994. *Changes in Land Use and Land Cover: A Global Perspective*. Cambridge: Cambridge University Press, 537 pp.

Morton, D.C., DeFries, R.S., Shimabukuro, Y.E., Anderson, L.O., Arai, E., Espirito-Santo, F.d.B., Freitas, R., and Morisette, J. 2006. Cropland expansion changes deforestation dynamics in the southern Brazilian Amazon. *Proceedings of the National Academy of Sciences of the USA*, 103, 14637–14641.

Mountrakis, G., Im, J., and Ogole, C. 2011. Support vector machines in remote sensing: A review. *ISPRS Journal of Photogrammetry and Remote Sensing*, 66, 247–259.

Pal, M. and Mather, P.M. 2005. Support vector machines for classification in remote sensing. *International Journal of Remote Sensing*, 26, 1007–1011.

Pandey, D.N. 2002. Sustainability science for tropical forests. *Conservation Ecology*, 6, 13.

Paola, J.D. and Schowengerdt, R.A. 1995. A review and analysis of backpropagation neural networks for classification of remotely sensed multi-spectral imagery. *International Journal of Remote Sensing*, 16, 3033–3058.

Pratt, W.K. 1974. Correlation techniques of image registration. *IEEE Transactions On Aerospace and Electronic Systems*, AES-10, 353–358.

Quinlan, J.R. 1993. *C4.5 Programs for Machine Learning*. The Morgan Kaufmann Series in Machine Learning. San Mateo, CA: Morgan Kaufmann Publishers, 302 pp.

Safavian, S.R. and Landgrebe, D. 1991. A survey of decision tree classifier methodology. *IEEE Transactions on Systems, Man, and Cybernetics*, 21, 660–674.

Schimel, D.S. 1995. Terrestrial biogeochemical cycles: Global estimates with remote sensing. *Remote Sensing of Environment*, 51, 49–56.

Singh, A. 1989. Digital change detection techniques using remotely-sensed data. *International Journal of Remote Sensing*, 10, 989–1003.

Song, C., Woodcock, C.E., Seto, K.C., Lenney, M.P., and Macomber, S.A. 2001. Classification and change detection using Landsat TM data: When and how to correct atmospheric effects? *Remote Sensing of Environment*, 75, 230–244.

Song, K. 2010. Tackling uncertainties and errors in the satellite monitoring of forest cover change. Dissertation Thesis. College Park. University of Maryland, 175 pp.

Stehman, S.V. and Czaplewski, R.L. 1998. Design and analysis for thematic map accuracy assessment: Fundamental principles. *Remote Sensing of Environment*, 64, 331–344.

Steininger, M.K., Desch, A., Bell, V., Ernst, P., Tucker, C.J., Townshend, J.R.G., and Killeen, T.J. 2001. Tropical deforestation in the Bolivian Amazon. *Environmental Conservation*, 28, 127–134.

Stow, D.A., Tinney, L.R., and Estes, J.E. 1980. Deriving land use/land cover change statistics from Landsat: A study of prime agricultural land. In *Proceedings of the 14th International Symposium on Remote Sensing of Environment* (pp. 1227–1237). Ann Arbor.

Sundaram, R.K. 1996. *A First Course in Optimization Theory* (pp. 357). New York: Cambridge University Press.

Teillet, P.M. and Fedosejevs, G. 1995. On the dark target approach to atmospheric correction of remotely sensed data. *Canadian Journal of Remote Sensing*, 21, 374.

Todd, W.J. 1977. Urban and regional land use change detected by using Landsat data. *Journal of Research by the US Geological Survey*, 5, 527–534.

Townshend, J.R.G., Justice, C.O., and McManus, J. 1992. The impact of misregistration on change detection. *IEEE Transactions on Geoscience and Remote Sensing*, 30, 1054–1060.

Tucker, C.J., Grant, D.M., and Dykstra, J.D. 2004. NASA's global orthorectified Landsat data set. *Photogrammetric Engineering and Remote Sensing*, 70, 313–322.

Vapnik, V.N. 1995. *The Nature of Statistical Learning Theory*. New York: Springer, 188 pp.

Vapnik, V.N. 1998. *Statistical Learning Theory. Adaptive and Learning Systems for Signal Processing, Communications, and Control* (pp. 736). New York: John Wiley & Sons, Inc.

Vermote, E.F. and Kotchenova, S. 2008. Atmospheric correction for the monitoring of land surfaces. *Journal of Geophysical Research Atmospheres*, 113, D23S90, doi:10.1029/2007JD009662.

Vicente-Serrano, S.M., Pérez-Cabello, F., and Lasanta, T. 2008. Assessment of radiometric correction techniques in analyzing vegetation variability and change using time series of Landsat images. *Remote Sensing of Environment*, 112, 3916–3934.

Zhu, G. and Blumberg, D.G. 2002. Classification using ASTER data and SVM algorithms: The case study of Beer Sheva, Israel. *Remote Sensing of Environment*, 80, 233.

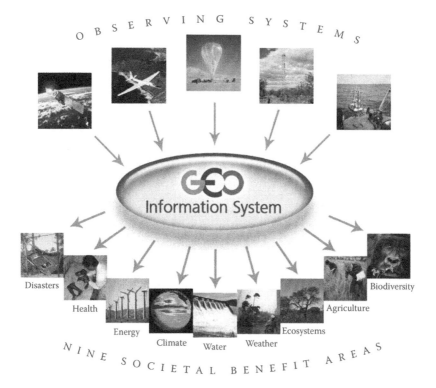

FIGURE 1.1 Nine areas of societal benefit of the Group on Earth Observations (GEO).

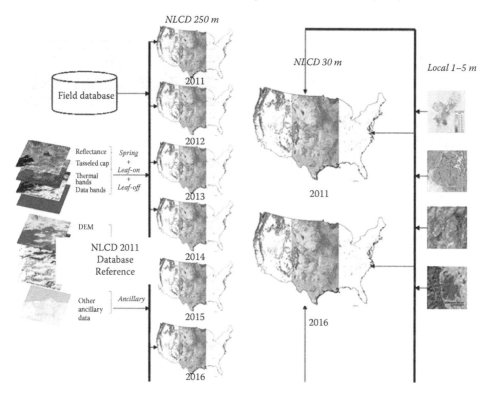

FIGURE 1.2 A potential product framework proposed for NLCD 2011 (Adapted from Xian, G., Homer, C., and Yang, L., 2011. Development of the USGS National Land-Cover Database over two decades. In: Weng, Q. H., ed., *Advances in Environmental Remote Sensing—Sensors, Algorithms, and Applications*. CRC Press, Boca Raton, FL, 525–543.).

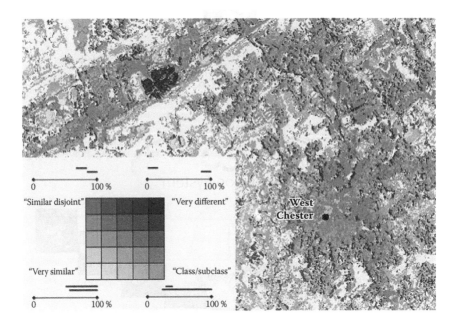

FIGURE 3.4 An example of semantic land-cover change map using a bivariate color scheme to represent different combinations of the semantic distance and overlap metric.

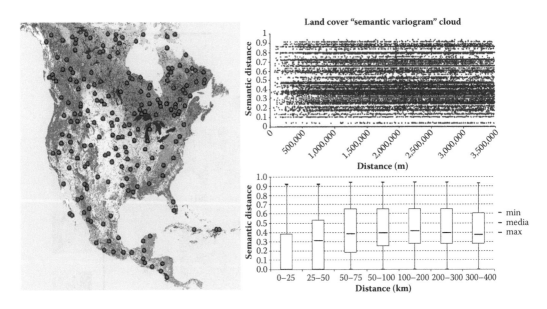

FIGURE 3.5 Random sample points ($n = 110$) across a North American land-cover dataset are plotted in a semantic variogram according to pair-wise comparison of spatial distance of the points and the semantic difference between the land-cover classes at these points. Box plots summarize the points for specified spatial distance intervals.

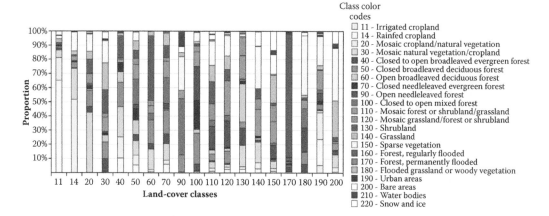

Class color
codes
☐ 11 - Irrigated cropland
☐ 14 - Rainfed cropland
☐ 20 - Mosaic cropland/natural vegetation
☐ 30 - Mosaic natural vegetation/cropland
■ 40 - Closed to open broadleaved evergreen forest
☐ 50 - Closed broadleaved deciduous forest
☐ 60 - Open broadleaved deciduous forest
■ 70 - Closed needleleaved evergreen forest
■ 90 - Open needleleaved forest
☐ 100 - Closed to open mixed forest
■ 110 - Mosaic forest or shrubland/grassland
■ 120 - Mosaic grassland/forest or shrubland
■ 130 - Shrubland
■ 140 - Grassland
☐ 150 - Sparse vegetation
■ 160 - Forest, regularly flooded
■ 170 - Forest, permanently flooded
■ 180 - Flooded grassland or woody vegetation
■ 190 - Urban areas
☐ 200 - Bare areas
■ 210 - Water bodies
☐ 220 - Snow and ice

FIGURE 5.1 Classification trajectories of the pixels that are not identically classified in the GlobCover 2005 and 2009 land-cover products. (From Bontemps, S. et al., GlobCover 2009—Products description and validation report, version 2.0, 17/02/2011. Available at: http://ionia1.esrin.esa.int/. With permission.)

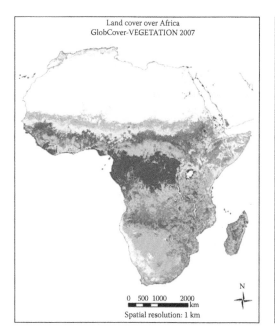
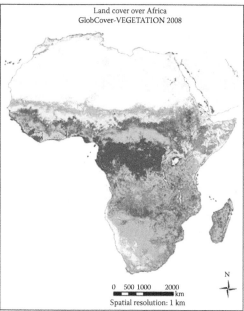

FIGURE 5.2 Land-cover results obtained by the automated GlobCover classification chain from 2007 to 2008 daily SPOT-Vegetation time series. (From Moreau, I., Méthode de cartographie globale de l'occupation du sol par télédétection spatiale: Analyse de la stabilité interannuelle de la chaîne de traitement GlobCover, mémoire de fin d'études, Université Catholique de Louvain, Faculté d'ingénierie biologique, agronomique et environnementale, 2009. With permission.)

TABLE 5.1

Illustration of the Proposed Concepts of Land-Cover Features and Land-Cover Conditions

Land-cover features (permanent aspect or stable elements of the landscape)	Land-cover condition (dynamic component of land cover)
Features' nature: built-up *Features' structure:* high density of building *Features' naturalness:* artificial *Features' homogeneity:* urban patterns made of a mixture of green areas, buildings, houses, and water channels	Seasonal behavior of the *green vegetation* (NDVI profile) *Snow cover* usually from December 15 to January 15 No *flooding* dynamic No *fire* dynamic

Possible denomination of this land cover according to the following:

A land-cover typology A: Urban area
A land-cover typology B: Residential area
A land-cover typology C: Impervious surface area

Land-cover features (permanent aspect or stable elements of the landscape)	Land-cover condition (dynamic component of the land cover)
Features' nature: tree cover *Features' structure:* high tree density (canopy cover of 92%) *Features' naturalness:* natural broadleaved, evergreen vegetation *Features' homogeneity:* homogeneous canopy (few clearings)	Slight seasonal behavior of the *green vegetation* (NDVI profile) No *snow* dynamic No *flooding* dynamic No *fire* dynamic

Possible denomination of this land cover according to the following:

A land-cover typology A: Closed evergreen forest
A land-cover typology B: Natural woody vegetation
A land-cover typology C: Dense broadleaved forest

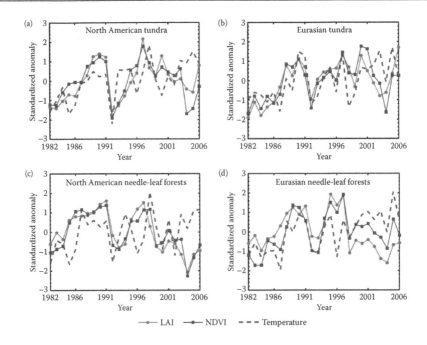

FIGURE 7.2 Standardized April-to-October anomalies of AVHRR LAI (green), GIMMS AVHRR NDVI (blue), and GISS temperature (red dashed line) for Eurasian and North American needle-leaf forests (panels [c] and [d]) and tundra (panels [a] and [b]) from 1982 to 2006. (From Ganguly, S. et al., *Rem. Sens. Environ.*, 112, 4318–4332, 2008a.)

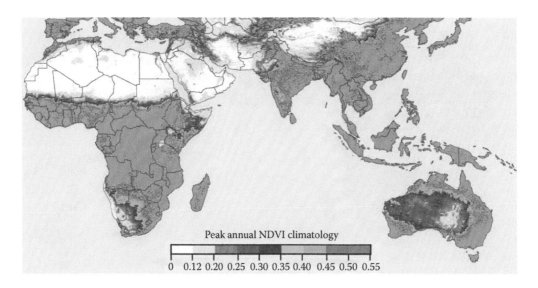

Peak annual NDVI climatology

0   0.12 0.20 0.25 0.30 0.35 0.40 0.45 0.50 0.55

FIGURE 7.3 Color map of peak annual NDVI climatology. Peak annual NDVI climatology was calculated by first estimating the 26-year (1981–2006) mean of monthly NDVI (monthly NDVI climatology) and then selecting the maximum value (per pixel, from 12-monthly climatological NDVI values). A spatial mask was applied on the color map based on peak annual NDVI climatology values in the range of 0.12–0.55. The NDVI data used is the AVHRR GIMMS NDVI product.

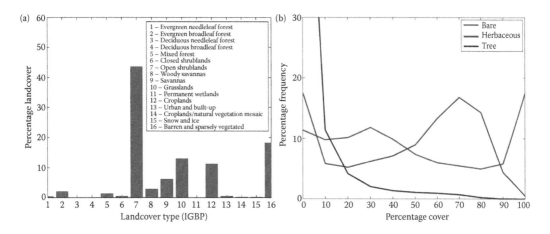

FIGURE 7.4 Percentage distribution of IGBP land-cover classes (panel [a]) and frequency distribution of bare (red), herbaceous (blue), and tree (black) cover from MODIS VCF map, expressed as percentage of total number of pixels (panel [b]) for the peak annual NDVI climatology range of 0.12–0.55.

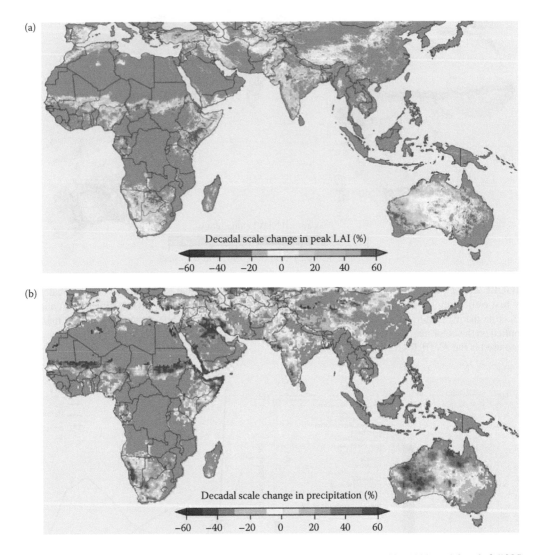

FIGURE 7.5 (a) Percentage change in mean peak annual LAI between decade 1 (1981–1990) and decade 2 (1995–2006). For each year in a decade, the peak LAI was selected (per pixel from 12 LAI values). The mean peak LAI was calculated for each decade. Finally, the percentage change was calculated as [100 × (mean peak LAI decade2 − mean peak LAI decade1)/(mean peak LAI decade1)]. A spatial mask was applied on the color map based on peak annual NDVI climatology values in the range of 0.12–0.55 (all values outside this range appear in gray—masked out). (b) Percentage change in mean peak annual precipitation (mm/year) between decade 1 (1981–1990) and decade 2 (1995–2006). Peak precipitation for each year was calculated by summing the precipitation in the three wettest months. The mean peak annual precipitation for each decade and percentage change were calculated as in (a).

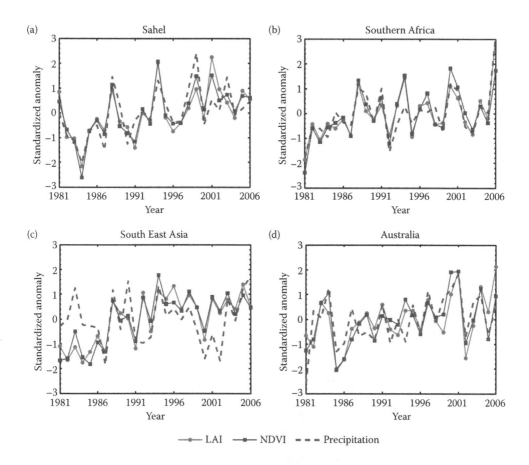

FIGURE 7.6 Standardized anomalies of annual peak AVHRR LAI (green line), annual peak AVHRR NDVI (blue line), and annual peak (three wettest month CRU + TRMM) precipitation (red dashed line) for the semi-arid regions (panels [a]–[d]) from 1981 to 2006.

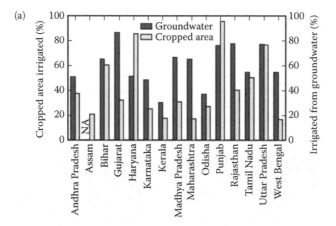

FIGURE 7.9 (a) Percentage of cropped area that is irrigated (blue bar) and percentage of irrigated land utilizing groundwater (green bar) for each of the major Indian states.

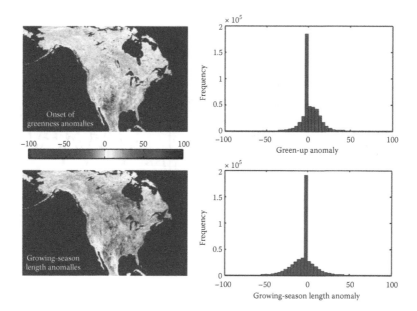

FIGURE 7.11 Anomalies in the timing of green-up onset and growing-season length for 2002 relative to the 2001–2006 mean. Histograms show the frequency of green-up and growing-season length anomalies. Details about processing the MODIS data and deriving the phenological parameters have been described in depth by Ganguly et al. (2010).

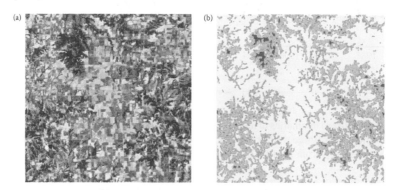

FIGURE 9.2 (a) Landsat 5 image, WRS2 path/row 024/032, centered on 91 10 21.5W, 39 59 8.7N with dimensions 26.3 km by 26.3 km. Near-infrared band 4 is shown in red, and visible red band 3 is shown in cyan. (b) Reference labels derived from a RapidEye forest/nonforest classification. Dark and light green are ≥50% forest cover. Yellow and orange are <50% forest cover. Dark green and yellow represent spatially homogeneous forest and nonforest labels, respectively. Light green and orange represent spatially heterogeneous forest and nonforest labels, respectively. These mixed pixels constitute a 120-m buffer along forest/nonforest interfaces. Forest accounts for 82.5% of the image and nonforest 17.5%.

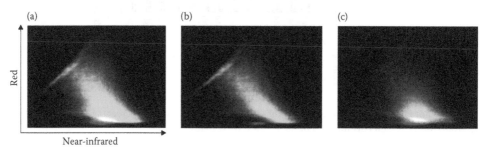

FIGURE 9.3 (a) All forest and nonforest data from Figure 9.2, (b) forest and nonforest pixels greater than 60 m from forest/nonforest interfaces (pure population), and (c) forest and nonforest pixels within a 120-m buffer along forest/nonforest interfaces (mixed population).

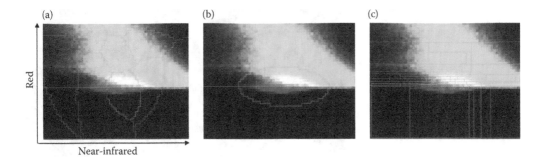

FIGURE 9.4 Results of (a) unsupervised clustering, (b) maximum likelihood, and (c) classification tree algorithms on partitioning the red/near-infrared feature space for forest (shown in red) and nonforest (shown in cyan). Green boundaries indicate forest, orange nonforest. For this test, all data were used as inputs.

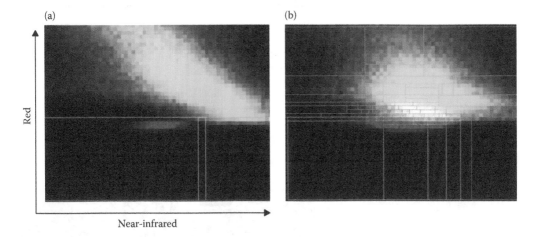

FIGURE 9.5 Example decision boundaries made using a classification tree for (a) core site training dataset and (b) mixed pixel training dataset. For each model, a 7% sample of forest and nonforest were drawn for model generation from the populations shown in Figure 9.2b. Cyan represents nonforest and red represents forest, based on Figure 9.2a.

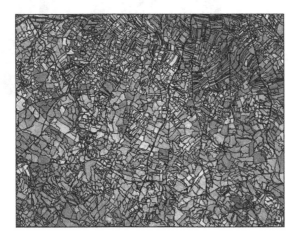

FIGURE 10.4 Segmented image using a scale parameter of 125.

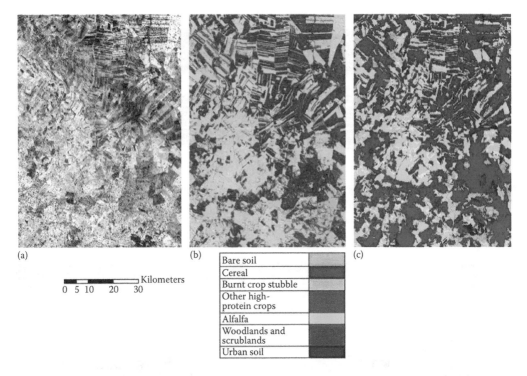

Bare soil	
Cereal	
Burnt crop stubble	
Other high-protein crops	
Alfalfa	
Woodlands and scrublands	
Urban soil	

(a) (b) (c)

Kilometers
0 5 10 20 30

FIGURE 10.5 Example of comparison between QuickBird image (a), supervised classification of the image formed by the principal component and the NDVI index (b), and the oriented-based classification (c).

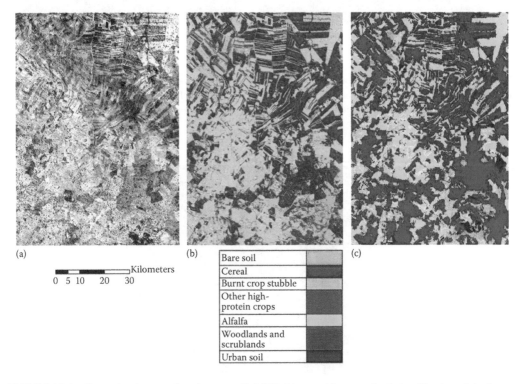

Bare soil	
Cereal	
Burnt crop stubble	
Other high-protein crops	
Alfalfa	
Woodlands and scrublands	
Urban soil	

(a) (b) (c)

Kilometers
0 5 10 20 30

FIGURE 10.6 Example of comparison between QuickBird image (a), supervised classification of the image formed by the principal component and the NDVI index (b), and the HTM classification (c).

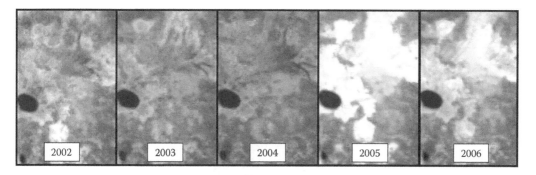

FIGURE 12.4 Example of wetland interannual spectral variability as seen in MODIS at 250-m spatial resolution, with bands displayed as red = band 2, green = band 6, and blue = band 1.

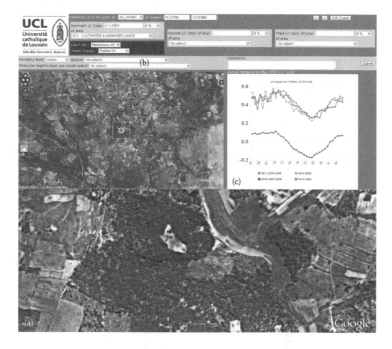

FIGURE 14.1 Web-based interface used for GlobCover 2009 reference data collection by the international expert network. Validation samples were automatically overlaid either in Virtual Earth or Google Earth (a), combo boxes to characterize the samples with LCCS classifiers were included (b), and SPOT-VGT NDVI and NDWI temporal profiles corresponding to the pixel displays as white square were provided (c).

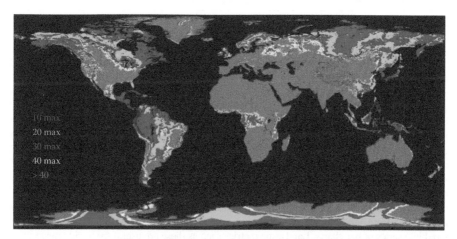

FIGURE 14.3 Number of valid MERIS FR surface reflectance observations for GlobCover 2005.

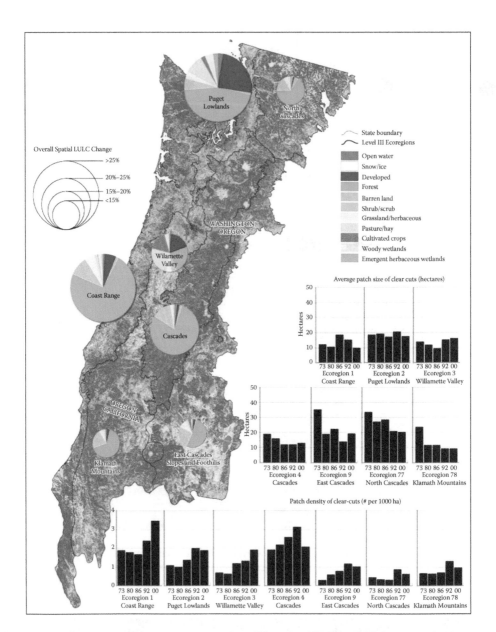

FIGURE 15.1 The area shown covers seven ecoregions (Omernik, 1987; EPA, 1999) in the Pacific Northwest of the western United States, falling in California, Oregon, and Washington. The ecoregions are the Coast Range, Puget Lowlands, Willamette Valley, Cascades, East Cascades, Slopes and Foothills, North Cascades, and Klamath Mountains. All the ecoregions are primarily forested, with varying levels of agriculture, urban, and other land uses. The map shows data from the 2001 National Land Cover Database (NLCD) (Homer et al., 2007) derived from Landsat thematic mapper (TM) and enhanced thematic mapper plus (ETM+) data at a spatial resolution of 30 m. The pie charts represent two sources of important LULC data. Land-cover composition is characterized by the relative size of each "wedge" and is based on NLCD data. The size of the pie charts reflects the amount of land that experienced changes in LULC as measured by the USGS Land Cover Trends project (Loveland et al., 2002). The extent of land area that changed at least once between 1973 and 2000 varies considerably across the seven ecoregions, including ecoregions with similar land-cover compositions. The changes in LULC reflect variability in the biophysical conditions, land ownership and management, and the impact of regional and national policy among other drivers. LULC change data, such as those presented here, are most readily obtained through examination of historical satellite imagery and aerial photographs. The size and density of forest clear-cuts for the seven ecoregions are displayed in a series of bar charts. Landscape metrics such as these are useful for a wide range of ecosystem assessments and are immediately available through examination of remotely sensed data.

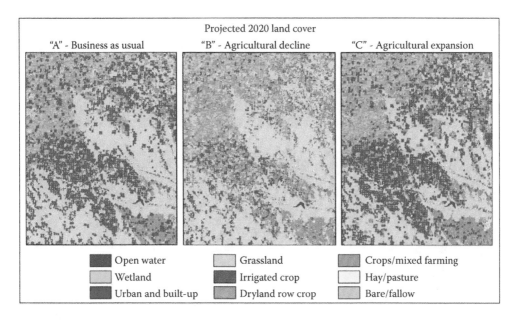

Projected 2020 land cover

"A" - Business as usual   "B" - Agricultural decline   "C" - Agricultural expansion

Open water   Grassland   Crops/mixed farming

Wetland   Irrigated crop   Hay/pasture

Urban and built-up   Dryland row crop   Bare/fallow

FIGURE 15.2 Scenarios are a vital component of LULC modeling, allowing the exploration of multiple possible futures and resultant impacts on ecological processes. Remote sensing both directly and indirectly informs the construction of viable LULC scenarios through (1) construction of regional landscape histories, (2) examination of LULC patterns, and (3) exploration of linkages between historical LULC change and socioeconomic and biophysical driving forces. Each of these three components was used to develop scenarios and model 2020 LULC for a portion of southwestern Kansas, in the central United States (Sohl et al., 2007). Scenario A depicts a business-as-usual scenario. Scenario B depicts a scenario of low precipitation and declining groundwater availability, leading to agricultural decline. Scenario C depicts a scenario of increased precipitation and a more efficient utilization of groundwater, leading to agricultural expansion. The modeled scenarios were used to examine the impacts of LULC change on regional weather and climate variability.

FIGURE 15.3 Two Landsat TM images acquired on August 29, 1987 (top) and August 12, 2010 (bottom). Both images are of the same region in northern California, covering parts of Humboldt and Del Norte counties. The images use visible and near-infrared bands to depict vegetation in hues of red. Dense old-growth conifer stands appear dark red, whereas recent clear-cuts appear bright. Dimensionally, the images are approximately 30 km from east to west and 13 km from north to south. The images span three major land ownership types. Redwood National Park is in the west and is most easily recognized by the large contiguous stand of old-growth redwoods found in Prairie Creek Redwoods State Park. In the eastern portion of the images is Six River National Forest (SRNF). SRNF is managed for multiple uses, including timber harvest. In the center of the image is a large swath of private land holdings along the Klamath River. Cutting on private lands generally occurs in relatively large, often contiguous patches, while SRNF is characterized by a smaller more dispersed pattern of cutting. No cutting is evident in the National Park. Cutting also seems to have accelerated in this area on both private and public lands. Satellite imagery, such as those presented here, are extremely useful for mapping and characterizing changes to landscapes, which provide the foundational understanding for LULC modeling efforts. In this example, land ownership is an important driver and constraint on LULC change and should be considered in any modeling effort.

TABLE 16.1

Twenty-Two Classes of the GlobCover Legend

Value	GlobCover legend	Color
11	Post-flooding or irrigated croplands	
14	Rainfed croplands	
20	Mosaic cropland (50%–70%)/natural vegetation (grassland, shrubland, forest) (20%–50%)	
30	Mosaic natural vegetation (grassland, shrubland, forest) (50%–70%)/cropland (20%–50%)	
40	Closed to open (>15%) broadleaved evergreen and/or semideciduous forest (>5)	
50	Closed (>40%) broadleaved deciduous forest (>5m)	
60	Open (15%–40%) broadleaved deciduous forest (>5m)	
70	Closed (>40%) needleleaved evergreen forest (>5m)	
90	Open (15%–40%) needleleaved deciduous or evergreen forest (>5m)	
100	Closed to open (>15%) mixed broadleaved and needleleaved forest (>5m)	
110	Mosaic forest/shrubland (50%–70%)/grassland (20%–50%)	
120	Mosaic grassland (50%–70%)/forest/shrubland (20%–50%)	
130	Closed to open (>15%) shrubland (<5m)	
140	Closed to open (>15%) grassland	
150	Sparse (>15%) vegetation (woody vegetation, shrubs, grassland)	
160	Closed (>40%) broadleaved forest regularly flooded—fresh water	
170	Closed (>40%) broadleaved semideciduous and/or evergreen forest regularly flooded—saline water	
180	Closed to open (>15%) vegetation (grassland, shrubland, woody vegetation) on regularly flooded or waterlogged soil—fresh, brackish or saline water	
190	Artificial surfaces and associated areas (urban areas >50%)	
200	Bare areas	
210	Water bodies	
220	Permanent snow and ice	

TABLE 16.3

Fourteen Classes of the GlobCorine Legend

Value	GlobCorine legend	Color
10	Urban and associated areas	
20	Rainfed cropland	
30	Irrigated cropland	
40	Forest	
50	Heathland and sclerophyllous vegetation	
60	Grassland	
70	Sparsely vegetated area	
80	Vegetated low-lying areas on regularly flooded soil	
90	Bare areas	
100	Complex cropland	
110	Mosaic cropland/natural vegetation	
120	Mosaic of natural (herbaceous, shrub, tree) vegetation	
200	Water bodies	
210	Permanent snow and ice	

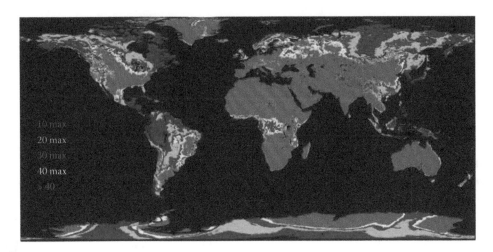

FIGURE 16.5 Number of valid observations obtained after 19 months of MERIS FRS acquisitions. Magenta areas are defined as well covered (>40 observations).

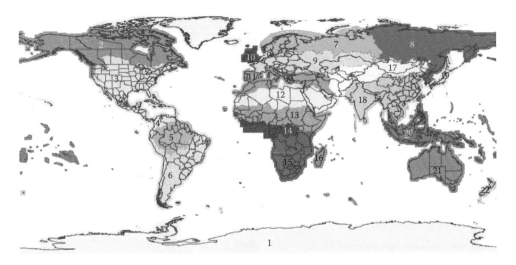

FIGURE 16.7 Overview of the 22 equal-reasoning areas used as stratification.

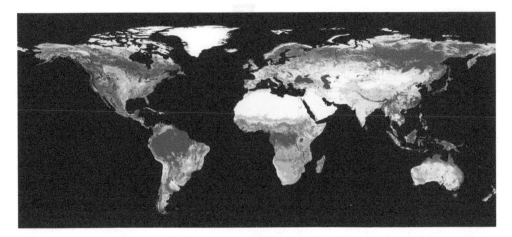

FIGURE 16.8 The GlobCover 2005 product as the first 300-m global land-cover map for the period December 2004–June 2006.

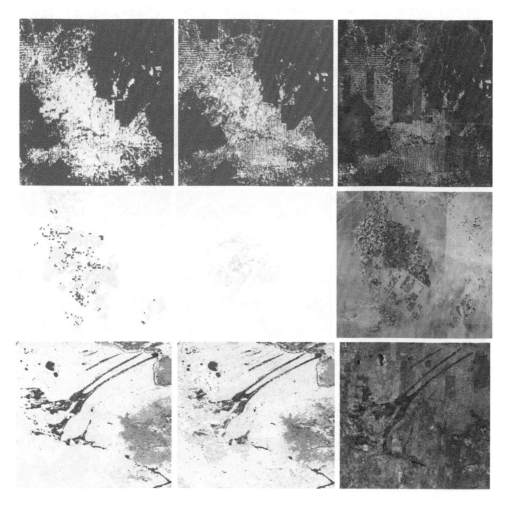

FIGURE 16.9 Improvement of the spatial detail due to the use of a 300-m spatial resolution. Deforestation clear-cuts in Amazonia (top), irrigated crops in Saudi Arabia's desert (center), and specific vegetation structure in Russia (bottom). GLC2000 (left), GlobCover (center), and Google Earth (right).

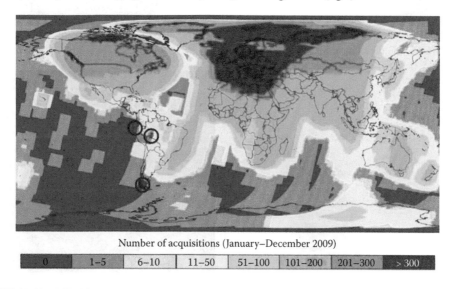

Number of acquisitions (January–December 2009)

| 0 | 1–5 | 6–10 | 11–50 | 51–100 | 101–200 | 201–300 | > 300 |

FIGURE 16.10 MERIS FRS density data acquisition over the year 2009.

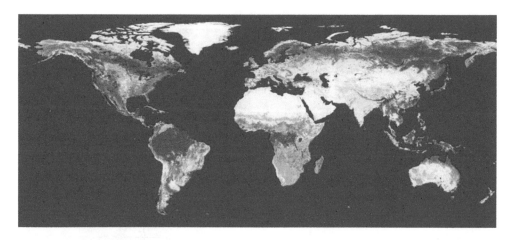

FIGURE 16.11 The GlobCover 2009 product as the first 300-m global land-cover map for the year 2009.

FIGURE 16.14 GlobCorine 2005 land-cover map.

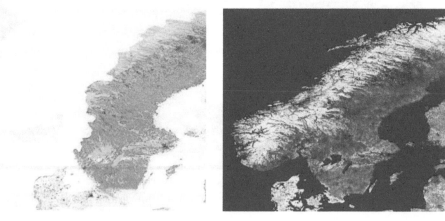

FIGURE 16.15 The classification of Norway (right), which was not covered by the reference database (left), proved to be spatially consistent with surrounding areas.

FIGURE 16.17 The GlobCorine 2009 product.

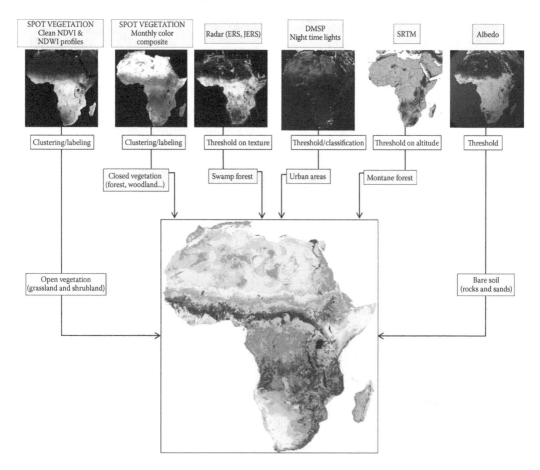

FIGURE 17.1 Datasets and main classification algorithms used in the production of the GLC2000 map of Africa. (From Mayaux, P. et al., *J. Biogeogr.*, 31, 861–877, 2004. With permission.)

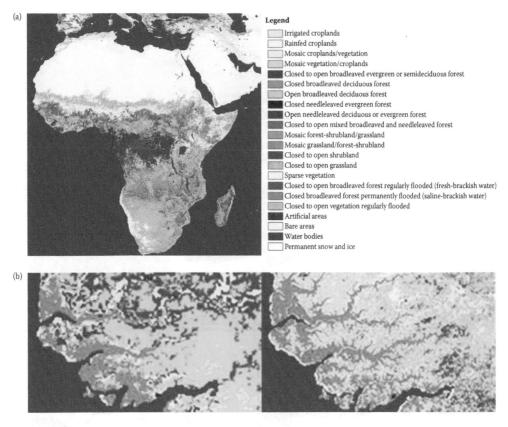

(a)

(b)

FIGURE 17.3 (a) (Top) Globcover classification over Africa (2005–2006) and legend; (b) (bottom) comparison of the GLC2000 map (left) with the Globcover map (right) over Senegal, Guinea-Bissau, and Gambia.

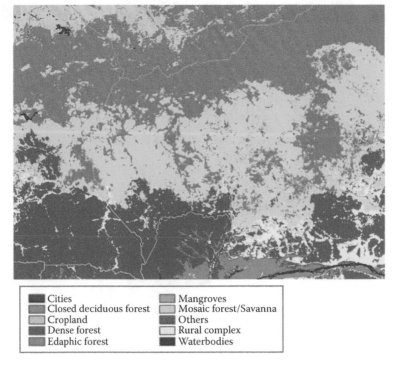

Cities
Closed deciduous forest
Cropland
Dense forest
Edaphic forest
Mangroves
Mosaic forest/Savanna
Others
Rural complex
Waterbodies

FIGURE 17.5 Detail of the fusion map over the northern part of the Congo Basin at the borders between Cameroon, Gabon, Congo, Central African Republic, and Democratic Republic of Congo.

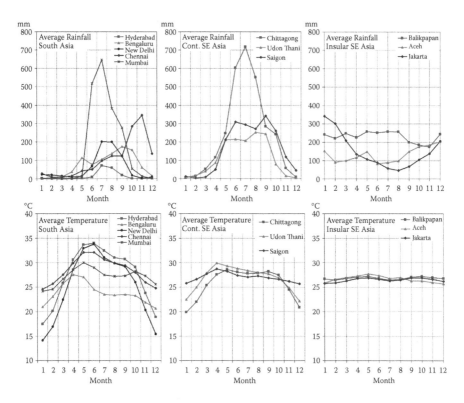

FIGURE 18.1 Rainfall and temperature patterns in the main subregions of tropical Asia. (From Arino, O. et al., *Eur. Space Agency Bull.*, 136, 24–31, 2008. With permission.)

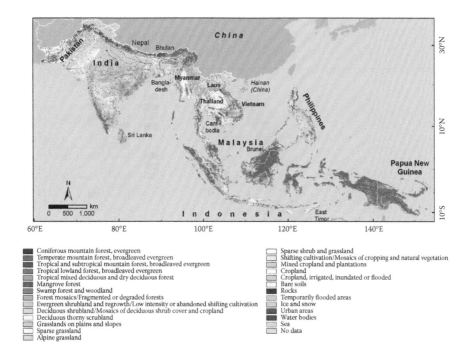

FIGURE 18.2 Land-cover map of the Lower Mekong Basin. (From FAO, *Forest Resources Assessment 1990—Tropical Countries*. FAO Forestry Paper 112, Rome, 1993. With permission.)

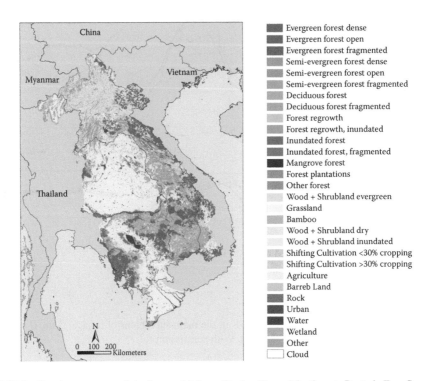

FIGURE 18.3 Land-cover map of the Lower Mekong Basin. (From Martimort, P. et al. *Eur. Space Agency Bull.* 131, 19–23, 2007. With permission.)

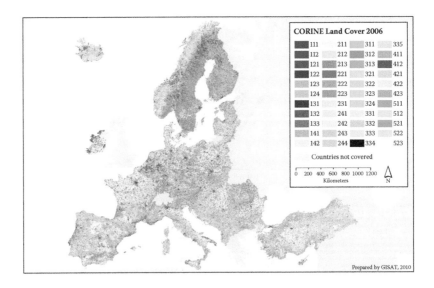

FIGURE 19.3 The spatial distribution of 44 CLC land-cover classes of Europe for the year 2006.

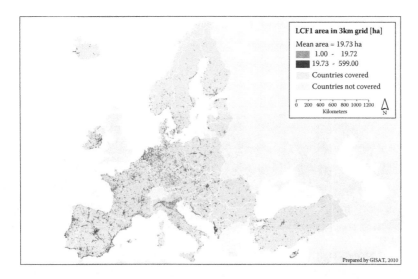

FIGURE 19.4 Spatial distribution of urbanization in European countries in 2000–2006.

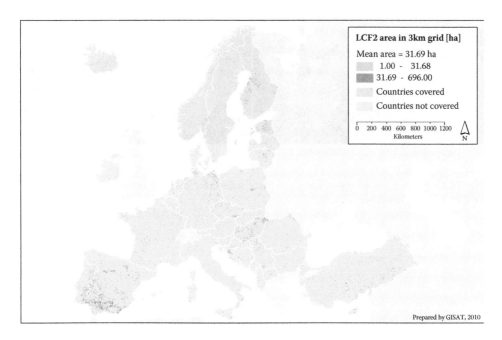

FIGURE 19.5 Spatial distribution of intensification of agriculture in European countries in 2000–2006.

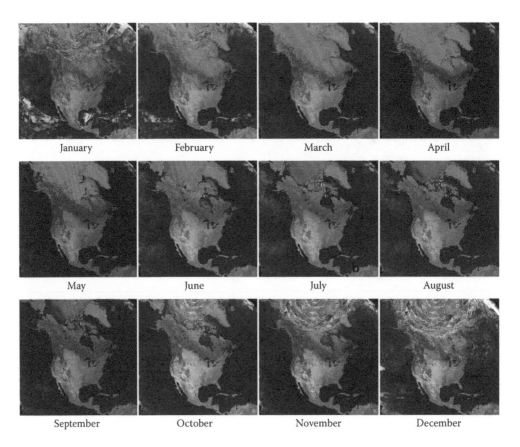

FIGURE 20.1 North America top-of-the-atmosphere reflectance monthly composites from MODIS/Terra 2005 at 250-m spatial resolution.

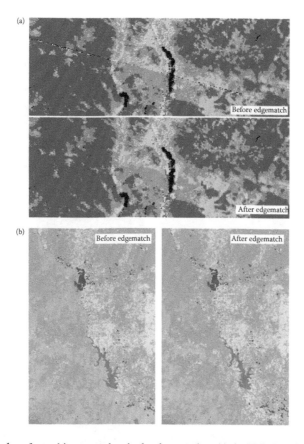

FIGURE 20.3 Examples of matching cross-border land-cover data (a) the U.S.–Mexico and (b) Canada–U.S. border before and after edge-matching procedure.

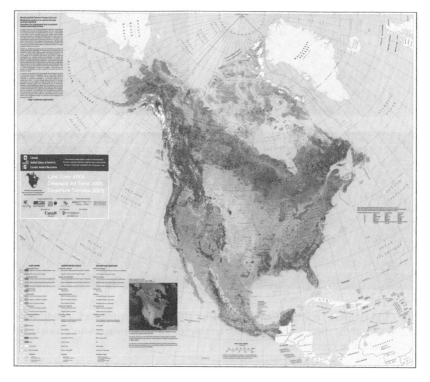

FIGURE 20.5 Land-cover map of North America 2005 at 250-m spatial resolution.

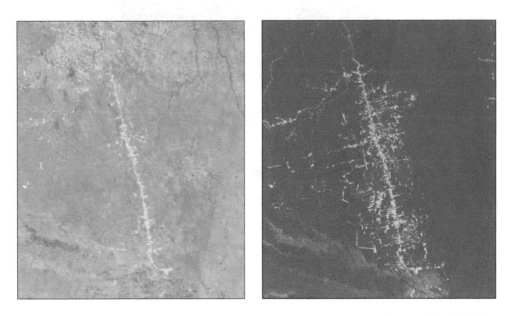

FIGURE 21.1 Trans-Amazonian highway (BR163) at the north of Pará state. Mosaic of SPOT VGT data (left) and mosaic of MERIS FR data (right).

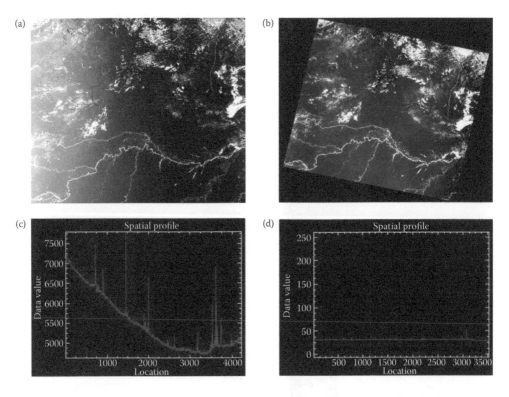

FIGURE 21.2 Top: MERIS image before (a) and after (b) applying cross-track illumination correction. Bottom: spatial profile of the spectral band 2 from a transect of the same image before (c) and after (d) applying cross-track illumination correction.

FIGURE 21.4 The final land-cover MERIS map.

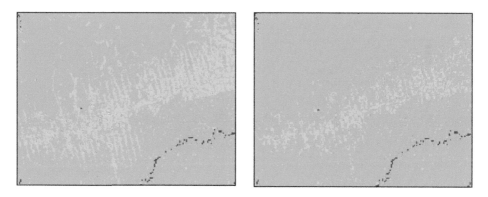

FIGURE 21.6 A 200 km by 150 km extract from the MERIS (left) and GLC2000 (right) maps along the Trans-Amazonian highway in Brazil. Agriculture is represented in gray and light green and forest in darker green.

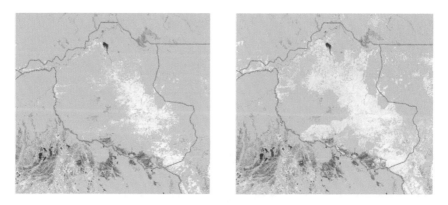

FIGURE 21.7 Extract of Rondônia showing the agricultural expansion in yellow from GLC2000 (a) and MERIS 2009/2010 (b). The forest cover is in green and savannahs in red.

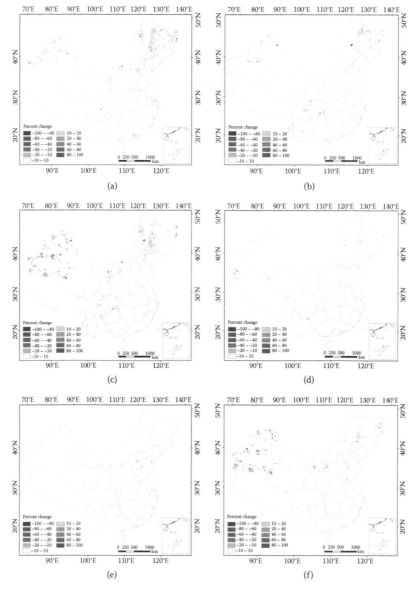

FIGURE 22.3 Spatial patterns of land-cover and land-use changes in China from the late 1980s to the late 1990s: (a) Cultivated land, (b) forest area, (c) grassland, (d) water area, (e) built-up area, and (f) unused land.

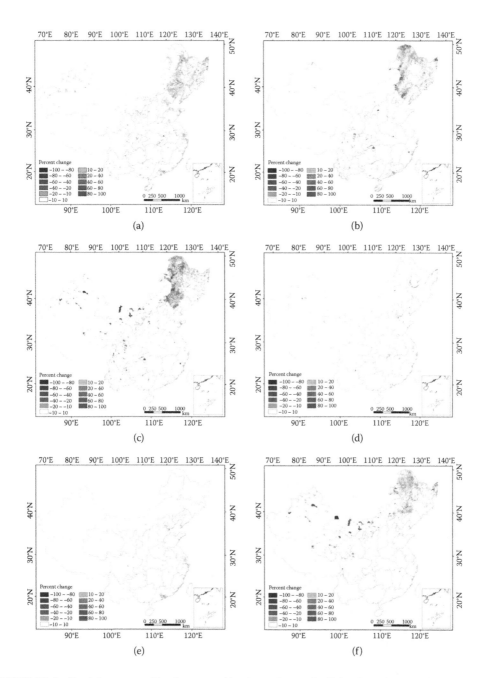

FIGURE 22.4 Spatial patterns of land-cover and land-use changes in China from the late 1990s to the mid-2000s: (a) Cultivated land, (b) forest area, (c) grassland, (d) water area, (e) built-up area, and (f) unused land.

FIGURE 23.3 Ecoregional distribution of the 20-km × 20-km sample blocks selected for the first nine completed ecoregions and the subsequent 10-km × 10-km sample blocks selected for the remaining 75 ecoregions of the conterminous United States.

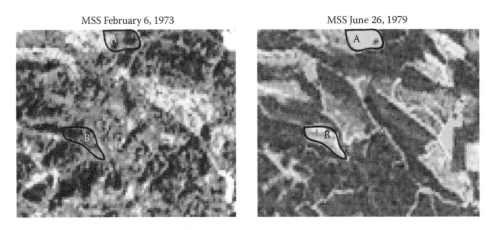

FIGURE 23.4 The definition for the "mechanically disturbed" LULC class accommodates a range of variance in land-cover conditions to support the conceptual intent of the project. In this sample block from the Ouachita Mountains ecoregion, Areas "A" and "B" were mature forests in 1973. The subsequent image for 1980 era reveals Area A as recently disturbed and unvegetated and Area B as vegetated but obviously altered since 1973.

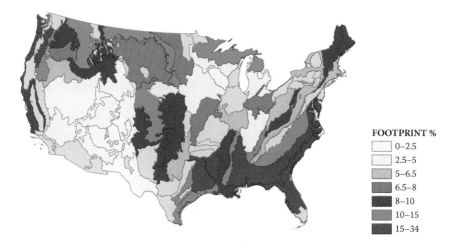

FOOTPRINT %

☐	0–2.5
☐	2.5–5
☐	5–6.5
☐	6.5–8
■	8–10
☐	10–15
■	15–34

FIGURE 23.5 Estimates of the total spatial extent, or footprint, of land-cover change for the 84 ecoregions of the conterminous United States.

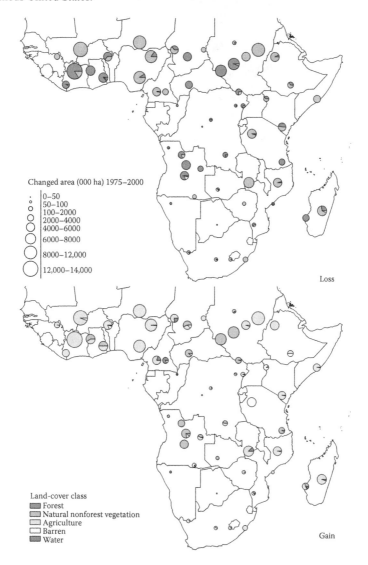

Changed area (000 ha) 1975–2000

·	0–50
◦	50–100
◦	100–2000
○	2000–4000
○	4000–6000
○	6000–8000
○	8000–12,000
○	12,000–14,000

Loss

Land-cover class
■ Forest
■ Natural nonforest vegetation
☐ Agriculture
☐ Barren
■ Water

Gain

FIGURE 24.2 Distribution and proportion of land-cover changes between 1975 and 2000. The top image represents the loss of land cover, whereas the bottom image shows the gain in land cover. The size of the pie chart corresponds to the extent of area changed.

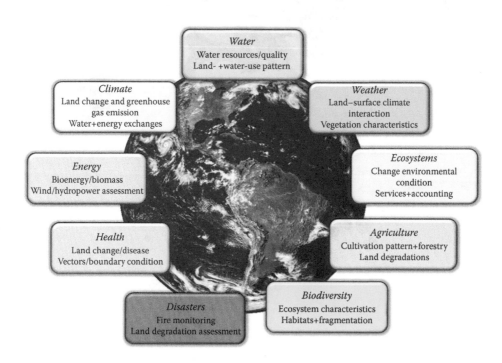

FIGURE 26.3 GEO areas of societal benefits and key land-cover observation needs emphasize the multitude of services from continuous and consistent global terrestrial observations.

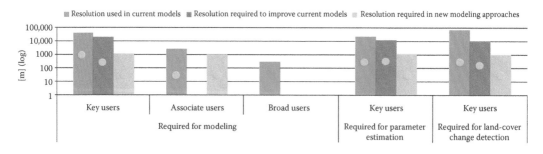

FIGURE 26.5 Spatial resolution median requirements (note *y*-axis in log-scale) from user surveys. The orange points indicate the minimum requirement.

14 Global Land-Cover Map Validation Experiences: Toward the Characterization of Quantitative Uncertainty

Pierre Defourny, Philippe Mayaux, Martin Herold, and Sophie Bontemps

CONTENTS

14.1 INTRODUCTION

Growing public awareness about satellite remote sensing can be attributed to the success of global geoportals that facilitate capturing of space-borne imagery from any place on the earth. Use of geoportals in land-cover mapping has made mapping an easy process. Moreover, technological advancements in mass computing, the ability to store and process large amounts of data, and global earth observation capabilities have opened the door to multiple initiatives in land-cover mapping on a global scale, and more can be expected in the near future. Land-cover maps may differ significantly depending on the quality of the input data, the classification algorithm used to produce the maps, spatial resolution, and the legend used.

Whereas the need for land-cover information is widely recognized, the quality requirements of the information are rather qualitative and do not rely on existing standards. Potential users are more concerned about the technical credibility and legitimacy of global land-cover products. As reported by Herold (2008), legitimacy is ensured if the process is perceived as fair and takes into account the concerns and insights of the relevant stakeholders (Clark et al., 2006).

In many global land-cover applications, the quality or accuracy of the maps is not considered. The rationale for this is that conventional sources of land-cover information are so generalized that anything is regarded as an improvement (Strahler et al., 2006). In principle, it is up to the potential user to determine the map's "fitness for use" in an application. Consequently, it is implied that no single figure can express the fitness for use for every user. However, it is clear that knowledge of the error structure of the land-cover data used will help users of land-cover information to improve the quality of information.

In the current context of multiple sources of global land information, it is important to rely on widely accepted guidelines and standards to assess data quality. Indeed, the constraints and limitations of specific remote-sensing products are often seen as being too technical and are rapidly forgotten by the users. Although there are several definitions of validation, the definition given by the Committee on Earth Observing Satellites Working Group on Calibration and Validation (CEOS-WGCV) should be adopted by the remote-sensing community for the sake of consistency. The committee defines validation as "The process of assessing, by independent means, the quality of the data products derived from the system outputs."

The report of the topical subgroup of the CEOS-Land Product Validation (CEOS-LVP) group on Global Land Cover Validation (Strahler et al., 2006) gave a specific set of recommendations for evaluation and accuracy assessment of global land-cover maps. To date, rigorous validation exercises of four different global land-cover maps have been completed: the map from the International Geosphere-Biosphere Program (IGBP) Data and Information System (DISCover) (Scepan, 1999), the GLC2000 map (Mayaux et al., 2006), the GlobCover 2005 product (Defourny et al., 2009), and more recently, the GlobCover 2009 product (Bontemps et al., 2011). As for the MODIS Land Cover product, it is yet to be validated with completely independent data sources. Only a procedure for internal consistency has hitherto been developed in training areas (Friedl et al., 2010).

Validation, as described in the CEOS-LVP report, is a whole process with various components, including quality control, qualitative assessment, cross-comparisons, confidence maps, and accuracy assessment. Recognizing that these components are truly a part of the validation process and are needed to get the whole picture, this chapter focuses particularly on the characterization of quantitative uncertainty. Whereas the overall validation process aims at reporting only on the fitness for use of a given product, and the accuracy assessment only on thematic accuracy, the characterization of uncertainty aims at documenting quantitatively the intrinsic product quality. On the basis of current experiences, this chapter seeks to contribute to further standardization of the characterization of uncertainty. As proposed in the Guide to the Expression of Uncertainty in Measurement (GUM) (JCGM, 2008), a readily implemented, easily understood, and generally accepted procedure is necessary for characterizing the quality of a result, that is, for evaluating and expressing its uncertainty.

First, a set of eight uncertainty components for a land-cover map is proposed and described briefly. The first seven components are also illustrated and discussed in the light of the four validation exercises effectively completed. The eighth component is addressed in a separate section dedicated to the accuracy assessment process. The most recent experience—corresponding to the GlobCover accuracy assessment—is further detailed as a case study to report on the recent findings and raise current issues.

14.2   UNCERTAINTY COMPONENTS

Current global land-cover maps assign a land-cover-type label to each area of the terrestrial surface. Except in the particular case of a continuous vegetation field (DeFries et al., 1995; Hansen et al., 2002), the already existing land-cover maps correspond to a hard classification run at the pixel level. To characterize the uncertainty of a final global land-cover product, eight components are proposed.

Induced by the earth observation data, the cartographic standards, and/or the image-processing algorithms, the components reflect the quality of spatiotemporal characteristics of land-cover products and the thematic definition as well as the performance of the classification method used in product generation. These eight components, which should be quantified, are listed in Table 14.1 and detailed in the following sections.

In the current land-cover validation exercises, only a few of these uncertainty components are quantified and/or considered. Basically, only the geolocation and the land-cover-type accuracy (standing for the two components generated by the preprocessing and classification algorithms) are assessed.

Furthermore, in principle, even the uncertainty of these uncertainty estimates should be addressed here, calling for standard deviation, confidence interval, or error bars for these estimates. It is clear that the focus of such an assessment is on quantifying the different components rather than investigating the possible reasons for the uncertainties. This clearly corresponds to a user's perspective, dealing only with the final product, unlike the remote-sensing community that is more interested in a discussion of the uncertainty sources.

It must also be mentioned that the uncertainty estimates of these components may concern the whole map or vary according to the land-cover type and the region. Finally, the impact of these uncertainties remains quite variable according to the landscape or the land cover of interest. For instance, the combined effect of the spatial resolution, the minimum mapping unit (MMU), and the geolocation accuracy depends strongly on the diversity and spatial patterns of the land surface.

14.2.1 Uncertainty Related to Spatial Resolution

Strictly speaking, the spatial resolution of a raster land-cover product should be referred to as the pixel size of the map. This should be equal to or larger than the spatial resolution of the

---

TABLE 14.1

List and Definition of Eight Uncertainty Components Proposed to Be Considered in the Validation Process of Global Land-Cover Maps

Uncertainty Component in the Land-Cover Product	Source
Spatial resolution of the land-cover map, defining the level of details for the class boundary delineation	Earth observation data: effective spatial resolution
Time span of the land-cover map, which usually corresponds to the acquisition period of the input data	Earth observation data: observation duration
Information gap, due to missing input data or areas not mapped for various reasons	Earth observation data: quantity of valid observation
Minimum mapping unit, corresponding to the minimum area required to be depicted on the map	Cartographic standards: map specification
Precision of the land-cover type definition, which is supposed to be exhaustive and mutually exclusive	Cartographic standards: legend quality
Thematic resolution, defined by the number of land-cover types included in the legend and effectively reported on the map—an associated thematic distance is sometimes used to describe the semantic proximity between classes	Cartographic standards and image-processing algorithms: legend level and discrimination capabilities
Geolocation accuracy, referring to the error in the geographic coordinates with regard to the absolute coordinate system selected for the map projection—this is often measured as the planimetric error	Image-processing algorithms and Earth observation data: geometric correction
Land-cover-type accuracy, which is usually referred to as the classification accuracy assessment	Image-processing algorithms and cartographic standards: map specifications matching

remote-sensing imagery used for mapping. The concept of spatial resolution is often simplified in that spatial resolution is considered to be equivalent to the ground projection of the pixel. As summarized by Duveiller and Defourny (2010), spatial resolution is actually a complex concept that depends mainly on three characteristics sensor: the ground-sampling distance, the modulation transfer function, and the signal-to-noise ratio. In addition, the relative geometric accuracy between repeated observations also influences the capacity to distinguish small features.

GlobCover land-cover products were produced at a 300-m spatial resolution, using a time series from the MERIS instrument in its Full spatial Resolution (FR), which is 290 m (along track) × 260 m (across track) at nadir for the 15 spectral bands. The pixel sizes of the GLC2000, MODIS Land Cover, and IGBP-DISCover maps also correspond to the nominal spatial resolution of their respective sensors. However, unlike the spatial resolution of push-broom instruments such as Medium Resolution Imaging Spectrometers (MERIS) and Systeme Pour 'Observation de la Terre (SPOT) VEGETATION, the spatial resolution of the scanning ones with a large field-of-view (e.g., AVHRR and MODIS) varies significantly from the subsatellite point to the image edge, making it more difficult to define the most appropriate output pixel size. Defined as the ratio of the intersection area between the nominal observation and the grid cell to the nominal area of the observation, the notion of "observation coverage" (Wolfe et al., 1998), more often called "obscov," was introduced for the MODIS whisk-broom scanner, and it allows for nice tackling of the aforementioned issue.

14.2.2 UNCERTAINTY RELATED TO TIME SPAN

Compilation of a multiyear dataset increases the quantity of the input data for the classification process and may also improve its quality. Use of a multiyear dataset, which most often comes from methodological choices, can also be a requirement for areas showing the most persistent cloud cover (Vancutsem et al., 2007).

The reference date or year defining the land-cover product should be documented as the actual time span of the input data. The extent of the observation period does introduce some uncertainty about the year of concern. This is even more critical for applications that aim at computing annual land-cover change rate, thus requiring the actual date of both epochs.

The IGBP-DISCover and GlobCover 2009 (Loveland et al., 2000; Bontemps et al., 2011) products have a time span of 1 year, but the GLC2000 and GlobCover 2005 maps (Bartholomé and Belward, 2005; Arino et al., 2008) rely on an observation period of 14 and 19 months, respectively. As for the most recent MODIS Land Cover product (Friedl et al., 2010), the input data span 12 months. A multiyear approach has been developed to stabilize the results and reduce the year-to-year spurious variability.

14.2.3 UNCERTAINTY DUE TO INFORMATION GAP

The reasons for missing data causing an information gap are numerous when working on a global scale. The most common is the extent to which satellites record along their orbit. In particular, polar and very high latitudes are hardly observed by orbiting instruments owing to low illumination conditions.

In the particular case of the GlobCover project, the lower temporal resolution of the MERIS sensor and the reception antenna footprints, combined with onboard storage capacity, produced gaps in the MERIS FR acquisition plan—independently of the cloud coverage issue. In some cases, the application of an inaccurate land–sea mask also resulted in the missing of some coastlines and isolated islands.

Such information gaps were usually filled by alternative sources of land-cover information. GLC2000 supplemented the GlobCover 2005 product for less than 2% of the continental areas. For the MODIS Collection 5 Land Cover products, Friedl et al. (2010) used information obtained from the same product for the previous year or, in some very rare cases, from the MODIS Collection 4 Land Cover output.

14.2.4 Uncertainty Related to Minimum Mapping Unit

From a map reading and user's perspective, the smallest area that can be depicted on a map is defined as the MMU. MMU is a cartographic concept, which corresponds to a simplification of the output to be expressed in square meters, hectares, square degrees, or the number of pixels, and which aims at facilitating and clarifying map reading. Undeniably useful when mapping outputs are printed on paper, it is also quite popular for on-screen visualization in spite of its zoom capabilities.

The introduction of MMU can also be explained on technical grounds. First, it removes the well-known salt-and-pepper effect obtained by pixel-based classification. Second, the geolocation error, usually below the pixel size, can still lead to a thematic error in landscapes with fine spatial patterns. The application of MMU can marginalize the resulting thematic error.

However, the use of land-cover products stored in raster formats hinders to some extent the actual implementation of MMU in transforming it in a simple filtering. Indeed, filters operate on windows, whereas a mapping unit may have an elongated form, stretching beyond the limits of a window. For each mapping unit, which is in fact a cluster of pixels, the surface area should be determined and then compared with the MMU size. New object-oriented classification techniques can easily fit in with the MMU concept.

The current practice in global land-cover mapping is to filter out the isolated pixels as in the case of GlobCover 2005. The 1-km IGBP-DIS and GLC2000 products were not filtered owing to their already coarser pixel size. Alternatively, GLC2000 somehow mimics the effect of MMU in terms of its accuracy assessment by taking into account the majority land-cover class within a 3 × 3 window.

14.2.5 Uncertainty Linked to the Precision of Land-Cover Type Definition

A major contribution of the Land-Cover Classification System (LCCS), developed by the Food and Agriculture Organization (FAO) of the United Nations (UN), has been to enhance the precision of the land-cover type definition in a standardized way. LCCS is a comprehensive, standardized, *a priori* classification system designed to meet specific user requirements and created for mapping products independent of the scale or means used for mapping (Di Gregorio and Jansen, 2000). According to this classification system, each class has to be defined by a set of quantitative and qualitative information called *classifiers*. LCCS classifiers are organized in a hierarchical way to ensure a high degree of precision in the class definition.

LCCS is characterized by two main phases: an initial dichotomous phase, which brings a division into the eight major land-cover types and a subsequent modular-hierarchical phase, where the set of classifiers and their hierarchical arrangement are tailored to the major land-cover type and are, to a large extent, specific for the concerned land-cover type. The two phases permit the use of the most appropriate classifiers and reduce the total number of impractical combinations of classifiers. Quantifying the combination of classifiers explicitly clarifies the class definition. However, land cover, especially on a global scale, can be very heterogeneous—in particular for a large MMU—and so several sets of classifiers are often to be used to define all land-cover types in the spatial unit.

GLC2000 and GlobCover typologies are based on LCCS, which also permits precise regional subclass definition. MODIS Collection 6 is also expected to migrate to this classification system to conform to international standards.

Although the adoption of LCCS gives an opportunity to significantly reduce the uncertainty in the definition of the legend, it does not prevent ambiguity or overlapping between classes; this clearly depends on the mixed or mosaic class definition. In GLC2000 (Mayaux et al., 2006), a thematic distance was calculated between all the classes, based on the mandatory and optional LCCS classifiers (Table 14.2).

The GlobCover 2005 experience also highlighted the fact that the translation of a set of classifiers into a given land-cover typology can be more ambiguous than expected. This problem occurred in its validation exercise (described in detail in Section 14.3) when experts used several land-cover types

TABLE 14.2
Matrix of Thematic Distance between GLC2000 Classes Based on LCCS Classifiers

	1	2	3	4	5	6	7	8	9	10	11	12	13	14	15	16	17	18	19	20	21
Tree cover, broad-leaved, evergreen	0.00																				
Tree cover, broad-leaved, deciduous, closed	0.17	0.00																			
Tree cover, broad-leaved, deciduous, open	0.21	0.13	0.00																		
Tree cover, needle-leaved, evergreen	0.13	0.29	0.33	0.00																	
Tree cover, needle-leaved, deciduous	0.25	0.17	0.21	0.13	0.00																
Tree cover, mixed leaf type	0.13	0.17	0.21	0.13	0.13	0.00															
Tree cover, regularly flooded, fresh	0.23	0.33	0.38	0.29	0.35	0.23	0.00														
Tree cover, regularly flooded, saline	0.17	0.21	0.25	0.29	0.29	0.29	0.03	0.00													
Mosaic: tree cover/other natural vegetation	0.19	0.23	0.27	0.19	0.19	0.13	0.29	0.32	0.00												
Shrub cover, closed-open, evergreen	0.13	0.29	0.33	0.25	0.38	0.25	0.29	0.26	0.19	0.00											
Shrub cover, closed-open, deciduous	0.31	0.23	0.27	0.31	0.19	0.19	0.35	0.39	0.13	0.19	0.00										
Herbaceous cover, closed-open	0.22	0.20	0.24	0.34	0.28	0.25	0.42	0.39	0.22	0.22	0.22	0.00									
Sparse herbaceous or sparse shrub cover	0.34	0.23	0.19	0.47	0.34	0.34	0.31	0.31	0.31	0.28	0.22	0.19	0.00								
Regularly flooded shrub and/or herbaceous cover	0.32	0.43	0.47	0.45	0.51	0.39	0.29	0.29	0.35	0.26	0.39	0.29	0.35	0.00							
Cultivated and managed areas	0.50	0.42	0.46	0.63	0.50	0.50	0.50	0.50	0.50	0.50	0.44	0.47	0.44	0.44	0.00						
Mosaic: Cropland/tree cover/other natural vegetation	0.21	0.31	0.35	0.27	0.33	0.21	0.31	0.38	0.34	0.41	0.47	0.44	0.56	0.41	0.35	0.00					
Mosaic: cropland/shrub and/or grass cover	0.33	0.31	0.35	0.40	0.33	0.27	0.41	0.47	0.38	0.44	0.38	0.34	0.47	0.41	0.29	0.09	0.00				
Bare areas	0.67	0.67	0.65	0.67	0.67	0.67	0.83	0.83	0.67	0.67	0.67	0.67	0.54	0.80	0.85	0.72	0.72	0.00			
Water bodies (natural and artificial)	0.92	0.92	0.92	0.92	0.92	0.92	0.75	0.75	0.92	0.92	0.92	0.92	0.92	0.75	0.92	0.92	0.92	0.38	0.00		
Snow and ice (natural and artificial)	0.92	0.92	0.92	0.92	0.92	0.92	0.75	0.75	0.92	0.92	0.92	0.92	0.92	0.75	0.92	0.92	0.92	0.38	0.15	0.00	
Artificial surfaces and associated areas	0.83	0.83	0.83	0.83	0.83	0.83	1.00	1.00	0.83	0.83	0.83	0.83	0.83	1.00	0.92	0.75	0.75	0.17	0.38	0.38	0.00

TABLE 14.3

Three Sets of Classifiers That Describe the Land Cover for an Observational Unit Out of the Validation Dataset

Land Cover 1	Land Cover 2	Land Cover 3
Natural and seminatural terrestrial vegetation	Cultivated and managed lands	Natural and seminatural terrestrial vegetation
Shrubs	Herbaceous	Trees
Open (70%–60%–20%–10%)	Rain-fed	Open to very open (40%–20%–10%)
5–0.3 m		>30–3 m (for trees)
Broad-leaved		Broad-leaved evergreen

to describe the same observational unit. Table 14.3 illustrates this situation with three land-cover types (each characterized by a set of LCCS classifiers), which are associated with a single validation unit. The fact that three land-cover types were used gives cause to consider mosaic classes (that are in fact mixed mapping units) in addition to "pure" GlobCover classes. In total, five GlobCover classes can correctly describe the land cover within the concerned observational unit:

- Closed to open (>15%) (broad-leaved or needle-leaved, evergreen or deciduous) shrubland (<5 m)—for land-cover type 1
- Rain-fed (cultivated and managed lands)—for land-cover type 2
- Closed to open (>15%) broad-leaved, evergreen or semideciduous forest (>5 m)—for land-cover type 3
- Mosaic natural vegetation (grassland/shrubland/forest) (50%–70%)/cropland (20%–50%)
- Mosaic cropland (50%–70%)/natural vegetation (grassland/shrubland/forest) (20%–50%)

Indicating the area proportion of the respective land-cover type identified within the same observational unit reduces the uncertainty related to the precision of land-cover type definition, as proved by the GlobCover 2009 accuracy assessment exercise. Regardless of this land-cover dominance issue, the ambiguity is of course more exacerbated when the expert is unable to provide information for all the required classifiers.

14.2.6 UNCERTAINTY DUE TO THEMATIC RESOLUTION

A precise definition of the land-cover typology as induced by the LCCS does not show the thematic resolution that prescribes the level of details used for the land-cover description.

The thematic resolution is first assessed by the number of land-cover classes included in the legend. Indeed, a global "forest–nonforest" map does not show the same thematic resolution that a typology of 16 or 22 classes does. The thematic resolution is also characterized by the semantic distance between the different classes of the legend. Indeed, the mapping detail level is not a necessary equivalent for all land-cover types. Most global maps differentiate several forest types but hardly separate different croplands or urban classes, which are of major interest for many applications. For instance, the vegetation-modeling community is eager to obtain the number of growing cycles for the different cropland areas as already reported by Thenkabail et al. (2010)—but at a 10-km aggregation level only. The thematic resolution of the global products is currently driven and much constrained by remote-sensing discrimination capabilities.

In GLC2000, Mayaux et al. (2006) complemented the overall accuracy value obtained for the GLC2000 map in weighting the class error with a similarity value derived from the LCCS classifiers. This aimed at balancing the misclassification between classes in terms of their thematic distance.

14.2.7 Uncertainty Related to Geolocation Accuracy

The ground location associated with each pixel must be known with high accuracy. The absolute geolocation accuracy defines the positioning error associated with the map. This limits the spatial matching between the global land-cover product and any ancillary data to be compared with, including the reference information used for the accuracy assessment.

In the context of GLC2000, the absolute planimetric error of the SPOT-VEGETATION products was considered to be about 300 m, corresponding to the planimetric error of the input mosaics used in the classification process. For the GlobCover products, a planimetric error assessment was specifically designed and completed. The MERIS products have demonstrated an absolute geolocation accuracy of 77 m RMS, which was found to be quite satisfactory for this ocean instrument and which permitted the use of the MERIS images at their full resolution of 300 m (Bicheron et al., 2011).

14.3 LAND-COVER ACCURACY ASSESSMENT

The accuracy of a land-cover type shows how much the classification diagnostic is in agreement with "ground truth." Collection of ground information is considered the best option for supporting the validation of remote-sensing products in general. However, for global land-cover products, organizing field surveys over thousands of large-sized plots is unrealistic. Most often, surrogates to ground truth are obtained from existing land-cover information, which are recognized as a suitable "reference" and which, of course, are independent of the product to be validated.

The accuracy of the land-cover type is then quantitatively assessed by comparing the land-cover type identified by the product and the "actual" land-cover type as determined by the reference dataset. Such a comparison provides a set of accuracy figures, such as the overall accuracy (with confidence interval), user's and producer's accuracy, and the kappa statistic. Quantitative assessment of the thematic accuracy must make use of an independent source for the reference dataset and derive accuracy figures in a sound and repeatable way, based on methodologies that are internationally acceptable and feasible from the point of view of cost and time.

As clearly expressed in the CEOS-LVP report (Strahler et al., 2006), accuracy assessment methodologies include three different steps: collecting reference data, elaborating the sampling strategy, and assessing the product's accuracy. These steps are detailed in the following sections, with special focus on the more recent GlobCover validation exercise.

14.3.1 Reference Data Collection

As already mentioned, the reference data collection for global products can rely only on the already existing expertise available all over the world. In the case of the IGBP-DISCover map (Scepan 1999) and the GLC2000 map (Mayaux et al., 2006), the reference dataset was built by visually interpreting a 50-m or 30-m resolution orthorectified Landsat color composite. In the GlobCover project, reference data were provided by a network of international experts using an online interface, summarizing land-cover information from different sources.

As recommended, the GlobCover validation plan was adopted from the very beginning of the project, before starting land-cover map production. An independent stakeholder developed the data collection tool and completed the data analysis for the accuracy assessment.

The GlobCover projects set up a network of international experts selected according to the following criteria: recognized expertise in land cover over relatively large areas, familiarity with interpreting remote-sensing imagery, commitment to performing the interpretation, mutual compatibility and association with well-known international networks. For the GlobCover 2005 validation exercise, 16 experts from all over the world were invited for six different 5-day workshops hosted by the Université catholique de Louvain (Louvain-la-Neuve, Belgium). The same network was again involved, though remotely, in the GlobCover 2009 validation. Indeed, on the basis of the GlobCover 2005 experience (Defourny et al., 2009), a specific web interface was proposed as a working environment.

Validation samples were automatically overlaid either in Virtual Earth or Google Earth, thus allowing rapid access to recent remote-sensing images with zooming capabilities. In addition, normalized difference vegetation index (NDVI) and normalized difference water index (NDWI) average profiles computed from the whole SPOT-VEGETATION archive (2000–2009) were displayed for each validation sample to complement the information provided by the high-resolution imagery with seasonal dynamics. Finally, the experts supported their work using any additional sources of information such as detailed maps. This GlobCover validation interface is illustrated in Figure 14.1.

To allow the use of the reference dataset beyond the scope of a given project, gathering information on the LCCS classifiers was strongly recommended to characterize the land cover of each validation sample irrespective of a given typology. Therefore, the GLC2000 as the GlobCover validation process required the experts to provide the classifier information (as shown in Figure 14.1), rather than the corresponding class in the legend. This information was then translated into the land-cover classes of the global product. Furthermore, a level of uncertainty was requested for each sample, allowing the expert to select between the certain, the reasonable, and the doubtful.

For a given sample, the expert was required to look not only at the sample point but also at a box that coincided with the so-called observational unit corresponding to 5 × 5 MERIS pixels (225 ha). The effective observational unit was not necessarily to be a square or a circle around the point. Some land-cover classes, notably lakes and wetlands, could indeed be elongated but could not be discarded owing to the shape of the observational unit.

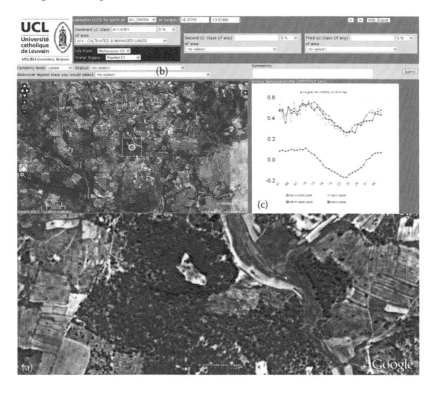

FIGURE 14.1 **(See color insert.)** Web-based interface used for GlobCover 2009 reference data collection by the international expert network. Validation samples were automatically overlaid either in Virtual Earth or Google Earth (a), combo boxes to characterize the samples with LCCS classifiers were included (b), and SPOT-VEGETATION NDVI and NDWI temporal profiles corresponding to the pixel displays as white square were provided (c).

14.3.2 Sampling Strategy

To ensure that each pixel had an equal chance of being sampled, the global product to be validated had to be projected into an equal-area projection. Most sampling designs relied on stratified random sampling.

For GLC2000, the global map was reprojected into an equal-area projection, and the stratification, which used an underlying grid of the Landsat World Reference 2 System, was based on the proportion of priority land-cover classes and on the landscape complexity. A set of 544 blocks dominated by one land-cover class (>80% of the area) allowed the selection of 1265 sample sites by a two-stage sampling (Mayaux et al., 2006).

For GlobCover, as there is no equal-area projection that does justice to the entire world, the world was divided into five regions (Africa, Australia and Pacific, Eurasia, North America, and South America) for which it is possible to apply an equal-area projection. Next, the GlobCover validation strategy opted for a stratified random sampling method based on the entire product. Stratification relied on the most recently available global land-cover map, that is, on the GlobCover 2005 beta product for GlobCover 2005 and on GlobCover 2005 for GlobCover 2009.

The GlobCover 2005 reference dataset contains 4258 samples. In 3167 cases, the experts were explicitly certain of the information they provided; in 797 cases, they were reasonably sure; and in 294 cases, they had some reservations. The distribution of the 3167 "certain" points is shown in Figure 14.2.

14.3.3 Accuracy Estimation

Literature on the subject has widely addressed the question of accuracy figures for land-cover maps. This question has also been specifically summarized in the CEOS-LVP report for global products (Strahler et al., 2006).

The confusion matrix remains the most popular standard. To build such a matrix, the reference dataset is crossed with the map, and land-cover codes are extracted for all the validation points. The overall accuracy weighted by the area proportions of the various land-cover classes is the most synthetic figure, although it does not reflect the whole story. The weighting factor corresponding to the area proportion of the given class is derived from the product that is projected into an equal-area projection.

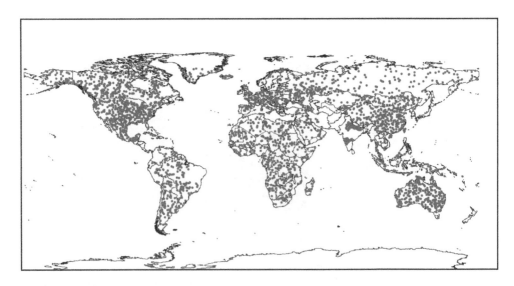

FIGURE 14.2 Geographic distribution of the "certain" points in the GlobCover 2005 reference dataset.

Table 14.4 reports the outcome of the accuracy assessment carried out for the existing global land-cover maps.

In spite of the diversity of sources for remote-sensing data, classification methods, and validation strategies, the overall accuracy figures weighted by area converge very much around 70%. This raises more general questions. Does the best effort of the producers increase thematic resolution and spatial resolution rather than thematic accuracy? Is this 70% ceiling the impact of an increasing strictness in the accuracy assessment (which, for instance, seeks to look at all landscapes and not at the homogeneous ones only)? What is the thematic resolution (i.e., the legend level of detail) that provides the best accuracy? Is there a natural limit around 70% of overall accuracy related to the quality of the reference dataset? Is the quality of the reference information hindering the accuracy improvement?

In all global validation experiences, the reliability assumption of the reference information has been unquestioned, as this corresponds to the best possible effort. However, as in the case of production of global land-cover maps, uncertainty sources are inherent in a validation exercise. On the basis of the GUM terminology (JCGM, 2008), four uncertainty components affecting the reference information can be discussed. They are presented in Table 14.5.

TABLE 14.4

GCOS Requirements and Accuracy of the Existing Global Land-Cover Maps

		Current Products			
GCOS Requirements		**AVHRR**	**Vegetation**	**MERIS**	**MODIS**
Class accuracy (max. error for individual classes)	15% omission/ commission per class	IGBP-DISCover 67% weighted across all classes	GLC2000 69% weighted across all classes	GlobCover 73% weighted across all classes	MODIS v5 (75% cross-validation accuracy)
Spatial resolution	250 m–1 km	1 km	1 km	300 m	500 m
Geometric accuracy	Better than 1/3 IFOV	–	300 m	70 m	–
Temporal resolution	1-year observing cycle	Yearly 1992	Yearly 2000	Yearly 2005 and 2009	Yearly 2002–2009
Stability	As class accuracy	Not specified No intercomparison possible			

TABLE 14.5

List and Definition of Four Uncertainty Components Affecting Land-Cover Accuracy Assessment

Uncertainty Component Related to Land-Cover Accuracy Assessment	Source
8a. Repeatability of the interpretation process that generates reference information	Interpretation process
8b. Reproducibility of the reference information as provided by the experts	Interpretation process
8c. Relevance of the reference information in terms of support	Land-cover data (reference and product to validate)
8d. Precision of the reference information translation into land-cover typology	Interpretation process, cartographic standards

According to the GUM, repeatability is defined as the closeness of the agreement between the results of successive measurements carried out under the same conditions of measurement. In the context of reference land-cover information, this corresponds to the reliability of the interpretation process by experts and answers the question: How many times will a sample point be identically interpreted under the same interpretation conditions (i.e., by the same expert using the same typology, the same imagery, and the same season)? Repeatability may be expressed quantitatively by the dispersion characteristics of the interpretation process. Unfortunately, no information seems to be available from the current global experiences about this uncertainty source. The only available indication that can be linked to this issue is the confidence level with which the experts accomplish the interpretation process. In the GlobCover project, this information was required from the experts, and the result was that for both GlobCover 2005 (3917 sample points) and GlobCover 2009 (3134 sample points) exercises, over 75% (respectively 80% and 77%) of the samples were labeled as certain by them.

Reproducibility of results is defined by the GUM (JCGM, 2008) as the closeness of the agreement between the results of measurements carried out under changed conditions of measurement. The reproducibility then quantifies the visual interpretation capability to identify the same land-cover type by different experts and/or by using high spatial resolution imagery from different sensors or from different seasons. The experts' influence on the interpretation result has been specifically studied in the framework of the GlobCover project. For the 2005 product validation, a set of 225 sample points was submitted to two experts, using the same dataset environment. Only 75% of the points were interpreted similarly by the two experts. In the GlobCover 2009 validation process, 1193 points from the GlobCover 2005 reference dataset were submitted to the experts for improvement (possibly owing to the availability of better high spatial resolution imagery on the proposed geoportals) or change detection. Only 2.8% of the sample points were declared as improved, whereas up to 30% of the points were considered as changed. These figures clearly call for further investigation as land-cover change rates at the global scale are much lower than this 30% proportion. What is to be questioned here is not the quality of the experts but only whether their work is feasible in a manner as reliable as expected.

In our context, the relevance of the reference information concerns the spatial matching between the two elements to be compared, that is, the sample point as described by the expert and the spatial unit of the map to be validated. Although this is not considered as an issue for those working at 1-km spatial resolution, and when homogeneous landscapes are mainly looked at, it becomes very critical for the validation of products with a subkilometer spatial resolution. At the same time, the matching of the information supports should, in principle, take into account the spatial resolution and the MMU effect. Studies carefully comparing different land-cover maps obtained at different spatial resolutions had previously addressed this issue by considering separately the homogeneous and heterogeneous land-cover patterns, with particular focus on class edges (Herold et al., 2008; Mayaux and Lambin, 1995; Smith et al., 2002). However, such strategies cannot be easily implemented in an accuracy assessment without the risk of introducing some bias.

Finally, as already noted (see Section 14.2.6), translating the reference information provided by experts into land-cover typology is not necessarily unambiguous. Since their reference information can be translated differently to describe a given sample point, all the translations must be taken into account to compile the confusion matrix. This introduces some uncertainty in the validation dataset and, therefore, in the overall accuracy measurement. To reduce this uncertainty for the GlobCover 2005 product, two overall accuracy figures were delivered: a first one of 73.14%, based on the 3167 samples acknowledged as certain by the experts, and a second one of 79.25%, obtained by using a subsample of 2115 homogeneous points, also acknowledged as certain. Indeed, there is almost no translation uncertainty for sample points covered by a single set of classifier, also called a homogeneous sample. Clearly, improving this translation precision is probably quite feasible but not straightforward.

Furthermore, it is worth mentioning that several combinations of land-cover types could not be transformed to a GlobCover mosaic class (6% of points were lost for this reason in GlobCover

2009). Indeed, a legend that will cover all these potential combinations is not desirable because the mosaic classes are often considered less informative and, therefore, less useful from the point of view of the end-user.

14.4   FURTHER IMPROVEMENTS

In this chapter, a set of uncertainty components has been listed (Tables 14.1 and 14.5). It documents the quality of the spatiotemporal characteristics and of the thematic definition of the land-cover products, the spatial matching between the land cover and validation products, the thematic relevancy of the validation data as well as the performance of the classification and interpretation methods implemented to generate the products and the validation datasets.

The existing validation exercises dedicated to global land-cover products have focused mainly on assessing the classification accuracy (Defourny et al., 2009; Mayaux et al., 2006; Scepan, 1999). The issues of geometric correction and geolocation accuracy assessment have been addressed for advanced very high resolution radiometer (AVHRR) (Moreno and Melia, 1993; Rosborough et al., 1994), SPOT VEGETATION (Sylvander et al., 2000), MODIS (Wolfe et al., 2002; Xiong et al., 2005), and MERIS (Bicheron et al., 2011). Several papers have attempted to deal with the uncertainty related to the definition and precision of the legend by comparing the existing global land-cover products (Herold et al., 2008; McCallum et al., 2006) or by deriving a better map from the existing ones for specific applications (Jung et al., 2006). Nevertheless, these studies have also demonstrated the extent of the thematic uncertainty and the associated negative impact on the possibility for their conjoint use. Clearly, there is room for improvement—better characterizing of the varying uncertainty sources affecting the land-cover products and thereby better documenting of their quality.

In classification accuracy assessment, the relevance of the supports (i.e., of the reference information and the global land-cover maps) should be improved to ensure that the same areas are effectively compared. For this, object-oriented techniques should be used to preprocess the images in order to interpret them for generating the reference information. Objects that guide the visual interpretation of the experts will make the integration of the MMU concept easier and support the characterization of the "dominance" when several land-cover types are needed to describe a single validation point. Yet, such object-based approaches will require the processing of all the images used for the accuracy assessment and thus the developing of automatic and consistent segmentation techniques.

Another critical point concerns the heterogeneous areas, where the combined effects of the spatial resolution, the MMU, and the geolocation accuracy often drastically increase the classification uncertainty. Current practices in the classification methodologies (e.g., filtering techniques to remove isolated pixels) should be coupled with specific effort in the validation process. For this, the definition of validation units complying with the spatial resolution, the MMU, and geolocation accuracy is a prerequisite. Then the quantitative evaluation of the dominance of the land-cover types (i.e., quantifying the area proportion of the land-cover types identified within the same validation unit) will allow a decrease in the uncertainty in the translation into land-cover typology.

Alternatively, accuracy assessment can be based on the mapping of very high spatial resolution extracts, allowing for detailed matching analysis. This can be advantageously coupled with global sampling exercises such as the FAO-Forest Resources Assessment. However, this type of imagery cannot be considered a unique validation support. Indeed, the most recent validation experiences have shown the critical importance of NDVI annual profiles—which are available only for lower spatial resolution time series—for land-cover identification (Defourny et al., 2009). Providing such information over time, in order to help generate reliable reference information, will also enable us to account for the time span of the land-cover product.

Furthermore, the current operational capacity to deliver global land-cover maps on a regular basis, as well as the respective findings of MODIS Collection 5 products and GlobCover 2005 and

2009 maps about the spurious interannual variability (Bontemps et al., 2011; Friedl et al., 2010), calls for a consideration of two additional uncertainty components (Table 14.6).

This twofold concern for consistency of global products—both in the spatial and temporal dimensions—is supported by the GCOS requirements document (GCOS, 2010) as well as by the climate users' land-cover requirements surveyed in the framework of the ESA Climate Change Initiative (Herold et al., 2011).

As for temporal consistency, the issue of interannual variability has been raised both for the MODIS Collection 5 and the GlobCover, but no formal analyses have been conducted so far to investigate this uncertainty source in detail.

Characterization of spatial consistency uncertainty was actually documented in the GlobCover 2005 and 2009 validation reports (Bicheron et al., 2008; Bontemps et al., 2011). In these reports, the classification accuracy was shown to be driven largely by the data coverage and the limited number of valid MERIS FR observations impacting the classification reliability. The number of valid observations available over a region (Figure 14.3) gives *a priori* information about the input-data quality and thus represents a valuable indicator of the output-product accuracy.

In the specific case of GlobCover, some regions of the world are particularly affected by a low number of MERIS FR acquisitions (Central and South America, northeastern America, the Korean peninsula, and eastern Siberia). Besides, when the data coverage is poor, a tendency to overestimate forest areas was noted. The fact that the GlobCover map quality varies according to the region of interest balances the positive figure of the overall accuracy assessment (73.14% using a validation

---

TABLE 14.6

Additional Uncertainty Components to Consider for Global Land-Cover Maps

Additional Uncertainty Component for Global Land-Cover Product	Source
Temporal consistency of the classification accuracy, corresponding to the reproducibility of the land-cover mapping of the same area, assuming an absence of land-cover change	Image-processing algorithms and earth observation data
Spatial consistency of the classification accuracy, describing the stability of the discrimination performances for the same land-cover class in different parts of the world	Image-processing algorithms and earth observation data

---

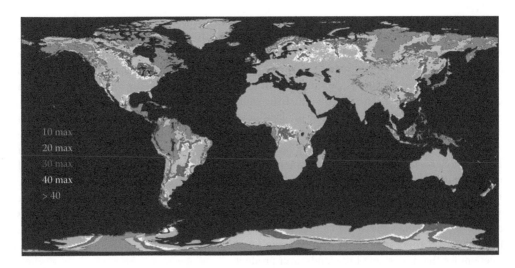

FIGURE 14.3   (See color insert.) Number of valid MERIS FR surface reflectance observations for GlobCover 2005.

dataset made of "certain" points and 79.25% when the points were both "certain" and "homogeneous"). Similarly, information about the confidence in the classification diagnostic (e.g., classification probability maps) also documents spatially the discrimination quality.

During the GLC2000 validation exercise, a confidence-building exercise (Mayaux et al., 2006) was conducted to document the spatial distribution of errors. This procedure, which also aimed at reducing macroscopic errors, resulted in drastic improvement in the quality of the final product as well as full endorsement by users.

Last but not least, the development of the voluntary information geographic community and the geowiki technology can significantly change the way ground-truth data is collected. Nevertheless, while such a strategy will multiply the human resources available and the capabilities of ground truthing on a global scale, other issues will have to be addressed. Since arriving at a common understanding of a land-cover legend is already a challenge, data collection will have to become creative in order to gather usable information.

14.5 PERSPECTIVES

Validation of global land-cover information is a challenge for several reasons. However, the need for quantitative uncertainty characterization has been clearly expressed by various user communities. This chapter systematically addresses the various sources of uncertainty embedded in any global land-cover map. A set of 10 uncertainty components is proposed and described. Each component is also illustrated or discussed in the light of the four previous global land-cover validation experiences. Furthermore, the uncertainty related to the reference information used as ground truth is also investigated on the basis of the GlobCover experiences. The definition of four subcomponents specifically related to the uncertainty in the reference information allows for documenting some working assumptions for global land-cover validation. Indeed, the reliability of the reference information has been rarely questioned in these global accuracy assessments. However, it needs to be explained why the four accuracies of the different global land-cover products curiously only reached an overall 70%, in spite of the significant improvements in technology and methodology.

This chapter calls for a more systematic and explicit characterization of all the uncertainty components associated with any global land-cover product, thereby providing a better documentation of the product and supporting better use of land-cover information.

Most recently, information technology development has been taken on board for global *in situ* data collection by pioneer projects and geowiki experiences. Based first on volunteered geographic information and now designed as a crowd-sourcing approach, these open data collection strategies still need to be assessed. Anyway, they not only enhance field observation and geographic coverage capabilities but also raise new methodological issues that have to be solved before entering this new world of global digital connectivity.

ACKNOWLEDGMENTS

The authors are most grateful to the GlobCover experts network comprising S. Bartalev (Space Research Institute, Russian Academy of Science), P. Caccetta (Commonwealth Scientific and Industrial Research Organisation, Australia), A.J.W. de Wit (Alterra, Wageningen, The Netherlands), C. Di Bella (Instituto Nacional de Tecnología Agropecuaria, Argentina), B. Gérard (International Livestock Research Institute, Addis Abeba, Ethiopia), C. Giri (United States Geological Survey— EROS Data Center, USA), V. Gond (CIRAD-Guyane—Université Laval), G.W. Hazeu (Alterra, Wageningen, The Netherlands), A. Heinimann (National Centre of Competence in Research North-South—Centre for Development and Environment, Austria), J. Knoops (INFRAM B.V., The Netherlands), G. Jaffrain (ETC-LUSI Technical Team, Barcelona, Spain), R. Latifovic (Canada Centre for Remote Sensing), H. Lin (Institute of Geographic Sciences and Natural Resources Research—Chinese Academy of Sciences), C.A. Mücher (Alterra, Wageningen, The Netherlands),

A. Nonguierma (Commission of Africa, Addis Abeba, Ethiopia), H.J. Stibig (EU-Joint Research Center), Y. Shimabukuro (INPE, Brazil), as well as to GlobCover validation team L. Schouten (INFRAM B.V., The Netherlands), E. Van Bogaert (Université catholique de Louvain, Belgium), and C. Vancutsem (Université catholique de Louvain, Belgium).

REFERENCES

Arino, O., Bicheron, P., Achard, F., Latham, J., Witt, R., and Weber, J.L. 2008. GlobCover: The most detailed portrait of the earth. *ESA Bulletin,* 136, 25–31.

Bartholomé, E. and Belward, A.S. 2005. GLC2000: A new approach to global land cover mapping from earth observation data. *International Journal of Remote Sensing*, 26(9), 1959–1977.

Bicheron, P., Amberg, V., Bourg, L., Petit, D., Huc, M., Miras, B., Brockmann, C., et al. 2011. Geolocation assessment of MERIS GlobCover orthorectified products. *IEEE Transactions on Geoscience and Remote Sensing*, 49, 2972–2982.

Bicheron, P., Defourny, P., Brockmann, C., Schouten, L., Vancutsem, C., Huc, M., Bontemps, S., et al., 2008. GlobCover: Products description and validation report, ESA GlobCover project. Available at: http://ionia1.esrin.esa.int/docs/GLOBCOVER2009_Validation_Report_2.2.pdf

Bontemps, P., Defourny, P., Van Bogaert, E., Kalogirou, V., and Arino, O. 2011. GlobCover 2009: Products description and validation report, ESA GlobCover project. Available at: http://ionia1.esrin.esa.int/docs/GLOBCOVER2009_Validation_Report_2.2.pdf

Clark, W., Mitchell, R.B., and Cash, D. 2006. Evaluating the influence of global environmental assets. In R.B. Mitchell, W. Clark, D. Cash, and N. Dickson (Eds.), *Global Environmental Assessments: Information and Influence*. Cambridge: MIT Press.

Defourny, P., Schouten, L., Bartalev, S., Bontemps, S., Caccetta, P., de Wit, A.J.W., Di Bella, C.M., et al. 2009. Accuracy assessment of a 300-m global land cover map: The GlobCover experience. In *33rd International Symposium on Remote Sensing of Environment—Sustaining the Millennium Development Goals*, May 4–8, 2009, Stresa, Italy.

DeFries, R., Field, C.R., Fung, I., Justice, C.O., Los, S., Matson, M.A., Matthews, E.A., et al. 1995. Mapping the land surface for global atmosphere-biosphere models: Towards continuous distributions of vegetation's functional properties. *Journal of Geophysical Research*, 100(D10), 20867–20882.

Di Gregorio, A. and Jansen, L.J.M. 2000. *Land Cover Classification System: Classification Concepts and User Manual*. Rome: FAO.

Duveiller, G. and Defourny, P. 2010. A conceptual framework to define the spatial resolution requirements for agricultural monitoring using remote sensing. *Remote Sensing of Environment*, 114, 2637–2650.

Friedl, M.A., Sulla-Menashe, D., Tan, B., Schneider, A., Ramankutty, N., Sibley, A., and Huang, X. 2010. MODIS Collection 5 global land cover: Algorithm refinements and characterization of new datasets. *Remote Sensing of Environment*, 114, 168–182.

GCOS—Global Climate Observing System. 2010. Implementation plan for the Global Observing System for Climate in Support of the UNFCCC, August 2010 (update), World Meteorological Organization. Available at: http://www.wmo.int/pages/prog/gcos/Publications/gcos-138.pdf

Hansen, M.C., Defries, R.S., Townshend, J.R.G., Sohlberg, R., DiMiceli, C., and Carroll, M. 2002. Towards an operational MODIS continuous field of percent tree cover algorithm: Examples using AVHRR and MODIS data. *Remote Sensing of Environment*, 83, 303–319.

Herold, M. 2008. Building saliency, legitimacy, and credibility towards operational global and regional land cover observation assessments. Habilitation thesis, University of Jena, Germany.

Herold, M., Mayaux, P., Woodcock, C.E., Baccini, A., and Schmullius, C. 2008. Some challenges in global land cover mapping: An assessment of agreement and accuracy in existing 1 km datasets. *Remote Sensing of Environment*, 112, 2538–2556.

Herold, M., van Groenestijn, A., Kooistra, L., Kalogirou, V., and Arino, O. 2011. ESA Land Cover CCI Project—User Requirements Document (URD), version 2.2 (23/02/2011). Available at: http://www.esa-landcover-cci.org/

JCGM—Joint Committee for Guides in Metrology—Working Group 1. 2008. Evaluation of measurement data—Guide to the expression of uncertainty in measurement, JCGM 100:2008. Available at: http://www.bipm.org/utils/common/documents/jcgm/JCGM_100_2008_E.pdf

Jung, M., Henkel, K., Herold, M., and Churkina, G. 2006. Exploiting synergies of global land cover products for carbon cycle modeling. *Remote Sensing of Environment*, 101(4), 534–553.

Loveland, T.R., Reed, B.C., Brown, J.F., Ohlen, D.O., Zhu, Z., Yang, L., and Merchant, J.W. 2000. Development of a global land cover characteristics database and IGBP DISCover from 1 km AVHRR data. *International Journal of Remote Sensing*, 21, 1303–1330.

McCallum, I., Obersteiner, M., Nilson, S., and Shvidenko, A. 2006. A spatial comparison of four satellite derived 1 km global land cover datasets. *International Journal of Applied Earth Observation and Geoinformation*, 8, 246–255.

Mayaux, P. and Lambin, E. 1995. Estimation of tropical forest area from coarse spatial resolution data— A 2-step correction function for proportional errors due to spatial aggregation. *Remote Sensing of Environment*, 53, 1–15.

Mayaux, P., Eva, H., Gallego, J., Strahler, A., Herold, M., Shefali, A., Naumov, S., et al. 2006. Validation of the Global Land Cover 2000 Map. *IEEE Transactions on Geoscience and Remote Sensing*, 44, 1728–1739.

Moreno, J.F. and Melia, J. 1993. A method for accurate geometric correction of NOAA AVHRR HRPT data. *IEEE Transactions on Geoscience and Remote Sensing*, 31, 204–226.

Rosborough, G.W., Baldwin, D., and Emery, W.J. 1994. Precise AVHRR image navigation. *IEEE Transactions on Geoscience and Remote Sensing*, 32, 644–657.

Scepan, J. 1999. Thematic validation of high-resolution global land-cover data sets. *Photogrammetric Engineering and Remote Sensing*, 65, 1051–1060.

Smith, J.H., Wickham, J.D., Stehman, S.V., and Yang, L. 2002. Impacts of patch size and land-cover heterogeneity on thematic image classification accuracy. *Photogrammetric Engineering and Remote Sensing*, 68, 65–70.

Strahler, A.H., Boschetti, L., Foody, G.M., Friedl, M.A., Hansen, M.A., Mayaux, P., Morisette, J.T., Stehman, S.V., and Woodcock, C.E. 2006. Global LandCover Validation: Recommendations for evaluation and accuracy assessment of global land cover maps. Office for Official Publications of the European Communities, Luxembourg.

Sylvander, S., Henry, P., Bastien-Thyri, Ch., Meunier, F., and Fuster, D. 2000. VEGETATION geometrical image quality. In *Proceedings of the VEGETATION 2000 Conference,* Belgirate, Italy, April 3–6, 2000, G. Saint (Ed.), CNES-Toulouse & JRC-Ispra (pp. 15–22). Available at: http://www.spot-vegetation.com/pages/vgtprep/vgt2000/sylvander.html

Thenkabail, P.S., Hanjra, M., Dheeravath, V., and Gumma, M. 2010. A holistic view of global croplands and their water use for ensuring global food security in the 21st century through advanced remote sensing and non-remote sensing approaches. *Remote Sensing*, 2, 211–261. Available at: http://www.mdpi.com/2072-4292/2/1/211

Vancutsem, C., Peckel, J.-F., Bogaert, P., and Defourny, P. 2007. Mean compositing, an alternative strategy for producing temporal syntheses. Concepts and performance assessment for SPOT VEGETATION times series. *International Journal of Remote Sensing*, 28, 5123–5141.

Wolfe, R.E., Roy, D.P., and Vermote, E. 1998. MODIS land data storage, gridding, and compositing methodology: Level 2 grid. *IEEE Transactions on Geoscience and Remote Sensing*, 36, 1324–1338.

Wolfe, R., Nishihama, M., Fleig, A.J., Kuyper, J.A., Roy, D.P., Storey, J.C., and Pratt, F.S. 2002. Achieving sub pixel geolocation accuracy in support of MODIS land science. *Remote Sensing of Environment*, 83, 31–49.

Xiong, X., Che, N., and Barnes, W. 2005. On Terra MODIS on-orbit spatial characterization and performance. *IEEE Transactions on Geoscience and Remote Sensing*, 43, 355–365.

15 Role of Remote Sensing for Land-Use and Land-Cover Change Modeling

Terry Sohl and Benjamin Sleeter

CONTENTS

15.1 INTRODUCTION

As the impacts of land-use and land-cover (LULC) change on carbon dynamics, climate change, hydrology, and biodiversity have been recognized, modeling of this transformational force has become increasingly important. Given the wide variety of applications that rely on the availability of LULC projections, modeling approaches have originated from a variety of disciplines, including geography, landscape ecology, economics, biology, and others. Initial modeling was often isolated within each discipline, but multidisciplinary modeling frameworks were developed as LULC modelers began to integrate the socioeconomic and biophysical components of LULC change. The empirical and theoretical basis for this work falls within land-use science, and this field documents both land-use and land-cover change, explains the coupled human–environment dynamics that produce the changes, and provides tools for producing spatially explicit LULC models (Mertens and Lambin, 1999; Rindfuss et al., 2004).

LULC models are data hungry, needing both historical and current land-cover maps coupled with data representing the driving forces of change. Availability of data, especially spatially and temporally consistent data representing those driving forces, is a primary challenge for LULC modeling (Parker et al., 2002; Tayyebi et al., 2008). Site-based observations can be used, but remote-sensing data have several characteristics, most notably repeated synoptic coverage with consistent observation at a relatively low cost, that make them ideal for modeling change. Direct observation and mapping of land cover through remote-sensing analysis are critical for identifying and quantifying the major processes of change. This raster (or grid-cell-based) view of the earth's surface offers simplicity, completeness, and efficient processing for analysis (Crews and Walsh, 2009). Empirical diagnostic models of LULC change can then be developed from these observations (Mertens and Lambin, 1999). However, to understand the driving forces of such observed change, these data must be linked to socioeconomic data.

Remote-sensing data play an increasingly important role in LULC modeling. Here we summarize the role of remote sensing in LULC modeling, the use of remote-sensing data in model construction, parameterization, and validation, and the challenges in linking remote-sensing data with analyses of LULC processes.

15.2   THE ROLE OF REMOTE SENSING IN LULC MODELING

Sohl et al. (2010) discussed the need to address several "foundational elements" of LULC modeling, including: (1) geographic context, (2) regional land-use history, (3) representation of drivers of change, and (4) representation of local land-use patterns. Heistermann et al. (2006) noted four classes of data needed for LULC models, three of which can be derived from or supported by remote-sensing data: (1) current and historical land-use data, (2) environmental data, and (3) scenario data. Several of these foundational elements can be addressed through the use and analysis of remotely sensed data. In the following section, we discuss remote-sensed information of relevance to LULC modeling and illustrate the discussion with a number of specific applications.

15.2.1   Information Obtained from Remote-Sensing Data Sources

Most LULC models attempt to untangle the driving forces behind anthropogenic land use, including socioeconomic and biophysical driving forces, but the resulting thematic classification produced by the models often focuses on resulting land covers or a mix of land-use and land-cover "classes." It is important to be clear about the definition of land use and land cover at this stage. Land use refers to how land is used by human beings, whereas land cover refers to the actual vegetative, structural, or other surface cover resulting from a given land use. Agriculture is a land use, but the crop "corn" is a land cover. Remote sensing simply measures the reflective response of the earth's surface, and so it can be used to directly observe the land cover for a given pixel. Land use must be inferred by linking the measured land cover with ancillary information such as socioeconomic data, field data, or "expert knowledge." Remote sensing excels at detecting surface cover type and condition and provides a number of landscape attributes that can be used by LULC models (Figure 15.1). These are outlined below:

15.2.1.1   Land Cover

Land cover refers to the actual surface cover for a given location (e.g., vegetation type, anthropogenic structure, etc.). Remote-sensing data have a long history of being used for deriving land-cover maps, even before the launch of the first Landsat platform in 1972. Aerial photography served as a primary source of information on land cover before the availability of satellite imagery, and it remains an important source of land-cover information even today (Akbari et al., 2003; Cots-Folch et al., 2007). Aerial photography, which was available before the launch of the first Landsat, remains a valuable tool for analyzing historical LULC change (Gerard et al., 2010; Thomson et al., 2007). With the advent of Landsat and other commercial remote-sensing satellites, land-cover mapping at all scales has flourished. Land-cover information at multiple spatial, thematic, and temporal resolutions has direct relevance to LULC forecast modeling. Consistent, broad-scale, multitemporal land-cover mapping programs such as the United States' National Land Cover Database (NLCD) (Homer et al., 2007; Vogelmann et al., 2001), USGS Land Cover Trends (Loveland et al., 2002), and LandFire (Rollins and Frame, 2006) projects, and Europe's CORINE Land Cover (CLC) databases (Büttner et al., 2002; Heymann et al., 1994) are particularly suited for LULC-modeling efforts.

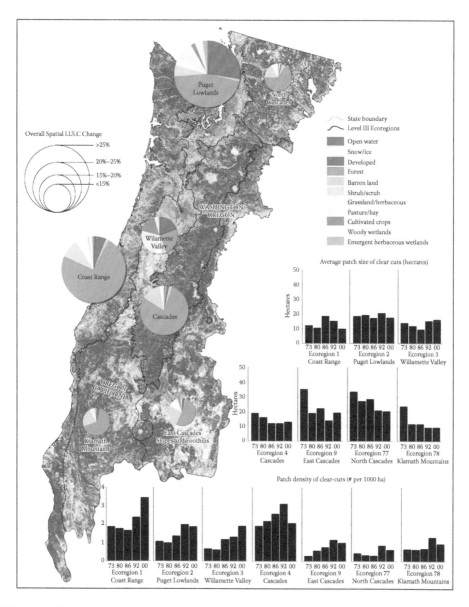

FIGURE 15.1 **(See color insert.)** The area shown covers seven ecoregions (Omernik, 1987) in the Pacific Northwest of the western United States, falling in California, Oregon, and Washington. The ecoregions are the Coast Range, Puget Lowlands, Willamette Valley, Cascades, East Cascades, Slopes and Foothills, North Cascades, and Klamath Mountains. All the ecoregions are primarily forested, with varying levels of agriculture, urban, and other land uses. The map shows data from the 2001 National Land Cover Database (NLCD) (Homer et al., 2007) derived from Landsat thematic mapper (TM) and enhanced thematic mapper plus (ETM+) data at a spatial resolution of 30 m. The pie charts represent two sources of important LULC data. Land-cover composition is characterized by the relative size of each "wedge" and is based on NLCD data. The size of the pie charts reflects the amount of land that experienced changes in LULC as measured by the USGS Land Cover Trends project (Loveland et al., 2002). The extent of land area that changed at least once between 1973 and 2000 varies considerably across the seven ecoregions, including ecoregions with similar land-cover compositions. The changes in LULC reflect variability in the biophysical conditions, land ownership and management, and the impact of regional and national policy among other drivers. LULC change data, such as those presented here, are most readily obtained through examination of historical satellite imagery and aerial photographs. The size and density of forest clear-cuts for the seven ecoregions are displayed in a series of bar charts. Landscape metrics such as these are useful for a wide range of ecosystem assessments and are immediately available through examination of remotely sensed data.

15.2.1.2 Land Use

Unlike land cover, which can be directly observed and monitored from remote-sensing data, land use typically must be inferred through a combination of remote-sensing observation, regional and local knowledge (including field observation), and other ancillary information that links a given land cover in a region with a given land use. As with the mapping of land cover, use of remote-sensing data to assist in the creation of spatially explicit maps of land use has had a long history. Marschner (1950) linked interpreted aerial photography with field notes and statistical summaries to produce the first continental scale, moderate-resolution map of major land uses in the United States. Marschner's basic paradigm is still widely utilized; it uses field surveys, regional knowledge, and other information to infer land use from land-cover observations made from remotely sensed imagery. For example, Brown et al. (2008a) used a combination of MODIS imagery, land-cover classification information from the 2001 NLCD (Homer et al., 2007), and county-level irrigation statistics to create a land-use map of irrigated cropland. Millette et al. (1995) advocated linkages between remote-sensing data, ground truthing, fieldwork, and personnel interviews to provide information on land-use and land-management practices for three villages in Nepal.

15.2.1.3 Landscape Pattern

Spatial patterns of LULC change are regionally unique and dependent on both physical and cultural factors (Gallant et al., 2004). Monitoring and characterizing spatial patterns of LULC change are vital for understanding and predicting LULC change (Petit et al., 2001). Within the field of landscape ecology, modeling of spatial patterns of LULC change needs to be improved (Wu et al., 2008), but this cannot be done without information on current and historical landscape patterns and the driving forces behind the patterns. Remote-sensing information is widely used to analyze landscape pattern. Typically land use and/or land cover is mapped from remote-sensing data and then processed using separate software such as FRAGSTATS (McGarigal et al., 2002) to analyze spatial patterns (Silva et al., 2008; Wang et al., 2009).

15.2.1.4 Landscape Condition

Remote sensing also has the ability to map and monitor changes in surface conditions, which are not related to a direct change in land cover or land use, most notably that of vegetation condition. Long-term datasets such as those provided by the Landsat sensor since the 1970s are particularly valuable for monitoring and understanding changes in vegetation condition (Vogelmann et al., 2009; Wallace et al., 2006). These are complemented by higher temporal resolution sensors such as MODIS, which can map within-season changes in condition (Brown et al., 2008b; Gu et al., 2008; Reeves et al., 2001). Active sensors such as LIDAR or RADAR have the ability to obtain measurements of land surface at any time or season and are also often used for monitoring landscape condition. Together, these sensors have the ability to assess a wide array of landscape condition metrics. Trends in the normalized difference vegetation index (NDVI) or other similar indices are often used as a proxy measure of vegetation condition (Al-Bakri and Taylor, 2003) and for analysis of the impact of drought (Liu and Kogan, 1996; Peters et al., 2002). Active sensors excel at measuring soil moisture (Njoku et al., 2002), canopy height, and forest structure (Lim et al., 2003; Means et al., 2000). Although not many LULC models directly forecast changes in landscape condition, this information can potentially be used within many LULC-modeling environments.

15.3 USE OF REMOTE-SENSING DATA IN LULC MODELING

Historical and current sources of remote-sensing information are obviously quite important for measuring and monitoring changes in landscape parameters. Table 15.1 provides a summary of major categories of spatially explicit LULC models, the majority of which rely directly on remote-sensing information. Data describing changes in land cover, land use, landscape condition, and/or landscape pattern are relevant to a number of associated modeling

TABLE 15.1

Most Commonly Used Methodologies for Producing Spatially Explicit LULC Projections

Category	Summary	Examples
Markov chain	• Probabilistic state-transition models, with LULC at time $t + 1$ strictly a function of LULC at time t • Transition rules for a given LULC type are often dependent on historical transition probabilities • Transition probabilities typically independent from status or dynamics of adjacent cells	• Muller and Middleton (1994) • Petit et al. (2001) • Coppedge et al. (2007) • Tang et al. (2007)
Geostatistical—empirical	• Development of suitability or probability maps for modeled LULC types to guide placement and location of LULC change • Regression-based analyses often used for development of probability surfaces • Artificial neural networks (ANNs) one subcategory	• GEOMOD—Hall et al. (1995); Pontius et al. (2001) • CLUE—Verburg et al. (1999b, 2008) • FORE-SCE—Sohl et al. (2007); Sohl and Sayler (2008) • Land Transformation Model—Pijanowski et al. (2002); Tang et al. (2005)
Cellular automata (CA)	• Spatial-temporal extension of Markov transition models • State-transition model with neighborhood component • Transition rules defined by current state of a cell, but also by status of neighboring cells	• SLEUTH—Claggett et al. (2004); Xibao et al. (2006) • Walsh et al. (2006) • Ozah et al. (2010)
Agent-based	• Recognizes and attempts to model the role of human decision-making in LULC change • Models behavior and interaction of "agents" (individuals, businesses, governmental bodies, or other entities with power to influence change) • Agents influence LULC change at a given location • LULC patterns emerge from interactions between human and natural processes	• MR. POTATOHEAD—Parker et al. (2006) • PALM—Matthews (2006) • SAMBA—Castella and Verburg (2007) • Valbuena et al. (2010)
Integrated	• Integration of multiple modeling approaches and/or frameworks • May include tightly coupled models with significant feedback or loose model coupling focusing on passing data between models • Advantage of potentially incorporating both spatial and aspatial modeling approaches • Typically include econometric or other economic modeling framework	• IMAGE—Alcamo et al. (1998); Strengers et al. (2004) • Verburg et al. (2008) • Jansson et al. (2008) • Moreira et al. (2009)

Note: This table is not all-inclusive. Aspatial modeling frameworks are not included, as they are less likely to directly incorporate remote-sensing data. Model names are noted under "Examples," where appropriate. For additional discussion of LULC modeling types, see Irwin and Geoghegan (2001), Agarwal et al. (2002), and Matthews et al. (2007). All of the aforementioned methodologies often rely on remote-sensing information.

fields including scenario development, driving-force analysis, model parameterization, and model validation.

15.3.1 Scenario Development

Scenarios of future land conditions are an important tool for a variety of research themes, including land-use impacts on greenhouse gas emissions and climate change (Strengers et al., 2004), biodiversity (Leadley et al., 2010; Sala et al., 2000), and hydrologic change and water availability (Ray et al., 2010; Wilk and Hughes, 2002) (Figure 15.2). The ability to blend thematically rich narratives describing future conditions with traditional quantitative results stimulates scenario users to think "outside the box" when considering complex human–environmental systems. Several large global environmental assessments have adopted a scenario-based approach. The Intergovernmental Panel on Climate Change (IPCC) defined scenarios as "images of the future that are neither projections nor forecasts" (Nakicenovic et al., 2000), whereas while the Millennium Ecosystem Assessment defined scenarios as "plausible and often simplified descriptions of how the future may develop based on a coherent and internally consistent set of assumptions about key driving forces and relationships" (Carpenter et al., 2005). Alcamo and Henrichs (2008) proposed the following definition: "A scenario is a description of how the future may unfold based on 'if-then' propositions and typically consists of a representation of an initial situation and a description of the key driving forces and changes that lead to a particular future state."

Regardless of the definition preferred, LULC scenarios require two things: knowledge of present conditions and an understanding of how drivers of change interact to create historical landscapes.

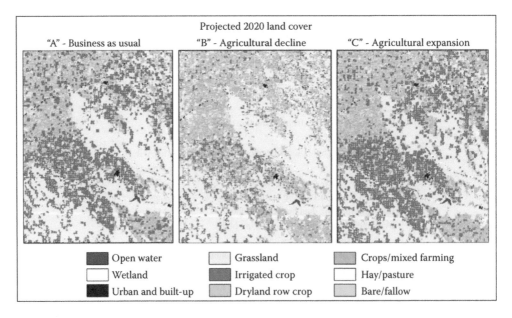

FIGURE 15.2 **(See color insert.)** Scenarios are a vital component of LULC modeling, allowing the exploration of multiple possible futures and resultant impacts on ecological processes. Remote sensing both directly and indirectly informs the construction of viable LULC scenarios through (1) construction of regional landscape histories, (2) examination of LULC patterns, and (3) exploration of linkages between historical LULC change and socioeconomic and biophysical driving forces. Each of these three components was used to develop scenarios and model 2020 LULC for a portion of southwestern Kansas, in the central United States (Sohl et al., 2007). Scenario A depicts a business-as-usual scenario. Scenario B depicts a scenario of low precipitation and declining groundwater availability, leading to agricultural decline. Scenario C depicts a scenario of increased precipitation and a more efficient utilization of groundwater, leading to agricultural expansion. The modeled scenarios were used to examine the impacts of LULC change on regional weather and climate variability.

LULC scenarios also critically need a baseline map, and irrespective of scale, remotely sensed data have a considerable advantage over survey and field-based methods for deriving such a product. Perhaps even more important is the use of remote-sensing data for developing LULC histories. LULC histories "expose the evolutionary patterns of a specific landscape by revealing its ecological stages, cultural periods, and keystone processes" (Marcucci, 2000). Specifically, LULC histories quantify LULC change over a sufficiently long timescale to illuminate relationships between driving forces such as population growth, economic development, and technological innovation and LULC change. The NLCD (Homer et al., 2007) and Land Cover Trends (Loveland et al., 2002; Sleeter et al., 2010) projects in the United States, the CORINE project in Europe (Büttner et al., 2002; Heymann et al., 1994), and the land-cover change component of Australia's National Carbon Accounting System (Furby, 2002; Waterworth et al., 2007) all rely on historical data from the Landsat archive to map and characterize LULC change. Identifying the rates and types of historical LULC change occurring within socioeconomic and biophysical settings offers the basic understanding from which we can construct alternative visions of the future. Combining the quantitative land-use histories with our understanding of the processes that drive change provides a powerful foundation from which we can develop alternative scenarios based on predefined assumptions about the interaction of various drivers of change.

Scenario development can also include simple projections of historical rates of LULC change. Verburg et al. (1999a) simulated land-use conversions in China using a scenario based on present land-use dynamics. Many LULC-modeling efforts focus on establishing a single reference condition, usually based on extrapolation of historical trends, while modifying certain LULC types to test the hypothesis about future impacts. For example, Kok and Winograd (2002) developed a "base" scenario established by extrapolating historical LULC trends, while using "optimistic" and "natural hazard" scenarios to test LULC response under extremely favorable and unfavorable conditions, respectively. While being relatively simple in design, these types of scenario are still dependent on LULC histories, which are based primarily on satellite observations.

15.3.2 Driving-Force Analysis

One of the great challenges in LULC modeling, and in remote sensing, is the ability to "socialize the pixel" (Geoghegan et al., 1998), linking social science analyses with remotely sensed data. Increasingly, spatially explicit LULC studies using remote sensing have an ultimate goal of not only analyzing the location and type of land-use change but also identifying the primary driving forces of that change (Chowdhury, 2006). Those LULC models that link remote-sensing observations with ground-based social data can greatly improve our understanding of the determinants of LULC change (Rindfuss and Stern, 1998). Generally, remote-sensing data are used in combination with ancillary information on the human decision-making process to understand the cause–effect relationship between driving forces and LULC change. Governmental policy, for example, can have a major impact on LULC change, but remote sensing cannot directly observe and monitor the policy. What remote sensing can do is examine the effects of the policy on land use, allowing LULC modelers to develop qualitative and quantitative relationships between a policy driver and impacts on LULC change. Remote sensing can be used to quantify the effects of a national policy such as the Conservation Reserve Program (CRP) of the United States, which pays farmers for converting environmentally sensitive lands to natural vegetative cover or examining land-use effects of local policies such as property taxation, zoning, or land ownership (Rindfuss and Stern, 1998) (Figure 15.3). Similarly, it can be used to deduce the effects of climate change on vegetation structure or condition (Ingram and Dawson, 2005; Stow et al., 2004).

Information on population distributions is used in a wide variety of LULC models. Census data can provide detailed information on population characteristics but are not available globally and vary greatly in consistency and accuracy. The LandScan database (Dobson et al., 2000) uses nighttime-lights data from the Defense Meteorological Satellite Program (DMSP) (Elvidge et al., 1997)

FIGURE 15.3 **(See color insert.)** Two Landsat TM images acquired on August 29, 1987 (top) and August 12, 2010 (bottom). Both images are of the same region in northern California, covering parts of Humboldt and Del Norte counties. The images use visible and near-infrared bands to depict vegetation in hues of red. Dense old-growth conifer stands appear dark red, whereas recent clear-cuts appear bright. Dimensionally, the images are approximately 30 km from east to west and 13 km from north to south. The images span three major land ownership types. Redwood National Park is in the west and is most easily recognized by the large contiguous stand of old-growth redwoods found in Prairie Creek Redwoods State Park. In the eastern portion of the images is Six River National Forest (SRNF). SRNF is managed for multiple uses, including timber harvest. In the center of the image is a large swath of private land holdings along the Klamath River. Cutting on private lands generally occurs in relatively large, often contiguous patches, while SRNF is characterized by a smaller more dispersed pattern of cutting. No cutting is evident in the National Park. Cutting also seems to have accelerated in this area on both private and public lands. Satellite imagery, such as those presented here, are extremely useful for mapping and characterizing changes to landscapes, which provide the foundational understanding for LULC modeling efforts. In this example, land ownership is an important driver and constraint on LULC change and should be considered in any modeling effort.

and remote-sensing-based land-cover and topography data in conjunction with census data and other ancillary data sources to produce global maps of population. Similarly, data on roads and other transportation infrastructure are widely used in LULC modeling, but reliable geographic information system (GIS) data are often not available. Remote sensing can provide either direct observation of transportation networks or can be used to deduce transportation network extents by mapping associated land-use changes (Bong et al., 2009; Lin et al., 2009). Such remote-sensing-based "proxy" datasets are invaluable for LULC models that require spatially explicit socioeconomic data for analyses and modeling of driving force. Other approaches attempted to establish more direct links between remote-sensing data and social science. McCracken et al. (1999) linked remote-sensing data with survey-based household property-level data to examine agricultural land-use strategies, and Silva et al. (2008) proposed a quantitative method to associate individual patches of land in a remote-sensing image with specific agents of change.

One of the most common uses of remote-sensing data for analyzing driving forces lies in regression-based approaches commonly used for empirical-/statistical-based modeling. Remote-sensing data are commonly used as sources of data for both the dependent variables (LULC type) and independent variables (topographic variables, climate variables, landscape structure information derived from LULC data, etc.) used in regression analyses meant to produce probability-of-occurrence or suitability maps for LULC and LULC change. For example, Verburg et al. (2006) used SPOT imagery along with data on slope, elevation, road access, population density, and market accessibility to develop probability surfaces for analyzing land-use change in the Philippines. Similarly, Sohl and Sayler (2008) used Landsat-derived LULC data from the NLCD project (Homer et al., 2007) and a series of spatially explicit independent variables to produce regression-based probability surfaces to model the placement of LULC change. Brown et al. (2002) used a generalized additive model rather than logistic regression to model forest change in Michigan, using a time series of LULC data derived from Landsat MSS to parameterize the model. Not only are the base LULC data from projects like these often derived from remote-sensing data, but so are many of the independent or ancillary variables. Information on landscape structure, topography, "distance to" measures (e.g., distance to roads), and other independent variables typically used in regression-based approaches often have a remote-sensing origin.

15.3.3 MODEL PARAMETERIZATION

Many modeling frameworks depend directly on consistent, historical remote-sensing data for model parameterization. The SLEUTH model is typically parameterized and calibrated with patterns of historical LULC change, with the historically derived parameters driving the modeling of future urban growth patterns (Claggett et al., 2004; Silva et al., 2002; Xibao et al., 2006). Model calibration is also built on a series of spatial metrics, and several of these parameters rely directly on remote-sensing data. Artificial neural network (ANN) models also typically rely on historically mapped LULC information, as well as other ancillary spatial data, for model parameterization and calibration. The Land Transformation Model, for example, develops spatially explicit predictor layers based on how neighborhood effects, patch size, distance measures, and site-specific characteristics affect LULC transitions, with these measures typically derived from remotely sensed data sources (Pijanowski et al., 2002; Tang et al., 2005). Other models, such as GEOMOD (Hall et al., 1995) and FORE-SCE (Sohl and Sayler, 2008) also use current local pattern information to parameterize and model future LULC change.

Many modeling frameworks rely on remote-sensing data for establishing transition rules and transition probabilities, with mapped historical LULC often driving model parameterization. Petit et al. (2001) used a temporal series of SPOT data to analyze historical LULC change in southeastern Zambia and provided transition probabilities for a Markov-chain model. Walsh et al. (2006) used remote sensing to parameterize a CA-based agricultural change model in Thailand, using time series land-cover information interpreted from Landsat TM and MSS images to define state and transition rules. Wu et al. (2006) used LULC information derived from Landsat data between 1986 and 2001 to establish transition probabilities in a first-order Markov-chain model for modeling urban development in Beijing, China. Coppedge et al. (2007) used aerial photography to analyze historical patterns of LULC change in the grasslands of Oklahoma, information that was used to populate transition probabilities and decision rules in a Markov-chain model. Tang et al. (2007) used both changed LULC area (derived from Landsat) and neighborhood LULC information to construct transition probabilities for a Markov-chain model. The FORE-SCE model used historical

LULC information from the USGS Trends project (Loveland et al., 2002) to establish conversion elasticity parameters, a parameter governing transition probability for a given LULC change (Sohl et al., 2007).

15.3.4 MODEL VALIDATION

Model validation remains an underdeveloped component of LULC-modeling science. The problem for LULC model validation lies more often with the availability of data rather than with suitable validation techniques, as several techniques and tools for validating LULC models have been developed (Chen and Pontius, 2010; Pontius and Petrova, 2010; Visser and de Nijs, 2006). Validation data obviously are not available for modeled future dates, so LULC modelers typically rely on modeling a historical period to perform model validation. Traditional accuracy assessment for one-point-in-time LULC classifications is often made by using aerial photography or another high-resolution remote-sensing source and developing rigorous, pixel-based accuracy assessments. Ray and Pijanowski (2010) used a similar procedure to validate model output for a backcasting application in the Muskegon River watershed in Michigan, using black-and-white aerial photography to interpret sampled validation points for assessing model output.

However, such pixel-by-pixel accuracy assessments are often less desirable choices for LULC-modeling applications, owing to path-dependence and the inherent stochasticity of LULC-modeling processes (Brown et al., 2005). LULC modelers often rely on the validation of landscape patterns rather than on pixel-by-pixel accuracy assessments where the model fit can be determined by the proportion of pixels correctly predicted in a local neighborhood (typically at multiple resolutions) or by comparison of generated landscape metrics between reference and modeled LULC. Brown et al. (2005) calculated edge density, patch size, and other landscape metrics from interpreted LULC to look at a model fit between an agent-based model and reference maps. Castella and Verburg (2007) used LULC maps interpreted from SPOT imagery to assess a CLUE-S model application in Vietnam, using a multiresolution neighborhood validation procedure (Costanza, 1989).

15.4 DISCUSSION

Remote-sensing data and analyses are a vital component of many LULC-modeling efforts. Given the reliance on remote-sensing data for informing LULC models, a primary concern for modelers is continued availability of consistent data. Heistermann et al. (2006) noted that a major problem in LULC modeling was the availability of spatially explicit time series data, while Rindfuss et al. (2004) stated that it was often impossible to obtain consistent, cost-effective, spatially and temporally relevant historical remote-sensing data for supporting land-change science. Crews and Walsh (2009) noted the extreme importance of continued development and maintenance of consistent satellite sensors and databases, with timely data delivery at reasonable prices. Sellers et al. (1995) noted that consistent land-cover data was one of the highest priorities for Land Science, with Herold et al. (2006) even advocating for adoption of a single standard land-cover legend. Even when synoptic, consistent data are available from remotely sensed sources, issues regarding data quality and consistency can strongly affect modeling results. Programs such as the USGS Land Cover Trends project (Loveland et al., 2002) or CORINE (Büttner et al., 2002) certainly have shown their ability to inform LULC modeling, but consistent databases of LULC change are often not widely available, and there is no guarantee of existing programs continuing. Continuance of the remote-sensing programs such as Landsat, as well as consistent LULC mapping and monitoring programs, is vital to the growth of land-change science and LULC modeling.

Continued improvements must also be made in linking remote sensing with the socioeconomic driving forces of LULC change. LULC modelers recognize the human dimension of change, but translating that recognition into sound, integrated theory has been a struggle. Linking pixel-based remote-sensing data with human decision making remains a primary challenge (Matthews et al.,

2007; Rindfuss et al., 2004). McNoleg (2003) stated that the issues were "unbridgeable" for social scientists to utilize raster data models and remote-sensing data, whereas Crews and Walsh (2009) stated that the problem was far too often choosing between "people and pixels." However, given their advantages, remote-sensing data remain too attractive to be dismissed, even for LULC models originating in social sciences. Remote-sensing platforms are built to measure physical phenomenon, and the typical unit of measure, the pixel, typically has no inherent meaning for the socioeconomic driving forces of LULC change. Although these socioeconomic driving forces are typically not directly measurable from remote-sensing platforms, LULC modelers do often develop proxy datasets from remote-sensing data that represent observable effects of social driving forces. Whether it is through the development of proxy datasets or through more direct links between remote-sensing data and the social sciences, it will be difficult for LULC modelers to move away from empirically based modeling systems to true process-based models without developing cost-effective, synoptic socioeconomic datasets that represent the social aspects of LULC change.

LULC modelers must continue to move from being passive users of remote-sensing data to vital team members for planning remote-sensing missions. The temporal, spatial, spectral, and radiometric resolutions of sensors directly affect our ability to map LULC change and monitor LULC processes. Spatial resolution of data alone has a very large impact on LULC-modeling processes and results (Silva and Clarke, 2002), with the driving forces of LULC change extremely dependent on scale. A model such as the aforementioned SLEUTH is extremely dependent on the availability of high-quality historical data on urban growth, but even SLEUTH users note the difficulty in accurate extraction of urban features from Landsat and other remote-sensing data sources (Claggett et al., 2004; Silva and Clarke, 2002). LULC modelers must stay engaged with the remote-sensing community to ensure availability of consistent, suitable remotely sensed data for analyzing LULC and LULC processes.

REFERENCES

Agarwal, C., Green, G.M., Grove, J.M., Evans, T.P., and Schweik, C.M. 2002. A review and assessment of land-use change models: Dynamics of space, time, and human choice. General Technical Report NE-297. Newton Square, Pennsylvania, U.S. Department of Agriculture, Forest Service, Northeastern Research Station. 61 pp.

Akbari, H., Shea Rose, L., and Taha, H. 2003. Analyzing the land cover of an urban environment using high-resolution orthophotos. *Landscape and Urban Planning*, 63, 1–14.

Al-Bakri, J.T. and Taylor, J.C. 2003. Application of NOAA AVHRR for monitoring vegetation conditions and biomass in Jordan. *Journal of Arid Environments*, 54, 579–593.

Alcamo, J., Leemans, R., and Kreileman, E. 1998. *Global Change Scenarios of the 21st Century. Results from the IMAGE 2.1 Model*. London: Pergamon & Elseviers Science, 296 pp.

Alcamo, J. and Henrichs, T. 2008. Towards guidelines for environmental scenario analysis. In J. Alcamo (Ed.), *Environmental Futures: The Practice of Environmental Scenario Analysis* (Chapter 2). Amsterdam: Elsevier.

Bong, D.B.L., Lai, K.C., and Joseph, A. 2009. Automatic road network recognition and extraction for urban planning. *International Journal of Engineering and Applied Sciences*, 5, 54–59.

Brown, D.G., Goovaerts, P., Burnicki, A., and Li, M.Y. 2002. Stochastic simulation of land-cover change using geostatistics and generalized additive models. *Photogrammetric Engineering and Remote Sensing*, 68(10), 1051–1061.

Brown, D.G., Page, S., Riolo, R., Zellner, M., and Rand, W. 2005. Path dependence and the validation of agent-based spatial models of land use. *International Journal of Geographical Information Science*, 19, 153–174.

Brown, J., Wardlow, B.D., Maxwell, S., Pervez, S., and Callahan, K. 2008a. National irrigated lands mapping via an automated remote sensing-based methodology. In *88th Annual Meeting, American Meteorological Society*. January 20–24, 2008, New Orleans, Louisiana.

Brown, J.F., Wardlow, B.D., Tadesse, T., Hayes, M.H., and Reed, B.C. 2008b. The vegetation drought response index (VegDRI): A new integrated approach for monitoring drought stress in vegetation. *GIScience and Remote Sensing*, 45(1), 16–46.

Büttner, G., Feranec, G., and Jaffrain, G. 2002. CORINE land cover update 2000. Technical guidelines. EEA Technical Report No 89. Available at: http://reports.eea.europa.eu/technical_report_2002_89/en

Carpenter, S., Pingali, P., Bennett, E., and Zurek, M. (Eds.). 2005. *Ecosystems and Human Well-Being, Volume 2, Scenarios*. Oxford: Island Press, pp. 145–172.

Castella, J.C. and Verburg, P.H. 2007. Combination of process-oriented and pattern-oriented models of land-use change in a mountain area of Vietnam. *Ecological Modelling*, 202, 410–420.

Chen, H. and Pontius, Jr., R.G. 2010. Diagnostic tools to evaluate a spatial land change projection along a gradient of an explanatory variable. *Landscape Ecology*, 25, 1319–1331.

Chowdhury, R.R. 2006. Driving forces of tropical deforestation: The role of remote sensing and spatial models. *Singapore Journal of Tropical Geography*, 27, 82–101.

Claggett, P.R., Jantz, C.A., Goetz, S.J., and Bisland, C. 2004. Assessing development pressure in the Chesapeake Bay Watershed: An evaluation of two land-use change models. *Environmental Monitoring and Assessment*, 94, 129–146.

Coppedge, B.R., Engle, D.M., and Fuhlendorf, S.D. 2007. Markov models of land cover dynamics in a southern Great Plains grassland region. *Landscape Ecology*, 22, 1383–1393.

Costanza, R. 1989. Model goodness of fit: A multiple resolution procedure. *Ecological Modelling*, 47, 199–215.

Cots-Folch, R., Aitkenhead, M.J., and Martinez-Casasnovas, J.A. 2007. Mapping land cover from detailed aerial photography data using textural and neural network analysis. *International Journal of Remote Sensing*, 28(7), 1625–1642.

Crews, K.A. and Walsh, S.J. 2009. Remote sensing and the social sciences. In T. Warner, M.D. Nellis, and G.M. Foody (Eds.), *The Sage Handbook of Remote Sensing* (pp. 437–445). London: Sage.

Dobson, J.E., Bright, E.A., Coleman, P.R., Durfee, R.C., and Worley, B.A. 2000. LandScan: A global population database for estimation populations at risk. *Photogrammetric Engineering and Remote Sensing*, 66(7), 849–857.

Elvidge, C.D., Baugh, K.e., Kihn, E.A., Kroehl, H.W., and Davis, E.R. 1997. Mapping city lights with nighttime data from the DMSP Operational Linescan System. *Photogrammetric Engineering and Remote Sensing*, 57(11), 1453–1463.

Furby, S.L. 2002. Land cover change: Specification for remote sensing analysis. National Carbon Accounting System Technical Report No. 9, Australian Greenhouse Office. Available at: http://pandora.nla.gov.au/pan/102841/20090717-1556/www.climatechange.gov.au/ncas/reports/pubs/tr09final.pdf

Gallant, A.L., Loveland, T.R., Sohl, T.L., and Napton, D.E. 2004. Using an ecoregion framework to analyze land-cover and land-use dynamics. *Environmental Management*, 34, s89–s110.

Gerard, F., Petit, S., Smith, G., Thomson, A., Brown, N., Manchester, S., Wadsworth, R., et al. 2010. Land cover change in Europe between 1950 and 2000 determined employing aerial photography. *Progress in Physical Geography*, 34(2), 183–205.

Geoghegan, J., Pritchard, L., Ogneva-Himmelberger, Y., Chowdhury, R.R., Sanderson, S., and Turner II, B.L. 1998. "Socializing the pixel" and "pixelizing the social" in land-use and land-cover change. In D. Liverman, E. Moran, R.R. Rindfuss, and P.C. Stern (Eds.), *People and Pixels: Linking Remote Sensing and Social Science* (Chapter 3, pp. 51–69). Washington, D.C.: National Academy Press.

Gu, Y., Hunt, E., Wardlow, B., Basara, J., Brown, J.F., and Verdin, J.P. 2008. Evaluation of MODIS NDVI and NDWI for vegetation drought monitoring using Oklahoma Mesonet soil moisture data. *Geophysical Research Letters*, 35, 5.

Hall, C.A.S., Tian, H., Pontius, G., and Cornell, J. 1995. Modelling spatial and temporal patterns of tropical land use change. *Journal of Biogeography*, 22, 753–757.

Heistermann, M., Muller, C., and Ronneberger, K. 2006. Land in sight? Achievements, deficits, and potentials of continental to global scale land-use modeling. *Agriculture, Ecosystems, and Environment*, 114, 141–158.

Herold, M., Latham, J.S., Di Gregorio, A., and Schmullius, C.C. 2006. Evolving standards in land cover characterization. *Journal of Land Use Science*, 1, 157–168.

Heymann, Y., Steenmans, Ch., Croissille, G., and Bossard, M. 1994. *CORINE Land Cover. Technical Guide.* EUR12585 Luxembourg: Office for Official Publications of the European Communities.

Homer, C., Dewitz, J., Fry, J., Coan, M., Hossain, N., Larson, C., Herold, N., McKerrow, A., VanDriel, J.N., and Wickham, J. 2007. Completion of the 2001 national land cover database for the conterminous United States. *Photogrammetric Engineering and Remote Sensing*, 73(4), 337–341.

Ingram, J.C. and Dawson, T.P. 2005. Climate change impacts and vegetation response on the island of Madagascar. *Philosophical Transactions of the Royal Society*, 363, 55–59.

Irwin, E.G. and Geoghegan, J. 2001. Theory, data, methods: Developing spatially explicit economic models of land-use change. *Agriculture, Ecosystems and Environment*, 85, 7–23.

Jansson, T., Bakker, M.M., Boitier, B., Fougeyrolla, A., Helming, J., van Meijl, H., and Verkerk, P.J. 2008. Linking models for land-use analysis: Experiences from the SENSOR project. In *12th Congress of the European Association of Agricultural Economists*. EAAE 2008, Ghent, Belgium.

Kok, K. and Winograd, M. 2002. Modelling land-use change for Central America, with special reference to the impact of hurricane Mitch. *Ecological Modelling*, 149, 53–69.

Leadley, P., Pereira, H.M., Alkemade, R., Fernandez-Manjarres, J.F., Proenca, V., Scharlemann, J.P.W., and Walpole, M.J. 2010. Biodiversity scenarios: Projections of 21st century change in biodiversity and associated ecosystem services. Secretariat of the Convention on Biological Diversity, Montreal. Technical Series no. 50, 132 pp.

Lin, X., Liu, Z., Zhang, J., and Shen, J., 2009. Combining multiple algorithms for road network tracking from multiple source remotely sensed imagery: A practical system and performance evaluation. *Sensors*, 9, 1237–1258.

Lim, K., Treitz, P., Wulder, M., St-Onge, B., and Flood, M. 2003. Lidar remote sensing of forest structure. *Progress in Physical Geography*, 27(1), 88–106.

Liu, W.T. and Kogan, F.N. 1996. Monitoring regional drought using the Vegetation Condition Index. *International Journal of Remote Sensing*, 17(14), 2761–2782.

Loveland, T.R., Sohl, T.L., Stehman, S.V., Gallant, A.L., Sayler, K.L., and Napton, D.E. 2002. A strategy for estimating the rates of recent United States land-cover changes. *Photogrammetric Engineering and Remote Sensing*, 68(10), 1091–1099.

Marschner, F.J. 1950. Major land uses in the United States [map, scale 1:5,000,000]: U.S. Dept. of Agriculture, Agricultural Research Service.

Marcucci, D.J. 2000. Landscape history as a planning tool. *Landscape and Urban Planning*, 49, 67–81.

Matthews, R.B. 2006. The People and Landscape Model (PALM): Towards full integration of human decision-making and biophysical simulation models. *Ecological Modeling*, 194(4), 329–343.

Matthews, R.B., Gilbert, N.G., Roach, A., Polhill, J.G., and Gotts, N.M. 2007. Agent-based land-use models: A review of applications. *Landscape Ecology*, 22, 1447–1459.

McCracken, S.D., Brondizio, E.S., Nelson, D., Moran, E.F., Siqueria, A.D., and Rodriguez-Pedraza, C. 1999. Remote sensing and GIS at farm property level: Demography and deforestation in the Brazilian Amazon. *Photogrammetric Engineering and Remote Sensing*, 65(11), 1311–1320.

McGarigal, K., Cushman, S.A., Neel, M.C., and Ene, E. 2002. FRAGSTATS: Spatial Pattern Analysis Program for Categorical Maps. Computer software program produced by the authors at the University of Massachusetts, Amherst. Available at: http://www.umass.edu/landeco/research/fragstats/fragstats.html

McNoleg, O. 2003. An account of the origins of conceptual models of geographic space. *Computers, Environment and Urban Systems*, 27(1), 1–3.

Means, J.E., Acker, S.A., Fitt, B.J., Renslow, M., Emerson, L., and Hendrix, C.J. 2000. Predicting forest stand characteristics with airborne scanning lidar. *Photogrammetric Engineering and Remote Sensing*, 66(11), 1367–1371.

Mertens, B. and Lambin, E. 1999. Modelling land cover dynamics: Integration of fine-scale land cover data with landscape attributes. *International Journal of Applied Earth Observation and Geoinformation*, 1(1), 48–52.

Millette, T.L., Tuladhar, A.R., Kasperson, R.E., and Turner II, B.L. 1995. The use and limits of remote sensing for analyzing environmental and social change in the Himalayan Middle Mountains of Nepal. *Global Environmental Change*, 5(4), 367–380.

Moreira, E., Costa, S., Aguiar, A.P., Camara, G., and Carneiro, T. 2009. Dynamical coupling of multiscale land change models. *Landscape Ecology*, 24, 1183–1194.

Muller, M.R. and Middleton, J. 1994. A Markov model of land-use change dynamics in the Niagara Region, Ontario, Canada. *Landscape Ecology*, 9(2), 151–157.

Nakicenovic, N., Alcamo, J., Davis, G., De Vrfies, B., Fenhann, J., Gaffin, S., Gregory, K., et al. 2000. *Special Report on Emissions Scenarios,* IPCC Special Reports, Cambridge University Press, Cambridge, 599 pp.

Njoku, E.G., Wilson, W.J., Yueh, S.H., Dinardo, S.J., Li, F.K., Jackson, T.J., Lakshmi, V., and Bolten, J. 2002. Observations of soil moisture using a passive and active low-frequency microwave airborne sensor during SGP99. *IEEE Transactions on Geoscience and Remote Sensing*, 40(12), 2659–2673.

Omernik, J.M. 1987. Ecoregions of the conterminous United States. *Annals of the Association of American Geographers*, 77, 118–125.

Ozah, A.P., Adesina, F.A., and Dami, A. 2010. A deterministic cellular automata model for simulating rural land use dynamics: A case study of Lake Chad basin. ISPRS Archive Vol. XXXVIII, Part 4-8-2-W9, *Core Spatial Databases—Updating, Maintenance, and Services—From Theory to Practice*, Haifa, Israel, 2010.

Parker, D.C., Berger, T., and Manson, S.M. (Eds.). 2002. Agent-based models of land-use and land-cover change. Report and review of an international workshop, Irvine.

Parker, D., Brown, D., Polhill, J.G., Manson, S.M., and Deadman, P. 2006. Illustrating a new 'conceptual design pattern' for agent-based models and land use via five case studies: the MR POTATOHEAD framework. In A.L. Paredes and C. H. Iglesias (Eds.), *Agent-based Modelling in Natural Resource Management* (pp. 29–62). Valladolid, Spain: Universidad de Valladolid.

Peters, A.J., Walter-Shea, E.A., Ji, L., Vina, A., Hayes, M., and Svoboda, M.D. 2002. Drought monitoring with NDVI-based standardized vegetation index. *Photogrammetric Engineering and Remote Sensing*, 68(1), 71–75.

Petit, C., Scudder, T., and Lambin, E. 2001. Quantifying processes of land-cover change by remote sensing: Resettlement and rapid land-cover changes in south-eastern Zambia. *International Journal of Remote Sensing*, 22(17), 3435–3456.

Pijanowski, B.C., Brown, D.G., Shellito, B.A., and Manik, G.A. 2002. Using neural networks and GIS to forecast land use change: A land transformation model. *Computers, Environment, and Urban Systems*, 26, 553–575.

Pontius, Jr., R.G., Cornell, J.D., and Hall, C.A.S. 2001. Modeling the spatial pattern of land-use change with GEOMOD2: Application and validation for Costa Rica. *Agriculture, Ecosystems and Environment*, 85, 191–203.

Pontius, Jr., R.G. and Petrova, S.H. 2010. Assessing a predictive model of land change using uncertain data. *Environmental Modelling and Software*, 25, 299–309.

Ray, D.K. and Pijanowski, B.C. 2010. A backcast land use change model to generate past land use maps: Application and validation at the Muskegon River watershed of Michigan, USA. *Journal of Land Use Science*, 5(1), 1–29.

Ray, D.K., Duckles, J.M., and Pijanowski, B.C. 2010. The impact of future land use scenarios on runoff volumes in the Muskegon River watershed. *Environmental Management*, 46(3), 351–366.

Reeves, M.C., Winslow, J.C., and Running, S.W. 2001. Mapping weekly rangeland vegetation productivity using MODIS algorithms. *Journal of Range Management*, 54, A90–A105.

Rindfuss, R.R. and Stern, P.C., 1998. Linking remote sensing and social science: The need and challenges. In D. Liverman, E.F. Moran, R.R. Rindfuss, and P.C. Stern (Eds.), *People and Pixels* (pp. 1–27). Washington, DC: National Academy Press.

Rindfuss, R.R., Walsh, S.J., Turner II, B.L., Fox, J., and Mishra, V. 2004. Developing a science of land change: Challenges and methodological issues. *Proceedings of the National Academy of Sciences of the USA*, 101(39), 13976–13981.

Rollins, M.G. and Frame, C.K. (Tech. Eds.). 2006. The LANDFIRE Prototype Project: Nationally consistent and locally relevant geospatial data for wildland fire management. Gen. Tech. Rep. RMRS-GTR-175. Fort Collins: U.S. Department of Agriculture, Forest Service, Rocky Mountain Research Station. 416 p.

Sala, O.E., Chapin, III, F.S., Armesto, J.J., Berlow, E., Bloomfield, J., Dirzo, R., Huber-Sanwald, E., et al. 2000. Global biodiversity scenarios for the year 2100. *Science*, 287(5459), 1770–1774.

Sellers, P.J., Meeson, B.W., Hall, F.G., Asrar, G., Murphy, R.E., Schiffer, R.A., Bretherton, F.P., et al. 1995. Remote sensing of land surface for studies of global change: Models—Algorithms—Experiments. *Remote Sensing of Environment*, 51, 3–26.

Silva, E.A. and Clarke, K.C. 2002. Calibration of the SLEUTH urban growth model for Lisbon and Porto, Portugal. *Computers, Environment and Urban Systems*, 26, 525–552.

Silva, M.P.S., Camara, G., Escada, M.I.S., and De Souza, R.C.M. 2008. Remote-sensing image mining: Detecting agents of land-use change in tropical forest areas. *International Journal of Remote Sensing*, 29(16), 4803–4822.

Sleeter, B. M., Wilson, T., Soulard, C., and Liu, J. 2010. Estimation of late 20th century landscape change in California. *Environmental Monitoring and Assessment*, 173(1), 251.

Sohl, T.L., Sayler, K.L., Drummond, M.A., and Loveland, T.R. 2007. The FORE-SCE model: A practical approach for projecting land cover change using scenario-based modeling. *Journal of Land Use Science*, 2(2), 103–126.

Sohl, T.L. and Sayler, K.L. 2008. Using the FORE-SCE model to project land-cover change in the southeastern United States. *Ecological Modelling*, 219, 49–65.

Sohl, T.L., Loveland, T.R., Sleeter, B.M., Sayler, K.L., and Barnes, C.A. 2010. Addressing foundational elements of regional land-use change forecasting. *Landscape Ecology*, 25, 233–247.

Stow, D.A., Hope, A., McGuire, D., Verbyla, D., Gamon, J., Huemmrich, F., Houston, S., et al. 2004. Remote sensing of vegetation and land-cover change in Arctic tundra ecosystems. *Remote Sensing of Environment*, 89, 281–308.

Strengers, B., Leemans, R., Eickhout, B., de Vries, B., and Bouwman, L. 2004. The land-use projections and resulting emissions in the IPCC SRES scenarios as simulated by the IMAGE 2.2 model. *GeoJournal*, 61, 381–393.

Tang, Z., Engel, B.A., Pijanowski, B.C., and Lim, K.J. 2005. Forecasting land use change and its environmental impact at a watershed scale. *Journal of Environmental Management*, 76, 35–45.

Tang, J., Wang, L., and Yao, Z. 2007. Spatio-temporal urban landscape change analysis using the Markov chain model and a modified genetic algorithm. *International Journal of Remote Sensing*, 15(10), 3255–3271.

Tayyebi, A., Delavar, M.R., Saeedi, S., Amini, J., and Alinia, H. 2008. Monitoring land use change by multi-temporal landsat remote sensing imagery. *The International Archives of the Photogrammetry, Remote Sensing and Spatial Information Sciences*, XXXVII, Part B7, Beijing.

Thomson, A.G., Manchester, S.J., Swetnam, R.D., Smith, G.M., Wadsworth, R.A., Petit, S., and Gerard, F.F. 2007. The use of digital aerial photography and CORINE-derived methodology for monitoring recent and historic changes in land cover near UK Natura 2000 sites for the BIOPRESS project. *International Journal of Remote Sensing*, 28(23), 5397–5426.

Valbuena, D., Verburg, P.H., Bregt, A.K., and Ligtenberg, A. 2010. An agent-based approach to model land-use at a regional scale. *Landscape Ecology*, 25(2), 185–199.

Verburg, P.H., Veldkamp, A., and Fresco, L.O. 1999a. Simulation of changes in the spatial pattern of land use in China. *Applied Geography*, 19, 211–233.

Verburg, P.H., DeKoning, G.H.J., Kok, K., Veldkamp, A., and Bouma, J. 1999b. A spatial explicit allocation procedure for modeling the pattern of land use change based upon actual land use. *Ecological Modelling*, 116, 45–61.

Verburg, P.h., Overmars, K.P., Huigen, M.G.A., de Groot, W.T., and Veldkamp, A. 2006. Analysis of the effects of land use change on protected areas in the Philippines. *Applied Geography*, 26, 153–173.

Verburg, P.H., Eickhout, B., and van Meijl, H. 2008. A multi-scale, multi-model approach for analyzing the future dynamics of European land use. *The Annals of Regional Science*, 42, 57–77.

Visser, H. and de Nijs, T. 2006. The map comparison kit. *Environmental Modelling and Software*, 21, 346–358.

Vogelmann, J.E., Howard, S.M., Yang, L., Larson, C.R., Wylie, B.K., and Van Driel, J.N. 2001. Completion of the 1990s national land cover dataset for the conterminous United States. *Photogrammetric Engineering and Remote Sensing*, 67, 650–662.

Vogelmann, J.E., Tolk, B., and Zhu, Z. 2009. Monitoring forest changes in the southwestern United States using multitemporal Landsat data. *Remote Sensing of Environment*, 113, 1739–1748.

Wallace, J., Behn, G., and Furby, S. 2006. Vegetation condition assessment and monitoring from sequences of satellite imagery. *Ecological Management and Restoration*, 7, S31–S36.

Walsh, S.J., Entwisle, B., Rindfuss, R.R., and Page, P.H. 2006. Spatial simulation modeling of land-use/land-cover change scenarios in northeastern Thailand: A cellular automata approach. *Journal of Land Use Science*, 1(1), 5–28.

Wang, Y., Mitchell, B.R., Nugranad-Marzilli, J., Bonynge, Zhou, Y., and Shriver, G. 2009. Remote sensing of land-cover change and landscape context of the National Parks: A case study of the Northeast Temperate Network. *Remote Sensing of Environment*, 13, 1453–1461.

Waterworth, R.M., Richards, G.P., Brack, C.L., and Evans, D.M.W. (2007). A generalized hybrid process-empirical model for predicting plantation forest growth. *Forest Ecology and Management*, 238, 231–243.

Wilk, J. and Hughes, D.A. 2002. Simulating the impacts of land-use and climate change on water resource availability for a large south Indian catchment. *Hydrological Sciences*, 47(1), 19–30.

Wu, Q., Li, H., Wang, R., Paulussen, J., He, Y, Wang, M., Wang, Bi., and Wang, Z. 2006. Monitoring and predicting land use change in Beijing using remote sensing and GIS. *Landscape and Urban Planning*, 78, 322–333.

Wu., X., Hu, Y., He., H.S., Bu, R., Onsted, J., and Xi, F. 2008. Performance evaluation of the SLEUTH model in the Shenyang metropolitan area of northeastern China. *Environmental Modeling and Assessment*, 13, 1–10.

Xibao, X., Feng, Z., and Jianming, Z. 2006. Modeling the impacts of different policy scenarios on urban growth in Lanzhou with remote sensing and cellular automata. *Geoscience and Remote Sensing Symposium 2006, IGARSS 2006*, Denver, CO, 1435–1438.

Section III

Application Examples

16 Operational Service Demonstration for Global Land-Cover Mapping
The GlobCover and GlobCorine Experiences for 2005 and 2009

Sophie Bontemps, Olivier Arino, Patrice Bicheron,
Christelle Carsten Brockmann, Marc Leroy,
Christelle Vancutsem, and Pierre Defourny

CONTENTS

16.1 INTRODUCTION

In view of the increasing concern over the functioning of the earth system, land-cover observation at a global scale is crucial in assessing the impacts of climate change, preserving biodiversity, and understanding biogeochemical cycling. The implementation plan for the Global Climate Observing

System (GCOS) in support of the United Nations Framework Convention on Climate Change (UNFCC) highlights the importance of regular land-cover assessment and includes land cover as an essential climate variable (ECV). The Group on Earth Observation (GEO) reports the contribution of land cover for all areas of societal benefits. However, unlike the case of other major earth observation domains, such as the oceans and the atmosphere, regular land-cover observation at a global scale is yet to be developed.

In the early nineties, the first global land-cover map derived from satellite remote sensing was produced at 1° spatial resolution by DeFries and Townshend (1994), using normalized difference vegetation index (NDVI) data recorded by the National Oceanic and Atmospheric Administration-Advanced Very High Resolution Radiometer (NOAA-AVHRR) and then at 8-km spatial resolution (DeFries et al., 1998). The International Geosphere–Biosphere Programme effort (IGBP–DIS) provided the first 1-km global map product derived from AVHRR data (Hansen et al., 2000; Loveland et al., 2000). The main challenge was acquiring and putting together such a global remote-sensing dataset. The legend and accuracy were very much constrained by the poor quality of AVHRR data (Loveland et al., 2000). However, this definitely confirmed the need for a consistent land-cover map for the whole world and raised the validation issue of such a 1-km global product.

With the availability of global remote-sensing datasets, global land-cover mapping has reached a new era. Large volumes of high-quality remotely sensed data have become available, provided by orbiting instruments such as NOAA-AVHRR, the Satellite Pour l'Observation de la Terre-Vegetation (SPOT-VGT), the MODerate Resolution Imaging Spectroradiometer (MODIS), and the MEdium Resolution Imaging Spectrometer (MERIS). These imagers provide near-daily multispectral imaging of the land surface at resolutions ranging from 250 to 1000 m. The frequent temporal coverage provides a continuous land-surface reflectance observation by minimizing interference from clouds, thus allowing the construction of global datasets in which nearly all points on the land surface are imaged on several occasions. This, in turn, has opened the door for global science data products derived from multispectral and multitemporal measurements.

With the launch of TERRA and AQUA satellites, the MODIS Land Cover product was expected to fulfill users' needs—thanks to advanced sensor capabilities. The first MODIS Land Cover map (Friedl et al., 2002) was generated by a supervised classification methodology that exploited a global database of training sites interpreted from high-resolution imagery in association with ancillary data. This tree-based classification was partly automatic but supervised, which required the definition of signatures for the 17 final classes. The most recent 500-m MODIS global land cover derived from collection 5 NBAR surface reflectance and Land Surface Temperature (LST) products (Friedl et al., 2010) was substantially different and improved than land cover derived earlier, in particular by using up to 1860 training sites and by refining the methods to postprocess the ensemble of decision tree results (sample bias and spatial prior probability adjustments). Unfortunately, no quantitative accuracy assessment has been made owing to the lack of an independent validation dataset.

Meanwhile, the Global Land Cover 2000 (GLC2000) project produced a new Global Land Cover database for the year 2000—thanks to an international partnership of about 30 research groups coordinated by the European Commission's Joint Research Centre (JRC) (Bartholomé and Belward, 2005). The project adopted an *ad hoc* processing strategy for the different regions of the world, but it followed a standardized land-cover approach based on the United Nations (UN) Land Cover Classification System (LCCS; Di Gregorio and Jansen, 2000) to ensure consistency of the various outputs. This initiative took advantage of the great quality of daily SPOT-VGT time series acquired during the year 2000 to differentiate 22 different land-cover classes at the global level. The overall accuracy of 68.6% (Mayaux et al., 2006) obtained for the GLC2000 global product confirmed that any land-cover mapping at a global scale is a challenging task.

The global land-cover community strongly pushed for improving global land-cover assessments because existing datasets were not yet fully satisfying. On the one hand, most global land-cover efforts were one-time exercises that did not allow repeated map production. On the other hand, their spatial resolution (1 km at best) was not fine enough to deal adequately with the large landscape heterogeneity. As regards the latter point, the recent availability of the MERIS instrument, with its fine spatial resolution, offered an opportunity to deal with landscape heterogeneity.

Building on the success of the GLC2000 project, the European Space Agency (ESA) launched in 2005 the GlobCover initiative in the framework of its Data User Element (DUE). The GlobCover product was intended for—but not limited to—European and international users such as the European Commission (EC), the European Environment Agency (EEA), the United Nations Environment Programme (UNEP), the UN Food and Agriculture Organization (FAO), the Global Observation of Forest and Land Cover Dynamics (GOFC-GOLD) program, and the IGBP.

The current needs of the global community are not only to supply a product but—equally important—to develop an operational service that enables the regular and timely delivery of consistent global land-cover maps. Such a service should also allow interactions with users and iterative improvements of follow-ups. Automation, timeliness, and transparency are the three main objectives and the most important prerequisites for a service to be included in the line of Global Monitoring for Environment and Security (GMES) Service Elements as defined by the European Union (EU). Meeting those requirements, the GlobCover and GlobCorine systems described in this chapter can be considered the precursors of a global and a regional land-cover service, respectively.

16.2 AN OPERATIONAL GLOBAL LAND-COVER SERVICE: THE GLOBCOVER EXPERIENCE

To meet the needs of the global land-cover community, the ESA-GlobCover initiative aimed at developing and demonstrating a global land-cover service that is able to produce a global land-cover map based on the 300-m MERIS time series. Therefore, unlike other global land-cover initiatives, the objective of the GlobCover initiative was not only to produce another global map from a new sensor but also to develop an automated processing system supporting such land-cover service.

This service development was conceived in such a way that the new GlobCover product could update, improve, and complement the other existing comparable global maps and, in particular, the GLC2000 map. As a result, the thematic legend had to be compatible with the LCCS in the GLC2000 project.

Started in April 2005, the ESA-GlobCover 2005 project was carried out by an international consortium (Medias-France, Brockmann Consult, Université catholique de Louvain, Noveltis, and Infram), which designed, implemented, and produced the first global GlobCover product at 300-m resolution. In 2010, the second GlobCover product derived from the 2009 MERIS time series was delivered, which successfully demonstrates the operational service capabilities of the system.

16.2.1 THE MERIS INSTRUMENT

On-board ENVISAT launched in 2002, MERIS is a wide field-of-view push-broom-imaging spectrometer measuring the solar radiation reflected by the earth in 15 spectral bands from 412.5 to 900 nm (Rast et al., 1999). Each of these 15 bands is programmable in position and in width. The instrument has a field of view of 68.5° and covers a swath width of 1150 km at a nominal elevation of 800 km, enabling a global coverage of the earth in 3 days. The wide field-of-view is shared between five identical optical cameras arranged in a fan-shaped configuration, with each camera covering a 14° field of view with a slight overlap (see Figure 16.1). An image is constructed using the push-broom principle: a narrow strip of the earth is imaged onto the entrance slit of the spectrometer, defining the across-track dimension, and the motion of the satellite provides the along-track dimension. The spectral dimension is achieved by imaging the entrance slit of each spectrometer via a dispersion grating onto a 2-D charged couple device.

The MERIS instrument resolution in full spatial resolution (FR) is 290 m (along-track) × 260 m (across-track) at nadir. Data at a coarser resolution are systematically generated on-board by spatially (across-track) and temporally (along-track) averaging a group of 4 × 4 pixels producing a reduced spatial resolution (RR) dataset with a 1160 × 1040 m resolution. The RR data are transmitted to the

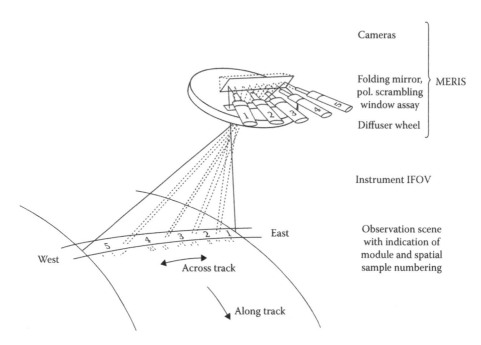

FIGURE 16.1 MERIS field-of-view, camera tracks, pixel enumeration, and swath dimension (From ESA, MERIS product handbook. 2011.)

ground on a global basis, whereas the FR data are limited to regional coverage, focusing on land surfaces and coastal areas.

For the purpose of GlobCover and GlobCorine processing chains, the level 1B MERIS Full Resolution full Swath (FRS) products, that is, calibrated top-of-atmosphere (TOA) gridded radiances over the full sensor's swath, were used as inputs.

16.2.2 GlobCover 2005

The GlobCover 2005 project aimed at developing the service infrastructure (expertise, software, and hardware) to produce a global land-cover map for the year 2005 using MERIS time series acquired in the FR mode. More precisely, the project included the development of the preprocessing and classification chain, the production of the GlobCover 2005 global land-cover map, and its validation.

This challenging project capitalized on a combination of several previous experiences. In particular, the experience gained in the GLC2000 project allowed tackling the main issues in producing a consistent land-cover map at a global scale, although land-cover in reality, is characterized by diversity, ambiguity, and continuum.

Five main challenges were identified:

- Data acquisition planning to ensure high temporal resolution of this global 300-m spatial resolution time series with 15 spectral bands.
- High-standard preprocessing chain required to produce consistent MERIS time series at 300-m spatial resolution.
- Land-cover classification methodology, which had to be automated, had to be global while being regionally tuned, and had to maintain 300-m spatial resolution throughout the whole process. The global scale of this mapping exercise forced the encompassing of the whole diversity of land-cover types, while the temporal dimension of the data analysis required an in-depth understanding of the related seasonality for the different bioclimatic regions. The key idea was to combine the spatial consistency of the classes' delineation obtained

from well-selected multispectral composites with the discrimination capacity of the temporal profile analysis. Before that, an *a priori* stratification of the world provided equal-reasoning regions to be processed separately. The great but much-controlled flexibility of this classification strategy allowed the defining of an automated process that tackled both the global consistency and the regional diversity of the land-cover characteristics.
- Handling and processing of a very large volume of data in a short time. The MERIS raw data volume was around 30 terabytes, which had to be fed into the preprocessing line. Besides the technical means necessary to physically transmit these data to the appropriate recipient, this implied a high level of collaboration between all the steps of the complete processing line. Data flow could also become a bottleneck if data were not transmitted quickly enough from their repository to the required process.
- Accuracy assessment of the GlobCover land-cover map at a global scale by independent validation.

16.2.2.1   Data Acquisition
The very first challenge was the global acquisition of a MERIS 300-m FRS time series, although the instrument was not initially designed to do so. Indeed, the data coverage was uneven owing to programmatic constraints. Therefore, ESA increased the MERIS FRS acquisition capacities, and the acquisition period was extended. The GlobCover product was then based on 19 months of global FRS MERIS level 1B product available from December 2004 to June 2006.

However, it has to be pointed out that despite this strategy, some parts of the world (such as Central and South America, northeast of America, Korean peninsula, and east Siberia) remained sparsely covered (Figure 16.2).

16.2.2.2   GlobCover Processing Chain
The GlobCover processing chain aimed at automatically delivering a land-cover map from MERIS FRS level 1B data. The processing system had two major modules (Figure 16. 3):

- A preprocessing module leading to global mosaics of land surface reflectance at 300-m spatial resolution in 13 spectral bands.
- A classification module leading to the final land-cover map at 300-m spatial resolution.

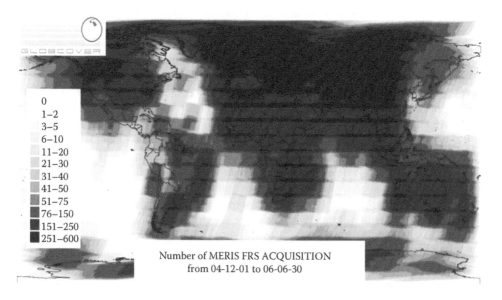

Number of MERIS FRS ACQUISITION
from 04-12-01 to 06-06-30

FIGURE 16.2   MERIS FRS density data acquisition from December 1, 2004 to June 30, 2006.

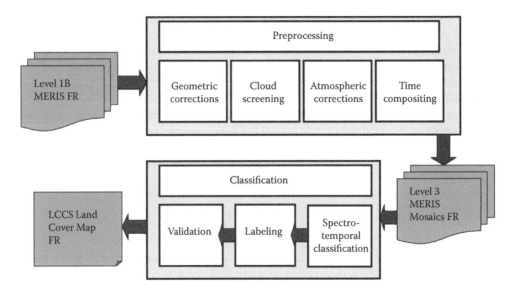

FIGURE 16.3 GlobCover processing line scheme.

16.2.2.2.1 Preprocessing Module

Overall, around 30 terabytes of MERIS FRS level 1B data were processed to produce the land surface reflectance mosaics. These mosaics were obtained from the MERIS FRS level 1B images with a series of preprocessing steps, including orthorectification of the input data to achieve at least 150-m geolocation accuracy, atmospheric corrections with a spectral normalization of the spectral bands most affected by the smile effect, cloud screening in the absence of short-wave infrared (SWIR) and thermal bands, shadow detection, land/water classification, projection, and temporal compositing.

Geometric corrections were done using the AMORGOS tool (Bourg et al., 2007), which provided geolocation information for every image pixels, whereas this information is only provided at tie points in MERIS full resolution raw data. The cartographic projection tool allowed the computation of radiances in a common grid (Plate-Carré coordinate reference system with a reference ellipsoid WGS84) and for all the MERIS FRS products used. The orthorectified images in the output of AMORGOS and of the projection tool demonstrated a relative geolocation accuracy of 52-m RMS and an absolute accuracy of 77-m RMS (Figure 16.4; Bicheron et al., 2011). These performances largely overcoming the initial specifications of 2-km accuracy were found very satisfactory for this ocean instrument and permitted the use of the MERIS images at their full resolution of 300 m.

Second, the atmospheric correction transformed the TOA radiances into surface reflectance values accounting for the effects of Rayleigh, aerosol scattering, and gaseous absorption. To this end, a neural network was used, relying on the so-called MOMO radiative transfer model based on Matrix Operator MOdel (Fischer and Grassl, 1991) already validated in the framework of the ESA Albedo Map project (Fisher et al., 2006). The aerosol correction was performed using a monthly aerosol optical depth product at 1-km spatial resolution, derived from a MERIS Reduced Resolution dataset from the years 2005 and 2006. The gaseous absorption correction used an ozone field from the European Centre for Medium-Range Weather Forecasts (ECMWF) and O_2 and H_2O fields derived from the MERIS data (ratios B11/B10 and B15/B14 for O_2 and H_2O, respectively).

For cloud screening, two methods were combined to achieve satisfactory results. The first one was based on the MOMO method already mentioned, and the second one used thresholds of reflectance on the bands at 443, 753, 760, and 865 nm. Results were validated using ground truth data from the synoptic network of meteorological stations. In addition, the cloud top height was

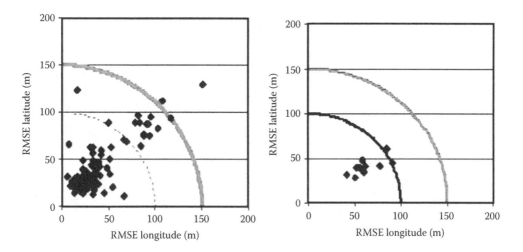

FIGURE 16.4 MERIS orthorectified images geolocation accuracy—relative (left) and absolute (right). (From Bicheron et al., *IEEE Trans. Geosci. Remote Sens.,* 49, 2972–2982, 2011. With permission.)

estimated for a better determination of cloud shadows, and the snow reflectance values were kept at their TOA level.

Finally, the cloud-free surface reflectance values were composited using the Mean Compositing strategy (Vancutsem et al., 2007a, 2007b), which reduced both the bidirectional reflectance distribution function (BRDF) effects and the possible remaining perturbations after atmospheric correction and cloud removal. The CYCLOPES method (Hagolle et al., 2004) was then applied for discarding the spurious values affected by undetected sources of noise (residual thin clouds, aerosols, shadows, etc.) through an iterative process. The daily images were composited on a 15-day basis, over a period of 2 months and 1 year (hereafter called "biweekly," "bimonthly," and "annual" composites, respectively).

As expected, the number of valid observations after all the preprocessing steps (in particular, that of cloud screening) was rather variable (Figure 16.5). As a result, combining the effect of poor acquisition and persistent cloud coverage, some areas (South America, northeast of America, central Siberia, northeast of Asia, Korea, Philippines, Malaysia, and Central Africa) showed a very low number of valid observations. These areas were not expected to be accurately classified in the land-cover product.

16.2.2.2.2 Classification Module

The classification process transforming the cloud-free land surface reflectance mosaics into a land-cover map was organized in four main steps (Figure 16.6). The global scale of this mapping exercise forced the encompassing of the whole diversity of land-cover types, whereas the temporal dimension of the data analysis required an in-depth understanding of the related seasonality for the different bioclimatic regions. The key idea was to combine the spatial consistency of the classes' delineation obtained from well-selected multispectral composites with the discrimination capabilities offered by the temporal profile analysis.

16.2.2.2.2.1 Stratification Before this classification process, an *a priori* stratification was applied to the world to delineate equal-reasoning areas to be processed separately. The stratification split the world into 22 equal-reasoning areas from an ecological and a remote-sensing point of view. The purposes were twofold: (1) reducing the land surface reflectance variability in the dataset to improve the classification efficiency and (2) allowing a regional tuning of the classification

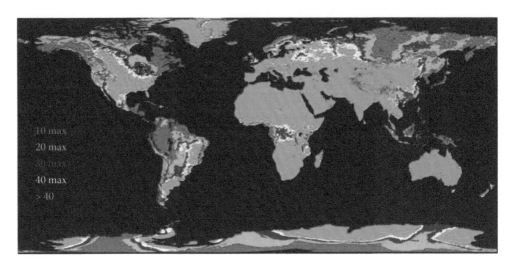

FIGURE 16.5 **(See color insert.)** Number of valid observations obtained after 19 months of MERIS FRS acquisitions. Magenta areas are defined as well covered (>40 observations).

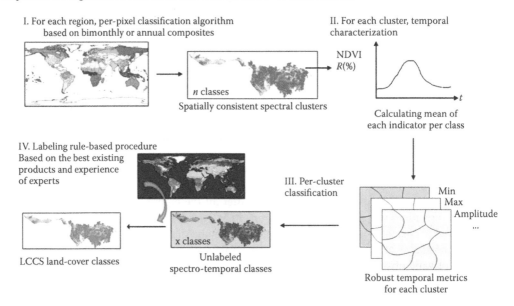

FIGURE 16.6 Scheme showing the principle of the classification algorithm starting with the bimonthly or the annual composites.

parameters to take into account the regional characteristics (vegetation seasonality, cloud coverage, etc.). Stratification offers the advantage of being based on natural limits directly derived from sharp boundaries observed in any remote-sensing dataset or through easy-to-classify and homogeneous land-cover areas. Figure 16.7 provides an overview of these areas.

16.2.2.2.2.2 Step I. Per-pixel classification algorithm   Spectral classification was made of both supervised and unsupervised classification algorithms. The supervised classification identified land-cover classes covering very small surfaces at the global scale such as irrigated crops, wetlands, and urban areas. The pixels classified through this process were masked out to run an unsupervised classification on the remaining pixels to create a large number of clusters (varying from 40 to 250) of spectrally similar pixels.

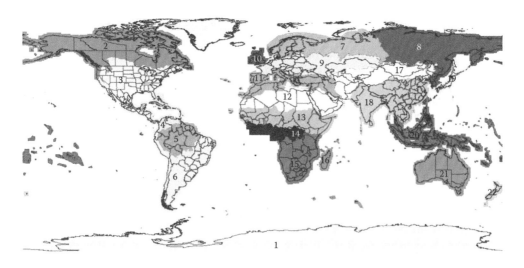

FIGURE 16.7 **(See color insert.)** Overview of the 22 equal-reasoning areas used as stratification.

16.2.2.2.2.3 Step II. Per-cluster temporal characterization The second step was a temporal characterization of the previously produced spectral clusters in the equal-reasoning areas that presented a high seasonality. In these strata, two phenological metrics (minimum and maximum of vegetation) were derived from the MERIS annual profiles and were spatially averaged for each spectral cluster.

16.2.2.2.2.4 Step III. Per-cluster classification algorithm Based on the temporal information characterizing each cluster (step 2), the third step merged the spectral clusters according to their similarity in the temporal space and defined a reduced number of spectro-temporal classes (varying from 50 to 70).

16.2.2.2.2.5 Step IV. Labeling-rule based procedure Finally, the labeling procedure transformed the spectro-temporal classes into land-cover classes defined using the LCCS. The labeling procedure was automated and based on a global reference land-cover database. Each spectro-temporal class was labeled according to the GlobCover land-cover legend based on the correspondence between this class and the reference land-cover classes. Several decision rules were defined with the help of international land-cover experts to automatically derive unique labels for each spectro-temporal class.

The global reference land-cover database was compiled from the GLC2000 product and a dozen of the existing national or regional land-cover maps. The reference maps were selected as the most accurate ones available for each region, with the highest spatial resolution and with a GlobCover-compatible legend.

16.2.2.2.2.6 Postclassification edition

Gap filling: As shown in Figure 16.5 and in spite of ESA's efforts, the data coverage of MERIS FR acquisitions was not complete, resulting in gaps in the data and therefore in the land-cover product. These gaps were filled out using the reference land-cover database (see in Step IV).

Flooded forest: The lack of SWIR band in the MERIS sensor hampered some discrimination. In particular, the class "Closed broadleaved forest regularly flooded with fresh water" appeared to be largely underestimated in the GlobCover classification and was therefore directly imported from the reference land-cover database.

Water bodies: A land/water mask was applied for producing the land surface reflectance products. Yet, this mask was not exhaustive—especially regarding inland water bodies—and had some geolocation inaccuracies. The SRTM Water Body Data (SWBD) was thus used to improve the delineation of "water bodies" in the GlobCover classification.

16.2.2.3 Results

The GlobCover 2005 land-cover map presented in Figure 16.8 is a globally consistent 300-m spatial resolution product, including 22 classes (Table 16.1) produced from MERIS FRS time series (December 2004–June 2006), using the GlobCover automated processing chain (Arino et al., 2007; Defourny et al., 2009a).

To produce a globally consistent land-cover map, the legend had to be determined by the level of information available, and that made sense at the scale of the entire world. From this point of view, defining the GlobCover legend using the LCCS proved to be highly suitable. Indeed, the LCCS had been designed as a hierarchical classification, which allowed adjusting the thematic detail of the legend to the amount of information available to describe each land-cover class, while following a standardized classification approach. In addition, it ensured compatibility with the GLC2000 global product.

The use of 300-m resolution data brings about considerable improvement in comparison with other global land-cover products at lower spatial resolution (Arino et al., 2008). Figure 16.9 provides a comparison between GLC2000 (1-km spatial resolution) and GlobCover (300-m spatial resolution) in Amazonia (Brazil), Saudi Arabia, and Russia.

One of the main issues the GlobCover project had to deal with was spatial coverage of the MERIS FR data. The use of a 19-month period (instead of a standard 12-month period over the year 2005 initially planned) contributed to mitigating this problem but did not solve it. Some areas over the globe remained underrepresented in the MERIS dataset, and that affected the quality of the land surface reflectance mosaics and finally the product. In areas of very low data coverage (about 2% of the continental areas), the pixel values were derived from the reference land-cover database, leading to possible discontinuity in the classification. A flag indicating whether the reference was used instead of the output of the GlobCover classification scheme was provided with the GlobCover map.

In addition, it has to be kept in mind that the identification of water bodies was largely based on the SWBD (cf. Section 16.2.2.2.2), which was based on year 2000 data and limited to −60° and +60° of latitude.

16.2.2.4 Validation

Apart from the production of a global land-cover map, the GlobCover initiative included an independent accuracy assessment (Defourny et al., 2009b). This effort was the first global exercise

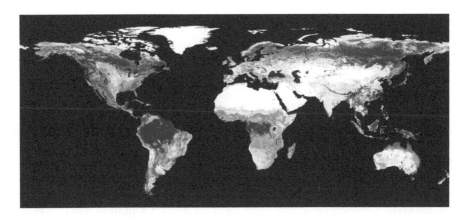

FIGURE 16.8 **(See color insert.)** The GlobCover 2005 product as the first 300-m global land-cover map for the period December 2004–June 2006.

TABLE 16.1

(See color insert.) Twenty-Two Classes of the GlobCover Legend

Value	GlobCover legend	Color
11	Post-flooding or irrigated croplands	
14	Rainfed croplands	
20	Mosaic cropland (50%–70%)/natural vegetation (grassland, shrubland, forest) (20%–50%)	
30	Mosaic natural vegetation (grassland, shrubland, forest) (50%–70%)/cropland (20%–50%)	
40	Closed to open (>15%) broadleaved evergreen and/or semideciduous forest (>5)	
50	Closed (>40%) broadleaved deciduous forest (>5m)	
60	Open (15%–40%) broadleaved deciduous forest (>5m)	
70	Closed (>40%) needleleaved evergreen forest (>5m)	
90	Open (15%–40%) needleleaved deciduous or evergreen forest (>5m)	
100	Closed to open (>15%) mixed broadleaved and needleleaved forest (>5m)	
110	Mosaic forest/shrubland (50%–70%)/grassland (20%–50%)	
120	Mosaic grassland (50%–70%)/forest/shrubland (20%–50%)	
130	Closed to open (>15%) shrubland (<5m)	
140	Closed to open (>15%) grassland	
150	Sparse (>15%) vegetation (woody vegetation, shrubs, grassland)	
160	Closed (>40%) broadleaved forest regularly flooded—fresh water	
170	Closed (>40%) broadleaved semideciduous and/or evergreen forest regularly flooded—saline water	
180	Closed to open (>15%) vegetation (grassland, shrubland, woody vegetation) on regularly flooded or waterlogged soil—fresh, brackish or saline water	
190	Artificial surfaces and associated areas (urban areas >50%)	
200	Bare areas	
210	Water bodies	
220	Permanent snow and ice	

implemented according to the Committee on Earth Observation Satellites (CEOS) Land Product Validation Subgroup recommendations (Strahler et al., 2006). The validation process had three different steps: elaborating the sampling strategy, collecting validation data, and assessing product accuracy. The validation strategy is described in detail in another chapter.

The validation dataset, containing in total 4258 points, was built with the support of international experts, who were asked to interpret points in LCCS classifiers. In 3167 cases, the experts were (explicitly) certain that the information they provided was correct. Only these points were considered in the validation step. Furthermore, to explore the effect of heterogeneous areas, the validation dataset was even further reduced to 2115 points by removing all the points for which the experts needed to define more than one land-cover type.

These two subsets of the validation dataset (made of "certain" and "certain and homogeneous" points, respectively) were then crossed with the GlobCover map to derive the confusion matrix. As the notion of dominance between land-cover types was not quantified for any validation sample, it was not taken into account in the validation process. Therefore, mosaic classes were positively

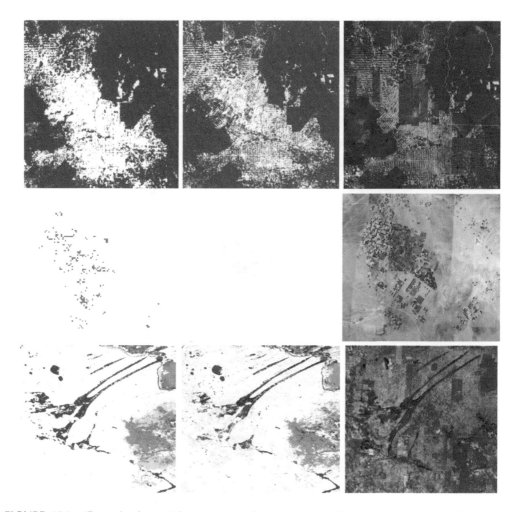

FIGURE 16.9 **(See color insert.)** Improvement of the spatial detail due to the use of a 300-m spatial resolution. Deforestation clear-cuts in Amazonia (top), irrigated crops in Saudi Arabia's desert (center), and specific vegetation structure in Russia (bottom). GLC2000 (left), GlobCover (center), and Google Earth (right).

TABLE 16.2
Accuracy of the GlobCover 2005 Land Cover Map

GlobCover Validation Dataset	Global Accuracy
3167 "certain" points	73.14%
2115 "certain" and "homogeneous" points	79.25%

validated even if only one of the classes making the mosaic matched the validation dataset. As recommended by the CEOS, the overall accuracy values derived from the confusion matrix were weighted by the area proportions of the various land-cover classes. Table 16.2 reports the results.

These final accuracy results documented the quality of the GlobCover product. The accuracy was higher than that of GLC2000 with spatial resolution improved by a factor 3.3, resulting in a product 10 times better than GLC2000 if the pixel area was considered.

This very positive figure must be balanced by the fact that the quality of the GlobCover map varies according to the region of interest. This can be explained by two factors: (1) the number of valid observations available over a region (Figure 16.5)—that gives *a priori* information about the input data quality and the expected classification reliability and (2) the quality of the reference data used for the automatic labeling. As for this latter concern, a reference dataset derived from medium- or low-resolution images will logically induce more inconsistencies and more mosaic classes in the classification result than a reference dataset derived from visual interpretation of high spatial resolution images.

16.2.3   GlobCover 2009

The GlobCover processing system including the preprocessing and classification modules was run again by ESA and the Université catholique de Louvain, exactly as they were defined in the 2005 project, to derive a new global land-cover map from 2009 MERIS FRS time series. The objective was to deliver the GlobCover 2009 product in 2010, thus demonstrating the operational service provided by the developed GlobCover chain.

16.2.3.1   Data Coverage

The GlobCover 2009 project benefited from 12 months of global MERIS FRS time series available from January 2009 to December 2009.

Just like in 2005, the global acquisition of a MERIS time series proved to be an important issue. The data coverage in 2009 was also uneven (Figure 16.10): regions such as Central and South America, western Canada, east Siberia, and the northern regions were covered by less than 50 observations for the whole year. As expected, the number of valid observations after the cloud screening was even more variable.

16.2.3.2   GlobCover 2009 Product

The GlobCover 2009 product (Arino et al., 2010) is presented in Figure 16.11. The GlobCover 2009 legend was identical to the GlobCover 2005 legend, thus counting 22 classes (Table 16.1) and being compatible with the GLC2000 product. The distribution of land-cover classes was highly similar to that associated with the GlobCover 2005 land-cover map, as illustrated in Figure 16.12. The quantitative accuracy assessment repeated from an updated version of the validation dataset by the same network provided lower accuracy figures. The overall accuracy values weighted by the class surface as computed for Table 16.2 were 67.5% using the 2190 samples that were heterogeneous and certain and 66.95% using the 1408 homogeneous samples that were considered certain. Unlike

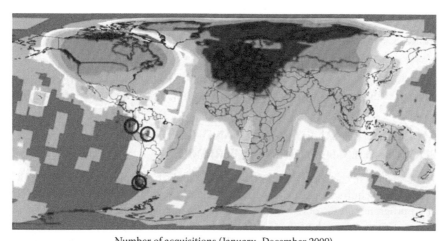

Number of acquisitions (January–December 2009)

| 0 | 1–5 | 6–10 | 11–50 | 51–100 | 101–200 | 201–300 | > 300 |

FIGURE 16.10   (See color insert.) MERIS FRS density data acquisition over the year 2009.

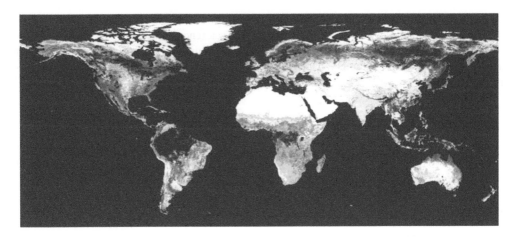

FIGURE 16.11 **(See color insert.)** The GlobCover 2009 product as the first 300-m global land-cover map for the year 2009.

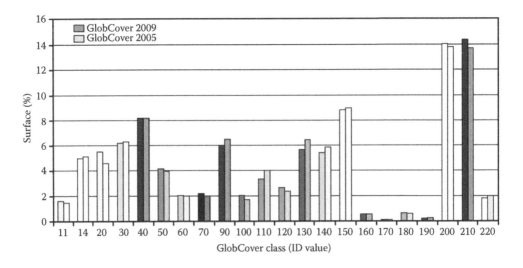

FIGURE 16.12 Comparison of class proportions between GlobCover 2005 and 2009 land-cover products.

for the GlobCover 2005, the analysis of validation results concluded that the large classes were less accurate than the others, reducing the 70.7% overall accuracy to 66.95% after area weighting.

16.3 AN OPERATIONAL REGIONAL LAND-COVER SERVICE: THE GLOBCORINE EXPERIENCE

Following the success of the GlobCover initiative, ESA and EEA decided to launch the GlobCorine initiative. EEA had acquired a unique experience in land-use database through the CORINE Land Cover (CLC) program and the derived information. The quality of the GlobCover 2005 product as well as the automated approach of the GlobCover processing chain prompted EEA to consider the MERIS time series as a great opportunity to address two major concerns over its CLC database: the spatial extent of the CLC products and their update frequency. The CLC database currently covers the EU countries and is updated on a 5-year basis with a delivery time of more than 2 years between the image acquisition and the derived results. Use of the MERIS time series coupled with a GlobCover-like processing system can allow a more frequent monitoring of some major land dynamics and result in a consistent mapping of the pan-European continent.

The GlobCorine study attempted to address these EEA concerns by making full use of the MERIS time series through a land-cover mapping service dedicated to the pan-European continent and based on the GlobCover findings. The pan-European area was defined as the 27 EU countries extending to the whole Mediterranean basin and western Russia. Like GlobCover, the GlobCorine system was first applied to the 2005 MERIS time series and then repeated for 2009.

16.3.1 GLOBCORINE 2005

16.3.1.1 Dataset

The main source of data for the GlobCorine project was the daily MERIS FRS composites as produced and delivered by the GlobCover processing chain. They were then processed in seasonal and annual surface reflectance composites from December 1, 2004 to June 30, 2006. As already mentioned, this dataset showed an uneven spatial and temporal coverage owing to constraints of the MERIS program. Nevertheless, the dataset available over most of Europe was significantly better than in many places of the world, allowing further fine-tuning of the classification module for this continent.

16.3.1.2 Methodology

The GlobCorine classification chain aimed at transforming the MERIS multispectral mosaics produced by the GlobCover preprocessing modules into a meaningful pan-European land-cover map in an automated way. As explicitly requested by ESA and EEA, the land-cover legend dedicated to the pan-European continent must be compatible as much as possible with the CLC aggregated legend (EEA, 2006).

The GlobCover classification module was thus adjusted, mainly by refining some methodological choices in three distinct areas: taking most advantage of the full MERIS spectral resolution, discriminating particular land-cover classes using specifically the temporal information of the MERIS time series, and adapting the reference land-cover database used in the labeling procedure to the GlobCorine legend. Finally, the GlobCorine classification module consisted of five main steps (Figure 16.13). It was also preceded by a stratification process, which split the pan-European continent into five equal-reasoning strata.

16.3.1.2.1 Steps I and II. Spectral classification and automated labeling procedure

The first step was similar to the GlobCover classification module, except that the algorithms were based on more and better selected spectral channels with up to 9 spectral bands for some strata. A supervised algorithm identified land-cover classes poorly represented at the pan-European scale (i.e., urban and wetland classes). An unsupervised algorithm was then applied on the remaining pixels to obtain clusters of spectrally similar pixels.

The second step was an automated labeling procedure that transformed the spectral clusters into land-cover classes according to the same procedure used in the GlobCover chain. The reference land-cover database was compiled from 2000 and 2006 CLC maps over Europe and from GlobCover 2005 map over North Africa and western Russia.

16.3.1.2.2 Steps III and IV. Temporal classification and automated labeling procedure

The third step used the temporal content of the MERIS time series to improve the discrimination of pixels labeled as cropland and mosaic classes. The application of an unsupervised classification on the 10-day NDVI profile allows disaggregating the mosaic classes into their pure components and splitting the rainfed from the irrigated croplands. A second labeling procedure then transformed these temporal classes into land-cover classes.

16.3.1.2.3 Step V. Merging of classifications

The land-cover classes obtained from steps II and IV were merged to produce the GlobCorine land-cover map.

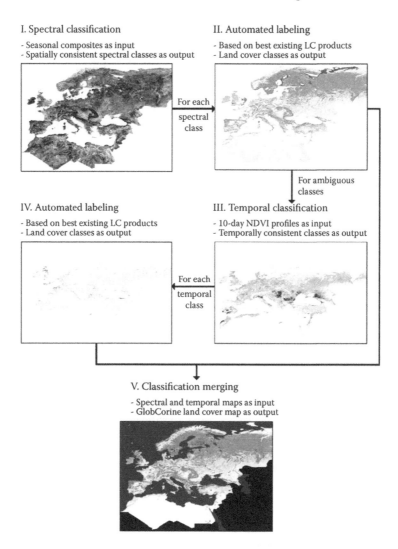

FIGURE 16.13 The five steps of the GlobCorine classification processing chain to be separately applied on each stratum.

16.3.1.3 Result

Figure 16.14 presents the GlobCorine land-cover map, which was the second 300-m spatial resolution land-cover map—after the GlobCover one—produced for the pan-European continent for the period December 2004–June 2006 (Bontemps et al., 2010).

The GlobCorine legend focuses on the CLC aggregated legend, which has demonstrated its capacity to capture the most important land-cover changes (EEA, 2006). Like GlobCover, the GlobCorine land-cover product was designed to be consistent at the pan-European scale. Its legend, which counts 14 classes, was therefore determined by the level of information available at this continental scale (Table 16.3).

With regard to the global GlobCover land-cover map, three major improvements were observed, which concerned the urban areas, the sparsely vegetated areas, and the significant reduction of mosaic classes. In addition, the high spatial consistency of the GlobCorine product was pointed out. Indeed, areas not covered by CLC2000 (i.e., by the European reference database) were coherently classified (Figure 16.15).

FIGURE 16.14 (See color insert.) GlobCorine 2005 land-cover map.

TABLE 16.3
(See color insert.) Fourteen Classes of the GlobCorine Legend

Value	GlobCorine legend	Color
10	Urban and associated areas	
20	Rainfed cropland	
30	Irrigated cropland	
40	Forest	
50	Heathland and sclerophyllous vegetation	
60	Grassland	
70	Sparsely vegetated area	
80	Vegetated low-lying areas on regularly flooded soil	
90	Bare areas	
100	Complex cropland	
110	Mosaic cropland/natural vegetation	
120	Mosaic of natural (herbaceous, shrub, tree) vegetation	
200	Water bodies	
210	Permanent snow and ice	

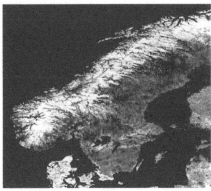

FIGURE 16.15 **(See color insert.)** The classification of Norway (right), which was not covered by the reference database (left), proved to be spatially consistent with surrounding areas.

16.3.1.4 Validation

The quantitative validation of the GlobCorine 2005 land-cover product aimed at assessing the accuracy of the 14 classes of the land-cover map from an independent validation database. A twofold validation exercise was achieved, based on the GlobCover validation dataset (restricted to the pan-European points) and on the CLC 2006 database.

16.3.1.4.1 Validation based on the GlobCover dataset

Over the pan-European region, 403 samples were extracted from the GlobCover validation database. The LCCS classifiers characterizing each sample were transformed into the GlobCorine legend. These 403 interpreted validation samples were then matched to the GlobCorine map, and a confusion matrix was built.

The overall accuracy was found to be 79.9%. When weighting the overall accuracy value by the area proportion of the various land-cover classes, the figure increased to 89.25%.

Nevertheless, these figures have to be used cautiously. First, there is a clear contribution of the mosaic classes in the high global accuracy figure. Indeed, their agreement with several classes increases the global accuracy. However, these classes are not easily interpretable, and they should therefore be avoided as much as possible. Second, the number of validation points highly varies between classes. The stratified sampling that generated the validation dataset was indeed achieved on a global scale, based on the GlobCover product. The stratification, which ensures that each class is representatively sampled at a global scale, is thus not necessarily valid at the GlobCorine pan-European scale. This slight bias has an influence on the overall accuracy value weighted by the class area, which is artificially increased.

16.3.1.4.2 Validation based on the CLC 2006 database

A second quantitative evaluation of the GlobCorine 2005 product was also achieved by the European Topic Centre on Land Use and Spatial Information (ETC-LUSI) using the CLC2006 data as validation dataset. It has to be stated that large countries like Finland, Germany, Greece, Italy, Norway, Spain, Sweden, and United Kingdom were missing in this validation dataset.

The CLC2006 dataset was resampled and translated into the GlobCorine legend. No sampling was defined, thus considering a pixel-to-pixel approach to validate the GlobCorine product. An overall agreement of 52.39% was found. This lower figure was mainly due to the presence of mosaic classes in the GlobCorine legend (classes 100, 110, and 120) which have a definition without strict equivalence in the CLC legend. When the mosaic classes were not considered, the global agreement between the GlobCorine product and the translated CLC2006 dataset increased up to 79.73%. In this case, overall class distributions were similar, except the grassland and urban areas that were clearly underestimated in the GlobCorine product (Figure 16.16).

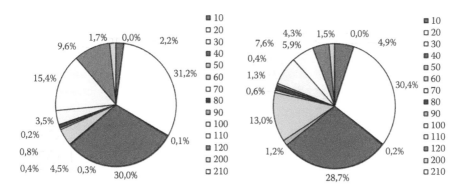

FIGURE 16.16 Proportions of each class of the GlobCorine legend in the GlobCorine 2005 land-cover map (left) and in the remapped CLC2006 dataset (right).

FIGURE 16.17 **(See color insert.)** The GlobCorine 2009 product.

16.3.2 GLOBCORINE 2009

To assess the operational service capabilities, a GlobCorine 2009 land-cover map was generated within a few months using exactly the same methodology (Figure 16.13) and the same legend (Table 16.3) as the ones developed for GlobCorine 2005. The only differences were in the input data:

- Twelve months of MERIS FRS time series (acquired from January 1 to December 31, 2009) were used instead of the 19 months from December 2004 to June 2006.
- The GlobCorine 2005 land-cover map was used as reference database instead of the previous database made of CLC 2000 and 2006 and GlobCover 2005.

Figure 16.17 presents the GlobCorine 2009 land-cover product.

The land-cover classes distribution was highly similar to the one associated with the GlobCorine 2005 land-cover map, as illustrated in Figure 16.18.

However, even if the proportions of the land-cover classes were similar between the 2005 and 2009 maps, the two products showed significant differences in the spatial distribution of land-cover classes. First, the classification of the MERIS time series from 2009 resulted in more compact

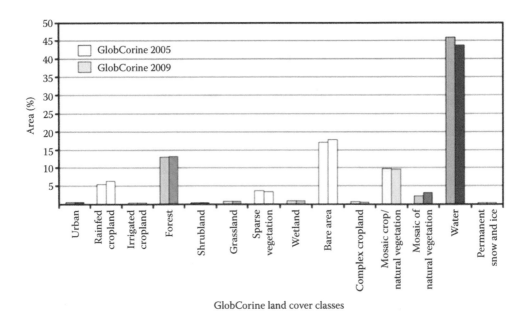

FIGURE 16.18 Comparison of class proportions between GlobCorine 2005 and 2009 land-cover products.

and homogeneous spatial patterns. According to the regions, this could have negative or positive impacts on the GlobCorine 2009 map. In addition, an increase of the "mosaic" class proportions was observed in the GlobCorine 2009 product.

The GlobCorine 2009 accuracy was assessed using the same validation dataset than the one used to validate the GlobCorine 2005 product. Based on the GlobCover validation dataset, the accuracy level was found to be 78% and 81.3% when the value was weighted by the area proportions of the various land-cover classes. Making a direct comparison with the CLC2006 database, the overall agreement was of 48.6%.

These results clearly demonstrated that the quality of the GlobCorine product is highly dependent on the input time series and on the land-cover database used for the labeling process, that is, GlobCover 2005 instead of the SPOT HRV-derived CLC database.

For the GlobCorine 2009 product, the number of valid observations available for the classification had significantly decreased compared with the GlobCorine 2005 product (Figure 16.10). This decrease had consequences on the consistency of the time series and the GlobCorine classification algorithm. In particular, the compositing period had to be adjusted, impacting negatively the discrimination between the different land-cover types according to their respective seasonal behavior. For instance, the "spring" composite over the northern region had to be based on 3 months in 2009 instead of 2 months in 2005.

16.4 LESSONS LEARNED

By delivering global and regional land-cover maps repeatedly within a short time, the ESA-supported GlobCover and GlobCorine initiatives paved the way for operational land-cover services. The GlobCover 2005 product was the very first 300-m global land-cover map derived from a time series acquired by Envisat's MERIS instrument for the period from December 2004 to June 2006 and successfully validated by an independent network of international experts. This GlobCover product, along with the three others, GlobCover 2009, GlobCorine 2005 and 2009, was made freely available through the ESA portal to the international community from http://ionia1.esrin.esa.int/ and http://ionia1.esrin.esa.int/globcorine/. This 300-m GlobCover 2005 product seems to meet some users' needs, as it has been downloaded by thousands since December 2008.

It is very important to keep in mind that the products delivered by the GlobCover and GlobCorine initiatives were strictly constrained by two requirements: only MERIS data could be used as surface reflectance input and the system must be fully automated. The latter requirement surely contributes to the operational capability of repeatability within a short time delivery. However, a trade-off between automation and interactive process is expected to further improve the output. Similarly, a multiple sensors approach should improve it in particular, thanks to the contribution of SWIR range.

The GlobCover service relies on a classification design that succeeds in being globally consistent but regionally tuned at the same time and that takes most advantage of the available data. Although being quite efficient, the GlobCover system has proved to be sensitive to the quality and amount of data used as input. Indeed, the MERIS surface reflectance composite as preprocessed by the GlobCover system is found to be unique wherever enough valid observations are acquired.

The concept of global and regional land-cover services has been demonstrated. The GlobCorine experience also illustrated how decision makers really prefer fresh and up-to-date, rather than detailed but outdated, information. Timeliness, defined as the time interval between satellite observation and product delivery, came out as a strong criterion and can be improved at the expense of details but not of quality.

To become an operational service according to common standards, both the satellite provision chain and the processing system should be consolidated to always ensure enough remote-sensing data provision and full processing capabilities in any case—thanks to system duplication. This calls for an even better coordination between space agencies to provide the long-term time series of various sensors with complementary spectral and spatial resolutions. It is indeed foreseen that global systematic acquisition capabilities of high-resolution imagery like Landsat and the forthcoming Sentinel-2 should be combined with medium-resolution time series made of daily global observation to deliver a product depicting most of the land-cover features. Development of these operational capabilities leading to yearly update calls for further conceptual and methodological research to enhance land-cover product consistency over time.

ACKNOWLEDGMENTS

The authors are most grateful to the GlobCover team including Franck Ranera (European Space Agency), José Ramos (European Space Agency), Vasileios Kalogirou (European Space Agency), Jean-François Pekel (Université catholique de Louvain), and Eric van Bogaert (Université catholique de Louvain). We also address a special thank to Jean-Louis Weber (European Environment Agency) for his support for the GlobCorine projects.

REFERENCES

Arino, O., Bicheron, P., Achard, F., Latham, J., Witt, R., and Weber, J-L. 2008. GlobCover the most detailed portrait of Earth. *ESA Bulletin*, 136, 25–31.

Arino, O., Gross, D., Ranera, F., Leroy, M., Bicheron, P., Brockmann, C., Defourny, P., et al. 2007. GlobCover: ESA Service for Global Land Cover from MERIS, In *Proceedings of the IEEE International Geoscience and Remote Sensing Society Symposium*, July 23–27, 2007, Barcelona, Spain.

Arino, O., Ramos, J., Kalogirou, V., Defourny, P., and Achard, F. 2010. GlobCover2009. In *Proceedings of the Living Planet Symposium*, June 27–July 2, 2010, Bergen, Norway, ESA SP-686.

Bartholomé, E. and Belward, A.S. 2005. GLC2000: A new approach to global land cover mapping from Earth Observation data. *International Journal of Remote Sensing*, 26, 1959–1977.

Bicheron, P., Amberg, V., Bourg, L., Petit, D., Huc, M., Miras, B., Brockmann, C., et al. 2011. Geolocation assessment of MERIS GlobCover orthorectified products. *IEEE Transactions on Geoscience and Remote Sensing*, 49, 2972–2982.

Bontemps, S., Defourny, P., Van Bogaert, E., Weber, J.L., and Arino, O. 2010. GlobCorine—A joint EEA-ESA project for operational land cover and land use mapping at pan-European scale. Proceedings of the 2010 European Space Agency Living Planet Symposium, June 28–July 2, 2010, Bergen, Norway.

Bourg, L. and Etanchaud, P. 2007. The AMORGOS MERIS CFI (Accurate MERIS Ortho Rectified Geolocation Operational Software). Software User Manual and Interface Control Document, PO-ID-ACRGS-003, February 2007.

Defourny, P., Bicheron, P., Brockman, C., Bontemps, S., Van Bogaert, E., Vancutsem, C., Pekel, J.F., et al. 2009a. The first 300 m global land cover map for 2005 using ENVISAT MERIS time series: A product of the GlobCover system. Proceedings of the 33rd International Symposium on Remote Sensing of Environment, May 4–8, 2009, Stresa, Italy.

Defourny, P., Schouten, L., Bartalev, S., Bontemps, S., Caccetta, P., de Witt, A., di Bella, C., et al. 2009b. Accuracy Assessment of a 300 m Global Land Cover Map: The GlobCover Experience. 33rd International Symposium on Remote Sensing of Environment, May 4–8, 2009, Stresa, Italy.

DeFries, R.S. and Townshend, J.R.G. 1994. NDVI-derived land cover classifications at a global scale. *International Journal of Remote Sensing*, 15, 3567–3586.

DeFries, R.S., Hansen, M., Townshend, J.R.G., and Sohlberg, R. 1998. 8-km Global Land Cover Data Set Derived from AVHRR. Global Land Cover Facility, University of Maryland Institute for Advanced Computer Studies, College Park, Maryland, USA.

Di Gregorio, A. and Jansen, L.J.M. 2000. Land cover classification system (LCCS): Classification concepts and user manual. GCP/RAF/287/ITA Africover-East Africa Project and Soil Resources, Management and Conservation Service, Food and Agriculture Organization.

EEA—European Environment Agency. 2006. Land accounts for Europe 1990-2000. EEA report 11/2006 prepared by Haines-Young, R. and Weber, J.-L. Available at: http://www.eea.europa.eu/publications/eea_report_2006_11

ESA—European Space Agency. 2011. MERIS product handbook. Issue 3.0. Available at: http://envisat.esa.int/handbooks/meris/ (accessed at August 1, 2011).

Fischer, J. and Grassl, J. 1991. Detection of cloud top height from backscattered radiances within the oxygen A band. Part I: Theoretical study. *Journal of Applied Meteorology*, 30, 1245–1259.

Fisher, J., Preusker, R., Muller, J.P., Schroeder, T., Brockmann, C., Zühle, M., and Formferra N. 2006. MERIS Land Surface Albedo/BRDF retrieval. Proceedings of the 2nd International Symposium on Recent Advances in Quantitative Remote Sensing (RAQRS II), October 25–29, 2006, Torrent (Valencia), Spain.

Friedl, M.A., McIver, D.K., Hodges, J.C.F., Zhang, X.Y., Muchoney, D., Strahler, A.H., Woodcock, C.E., et al. 2002. Global land cover mapping from MODIS: Algorithms and early results. *Remote Sensing of Environment*, 83, 287–302.

Friedl, M.A., Sulla-Menashe, D., Tan, B., Schneider, A., Ramankutty, N., Sibley, A., and Huang, X. 2010. MODIS Collection 5 global land cover: Algorithm refinements and characterization of new datasets. *Remote Sensing of Environment*, 114, 168–182.

Hagolle, O., Lobo, A., Maisongrande, P., Cabot, F., Duchemin, B., and De Peyrera, A. 2004. Quality assessment and improvement of temporally composited products of remotely sensed imagery by combination of VEGETATION 1 and 2 images. *Remote Sensing of Environment*, 94 (2), 172–186.

Hansen, M.C., DeFries, R.S., Townshend, J.R.G., and Sohlberg, R. 2000, Global land cover classification at 1 km spatial resolution using a classification tree approach. *International Journal of Remote Sensing*, 21, 1331–1364.

Loveland, T.R., Reed, B.C., Brown, J.F., Ohlen, D.O., Zhu, Z., Yang, L., and Merchant, J.W. 2000. Development of a global land cover characteristics database and IGBP DISCover from 1 km AVHRR data. *International Journal of Remote Sensing*, 21, 1303–1365.

Mayaux, P., Eva, H., Gallego, J., Strahler, A., Herold, M., Shefali, A., Naumov, S., et al. 2006. Validation of the Global Land Cover 2000 Map. *IEEE Transactions on Geoscience and Remote Sensing*, 44, 1728–1739.

Rast, M., Bezy, J.L., and Bruzzi, S. 1999. The ESA Medium Resolution Imaging Spectrometer MERIS—A review of the instrument and its mission. *International Journal of Remote Sensing*, 20, 1681–1702.

Strahler, A.H., Boschetti, L., Foody, G.M., Friedl, M.A., Hansen, M.A., Mayaux, P., Morisette, J.T., Stehman, S.V., and Woodcock, C.E. 2006. Global Land Cover Validation: Recommendations for evaluation and accuracy assessment of global land cover maps. Office for Official Publications of the European Communities, Luxembourg. Available at: http://nofc.cfs.nrcan.gc.ca/gofc-gold/Report%20Series/GOLD_25.pdf

Vancutsem, C., Bicheron, P., Cayrol, P., and Defourny, P. 2007a. Performance assessment of three compositing strategies to process global ENVISAT MERIS time series. *Canadian Journal of Remote Sensing*, 33, 492–502.

Vancutsem, C., Pekel, J.F., Bogaert, P., and Defourny, P. 2007b. Mean compositing, an alternative strategy for producing temporal syntheses. Concepts and performance assessment for SPOT VEGETATION times series. *International Journal of Remote Sensing*, 28, 5123–5141.

17 Continental and Regional Approaches for Improving Land-Cover Maps of Africa

Philippe Mayaux, Christelle Vancutsem, Jean-François Pekel,
Carlos de Wasseige, Pierre Defourny, Matthew C. Hansen,
and Landing Mane

CONTENTS

17.1 INTRODUCTION

Land-cover information provides essential information for global scientific applications and regional environmental policies. It establishes the boundary conditions for general circulation models used for simulating climate and for land-surface process models used for studying earth system energy, water, and material transport. The accuracy with which such maps depict actual land cover at some specified time can influence the reliability of the scenarios the models generate.

Policy users also need information on the state of land cover to formulate sustainable development policies and strategies at scales ranging from local projects to the global perspective of multilateral environmental agreements such as the UN Framework Convention on Climate Change (UNFCCC), the UN Convention to Combat Desertification (UNCCD), the Convention on Biological Diversity (CBD), and the Ramsar Wetlands Convention. The reporting mechanisms under the terms of multilateral environmental agreements include land cover as main parameter to assess. In particular, the prominent role of forests in carbon cycle was underlined during the recent negotiations on climate from Copenhagen to Durban and the mechanisms put in place by the Convention of Parties (REDD+, CDM) require detailed information on land cover and land-cover changes.

Land-cover information is also needed to measure the impact and effectiveness of management actions associated with sustainable development policies. Addressing of issues such as sustainable management and use of forests and other land resources in developing countries, forest conservation and restoration, extension of croplands, desertification, or watershed degradation will substantially depend on the availability of accurate baseline land-cover information (United Nations, 2002).

Thus, the need to document the extent and condition of ecosystems is well recognized. This is especially true in tropical areas, where land-cover change has occurred at an unprecedented rate in recent decades. Geospatial representations of African land cover in forms we would recognize as maps have been produced since the sixteenth century at least, but viewing (let alone mapping) the entire land mass in a consistent and uniform way and representing actual land cover over a fixed, contiguous period of time was unimaginable until the end of the twentieth century. In this chapter, we present two continental land-cover maps of Africa, GLC2000 and GlobCover, and a regional multisource map of the Congo Basin countries.

17.2   PREVIOUS LAND-COVER MAPS OF AFRICA

Several continental cartographic studies have been undertaken (Table 17.1). The first ones were based on the compilation of national and local maps enriched by consultation with many experts (Olson et al., 2001; White, 1983). By the end of the 1980s, the International Geosphere Biosphere Programme (IGBP) showed a clear requirement for global land-cover maps to support global change research. Loveland et al. (1999) published the IGBP land-cover map based on 1-km resolution data collected between 1992 and 1993 from the Advanced Very High Resolution Radiometer (AVHRR). This product has been widely used in global change research and for supporting the work of groups such as nongovernmental conservation organizations and development assistance programs. However, the latter two groups of users showed a clear requirement for better spatial and thematic detail as they exploited the map effectively at regional/continental scales, rather than as a single global dataset.

17.3   THE GLOBAL LAND-COVER 2000 MAP OF AFRICA

The Joint Research Centre (JRC) decided to produce a global land-cover map in partnership with 30 institutions, using SPOT–4 VEGETATION daily images for the year 2000 as primary data source (Bartholomé and Belward, 2005). A number of different types of remotely sensed data are available for vegetation mapping at continental scale; each of these sources has its own potential application. Previous maps were derived from single source data, whereas the GLC2000 map used four sets of satellite information: SPOT VEGETATION daily images for the entire year 2000, ERS SAR data,

---

TABLE 17.1
Previous Land-Cover Maps of Africa

Title	Global/Africa	References	Methods
Vegetation of Africa	Africa	White (1983)	Consultation of experts and compilation of local information
IGBP DISCover	Global/Africa	Loveland et al. (1999)	Satellite-based analysis
Global land-cover classification	Global	Hansen et al. (2000)	Satellite-based analysis
Terrestrial ecosystems (WWF)	Global/Africa	Olson et al. (2001)	Consultation of experts and compilation of local information
Vegetation continuous fields	Global	Hansen et al. (2005)	Satellite-based analysis
GLC 2000	Global/Africa	Mayaux et al. (2004)	Satellite-based analysis by continent and global aggregation
GlobCover	Global/Africa	Defourny et al. (2006)	Satellite-based analysis
MODIS land cover	Global	Friedl et al. (2010)	Satellite-based analysis

Note: The column Global/Africa informs if the map was produced at global level (Global), only on Africa (Africa), or if it was fine-tuned for Africa and then integrated into a global product (Global/Africa).

JERS SAR data, and DMSP nighttime lights. The Digital Elevation Model was also used for mapping mountainous ecosystems. Each of the sources of data used, outlined below, contributes to mapping a specific ecosystem or land cover, seasonality, or water regime (Figure 17.1). The classification methods include unsupervised clustering and interactive labeling of seasonal profiles and monthly average composites (Cabral et al., 2003; Vancutsem et al., 2007), classification based on radar texture (Mayaux et al., 2002), and unsupervised classification based on nighttime lights. The continental legend of the GLC2000 map includes 27 classes: 9 with a dominant tree layer, 8 with a dominant shrub or grass layer mixed with agricultural field/land, 4 agricultural classes, 4 classes of bare soil and deserts, cities, and water. These 27 classes were recombined into 15 classes at the global level. Note that the global product contains 21 classes, but 6 classes, for example, needle-leaf forests, snow, and ice, were not represented in the African map.

The thematic accuracy of the GLC2000 map was computed on 544 points at the global level, with a value of 68.4%, with very high producer and user accuracy for forested classes (Mayaux et al., 2006). When computed on the 164 points falling in Africa, the accuracy was 82.4%, although the number of points did not allow for providing a good confidence interval to the estimator.

The strength of the GLC2000 project was in the partnership with regional experts. More than 30 research teams participated in production of the land-cover map from SPOT VEGETATION data or validation of the maps over Africa. This offered a number of technical and political advantages. The project teams had an experience of mapping their region, and this ensured that optimum image classification methods were used, that the legend was regionally appropriate, and that there was access

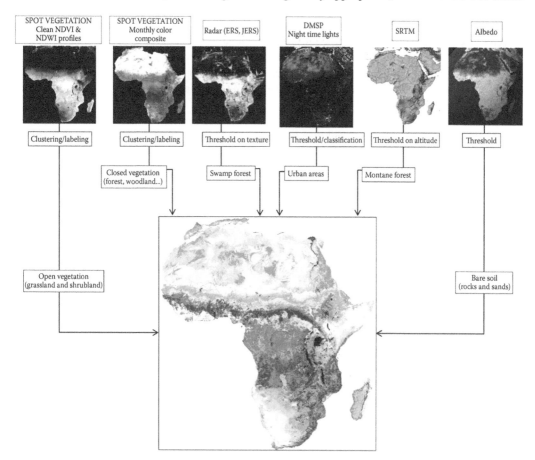

FIGURE 17.1   (See color insert.) Datasets and main classification algorithms used in the production of the GLC2000 map of Africa. (From Mayaux, P. et al., *J. Biogeogr.*, 31, 861–877, 2004. With permission.)

to reference material. The presence of national teams in producing the information weakened the reluctance of some nations to accept assessments of their territory such as land-cover maps made by third parties. Finally, there was clear capacity building, with the involvement of scientists from the developing world in international projects.

17.4 THE GLOBCOVER MAP

As a natural evolution of GLC2000, the European Space Agency (ESA) exploited the full potential of the Medium Resolution Imaging Spectrometer (MERIS) fine resolution (300 m) and demonstrated a service that produced automatically a global land-cover map in a consistent manner (Defourny et al., 2006). For this purpose, a system made up of two components was developed: a first component dealing with data preprocessing and a second component providing an automatic classification, including the transformation of composites of surface reflectance into classes satisfying the land-cover classification system (LCCS) nomenclature.

The land-cover mapping approach (Figure 17.2) combines the high spatial consistency of class delineation obtained from multispectral composite(s) with the good land-cover discrimination provided by temporal profile analysis. The overall classification performance relies on four steps: (1) a stratification that splits the world into 22 equal-reasoning regions based on bioclimatic, land-cover, and satellite observation conditions and that allows optimization of the data and classification parameters for each region; (2) a classification algorithm to define homogenous land-cover classes based on one (or at the most two) multispectral reflectance composite(s); (3) a land-cover discrimination algorithm with iterative multitemporal clustering steps: and (4) a labeling procedure built on reference classifications such as the GLC2000 regional products and Africover maps and then adjusted to MERIS mapping capabilities with the support of international experts.

The GlobCover legend (Figure 17.3a) comprises 22 land-cover classes, including croplands (irrigated and rainfed), wetlands, forest, savannah (shrubland, grassland, and sparse vegetation), artificial surfaces, water bodies, and bare soils.

The first validated and calibrated product delivered in September 2008 covered a period of 19 months (between December 2004 and June 2006) and was composed of the MERIS mean composite products (bimonthly and annual; Vancutsem et al., 2007) and a land-cover map at 300 m (Figure 17.3b). A second land-cover map was produced in 2010, using MERIS data from 2009.

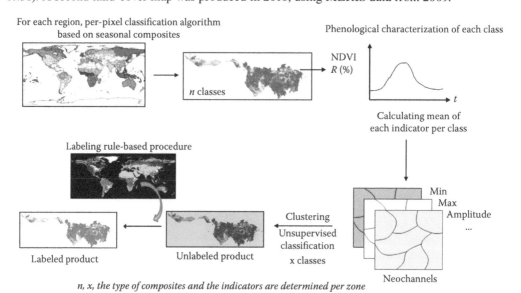

n, x, the type of composites and the indicators are determined per zone

FIGURE 17.2   GlobCover land-cover mapping methodology.

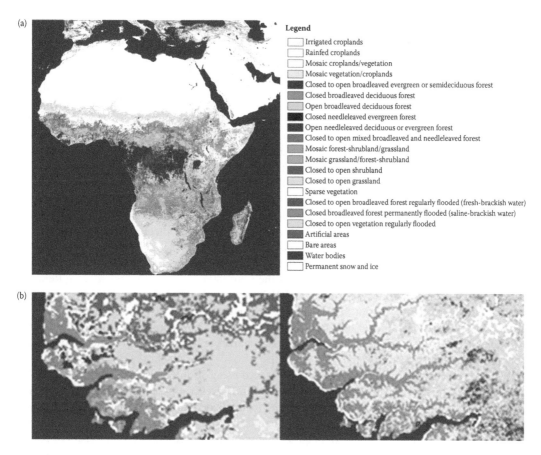

Legend

☐ Irrigated croplands
☐ Rainfed croplands
☐ Mosaic croplands/vegetation
☐ Mosaic vegetation/croplands
■ Closed to open broadleaved evergreen or semideciduous forest
■ Closed broadleaved deciduous forest
☐ Open broadleaved deciduous forest
■ Closed needleleaved evergreen forest
■ Open needleleaved deciduous or evergreen forest
■ Closed to open mixed broadleaved and needleleaved forest
■ Mosaic forest-shrubland/grassland
■ Mosaic grassland/forest-shrubland
■ Closed to open shrubland
☐ Closed to open grassland
☐ Sparse vegetation
■ Closed to open broadleaved forest regularly flooded (fresh-brackish water)
■ Closed broadleaved forest permanently flooded (saline-brackish water)
☐ Closed to open vegetation regularly flooded
■ Artificial areas
☐ Bare areas
■ Water bodies
☐ Permanent snow and ice

FIGURE 17.3   (See color insert.) (a) (Top) GlobCover classification over Africa (2005–2006) and legend; (b) (bottom) comparison of the GLC2000 map (left) with the GlobCover map (right) over Senegal, Guinea-Bissau, and Gambia.

The first validation exercise (February 2008) based on the expertise of an international network of regional experts established a reference dataset of almost 4000 points. The overall accuracy of the GlobCover land cover was found to be 73.3%. However, considering only the points about which the experts were very confident in their interpretation, the following results for homogeneous land cover were achieved. For the principal classes, the user accuracy was as follows: 82.7% for cultivated and managed terrestrial land, 69.5% for natural and seminatural terrestrial vegetation, 19% for natural and seminatural aquatic vegetation, 63.6% for artificial surfaces, 88.1% for bare areas, and 74.1% for water, snow, and ice. The producer accuracy was 69.6, 87.8, 19.0, 43.8, 77.1, and 82.2, respectively, for the aforementioned classes, leading to an overall accuracy of 77.9%.

A qualitative assessment was realized as well. Different institutions such as JRC, FAO, and GOFC-GOLD highlighted the improvement in the spatial detail and coverage and in thematic content in many areas, compared with GLC2000. As illustrated in Figure 17.3b, the thin and linear features such as mangroves were particularly well delineated with GlobCover—thanks to the improvement in spatial resolution.

17.5   A REGIONAL SYNTHESIS OVER THE CONGO BASIN

Although continental land-cover maps are much more adapted to users' needs than global products, they can be improved in specific regions by refining the legend and using more appropriate data sources. The most important part of Africa for climate regulation and biodiversity conservation is undoubtedly Central Africa, in particular the Congo Basin.

The various partners of the Congo Basin Forest Partnership have set up the Observatory for the Forests of Central Africa (OFAC in French for "Observatoire des Forêts d'Afrique Centrale"), which aims at pooling the knowledge and data necessary to monitor the ecological, environmental, and social services provided by Central Africa's forests (de Wasseige et al., 2009).

Recent vegetation maps are compiled and cross-validated by OFAC partners. A promising approach is the combination of medium-resolution maps (Landsat-derived), which give the best possible details of the forest–nonforest interface, and coarse-resolution maps (SPOT VEGETATION and MODIS-derived), which are able to depict the ecological types depending on the seasonality.

For the 2008 State of the Forest Report, all of the available data and state-of-the-art methods were used to deliver the most recent and best-area estimates currently available from satellite remote sensing. This map covers a total area of 5,450,000 square kilometers.

Forest area for the Congo Basin was estimated from five complementary sources provided by the South Dakota State University (SDSU), the Université Catholique de Louvain (UCL), and the EC JRC. Based on the GlobCover map, a new forest map including edaphic forests was produced at UCL using 300-m resolution MERIS data for the year 2005–2006 for Central African Republic (CAR) and Democratic Republic of Congo (DRC). For the four coastal countries of the Congo Basin, 1-km daily observations of SPOT-Vegetation acquired over the last 9 years provided an even clearer mosaic, allowing a better forest/no-forest delineation. This map contained four main land-cover classes: dense rain forest, swamp forest, rural complex, and nonforest. For a detailed description of the classes, see Mayaux et al. (1999).

Wall-to-wall mapping of forest cover was performed using 30-m Landsat data for the year 2000, covering the major part of the Congo Basin at SDSU. However, this forest map did not exhaustively map the entire Basin. To derive an estimate that included all lands of the Congo Basin, the Landsat-derived map product was used to calibrate data from the MODerate Resolution Imaging Spectroradiometer (MODIS) sensor in mapping humid tropical forest areas. Eight years of 250-m MODIS data were used as inputs to overcome atmospheric contamination. In the dry domain, the GLC2000 map, presented earlier in this chapter, was resampled at 300 m and used as a reference. Finally, the SRTM 90-m digital elevation was used to classify forest types according to an altitudinal gradient.

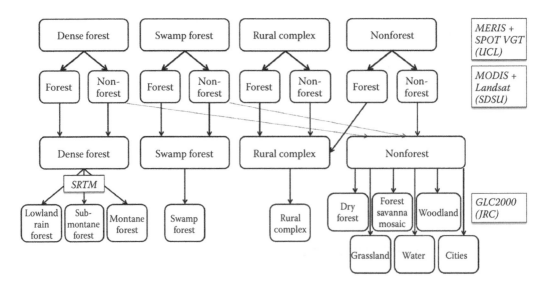

FIGURE 17.4 Scheme of integration of various maps on the Congo Basin.

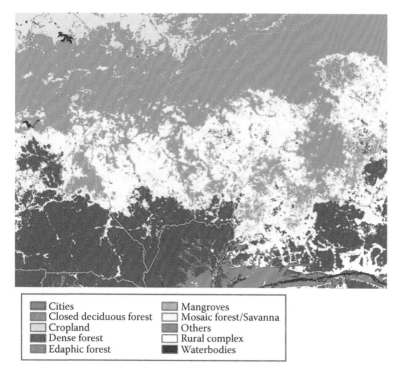

■ Cities	■ Mangroves
■ Closed deciduous forest	□ Mosaic forest/Savanna
□ Cropland	■ Others
■ Dense forest	□ Rural complex
■ Edaphic forest	■ Waterbodies

FIGURE 17.5   (See color insert.) Detail of the fusion map over the northern part of the Congo Basin at the borders between Cameroon, Gabon, Congo, Central African Republic, and Democratic Republic of Congo.

When a combination of various maps (Figure 17.4) consists of a thematic refinement (e.g., forest in SDSU and swamp forest in UCL), integration of the various products is quite straightforward. But complex decision rules based on thematic proximity and spatial pattern (e.g., distance to river network, to roads) are necessary for resolving conflicting situations, as in the case where SDSU indicates nonforest and UCL map indicates dense forest. It must be underlined that this situation occurs only rarely. The result of the map is presented in Figure 17.5.

17.6   CONCLUSIONS

- Low-scale vegetation maps are often blamed for not responding to foresters' or park managers' requirements. Indeed, forest and protected area management cannot be done by analyzing such maps, which are targeted at totally different "customers," but low-scale maps can provide a regional perspective for local studies about conservation of biological resources or national forest inventories. Moreover, such maps can help correct discrepancies in official statistics that come from various sources. For instance, it is quite surprising that, according to the last FAO Forest Resource Assessment (FAO, 2001), Mali has more than 7 million hectares of "dense forest" (with more than 40% of tree cover), that is, twice more than Cote d'Ivoire. In the same way, Zambia has 40 million hectares of closed forest, that is, 40 times more than Angola, which has very similar ecological conditions. Careful comparison of national statistics with the continental maps derived from satellite datasets can help eliminate many inconsistencies between contiguous countries.
- Although maps providing biophysical information such as the tree cover percentage or the leaf area index will be more and more used by the scientific community working for global change, the maps reducing the continuous reality into discrete classes will continue to be

used by ecologists and decision makers for understandable reasons of facility. They are just starting to integrate the spatial data in their analyses. Therefore, it would be too audacious to provide required information in a legend that they are not used to working with.

- The spatial resolution of the maps (300-m to 1-km pixel resolution) does not allow accurate determination of land-cover trends. For many classes, the spatial fragmentation of the land cover leads to an overestimation/underestimation of land-cover classes depending on the spatial arrangement of that class. In Africa, there is a specific problem with the agricultural areas that are often mixed with the natural grasslands or shrublands. However, for most of the continent, this resolution yields good results, taking into account the mean size of vegetation communities.

- Detection of agriculture in Africa from remote-sensing data is quite problematic owing to the farming system and the spatial pattern of croplands. The fields are small and mixed with savannas and fallows, which preclude a reliable mapping at 1-km spatial resolution. On the other hand, the low intensification level of agricultural techniques induces spectral or temporal properties of agriculture close to the surrounding natural vegetation. However, large pure cropland areas were mapped in the Sahelian belt, in Ethiopia, east Africa, and southern Africa in areas of intensive agriculture.

- The quality of the land-cover information available now is far better than that of the information that was available 10 years ago. However, new projects such as the ESA Climate Change Initiative aim at improving and updating the existing land-cover products by reprocessing historical datasets since the early nineties.

- The current land cover information is the result of a historical compromise between different beneficiaries and users, but it also represents the first element of a holistic sustainable management of natural resources. The maps of current and future land-cover of Africa can now be combined with databases on ecosystem services, such as carbon sequestration, conservation of biodiversity, or regulation of water cycle.

REFERENCES

Bartholomé, E. and Belward, A.S. 2005. GLC2000: A new approach to global land cover mapping from earth observation data. *International Journal of Remote Sensing*, 26(9–10), 1959–1977.

Cabral, A., de Vasconcelos, M.J.P., Pereira, J.M.C., Bartholomé, É., and Mayaux, P. 2003. Multitemporal compositing approaches for SPOT-4 VEGETATION data. *International Journal of Remote Sensing*, 24, 3343–3350.

Defourny, P., Vancutsem, C., Pekel, J-F., Bicheron, P., Brockmann, C., Nino, F., Schouten, L., and Leroy, M. 2006. GlobCover: A 300m global land cover product for 2005 using ENVISAT MERIS time series. *Proceedings of ISPRS Commission VII Mid-Term Symposium: Remote Sensing: From Pixels to Processes*, 8–11 May 2006, Enschede, the Netherlands.

de Wasseige, C., Devers, D., de Marcken, P., Eba'a Atyi, R., Nasi, R., and Mayaux, Ph. 2009. *The Forests of the Congo Basin—State of the Forest 2008*. Luxembourg: Publications Office of the European Union.

FAO. 2001. *Global Forest Resources Assessment 2000 Main Report*. FAO Forestry paper 140, 479 pp, Food and Agriculture Organization of the UN, Rome.

Friedl, M.A., Sulla-Menashe, D., Tan, B., Schneider, A., Ramankutty, N., Sibley, A., and Huang X. 2010. MODIS Collection 5 global land cover: Algorithm refinements and characterization of new datasets. *Remote Sensing of Environment*, 114(8), 168–182, ISSN 0034-4257, doi: 10.1016/j.rse.2009.08.016.

Hansen, M.C., Defries, R.S., Townshend, J.R.G., and Sohlberg, R. 2000. Global land cover classification at 1 km spatial resolution using a classification tree approach. *International Journal of Remote Sensing*, 21 (6–7), 1331–1364.

Hansen, M.C., Townshend, J.R.G., Defries, R.S., and Carroll, M. 2005. Estimation of tree cover using MODIS data at global, continental and regional/local scales. *International Journal of Remote Sensing*, 26(19), 4359–4380.

Loveland, T.R., Estes, J.E., and Scepan, J. 1999. Introduction: Special issue on global land cover mapping and validation. *Photogrammetric Engineering and Remote Sensing*, 65(9), 1011–1012.

Mayaux, P., Richards, T., and Janodet, E. 1999. A vegetation map of Central Africa derived from satellite imagery. *Journal of Biogeography*, 26, 353–366.

Mayaux, P., De Grandi, G.F., Rauste, Y., Simard, M., and Saatchi, S. 2002. Large scale vegetation maps derived from the combined L-band GRFM and C-band CAMP wide area radar mosaics of Central Africa. *International Journal of Remote Sensing*, 23, 1261–1282.

Mayaux, P., Bartholomé, E., Fritz, S., and Belward, A. 2004. A new land-cover map of Africa for the year 2000. *Journal of Biogeography*, 31, 861–877.

Mayaux, P., Strahler, A., Eva, H., Herold, M., Shefali, A., Naumov, S., Dorado, A., et al. 2006. Validation of the global land cover 2000 map. *IEEE-Transactions on Geoscience and Remote Sensing*, 44(7), 1728–1739.

Olson, D.M., Dinerstein, E., Wikramanaya, E.D., Burgess, N.D., Powell, G.V.N., Underwood, E.C., D'amico, J.A., et al. 2001. Terrestrial ecoregions of the world: A new map of life on earth. *Bioscience*, 51, 933–938.

United Nations. 2002. Report of the World Summit on Sustainable Development, Johannesburg, South Africa, August 26–September 4, 2002, 178 pp, New York, ISBN 92-1-104521-5.

Vancutsem, C., Bicheron, P., Cayrol, P., and Defourny, P. 2007. An assessment of three candidate compositing methods for global MERIS time series. *Canadian Journal of Remote Sensing*, 33(6), 492–502.

White, F. 1983. *The Vegetation of Africa: A Descriptive Memoir to Accompany the UNESCO/AEFTAT/UNSO Vegetation Map of Africa*. Paris: UNESCO.

18 Land-Cover Mapping in Tropical Asia

Hans-Jürgen Stibig and Chandra P. Giri

CONTENTS

18.1 CLIMATIC CONDITIONS AND LAND COVER OF TROPICAL ASIA

Tropical Asia stretches from the Indian subcontinent in the west through the mainland of continental Southeast Asia to the islands of Sumatra, Borneo, and Java in the south and to New Guinea in the Far East. The region is characterized by mountains, plains, and deltas. The elevations of the mountains of northern Myanmar or the islands of New Guinea, for example, reach more than 4000 m (asl), and the highest elevations of Sumatra and Borneo are more than 2500 m (asl). On the mainland, there are huge plains and deltas formed by large rivers, including the Ganges, Irrawaddy, and Mekong. In insular Southeast Asia, flatlands along the coastal zones or in river plains as for example formed by the Fly River (Papua New Guinea) are often covered by extensive areas of swampland. The climate of tropical Asia is dominated by the regime of the monsoon winds, causing a typical annual pattern of a dry and a rainy season on the mainland, with notable differences in precipitation and temperatures. South Asia (Pakistan, India, Sri Lanka, Bangladesh, Nepal, and Bhutan) has very large ecoclimatic amplitude, ranging from alpine conditions in the Himalayas to hot and arid zones in the west and humid tropical climate in the south and the east of the subregion. Continental Southeast Asia (Myanmar, Thailand, Laos, Cambodia, and Vietnam) shows a rather homogenous seasonal monsoon pattern, characterized by a distinct dry season from December to April and rainfall between May and October. In contrast, insular Southeast Asia (Malaysia, Indonesia, East Timor, and the Philippines, including Papua New Guinea) displays a typical equatorial climate pattern, where rainfall is rather evenly distributed throughout the year and annual variations of temperatures are much lower than those on the continent. The local seasonal patterns in this subregion are largely influenced by the geographical location in relation to the hemisphere and to coastal zones (Worldclimate, 2011; Figure 18.1).

Tropical forests are one of the main land-cover components in the region. The mixed deciduous forests of the mainland are famous for precious timber species (e.g., *teak* and *rosewood*). Insular Southeast Asia is known for its moist-evergreen tropical forests, including the highly productive *Dipterocarp* forests and the peat swamp forests in the lowlands of Sumatra, Borneo, and New Guinea. The region harbors 42% of the mangrove forests of the world (Giri et al., 2011), including the largest remaining tract of mangrove forests of the world in the Sundarbans (Bangladesh, India; Collins et al., 1991; Whitmore, 1984a). Many hilly and mountainous zones have been shaped by

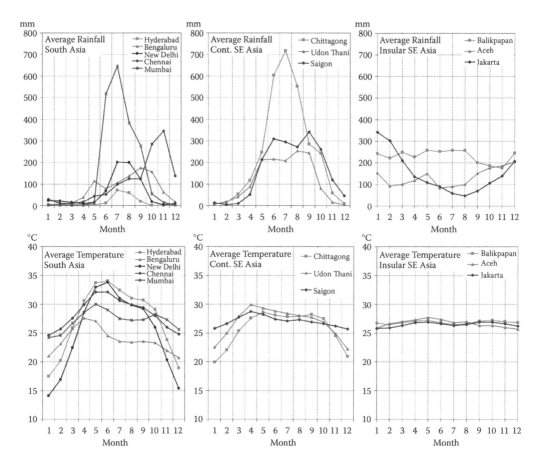

FIGURE 18.1 (See color insert.) Rainfall and temperature patterns in main subregions of tropical Asia. (From Arino, O. et al., *Eur. Space Agency Bull.,* 136, 24–31, 2008. With permission.)

shifting cultivation, a cycle of tree slashing, burning, and cropping, followed by a longer period of fallow. Shifting cultivation has caused notable forest loss in the past, creating typical mosaic patterns of fields, patches of regrowth, and forest remnants. Flatlands and deltas are mostly used for agriculture, most prominently for rice cultivation but also for cash crop plantations such as oil palm plantations in insular Southeast Asia (ADB and UNEP, 2004; APFC, 2009; Collins et al., 1991). The landscape of the region has changed notably during the last few decades, characterized by agricultural expansion (domestic crops, cash crops), urban and infrastructural development (road construction, hydropower), and an estimated loss of forest cover of about 74 million ha in the period from 1980 to 2010 (FAO, 2010). Insular Southeast Asia, particularly Borneo, has been heavily affected by extensive fires in 1982/1983 and 1997/1998, destroying millions of hectares of forests and causing severe air pollution and health threats to the population.

18.2 REGIONAL FOREST AND LAND-COVER MAPPING

The need for assessing forest and land cover at regional scales has been realized in the past in view of the magnitude of change, specifically in forest cover since the 1980s (e.g., Blasco et al., 1996). Satellite remote sensing offers the unique possibility of documenting land cover at regional or subregional levels consistently and uniformly across country boundaries and at a reasonable cost and effort. In tropical Asia, the mapping and monitoring of forest cover has received specific attention, reflecting the importance of the region's forests for the environment and for sustainable development.

The regional forest and land-cover maps of tropical Asia compiled in the 1980s were based on satellite remote sensing in combination with aerial photography, existing maps, and field knowledge. The use of satellite remote sensing was stimulated by the availability of the Landsat satellite series, with the initial application of visual interpretation and manual mapping techniques. Standard false color composites produced from the MSS (multispectral scanner, spatial resolution ~80 m) infrared, red, and green spectral bands were typically interpreted at scales ranging between 1:1 million and 1:250000. Starting from the mid-1980s, land-cover maps were prepared using new satellite sensors, including Landsat TM (Thematic Mapper, spatial resolution ~30 m) with additional spectral bands in the shortwave infrared range, the SPOT (Satellite Pour l' Observation de la Terre) HRV (high resolution visible) instrument (spatial resolution ~20 m), and for the Indian sub-continent, increasingly the IRS (Indian remote sensing) LISS (linear imaging self-scanning) sensors (spatial resolution ~ 36 m and 23.5 m).

The "Vegetation Map of Malesia" (Whitmore, 1984b) provided the first comprehensive overview of forest and other land cover of insular Southeast Asia, displaying the distribution and approximate extent of the humid tropical forest at a scale of 1:5 million and covering the area from peninsular Malaysia, Sumatra, and Borneo to the west of New Guinea. Evergreen and monsoon lowland and mountain forest types, as well as specific formations such as heath, limestone, mangrove, peat swamp, and swamp forests, were mapped; other lands were grouped in one class. The forest limits had been obtained by manual delineation from Landsat MSS imagery combined with other map information. Whitmore highlighted the fact that "much forest has been cleared" when compared to the situation in the 1950s.

For continental Asia, regional mapping of tropical vegetation was done at a scale of 1:5 million in the beginning of the 1990s (Blasco et al., 1996). A complete set of Landsat MSS imagery and a mosaic of Landsat TM data from 1991 were used for visual interpretation and for manual updating of the limits of the main vegetation types from existing vegetation maps. The "Vegetation Map of Tropical Continental Asia" discriminated eight tropical forest-cover types (lowland evergreen, semievergreen, peat swamp, swamp, mangrove, montane rain, dry evergreen, and dry deciduous forests) and categories of dry and mixed deciduous woodlands, thickets, and grasslands; three mosaic classes including cropland components were mapped. Bioclimatic and ecofloristic criteria were taken into account for the classification. A system for classification and mapping of vegetation types based on ecofloristic zoning and on floristic, ecological, and phenological parameters had already been proposed by Blasco and Legris (FAO, 1989), aiming at a regional standard for vegetation-cover classification for the whole region of tropical Asia.

A comprehensive regional overview of forest cover was then compiled for the "Conservation Atlas of Tropical Forests: Asia and the Pacific" (Collins et al., 1991). The extent of tropical rain and monsoon forests was mapped uniformly for all countries of tropical Asia (including PNG), discriminating lowland, montane, swamp, and mangrove forest types and indicating heavily degraded forest areas. The main data source was national land-cover maps derived from visual interpretation of remote sensing data, including aerial photography, Landsat and SPOT imagery, and radar data in specific cases.

Since the 1990s, a continental overview of land cover was also generated in the context of global mapping initiatives. These maps were based on digital classification of satellite imagery of sensors of coarse spatial resolution (1 km–250 m), such as NOAA AVHRR (Hansen et al., 2000; Loveland et al., 1999), SPOT VEGETATION (Bartholomé and Belward, 2005), Terra MODIS (Hansen et al., 2005), and ENVISAT MERIS (Arino et al., 2008). These studies used coarse spatial resolution satellite data but with daily or almost daily acquisition of imagery, with a large swath needed for continental applications. The thematic legends of these maps were tuned for global applications; however, they provided a useful overview of the main land-cover pattern of tropical Asia at regional scales.

NOAA AVHRR imagery of 1-km spatial resolution from 1985/1986 and 1992/1993 was employed for land-cover assessment across continental tropical Asia (except India) in the context of the UNEP (United Nations Environmental Program) Land Cover Assessment and Monitoring project (Giri et al., 2003; UNEP EAP-AP, 1995). Techniques of NDVI (normalized difference vegetation index) compositing and NDVI slicing were used to select cloud-free pixels. Clustering and digital classification were applied to the visible, near infrared, and thermal bands of the AVHRR sensor, resulting in a map of main vegetation and cropland classes. Forest and vegetation types were further differentiated at the country level, adapting the land-cover legend as best as possible to the conditions of the individual countries. This resulted in a variation of national legends, displaying apart from agricultural and water surfaces, for example, for Laos, categories such as "moist mixed deciduous forests," "dry mixed deciduous forests," "scrubland," "savannah," and "woody or shrubby vegetation," whereas for Vietnam, the land-cover classes that were mapped consisted of "evergreen," "deciduous," and "mangrove forests" and "marshes" and "scrubland."

In the Joint Research Centre's TREES and GLC2000 (Global Land Cover 2000) projects, land-cover mapping of tropical Asia was implemented at the subregional level based on SPOT VEGETATION imagery (1-km spatial resolution) from 1998 to 2000 (Stibig et al., 2007). Image mosaics were generated, selecting pixels from the 10-day standard SPOT VEGETATION composites by minimum values in the near and shortwave infrared channels. The 10-day standard products were produced by pixel compositing based on the minimum NDVI pixel selection criterion. For continental Southeast Asia, all acquisitions of two dry seasons, and for insular Southeast Asia, all acquisitions of two complete years were needed to cope with cloud cover and to obtain subregional image composites of sufficient quality. These image composites served as an input to unsupervised classification, assigning land-cover classes after further subregional stratification in major landscape strata. For South Asia, a different approach was chosen. The approach was based on monthly NDVI mosaics, which were combined to a 9-month NDVI mosaic as an input to digital classification. The LCCS land-cover classification system (Di Gregorio and Jansen, 2000) was applied to describe individual classes in each subregional data set and to finally aggregate the classes to one uniform legend for the whole of tropical Asia. The continental land-cover map (Figure 18.2)

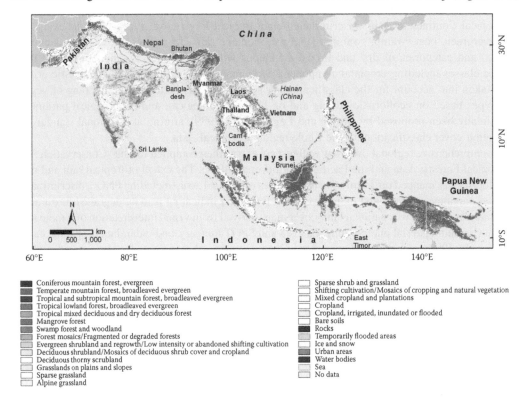

FIGURE 18.2 (See color insert.) Land-cover map of South and Southeast Asia, including PNG (GLC2000). (From FAO, *Forest Resources Assessment 1990—Tropical Countries*. FAO Forestry Paper 112, Rome, 1993. With permission.)

discriminated between boreal and temperate mountain forests, evergreen and mixed tropical low-land forests, and mangrove and swamp forests. Nonforest classes consisted of shrubland, grassland, cropland mosaics, and permanent cropland, separating upland and inundated cropland. The forest cover of tropical Asia was estimated to be about 310 million ha and cropland to be about 350 million ha; using Landsat TM data as a reference, the mapping accuracy for these two land-cover categories was assessed to be about 72%.

Subregional land-cover mapping initiatives are also important in providing insight to the regional or continental analysis. For example, the "Land cover map of the lower Mekong basin" (Laos, Cambodia, Thailand, Vietnam) was produced based on complete coverage of Landsat TM images of the years 1993 and 1997 (Figure 18.3). Thanks to the Mekong River Commission (MRC) program and the cooperation of the forestry departments of the four Mekong countries, forest and land cover was interpreted and delineated manually at a scale of 1:250000. The subregional map displayed evergreen, semievergreen, and deciduous forest-cover types, thereby indicating tree-cover density and fragmentation. Other land-cover classes mapped were woodlands, grasslands, and bamboo-dominated areas, as well as croplands, including two intensities of shifting cultivation mosaics and one category of permanent agriculture. Owing to the spatial and thematic detail, the land-cover map was used as an input to Mekong basin-wide programs developed by the MRC and as a national reference, for example, in Cambodia (MRC, 2003; Figure 18.3).

In South Asia, the forest-cover monitoring program implemented for India needs to be seen in a regional context due to the sheer size of the subcontinent. The forest cover of India has been mapped

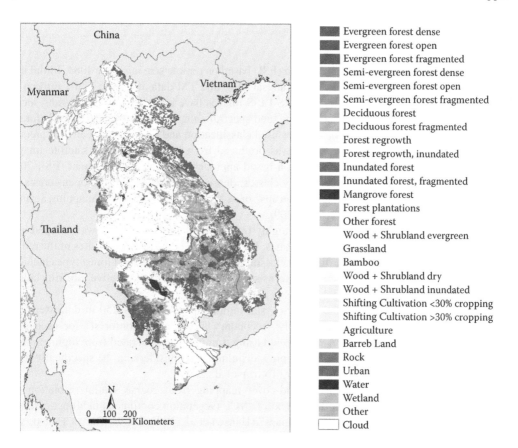

FIGURE 18.3 **(See color insert.)** Land-cover map of the Lower Mekong Basin. (From Martimort, P. et al. *Eur. Space Agency Bull.* 131, 19–23, 2007. With permission.)

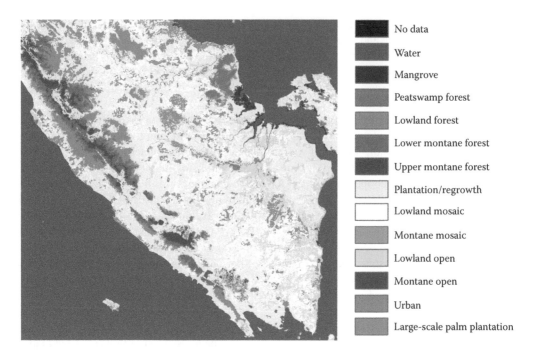

■	No data
■	Water
■	Mangrove
■	Peatswamp forest
■	Lowland forest
■	Lower montane forest
■	Upper montane forest
□	Plantation/regrowth
□	Lowland mosaic
■	Montane mosaic
■	Lowland open
■	Montane open
■	Urban
■	Large-scale palm plantation

FIGURE 18.4 2010 Land-cover map of insular Southeast Asia: details of Sumatra. (Courtesy of CRISP, National University of Singapore, Singapore.)

and monitored by the Forest Survey of India (FSI) based on remote sensing since 1984. Initial mapping started with visual interpretation of Landsat MSS and TM data at a scale of 1:1 million and 1:250000, respectively. Since the 1990s, IRS LISS images have been used. Preprocessing such as radiometric correction, geometric correction, and spectral enhancement was performed before the classification. A hybrid approach of unsupervised classification and visual on-screen interpretation was applied. The 2009 forest-cover map was produced from IRS LISS III P6 satellite imagery (~23.5-m spatial resolution) using a polygon-based approach for change assessment (FSI, 2009). Forest cover was mapped into three density classes: dense, moderate-dense, and open forests. All other land covers were mapped either as "scrubs" or as "nonforest." The overall mapping accuracy was assessed to be more than 92% (FSI, 2009).

The land cover of northeast India, with highly fragmented land-cover types, was mapped using IRC-C WiFS (188-m spatial resolution) mosaics. Monthly NDVI data and composites of the red and near infrared WiFS channels were digitally classified using K-means. Several forest types including degraded forests and typical land-cover patterns such as "abandoned" and "active" shifting cultivation could be mapped at satisfying detail (Roy and Joshi, 2002).

Similarly, for the mainland of Southeast Asia, multitemporal MODIS 250-m data were used to determine the fractions of mature forest, secondary forest, and "nonforest" for each pixel. A supervised regression tree model was applied using training data generated from high-resolution (ASTER, SPOT) satellite images. The mapping technique was considered to be specifically useful in inferring human imprints on forest cover (Tottrup et al., 2007).

For Indonesia, a new approach of land-cover mapping from coarse spatial resolution data (AVHRR, MODIS) was applied in the context of "VCF (vegetation continuous field) per cent tree-cover maps" and "forest-change indicator maps" (Hansen et al., 2003, 2009). The VCF maps were based on multitemporal metrics derived from coarse-resolution satellite data. Based on a continuous training data set, tree-cover information from high-resolution imagery was aggregated to the coarse spatial resolution data; the percentage of tree cover in the coarse spatial resolution pixel was then predicted by a regression tree algorithm. The MODIS "forest-change indicator map" was produced by a classification tree-bagging procedure using MODIS-scale "change" and "no-change" training sites and relating expert-interpreted forest-cover loss to the MODIS data. Although not displaying classical land-cover classes, these products provide a basis for further land-cover analysis and mapping. Both maps were used for stratification purposes, that is, selecting a stratified sample of Landsat image subsets for quantitative forest-change assessment. The study showed that the annual forest-cover loss in Indonesia for the period 2000–2005 was estimated at 0.71 million ha (Hansen et al., 2009).

The land cover of insular Southeast Asia was furthermore documented for the year 2010 based on the MODIS Daily Surface Reflectance Product at 250-m spatial resolution. Starting from unsupervised classification and five main land-cover categories, further refinement was done using ancillary map and elevation data, employing manual delineation techniques. The final 12 land-cover classes included lowland forests, lower and upper montane forests, mangrove and peat-swamp forests, regrowth, plantations, as well as land-cover mosaics and cropland categories. Large-scale palm plantations were mapped as a separate class using supplementary ALOS PALSAR data (Miettinen et al., 2010; Figure 18.4).

Because of the regional impact of the extensive burning of tropical forests in insular Southeast Asia in 1982/1983 and in 1997/1998, specific attention was drawn to the mapping of fires and forest burning in Borneo. For assessing the impact of the devastating 1997/1998 fires in the forests and the vegetation of Kalimantan, a multiscale remote sensing analysis was performed. Coarse and high-resolution optical imagery and radar satellite imagery were used along with ground and aerial surveys. By applying "difference detection techniques" to pairs of SAR (synthetic aperture radar) data of the Active Microwave Instrument (AMI, 25-m spatial resolution) of the European Remote Sensing Satellite (ERS-2), the area burnt in 1997/1998 was assessed and estimated at about 5.2 million ha (Siegert et al., 2001). SAR AMI imagery, Landsat TM data, and ground measurements of peat were also the basis for determining carbon emissions from peat and forest fires in Indonesia. Extrapolating the data for the whole region, the emissions caused by the burning of peat and vegetation in 1997 were estimated to range between 0.81 and 2.57 Gt (Page et al., 2002).

In Borneo (Indonesia), MODIS 250-m data from the years 2002 and 2005 were used to study forest change and the role of fires. Starting from eight initial land-cover classes, the forest cover was further stratified into three elevation categories based on SRTM elevation data. The final land-cover categories comprised lowland, upper and mountain evergreen forest, peat swamp and fresh-water swamp forest, mangrove forest, degraded forest and regrowth, cultivation forest mosaics, soils, grassland, and agricultural land. This land-cover information was combined with MODIS-derived hot-spot products of active fires at 1-km spatial resolution. The study estimated that the annual deforestation rate in Borneo was 1.7%; for the carbon-rich peat swamp forests, the rate was 2.2%. Additionally, 98% of all forest fires were detected within the 5-km buffer zone from the forest edges (Langner et al., 2007).

Continental-scale land-cover mapping in tropical Asia also benefits from the ongoing development of cloud-independent radar sensors. The Japanese Aerospace Exploration Agency (JAXA) has begun to provide mosaics of polarimetric ALOS PALSAR (phased array L-band synthetic aperture radar) imagery (spatial resolution 10–100 m) covering the region at least twice a year. Successful application of PALSAR data has been demonstrated in the land-cover mapping of Borneo. Mixture modeling and Markov Random Field classification techniques were applied to map seven forest types, two categories of woodland, two of shrubland, and two of grassland, as well as three cropland classes. In particular, vegetation formations influenced by water (daily or seasonally inundated or flooded), such as *Nipah* mangrove forests, open and very open peat swamp forests, and oil palm plantations, could be mapped as separate categories. An overall mapping accuracy of 85% was achieved compared to reference data from satellite imagery of high spatial resolution (Hoekman et al., 2010).

18.3 PERSPECTIVES

Reliable and timely information on land cover and land-cover change of tropical Asia is increasingly needed to better address evolving regional processes and environmental threats. For example, increasing and competing pressure on land use due to growing population and rising international demand for products from the region (e.g., palm oil) is contributing to the loss of natural forests, loss of biodiversity, and land degradation. Such information is needed to better manage natural resources of the region, including basin-wide watershed management (e.g., MRC, 2003; UNEP, 2002, 2007). The regular monitoring of forest cover will remain a key topic in tropical Asia not only because of environmental and biodiversity-related aspects but also because of the contribution of tropical deforestation to global carbon emissions (e.g., Achard et al., 2010; APFC, 2009; IPPC, 2000). Improved land-cover information can be expected from the ongoing development in satellite remote sensing. Coarse spatial resolution satellite data (e.g., MODIS 250 m or MERIS 300 m), covering the whole region with almost daily acquisitions, are available. This has increased the possibility of acquiring cloud-free imagery, particularly in insular Southeast Asia. Coarse spatial resolution satellite data together with high spatial resolution satellite data can provide a basis for regular and short-term forest and land-cover monitoring (e.g., Broich et al., 2011).

At the same time, availability and accessibility of free and new satellite data provide an opportunity for wall-to-wall regional land-cover mapping at high spatial detail. For example, Landsat satellite data are freely available (http://glovis.usgs.gov). Similarly, the SENTINEL-2 mission [European Space Agency (ESA) and European Commission, planned 2013] will provide multispectral imagery of 10-m spatial resolution with a swath of 290 km; the revisiting time will be only 5 days when the pair of satellites is operational (Martimort et al., 2007). There is increasing availability of satellite imagery of high spatial resolution within the region, for example, from the National Remote Sensing Centre of India (e.g., IRS, CARTOSAT satellites; NRSC, 2011) or from the Geo-Informatics and Space Technology Development Agency of Thailand (e.g., THEOS satellite; GISTDA, 2011). For frequently cloud-covered tropical Asia, progress can also be expected from new, cloud-independent microwave sensors, such as ALOS-2 (JAXA, planned 2013) with L-band imagery of very high (1 m–3 m) and high (3 m–10 m) spatial resolution (JAXA, 2011) or SENTINEL-1 (ESA, planned 2012) with a pair of C-band SAR sensors at fine spatial resolution (5 m × 20 m) and about a 6-day revisit time and biweekly global coverage (Attema et al., 2007). BIOMASS (ESA, planned ~ 2016) would add a P-band SAR, enhancing forest mapping and biomass measurement, both of specific interest for the region (ESA, 2008).

Recent advancement in computer technology and image-processing methodology provides an opportunity to analyze a large volume of data using complex algorithms. In future, it is possible to generate land-cover information at higher spatial and thematic resolutions on an operational basis.

REFERENCES

Achard, F., Stibig, H.-J., Eva, H.D., Lindquist, E., Bouvet, A., Arino, O., and Mayaux, P. 2010. Estimating tropical deforestation. *Carbon Manage*, 1(2), 271–287.

ADB and UNEP. 2004. *Greater Mekong Sub-Region Atlas of the Environment*. Asian Development Bank and United Nations Environmental Program. Manila, Philippines: ADB, p. 216.

APFC. 2009. *The Future of Forests in Asia and the Pacific: Outlook 2020*. Asia-Pacific Forestry Commission. Bangkok, Thailand: FAO, p. 600.

Arino, O., Bicheron, P., Achard, F., Latham, J., Witt, R., and Weber J.L. 2008. The most detailed portrait of Earth. *Euro Space Agency Bull*, 136, 24–31.

Attema, E., Bargellini, P., Edwards P., Guido, L., Svein, L., Ludwig, M., Betlem, R-T., et al. 2007. Sentinel-1 The Radar Mission for GMES Operational Land and Sea Services. *Euro Space Agency Bull*, 131, 11–17. Available at: http://www.esa.int/esapub/bulletin/bulletin131/bul131a_attema.pdf

Bartholomé, E. and Belward, A.S. 2005. GLC2000: A new approach to global land-cover mapping from earth observation data. *Int J Rem Sens*, 26(9), 1959–1977.

Blasco, F., Bellan, M.F., and Aizpuru, M. 1996. A vegetation map of tropical continental Asia at scale 1:5 million. *J Vegetat Sci*, 7, 623–634.

Broich, M., Hansen, M.C., Stolle, F., Potapov, P.V., Margono, B.A., and Adusei, B. 2011. Remotely sensed forest cover loss shows high spatial and temporal variation across Sumatera and Kalimantan, Indonesia 2000–2008. *Environ Res Lett*, 6(1), 9.

Collins, N.M., Sayer, J.A., and Whitmore, T.C. 1991. *The Conservation Atlas of Tropical Forests: Asia and the Pacific*. London, UK: Macmillan Press, p. 256.

Di Gregorio, A. and Jansen, L.J.M. 2000. *Land Cover Classification System—LCCS: Classification Concepts and User Manual*. Rome, Italy: FAO, p. 179.

ESA. 2008. BIOMASS—To observe global forest biomass for a better understanding of the carbon cycle: Report for assessment. *European Space Agency Special Publication* 1313/2. Available at: http://esamultimedia.esa.int/docs/SP1313-2_BIOMASS.pdf

FAO. 1989. *Classification and Mapping of Vegetation Types in Tropical Asia*. Rome, Italy: FAO, p. 169.

FAO. 2010. *Global Forest Resources Assessment 2010—Main Report*. FAO Forestry Paper 163. Rome, Italy: FAO, p. 340.

FSI. 2009. *India State of Forest Report 2009*. Dehradun, India: Forest Survey of India, p. 340.

Giri, C., Defourny, P., and Shrestha, S. 2003. Land cover characterization and mapping of continental Southeast Asia using multi-resolution satellite sensor data. *Int J Rem Sens*, 24(21), 4181–4196.

Giri, C., Ochieng, E., Tieszen, L.L., Zhu, Z., Singh, A. et al. 2011. Status and distribution of mangrove forests of the world using earth observation satellite data. *Global Ecol Biogeogr*, 20(1): 154–159.

GISTDA. 2011. *Geo-Informatics and Space Technology Development Agency*. Available at: http://www.gistda.or.th/gistda_n/

Hansen, M.C., DeFries, R.S., Townshend, J.R.G., and Sohlberg, R.A. 2000. Global land cover classification at 1 km spatial resolution using a classification tree approach. *Int J Rem Sens*, 21(6–7), 1331–1364.

Hansen, M.C., DeFries, R.S., Townshend, J.R.G., Caroll, M., Dimiceli, C., and Sohlberg, R.A. 2003. Global percent tree cover at a spatial resolution of 500 meters: First results of the MODIS vegetation continuous fields algorithm. *Earth Interact*, 7(10), 1–15.

Hansen, M.C., Townshend, J.R.G., DeFries, R.S., and Carroll, M. 2005. Estimation of tree cover using MODIS data at global, continental and regional/local scales. *Int J Rem Sens*, 26(19), 4359–4380.

Hansen, M.C., Stehman, S.V., Potapov, P.V., Arunarwati, B., Stolle, F., and Pittman, K. 2009. Quantifying changes in the rates of forest clearing in Indonesia from 1990 to 2005 using remotely sensed datasets. *Environ Res Lett*, 4, 12.

Hoekman, D., Vissers, M., and Wielaard, N. 2010. PALSAR wide-area mapping of Borneo: Methodology and map validation. *IEEE J Select Topics Appl Earth Observ Rem Sens*, 3(4), 605–617.

IPPC. 2000. *Land Use, Land Use Change, and Forestry*. Special Report. International Panel on Climate Change. Cambridge, UK: Cambridge University Press, p. 377.

JAXA. 2011. ALOS-2: The Advanced Land Observing Satellite-2. Available at: http://www.jaxa.jp/pr/brochure/pdf/04/sat29.pdf

Langner, A., Miettinen, J., and Siegert, F. 2007. Land Cover Change 2002–2005 in Borneo and the role of fire derived from MODIS imagery. *Global Change Biol*, 13, 2329–2340.

Loveland, T.R., Zhu, Z., Ohlen, D.O., Brown, J.F., Reed, C., and Yang, L. 1999. An analysis of the IGBP global land-cover characterization process. *Photogramm Eng Rem Sens*, 65, 9, 1021–1032.

Martimort, P., Berger, M., Carnicero, B., Umberto, D.B., Valérie, F., Ferran, G., Pierluigi, S., et al. 2007. Sentinel-2: The optical high-resolution mission for GMES operational services. *Euro Space Agency Bull*, 131, 19–23. Available at: http://www.esa.int/esapub/bulletin/bulletin131/bul131b_martimort.pdf

Miettinen, J., Shi, C., Tan, W.J., and Liew, S.C. 2010. 2010 land cover map of insular Southeast Asia in 250m spatial resolution, *Rem Sens Lett*, 3(1), 11–20.

MRC. 2003. *People and the Environment Atlas of the Lower Mekong Basin*. CD ROM . Phnom Penh, Cambodia: Mekong River Commission.

NRSC. 2011. *National Remote Sensing Centre*. Available at: http://www.nrsc.gov.in/index.html

Page, S.E., Siegert, F., Rieley, J.O., Boehm, H-D., Jaya, A., and Limin, S. 2002. The amount of carbon released from peat and forest fires in Indonesia during 1997. *Nature*, 420, 61–65.

Roy, P.S., and Joshi, P.K. 2002. Forest Cover Assessment in northeast India—The potential of temporal wide swath satellite sensor data (IRS-1 WiFS). *J Rem Sens*, 23(22), 4881–4896.

Siegert, F., Rücker, G., Hinrichs, A., and Hoffmann, A. 2001. Increased damage from forest fires in logged over forests during droughts caused by El Niño. *Nature*, 414, 437–440.

Stibig, H.J., Belward, A.S., Roy, P.S., Rosalina-Wasrin, U., Agrawal, S., Joshi, P.K., Hildanus, et al. 2007. A land-cover map for South and Southeast Asia derived from SPOT-VEGETATION data. *J Biogeogr*, 34, 625–637.

Tottrup, C., Rassmussen, M.S., Eklundh, L., and Jönsson, P. 2007. Mapping fractional forest cover across the highlands of mainland Southeast Asia using MODIS data and regression tree modelling. *Int J Rem Sens*, 28(1), 23–46.

UNEP EAP-AP. 1995. *Land Cover Assessment and Monitoring*, Vol. 1a-10a. Bangkok, Thailand: UNEP.

UNEP. 2002. *Global Environment Outlook-3. Past, Present and Future Perspectives*. Report, United Nations Environment Programme, Nairobi. London, UK: Earthscan Publications, p. 540.

UNEP. 2007. *Global Environment Outlook-4. Environment for Development*. Report, United Nation as Environment Programme, Nairobi. Malta: Progress Press, p. 426.

Whitmore, T.C. 1984a. *Tropical Rain Forests of the Far East*. Oxford, UK: Clarendon Press, p. 352.

Whitmore, T.C. 1984b. A vegetation map of Malesia at scale 1:5 million. *J Biogeogr*, 11, 461–471.

Worldclimate. 2011. *Worldclimate*. Available at: http://www.worldclimate.com/

19 Land Cover and Its Change in Europe: 1990–2006

Jan Feranec, Tomas Soukup, Gerard Hazeu, and Gabriel Jaffrain

CONTENTS

19.1 INTRODUCTION

The foundation for progressive monitoring of the European land cover (LC) and its changes was laid by the Co-Ordination of Information on the Environment (CORINE) program approved by the European Commission on June 27, 1985. The aim was to provide compatible environmental data for the European countries (Heymann et al., 1994). Nowadays, there are numerous activities aimed at meeting this goal. Among them is the CORINE Land Cover (CLC) project, which seeks to generate a digital database of the European LC/land use (LU) and its changes. So far, three projects have been realized: CLC1990, Image 2000 & CLC2000 (I&CLC2000), and CLC2006 (under the Global Monitoring for Environment and Security—GMES umbrella). The first project was launched by the European Commission in 1985, the second by the European Environment Agency (EEA) and Joint Research Centre (JRC) of the European Commission in 2000, and the third was part of the implementation of GMES in the 2006 fast-track service on land monitoring (Büttner et al., 2004; EEA-ETC/LUSI, 2007; Feranec et al., 2007a; Steenmans and Perdigao, 2001). More, recently, there are ongoing preparatory activities for the next update in 2012 under GMES Initial Operation (GIO) Land framework.

Thanks to these activities, a complete picture of LC and its changes in Europe can be provided in a consistent way. With more countries joining the activity, the CLC coverage has evolved from 1990 onward and has gradually expanded (see coverage for databases for different reference years in Table 19.2). Nevertheless, for some countries the most recent data (CLC2006 data) were still

TABLE 19.1

Evolution of the CLC Projects

	CLC1990 Specifications	CLC2000 Specifications	CLC2006 Specifications
Satellite data	Landsat-4/5 TM single date (in a few cases Landsat MSS, as well)	Landsat-7 ETM single date	SPOT-4 and/or IRS LISS III two dates
Time consistency	1986–1998	2000 ± 1 year	2006 ± 1 year
Geometric accuracy of satellite images	≤50 m	≤25 m	≤25 m
CLC minimum mapping unit	25 ha	25 ha	25 ha
Geometric accuracy of CLC data	100 m	Better than 100 m	Better than 100 m
Thematic accuracy	≥85% (not validated)	≥85% (validated; see Büttner and Maucha, 2006)	≥85%
Change mapping	NA	Boundary displacement minimum 100 m; change area for existing polygons ≥ 5 ha; isolated changes ≥ 25 ha	Boundary displacement minimum 100 m; *all* changes > 5 ha have to be mapped
Production time	10 years	4 years	1.5 years
Documentation	Incomplete metadata	Standard metadata	Standard metadata
Access to the data	Unclear dissemination policy	Free access	Free access
Number of European countries involved	27	39	37

Source: EEA-ETC/LUCI, CLC2006 technical guidelines, EEA Technical Report 17. Office for Official Publications of the European Communities, Luxembourg, 2007. Available at: http://www.eea.europa.eu/publications/technical_report_2007_17. With permission.

not available when the manuscript of this chapter was being prepared (Greece, Switzerland, and Great Britain). Tables 19.1 and 19.2 demonstrate the basic characteristics of the aforementioned projects. The CLC project was also realized in the French overseas department (including Guyana, Guadeloupe, Martinique, and Isle of la Reunion) and was supervised by the Institut Geographique National France International (IGN FI) in 2008. Furthermore, IGN FI has implemented several CLC projects in different biogeographical regions in collaboration with national institutions in Burkina Faso, Colombia, and Central America, in which the CLC nomenclature has been followed and adapted according to the dominant agricultural and natural landscape of those regions (Jaffrain et al., 2005).

Five data layers were derived under the aforementioned projects: CLC1990 (frequency and area of LC classes in 27 states from the 1990s), CLC2000 (±1 year in 39 states), CLC1990/2000 (LC changes for 10 years), CLC2006 (±1 year in 37 states), and CLC2000/2006 (LC changes for a 6-year period). The methodology of deriving these data and the characterization of LC in Europe and its changes during 1990–2006 are the theme of this chapter.

19.2 CORINE LAND COVER NOMENCLATURE

The CLC nomenclature is based mainly on physiognomic attributes and spatial relationships of landscape objects, for instance, the attribute of association. The natural, modified/cultivated, and artificial landscape objects are characterized by physiognomic attributes such as shape, size, color,

TABLE 19.2
Participants in the CLC Projects

Country	CLC1990	Change 1990/2000	CLC2000	Change 2000/2006	CLC2006
Albania	No	No	Yes	Yes	Yes
Austria	Yes	Yes	Yes	Yes	Yes
Belgium	Yes	Yes	Yes	Yes	Yes
Bosnia/Herzegovina	No	No	Yes	Yes	Yes
Bulgaria	Yes	Yes	Yes	Yes	Yes
Serbia	Yes	Yes	Yes	Yes	Yes
Cyprus	No	No	Yes	Yes	Yes
Czech Republic	Yes	Yes	Yes	Yes	Yes
Germany	Yes	Yes	Yes	Yes	Yes
Denmark	Yes	Yes	Yes	Yes	Yes
Estonia	Yes	Yes	Yes	Yes	Yes
Spain	Yes	Yes	Yes	Yes	Yes
Finland	No	No	Yes	Yes	Yes
France	Yes	Yes	Yes	Yes	Yes
Greece	Yes	Yes	Yes	No[a]	No[s]
Croatia	Yes	Yes	Yes	Yes	Yes
Hungary	Yes	Yes	Yes	Yes	Yes
Switzerland	No	No	No[a]	No[a]	No[a]
Ireland	Yes	Yes	Yes	Yes	Yes
Iceland	No	No	Yes	Yes	No[a]
Italy	Yes	Yes	Yes	Yes	Yes
Kosovo	No	No	Yes	Yes	Yes
Liechtenstein	Yes	Yes	Yes	Yes	Yes
Lithuania	Yes	Yes	Yes	Yes	Yes
Luxembourg	Yes	Yes	Yes	Yes	Yes
Latvia	Yes	Yes	Yes	Yes	Yes
Monte Negro	Yes	Yes	Yes	Yes	Yes
Macedonia FYR	No	No	Yes	Yes	Yes
Malta	Yes	Yes	Yes	Yes	Yes
Netherlands	Yes	Yes	Yes	Yes	Yes
Northern Ireland	No	Yes	Yes	Yes	Yes
Norway	No	No	Yes	Yes	Yes
Poland	Yes	Yes	Yes	Yes	Yes
Portugal	Yes	Yes	Yes	Yes	Yes
Romania	Yes	Yes	Yes	Yes	Yes
Sweden	No	No	Yes	Yes	Yes
Slovenia	Yes	Yes	Yes	Yes	Yes
Slovakia	Yes	Yes	Yes	Yes	Yes
Turkey	No	No	Yes	Yes	Yes
Great Britain	No	Yes	Yes	No[a]	No[a]
Total	**27**	**29**	**39**	**37**	**37**

[a] In process, but not available for the study.

TABLE 19.3

Physiognomic Attributes Relevant for Identification of CLC Classes

Urban fabric areas	Size, shape, and density of the buildings, share of supplementing parts of the class (e.g., square, width of the streets, gardens, urban greenery parking lots), character of transport network, size and character of neighboring water bodies, arrangement of infrastructure, size of quays, character of the runway surfaces, state of the dumps, and arrangement and share of playgrounds and sport halls
Agricultural areas	Share of dispersed greenery within agricultural land, arrangement and share of areas of permanent crops, relationships of grasslands with urban fabric, occurrence of dispersed houses (cottages), arrangement and share of agricultural land (arable land), grasslands, permanent crops and natural vegetation (mainly trees and bushes), irrigation channel network
Forest and seminatural areas	Character (composition), developmental stage and arrangement of vegetation (mainly trees and bushes), share of grass and dispersed greenery (composition density)
Wetlands	Character of substrate, water, and vegetation
Water bodies	Character (shape) of water bodies

Source: Feranec, J. et al., *Land Use Policy*, 24, 234–247, 2007a. With permission.

texture, and pattern (Feranec, 1999). These attributes (see Table 19.3) are crucial for identifying LC classes on satellite images by visual or computer-aided visual interpretation (Feranec et al., 2007a).

LC represents the biophysical state of the real landscape, which means that it consists of natural and also modified and artificial objects, whereas LU refers to the purpose for which land is used (function) (Meyer and Turner, 1994). On the one hand, artificial surfaces, intensively used arable land, or permanent crops are LC, but the term also indicates their LU, that is, their societal function. On the other hand, the nature and appearance of the natural or seminatural objects do not mean that they are not used or that they have no function. This explains why the nomenclature does not consistently separate LC from LU. However, the functional aspect is difficult to identify visually, especially from remote-sensing data.

The CLC nomenclature (Table 19.4) is hierarchical with three class levels characterized by the following attributes:

- The first level contains 5 items and addresses the major categories of LC in Europe.
- The second level contains 15 items and addresses scales 1:500,000–1:1,000,000.
- The third level contains 44 items and addresses the scale 1:100,000 (Heymann et al., 1994).

19.3 METHODOLOGY

The CLC1990 data layer represents the LC of the 27 European countries in the 1990s, which was produced by the method of visual interpretation of single-date images of Landsat TM (in a few cases, Landsat MSS) and SPOT2/3 (in France; see Table 19.1). Generation of the CLC2000 and CLC2006 data layers was based on the *updating* approach (the opposite of *backdating*; see Figure 19.1) by computer-aided visual interpretation of single-date Landsat7 ETM images and two-date SPOT4 and/or IRS LISS III images (see Table 19.1). The primary purpose of *updating* was to minimize the chance of introducing inaccuracies of changes into the data layer. The independent generation of data layers for two-time horizons may result in inaccurate drawing of the same LC class borders in one or both data layers. Consequently, such technical differences between both data layers do not reflect real changes (Comber et al., 2004; de Zeeuw and Hazeu, 2001; Feranec et al., 2007a; Foody, 2002; Fuller et al., 2003; Khorram, 1999; van Oort, 2005). Instead of generating a completely new data layer, the approach used the copy of the *corrected* CLC1990 and CLC2000 *template data layers* as the initial CLC2000 and CLC2006 data layers for updating (Feranec et al., 2007a).

TABLE 19.4
The CLC Nomenclature

1 Artificial surfaces	**3 Forest and seminatural areas**
1.1 Urban fabric	*3.1 Forests*
1.1.1 Continuous urban fabric	3.1.1 Broad-leaved forests
1.1.2 Discontinuous urban fabric	3.1.2 Coniferous forests
1.2 Industrial, commercial, and transport units	3.1.3 Mixed forests
1.2.1 Industrial or commercial units	*3.2 Scrub and/or herbaceous vegetation associations*
1.2.2 Road and rail networks and associated land	3.2.1 Natural grasslands
1.2.3 Port areas	3.2.2 Moors and heathland
1.2.4 Airports	3.2.3 Sclerophyllous vegetation
1.3 Mine, dump, and constructions sites	3.2.4 Transitional woodland-scrub
1.3.1 Mineral extraction sites	*3.3 Open spaces with little or no vegetation*
1.3.2 Dumpsites	3.3.1 Beaches, dunes, sands
1.3.3 Construction sites	3.3.2 Bare rocks
1.4 Artificial, nonagricultural vegetated areas	3.3.3 Sparsely vegetated areas
1.4.1 Green urban areas	3.3.4 Burnt areas
1.4.2 Sport and leisure facilities	3.3.5 Glaciers and perpetual snow
2 Agricultural areas	**4 Wetlands**
2.1 Arable land	*4.1 Inland wetlands*
2.1.1 Nonirrigated arable land	4.1.1 Inland marshes
2.1.2 Permanently irrigated land	4.1.2 Peat bogs
2.1.3 Rice fields	*4.2 Maritime wetlands*
2.2 Permanent crops	4.2.1 Salt marshes
2.2.1 Vineyards	4.2.2 Salines
2.2.2 Fruit trees and berry plantations	4.2.3 Intertidal flats
2.2.3 Olive groves	**5 Water bodies**
2.3 Pastures	*5.1 Inland waters*
2.3.1 Pastures	5.1.1 Water courses
2.4 Heterogeneous agricultural areas	5.1.2 Water bodies
2.4.1 Annual crops associated with permanent crops	*5.2 Marine waters*
2.4.2 Complex cultivation patterns	5.2.1 Coastal lagoons
2.4.3 Land principally occupied by agriculture, with significant areas of natural vegetation	5.2.2 Estuaries
2.4.4 Agro-forestry areas	5.2.3 Sea and ocean

Source: Heymann, Y. et al., *CORINE Land Cover. Technical Guide*, Office for Official Publications European Communities, Luxembourg, 1994; Bossard, M. et al., CORINE Land-Cover Technical Guide—Addendum 2000, Technical Report 40. European Environment Agency, Copenhagen, 2000. Available at: http://www.eea.europa.eu/data-and-maps/data/corine-land-cover-clc1990-250-m- version-06-1999/corine-land-cover-technical-guide-volume-2. With permission.

All needed modifications were done on these initial CLC2000 and CLC2006 data layers, which were altered only locally in areas of the identified LC changes. The common boundaries of all unchanged areas were maintained without any modifications. The basic principle of this approach is shown in Figure 19.1 (Feranec et al., 2007a). The referential layer, copy of which (the template) is modified by updating or backdating subject to the changes of LC shape, for instance, in the $T + 1$ time horizon or $T - 1$, and so on, is in red.

Figure 19.2 shows an example of CLC2000 layer derivation. The CLC2006 layer was derived in a similar way. This approach reduced to a minimum the generation of spatial discrepancies during

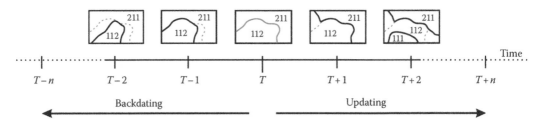

FIGURE 19.1 Basic principles of updating and backdating. (From Feranec, J. et al., Photo-to-photo interpretation manual (Revised), BIOPRESS Document, Biopress-d-13-1.3, Institute of Geography, Slovak Academy of Sciences, Bratislava, 2005. With permission.)

identification and interpretation of CLC classes. The approaches A and B shown in this figure were specified by Feranec et al. (2007a). Note that owing to the focus on changes, most of the countries applied approach B for deriving the CLC2006 layer.

Identification respected the requirement of spatial characteristics of the generated data layers in terms of the CLC project methodology (Heymann et al., 1994); for example, the resulting new area would comply with the criteria of the minimum area of 25 ha and minimum width of 100 m (see Table 19.1).

The identified change in the CLC2000 and CLC2006 data layers was accepted only if the total change area was larger than 5 ha and the width of change was ≥ 100 m (see Table 19.1). CLC change shows a categorical change, when one LC class or its part(s) is replaced by other class(es) (cf. Coppin et al., 2004).

All technical details of these approaches are quoted in technical guidelines (EEA-ETC/TE 2002), CLC2006 technical guidelines (EEA-ETC/LUSI 2007), and in the works of Feranec et al. (2007a, 2007b).

19.3.1 Conversion of LC Change

As LC is indivisible from the landscape, it reflects its states in different stages of development. This is why LC changes can be regarded as a relevant information source about processes (flows) in the landscape. The methodology applied in Haines-Young and Weber's study (2006: 9) categorizes LC changes into LC flows (LCF) on the basis of the second CLC data level and presents the spatial aspects of LC changes through LCF intensity maps. The analysis in this chapter of LC changes in European landscapes between 1990 and 2006 makes use of this approach.

The main LCFs for the second level of CLC classes have been derived by means of the conversion table (Table 19.5). This table, or in other words "matrix of flows," groups LC changes of the same type. There are $15 \times 14 = 210$ possible combinations of one-to-one changes between the 15 CLC classes at the second level (Feranec at al., 2010).

The table summarizes aggregation of LC changes at the second CLC level with codes 11–52 (see Table 19.4 for explanation). The CLC changes that took place between two time horizons were grouped into seven processes going on in landscape—(1) urbanization (industrialization), (2) intensification of agriculture, (3) extensification of agriculture, (4) afforestation, (5) deforestation, (6) construction and management of water bodies, (7) other changes (recultivation, dumpsites, unclassified changes, etc.)—were not taken into consideration in the LCF context.

The changes, grouped into LCFs, represent seven major LU processes (see Table 19.5):

- (LCF1) Urbanization: a flow that represents the change of agricultural (classes 21, 22, and 23) and forestland (classes 31, 32, and 33), wetlands (classes 41 and 42), and water bodies (51 and 52) into urbanized land (construction of buildings for living, education, healthcare,

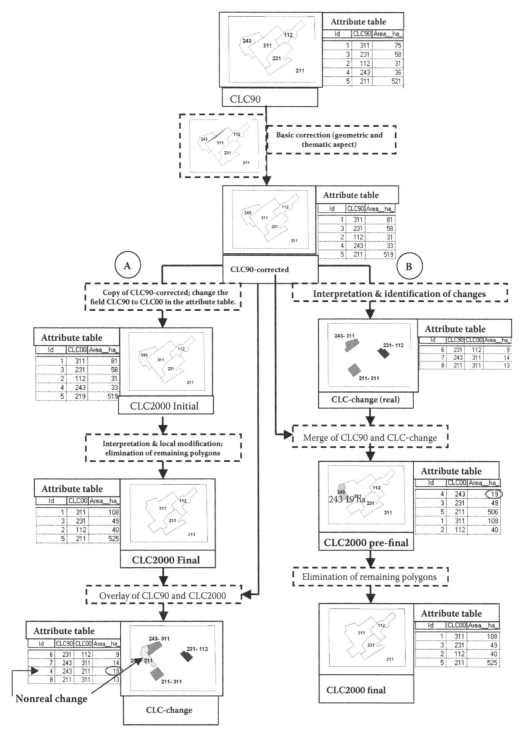

FIGURE 19.2 Generation of the CLC2000 by the computer-aided visual interpretation method: two different approaches—A and B. (From Feranec, J. et al., *Land Use Policy*, 24, 234–247, 2007a. With permission.)

TABLE 19.5

Conversion Table

		2000s Classes													
	11	12	13	14	21	22	23	24	31	32	33	41	42	51	52
11	0	7	7	7	7	7	7	7	7	7	7	7	7	7	7
12	7	0	7	7	7	7	7	7	7	7	7	7	7	7	7
13	7	7	0	7	7	7	7	7	7	7	7	7	7	6	7
14	7	7	7	0	7	7	7	7	7	7	7	7	7	6	7
21	1	1	1	1	0	2	3	3	4	4	7	7	7	6	7
22	1	1	1	1	3	0	3	3	4	4	7	7	7	6	7
23	1	1	1	1	2	2	0	2	4	4	7	7	7	6	7
24	1	1	1	1	2	2	3	0	4	4	7	7	7	6	7
31	1	1	1	1	5	5	5	5	0	5	5	5	7	6	7
32	1	1	1	1	2	2	2	2	4	0	5	7	7	6	7
33	1	1	1	1	2	2	2	2	4	4	0	7	7	6	7
41	1	1	1	1	2	2	2	2	4	4	7	0	7	6	7
42	1	1	1	1	2	2	2	2	4	4	7	7	0	6	7
51	1	1	1	1	7	7	7	7	4	4	7	7	7	0	7
52	1	1	1	1	7	7	7	7	4	4	7	7	7	7	0

(Row labels under "1990s Classes")

Note: 1—Urbanization (industrialization), 2—intensification of agriculture, 3—extensification of agriculture, 4—afforestation, 5—deforestation, 6—construction and management of water bodies, 7—other changes (recultivation, dump sites, unclassified changes, etc., were not taken into consideration in LCF context).

Source: Feranec, J., *Applied Geography*, 30, 19–35, 2010.

recreation, and sports) as well as industrialized land (construction of facilities for production, all forms of transport, and electric-power generation; see Table 19.5).

- (LCF2) Intensification of agriculture: a flow that represents the transition of LC types associated with lower intensity use (e.g., from the natural area—classes 32, 33, except forest class 31 and wetland—class 4) into higher intensity use (see Table 19.5).
- (LCF3) Extensification of agriculture: a flow that represents the transition of the LC type, associated with a higher intensity use (classes 21 and 22) to lower intensity use (classes 23 and 24; see Table 19.5).
- (LCF4) Afforestation: a flow that represents forest regeneration—establishment of forests by planting and/or natural regeneration (change of classes 21, 22, 23, 24, 33, 41, 42 into classes 31 and 32; see Table 19.5).
- (LCF5) Deforestation: a flow involving forestland (class 31) changes into another LC or damaged forest (classes 21, 22, 23, 24, 32, 33, and 41, e.g., the tree canopy falls below a minimum percentage threshold of 30%; see Table 19.5).
- (LCF6) Construction and management of water bodies: a flow involving the change of mainly agricultural (classes 21, 22, 23, and 24) and forestland (classes 31 and 32) into water bodies (see Table 19.4).
- (LCF7) Other changes: changes caused by various anthropic activities such as recultivation of former mining areas, dumpsites, unclassified changes, etc. (see Table 19.5). More detailed characteristics of LCFs have been quoted by Feranec et al. (2010).

Individual areas of LC change are mostly too small to be presented in the form of an overview map at national or even European levels; therefore, an LCF mean value summarizing individual changes in a regular grid pattern is often used (cf. Feranec et al., 2000). Moreover, this approach allows presenting the intensity of particular LCFs. In this chapter, the 3 × 3 km grid has been selected for presentation. The mean LCF value in Figures 19.4 through 19.10 is calculated by summing up all areas within the 3 × 3 km squares that are characterized by a specific LCF divided by the number of 3 × 3 km squares in which such changes took place.

The mean LCF value is only a theoretical value, which facilitates marking the squares with specific changes that occurred. For instance, the LCF mean value for urbanization is 19.73 ha. Figure 19.4 indicates which squares have total areas of changes above or below the mean value of that specific LCF. It is important to emphasize that the area of the square is 900 ha (3 × 3 km). It means that within these square plots, a contrary process may have taken place as well; in other words, in the same square a more pronounced intensification or other processes may have occurred (cf. Feranec et al., 2010).

CLC data adapted in this way are considered suitable for quick presentation and assessment of European LC change.

19.4   LAND COVER OF EUROPE

The most recent CLC2006 (see Table 19.6 and Figure 19.3) data layer, an LC seamless mosaic of 37 countries, provides the topical view of the European LC at the scale 1:100,000 (Table 19.2). Statistics of CLC2006 classes and the map containing *CLC2006 of Europe* (Figure 19.3) are important contributions to the knowledge of the present structure of European landscapes.

TABLE 19.6

Statistical Characteristics of the CLC1990, CLC2000, and CLC2006 Data Layers of European Countries

CLC Classes	1990		2000		2000[a]		2006	
2nd Level	Total Area (in ha)	Share (in %)	Total Area (in ha)	Share (in %)	Total Area (in ha)	Share (in %)	Total Area (in ha)	Share (in %)
11	13,165,947	3.57	13,600,646	3.69	14,191,243	2.62	14,434,426	2.66
12	2,201,588	0.60	2,483,112	0.67	2,611,810	0.48	2,815,939	0.52
13	758,133	0.21	829,062	0.22	890,168	0.16	1,017,529	0.19
14	956,995	0.26	1,045,460	0.28	959,574	0.18	1,010,670	0.19
21	108,093,487	29.29	107,435,829	29.11	123,482,430	22.77	123,075,745	22.69
22	10,183,129	2.76	10,217,414	2.77	10,578,022	1.95	10,698,511	1.97
23	36,540,946	9.90	36,403,505	9.87	32,242,530	5.94	32,132,941	5.92
24	48,518,949	13.15	48,396,919	13.12	62,917,859	11.60	62,785,819	11.58
31	92,844,629	25.16	92,902,765	25.18	160,845,794	29.65	158,610,466	29.24
32	41,178,939	11.16	41,182,765	11.16	73,193,247	13.49	75,371,854	13.90
33	5,168,365	1.40	5,124,121	1.39	34,207,227	6.31	34,142,735	6.29
41	3,246,276	0.88	3,140,234	0.85	10,795,202	1.99	10,745,915	1.98
42	1,406,269	0.38	1,409,255	0.38	1,201,575	0.22	1,207,431	0.22
51	3,903,355	1.06	4,004,217	1.09	13,540,085	2.50	13,605,441	2.51
52	844,999	0.23	836,702	0.23	760,320	0.14	761,664	0.14
Total	369,012,006	100.00	369,012,006	100.00	542,417,086	100.00	542,417,086	100.00

[a] New countries have joined the CLC program, so the total area mapped in 2000–2006 was considerably enlarged (see Table 19.2).

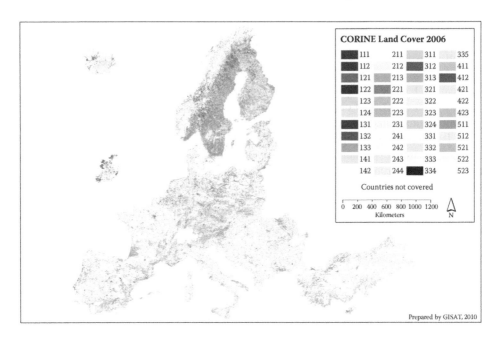

FIGURE 19.3   (See color insert.) The spatial distribution of 44 CLC classes of Europe for the year 2006.

CLC classes defining the agricultural landscape clearly dominate in Europe (classes 21, 22, 23, and 24; see Table 19.4). They cover 228,693,016 ha in 37 countries and represent 42.2% (Table 19.6) of their surface. Areas of these classes are most numerous in lowland and hilly landscapes. Figure 19.3 gives an idea of the distribution of agricultural landscape areas.

Forests (class 31) are at the second position (they occupy 158,610,466 ha; it means 29.2% of the area of 37 countries [Table 19.6]). They occur in mountain ranges (Figure 19.3), but also in lowland and hilly landscapes. This class does not include *transitional woodland/shrubs* (324) represented either by woodland degradation or by forest regeneration/colonization, which along with *natural grasslands* (321), *moors and heathland* (322), and *sclerophyllous vegetation* (323) are part of shrub *and/or herbaceous vegetation associations* with an overall surface of 75,371,854 ha (13.9%). Classes 322 and 323 represent bushy vegetation in the "climax" stage of development.

Open spaces with little or no vegetation (class 33), which cover, for example, beaches, dunes, sand plains, bare rocks, sparsely vegetated areas, burnt areas, glaciers, and perpetual snow, occupy 34,142,735 ha (6.3%), and they are another class in the European LC mosaic that occurs in the range from high mountains to the sea or coasts (see Table 19.6 and Figure 19.3).

Artificial surfaces (1) including the areas of classes *urban fabric* (11); *industrial, commercial, and transport units* (12); *mine, dump, and construction sites* (13); and *artificial, nonagricultural vegetated areas* (14) (see Table 19.6 and Figure 19.3) with a total area of 19,278,564 ha (3.6%) are also part of the European LC.

The two last classes are *water bodies* (5) occupying 14,367,105 ha (2.6%) and *wetlands* (4) with a surface of 11,953,346 ha (2.2%) (see Table 19.6 and Figure 19.3). The third level CLC nomenclature is explained in Table 19.4.

19.5   CHANGES IN EUROPEAN LC: 1990–2000–2006

In the wake of various socioeconomic and natural processes, landscape constantly changes, and most of these changes become visible precisely through LC changes. Their transformation into LCFs makes their interpretation easier in the wider context of processes taking place in landscape.

Approximate areas of LCFs are demonstrated through LC changes, particularly their LCF values and their spatial distribution in 37 European countries during 2000–2006 by maps (Figures 19.4 through 19.10).

19.5.1 URBANIZATION (LCF1)

The urbanization process manifests itself in this approach by enlargement of artificial surfaces (CLC classes 11×, 12×, 13×, and 14×), that is, construction of new buildings, industries, etc. (Feranec et al., 2010). Results in Table 19.7 represent an acceleration of the enlargement of urban fabric with an average of 16,338 ha per year during 2000–2006. The mean LCF1 value during 2000–2006 reached 19.73 ha, and its spatial distribution is represented in Figure 19.4. LCF1 distinctly manifests itself,

TABLE 19.7
European Landscape Changes in 1990–2006

	Area of Change (in ha)				Difference	Difference
LCF	**1990–2000**	**Yearly in 1990–2000**	**2000–2006**	**Yearly in 2000–2006**	**(in ha/year)**	**(in %/year)**
Urbanization	980,620	98,062	686,397	114,400	16,338	+17
Intensification of agriculture	1,500,111	150,011	467,969	77,995	−72,016	−48
Extensification of agriculture	1,302,440	130,244	154,662	25,777	−104,467	−80
Afforestation	2,651,582	265,158	1,540,977	256,830	−8329	−3
Deforestation	2,079,170	207,917	3,405,409	567,568	359,651	+173
Construction and management of water bodies	121,199	12,120	93,872	15,645	3525	+29
Other changes	215,428	21,543	222,901	37,150	15,607	+72

Note: + indicates increase and −indicates decrease. Total mapped area has changed in periods (see Table 19.6).

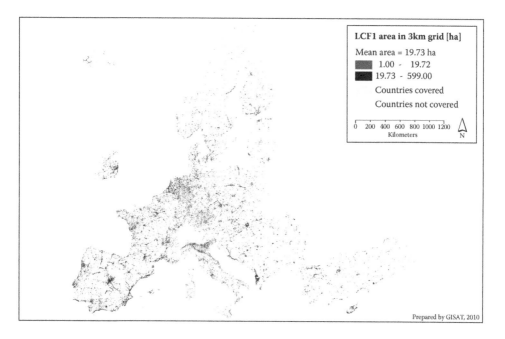

FIGURE 19.4 **(See color insert.)** Spatial distribution of urbanization in European countries in 2000–2006.

for instance, in the eastern part of Ireland, western part of The Netherlands, along the River Po in northern Italy, in central and eastern Spain, western Albania, the northern parts of Hungary, and so on. Results obtained under the BIOPRESS, LACOAST, and MOLAND/MURBANDY (Gerard et al., 2010) projects also confirm enlargement of urban sprawl in Europe.

Conversion of landscape into areas of artificial surfaces is most prominent in eastern Ireland and in western part of the Netherlands, along the River Po in Italy, and in central and eastern Spain.

19.5.2 Intensification of Agriculture (LCF2)

Table 19.5 contains changes of CLC classes resulting from agricultural intensification. The extent of changes contributing to the intensification decreased during 2000–2006, when compared with the years 1990–2000, on an average with 72,016 ha a year (see Table 19.7). The mean LCF2 value reached 31.69 ha, and its spatial distribution is evident in southwestern Finland, central part of Estonia, northwestern Germany, northeastern Hungary, southern Spain, and so on (Figure 19.5).

Changes in favor of intensification of agriculture are most prominent in southwestern Finland, central part of Estonia, northwestern Germany, northeastern Hungary, and southern Spain.

19.5.3 Extensification of Agriculture (LCF3)

Changes in agricultural landscape in favor of extensification are shown in Table 19.5. Areas that changed in favor of extensification of agriculture strongly decreased by an average of 104,467 ha (see Table 19.7) a year compared with the 1990–2000 period. The mean LCF3 value reached 31.02 ha, and its spatial distribution is clearly evident in southwestern, central, and northeastern parts of the Czech Republic and less in the eastern part of Hungary, southern part of Norway, southern part of Spain, and so on (see Figure 19.6).

These changes are most prominent in the southwestern, central, and northeastern parts of the Czech Republic and less in the eastern part of Hungary, southern part of Norway, and southern part of Spain.

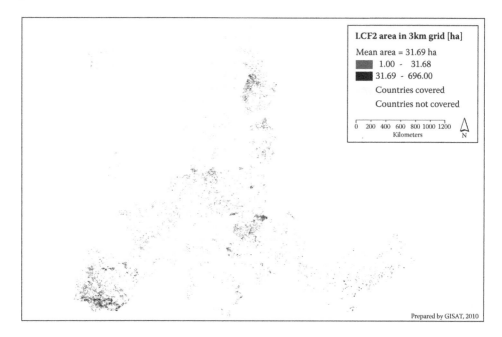

LCF2 area in 3km grid [ha]

Mean area = 31.69 ha
- 1.00 - 31.68
- 31.69 - 696.00

Countries covered
Countries not covered

0 200 400 600 800 1000 1200
Kilometers

N

Prepared by GISAT, 2010

FIGURE 19.5 (**See color insert.**) Spatial distribution of intensification of agriculture in European countries in 2000–2006.

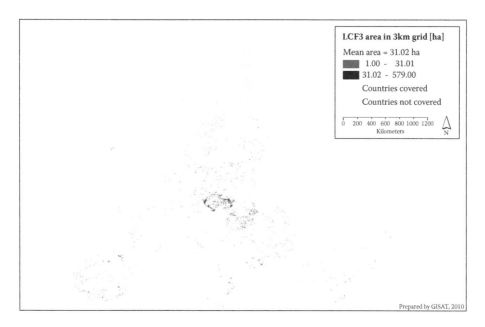

FIGURE 19.6   Spatial distribution of extensification of agriculture in European countries in 2000–2006.

19.5.4   Afforestation (LCF4)

Forest regeneration manifested itself during 2000–2006 as the most stable one. Areas that changed into "forest" per year were nearly constant (265,158–256,830 ha a year) throughout both periods. Areas that changed in favor of forestation during 2000–2006 only decreased by a mean of 8329 ha a year (see Table 19.7) compared with the 1990–2000 period. It must be stressed that afforestation culminated during 1990–2000 when the LC changes attributed to the LCF afforestation were the most extensive. During 2000–2006, these changes were second to the most extensive changes that could be related to deforestation. The mean LCF4 value reached 28.89 ha, and its spatial distribution is observable in almost all the territories of Norway, Finland, and Ireland, the northeastern part of the Czech Republic, eastern and southeastern Hungary, southwestern part of France, northern Spain, southern half of Portugal, northwest of Turkey, and so on (see Figure 19.7).

This type of change in the forest landscape occurred especially in the whole of Norway, Finland, and Ireland, in the northeastern part of the Czech Republic, eastern and southeastern Hungary, southwestern part of France, northern Spain, southern half of Portugal, and northwest of Turkey.

19.5.5   Deforestation (LCF5)

Deforestation is connected with logging and natural catastrophes (strong winds, etc.) or emissions that cause forests to die (Feranec et al., 2010). Changed areas that contributed to deforestation enlarged by a mean of 359,651 ha a year compared with the 1990–2000 period and were the most extensive in Europe during 2000–2006 (see Table 19.7). The mean LCF5 was 31.77 ha, and its spatial distribution is evident in the whole of Norway, Finland, Estonia, Latvia, Ireland, and Portugal, southeastern part of Sweden, northeastern, eastern, and southern parts of the Czech Republic, northern and central part of Slovakia, southern and eastern parts of Hungary, northeastern part of Romania, southwestern and northern parts of France, and so on (see Figure 19.8).

Deforestation dominates the whole of Norway, Finland, Estonia, Latvia, Ireland, and Portugal, southeastern part of Sweden, northeastern, eastern, and southern parts of the Czech Republic, northern and central part of Slovakia, southern and eastern parts of Hungary, northeastern part of Romania, and southwestern and northern parts of France.

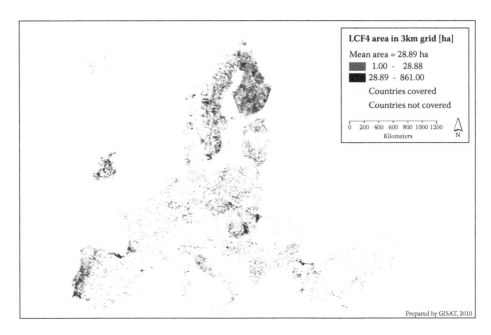

FIGURE 19.7   Spatial distribution of afforestation in the European countries in 2000–2006.

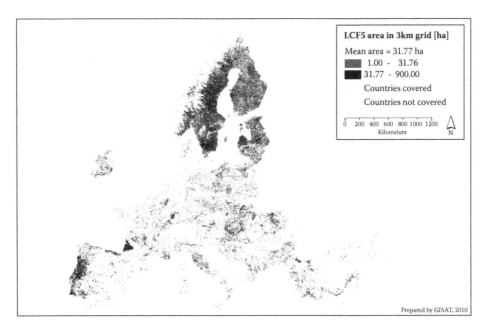

FIGURE 19.8   Spatial distribution of deforestation in European countries in 2000–2006.

19.5.6   CONSTRUCTION AND MANAGEMENT OF WATER BODIES (LCF6)

The results suggest that the area of water bodies increased in the European countries assessed. Areas that changed in favor of water bodies increased by an average of 3525 ha a year for the 2000–2006 period compared with the 1990–2000 period (see Table 19.7). The mean LCF6 value was 34.74 ha, and its spatial distribution is marked in the south of Iceland, The Netherlands, Hungary, eastern part of Germany, central and southwestern parts of Poland, south of Portugal, and so on (Figure 19.9).

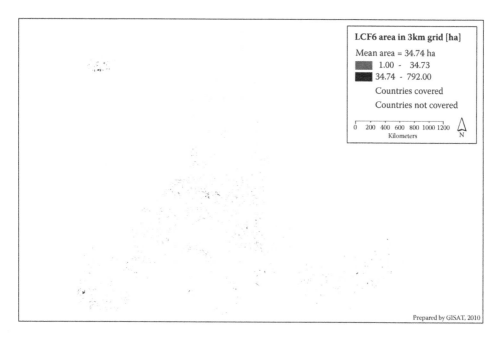

FIGURE 19.9 Spatial distribution of construction of water bodies in European countries in 2000–2006.

Landscape changes in favor of construction of water bodies are evident especially in Iceland, The Netherlands, Hungary, eastern part of Germany, central and southwestern parts of Poland, and south of Portugal.

19.5.7 OTHER CHANGES (LCF7)

Table 19.5 demonstrates other changes: for instance, LC changes in favor of recultivation of former mining areas and dumpsites, unclassified changes, and so forth. Areas that contributed to other changes during 2000–2006 increased by a mean of 15,607 ha a year (see Table 19.7). The mean LCF7 value was 25.57 ha, and its spatial distribution is observable in the southern part of Iceland, eastern part of Ireland, Denmark, Germany, northwestern part of the Czech Republic, southern part of Poland, western part of Portugal, southern and eastern parts of Spain, central Turkey, and so on (see Figure 19.10).

Other changes dominate especially in the southern part of Iceland, eastern part of Ireland, Denmark, Germany, northwestern part of the Czech Republic, southern part of Poland, western part of Portugal, southern and eastern parts of Spain, and central part of Turkey.

19.6 CONCLUSIONS

Results in this chapter show the European LC and its changes during the 1990–2000–2006 periods. The CLC2006 second level data were used to illustrate the occurrence and areas of 15 LC classes. In terms of size, CLC classes defining the agricultural landscape dominate (42.2% of the total area of the countries concerned) followed by forests (29.2%), shrub and/or herbaceous vegetation associations (13.9%), open spaces with little or no vegetation (6.3%), artificial surfaces (3.6%), water bodies (2.6%), and wetlands (2.2%) (see Table 19.6). Spatial distribution of these classes is illustrated in Figure 19.3.

LC changes transformed into LCFs show acceleration of urbanization, that is, enlarged urbanized areas by 17% in 2000–2006 compared with 1990–2000, decrease of LC areas in favor of

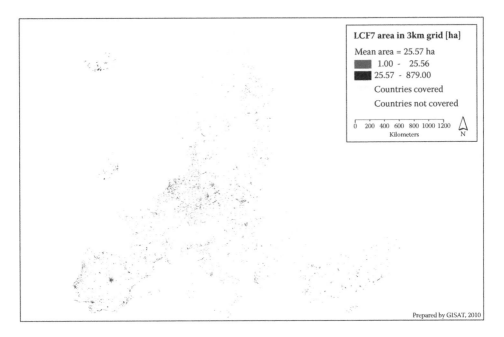

FIGURE 19.10   Spatial distribution of other changes in European countries in 2000–2006.

intensification of agriculture by 48%, decrease of LC areas in favor of extensification of agriculture by −80%, decrease of LC areas in favor of afforestation by −3%, a very marked enlargement of LC areas in favor of deforestation by 173%, enlargement of LC areas in favor of water bodies by 29%, and enlargement of LC areas in favor of other changes by 72% (see Table 19.7). Figures 19.4 through 19.10 show the spatial distribution of individual types.

Comparison of processes in European landscapes during the 1990–2000 and 2000–2006 periods shows that acceleration of those processes associated with deforestation, other changes, urbanization, and construction of water bodies was the most conspicuous. In turn, rates of processes involving extensification of agriculture, intensification of agriculture, and afforestation decreased during those periods (see Table 19.7).

ACKNOWLEDGMENTS

We would like to acknowledge the European Environment Agency (EEA) for its vision and effort in support of CORINE landcover activities in Europe. We would also like to thank Hana Contrerasova for the translation of Chapter 19 from Slovak to English.

REFERENCES

Bossard, M., Feranec, J., and Otahel, J. 2000. CORINE Land Cover Technical Guide—Addendum 2000, Technical Report 40. Copenhagen: European Environment Agency. Available at: http://www.eea.europa.eu/publications/tech40add

Büttner, G., Feranec, J., Jaffrain, G., Mari, L., Maucha, G., and Soukup, T. 2004. The CORINE land cover 2000 project. In R. Reuter (Ed.), *EARSeL eProceedings*, 3(3) (pp. 331–346). Paris: EARSeL.

Büttner, G. and Maucha, G. 2006. The thematic accuracy of CORINE land cover 2000—Assessment using LUCAS (land use/cover area frame statistical survey). Technical report, 7. Copenhagen: European Environment Agency: http://reports.eea.europa.eu/technical_report_2006_7/en

Comber, A., Fisher, P., and Wadsworth, R. 2004. Integrating land-cover data with different ontologies: Identifying change from inconsistency. *International Journal of Geographical Information Science*, 18, 691–708.

Coppin, P., Jonckheere, J., Nackaerts, K., Muys, B., and Lambin, E. 2004. Digital change detection methods in ecosystem monitoring: A review. *International Journal of Remote Sensing*, 25, 1565–1596.

De Zeeuw, C.J. and Hazeu, G.W. 2001. Monitoring land-use changes using geo-information: Possibilities, methods and adapted techniques, Alterra-rapport 214/CGI-Report 9. Wageningen, Alterra.

EEA-ETC/TE. 2002. *CORINE Land Cover Update 2000. Technical Guidelines*. Copenhagen: European Environment Agency. Available at: http://terrestrial.eionet.eu.int

EEA-ETC/LUSI. 2007. CLC2006 technical guidelines. EEA Technical Report 17. Luxembourg: Office for Official Publications of the European Communities. Available at: http://www.eea.europa.eu/publications/technical_report_2007_17

Feranec, J. 1999. Interpretation element "Association": Analysis and definition. *International Journal of Applied Earth Observation and Geoinformation*, 1, 64–67.

Feranec, J., Cebecauer, T., and Otahel, J. 2005. *Photo-to-Photo Interpretation Manual* (Revised), BIOPRESS Document, Biopress-d-13-1.3. Bratislava: Institute of Geography, Slovak Academy of Sciences.

Feranec, J., Hazeu, G., Christensen, S., and Jaffrain, G. 2007a. Corine land-cover change detection in Europe (case studies of the Netherlands and Slovakia), *Land Use Policy*, 24, 234–247.

Feranec, J., Hazeu, G., Jaffrain, G., and Cebecauer, T. 2007b. Cartographic aspects of land cover change detection (over- and underestimation in the I&CORINE Land Cover 2000 Project). *Cartographic Journal*, 44, 44–54.

Feranec, J., Jaffrain, G., Soukup, T., and Hazeu, G. 2010. Determining changes and flows in European landscapes 1990–2000 using CORINE land cover data. *Applied Geography*, 30, 19–35.

Feranec, J., Suri, M., Otahel, J., Cebecauer, T., Kolar, J., Soukup, T., Zdenkova, D., et al. 2000. Inventory of major landscape changes in the Czech Republic, Hungary, Romania and Slovak Republic. *International Journal of Applied Earth Observation and Geoinformation*, 2, 129–139.

Foody, G.M. 2002. Status of land cover classification accuracy assessment. *Remote Sensing of Environment*, 80, 185–201.

Fuller, R.M., Smith, G.M., and Devereux, B.J. 2003. The characterisation and measurement of land cover change through remote sensing: Problems in operational applications. *International Journal of Applied Earth Observation and Geoinformation*, 4, 243–253.

Gerard, F., Petit, S., Smith, G., Thomson, A., Brown, N., Manchester, S., Wadsworth, R., et al. 2010. Land cover change in Europe between 1950 and 2000 determined employing aerial photography. *Progress in Physical Geography*, 34, 183–205.

Haines-Young, R. and Weber, J.-L. 2006. Land accounts for Europe 1990–2000. Towards integrated land and ecosystem accounting. EEA Report 11. Copenhagen: European Environment Agency.

Heymann, Y., Steenmans, Ch., Croissille, G., and Bossard, M. 1994. *CORINE Land Cover. Technical Guide*. Luxembourg: Office for Official Publications European Communities.

Jaffrain, G., Boussim, J., and Diallo, A. 2005. *Technical Guide for Land-Cover Database "BDOT" in Burkina Faso*. Paris: IGN FI–IGB.

Khorram, S. 1999. *Accuracy Assessment of Remote Sensing-Derived Change Detection*. Bethesda, MD: American Society for Photogrammetry and Remote Sensing.

Meyer, W.B. and Turner II, B.L. 1994, *Changes in Land Use and Land Cover: A Global Perspective*. Cambridge: Cambridge University Press.

Steenmans, Ch. and Perdigao, V. 2001. Update of the CORINE Land Cover Database. In G. Groom and T. Reed (Eds.), *Strategic Landscape Monitoring for the Nordic Countries* (pp. 101–107). Copenhagen: Nordic Council of Ministers.

Van Oort, P.A.J. 2005. Improving land cover change estimates by accounting for classification errors. *International Journal of Remote Sensing*, 26, 3009–3024.

20 North American Land-Change Monitoring System

Rasim Latifovic, Colin Homer, Rainer Ressl, Darren Pouliot, Sheikh Nazmul Hossain, René R. Colditz, Ian Olthof, Chandra P. Giri, and Arturo Victoria

CONTENTS

20.1 INTRODUCTION

Global- and continental-scale land-cover and land-cover change information is required to better understand land-surface processes that characterize environmental, social, and economic aspects of sustainability. Land cover plays an important role in a number of environmental issues, such as migratory wildlife, water supply, landscape biochemical processes, climate change, and pollution, which may influence ecosystem and human health across borders. The scientific requirement for global and continental land-cover information has long been articulated, especially at the international level by the International Geosphere Biosphere Programme (IGBP), World Climate Research Programme (WCRP), and Global Monitoring for Environment and Security (GMES). Global and continental land-cover information is needed to implement the UN Millennium Development Goals (MDGs), the UN Framework Convention on Climate Change (FCCC), the UN Convention on Biological Diversity (CBD), the UN Convention to Combat Desertification (CCD), and the UN

Forum on Forests (UNFF). Recently, a list of essential climate variables (ECVs) endorsed by the GCOS (Global Climate Observing System) and CEOS (Committee on Earth Observation Satellites) science community included land cover as an essential terrestrial variable (CEOS, 2006; GCOS, 2006; GTOS, 2008). The ECV specifications are provided in the GCOS-107 report, in which key requirements for land cover undergo annual updates at 0.25–1-km spatial resolution and 5-year updates at 10–30-m spatial resolution. A new collaborative model for land-cover mapping that provides the information needed by these international initiatives is clearly required. Harmonization of land-cover mapping procedures across national borders, increasing flexibility of land-cover classifications, engagement of local expertise, and efficient integration of land-cover and other geospatial data are some of the challenges that must be addressed to achieve this goal.

At present, land-cover monitoring at the North American scale with the consistency and update frequency required under GCOS is not being systematically maintained. Thus, the joint North American Land Change Monitoring System (NALCMS) project was initiated at the North America Land Cover Summit held in Washington in 2006 by Natural Resources Canada/Canada Centre for Remote Sensing (NRCan/CCRS), the United States Geological Survey (USGS), and Mexico's National Institute for Statistics and Geography (Instituto nacional de estadística y geografía—INEGI). Other organizations involved in this initiative were Mexico's National Commission for the Knowledge and Use of Biodiversity (Comisión nacional para el conocimiento y uso de la biodiversidad—CONABIO) and the National Forestry Commission (Comisión nacional forestal—CONAFOR). The project is supported by the Commission for Environmental Cooperation (CEC), an international organization set up by Canada, Mexico, and the United States under the North American Agreement on Environmental Cooperation to promote environmental collaboration among the three countries. Target users of the information provided by the NALCMS include decision makers, international organizations such as the United Nations Environment Programme (UNEP), nongovernmental conservation organizations, and scientific researchers in the domains of climate change, carbon sequestration, biodiversity loss, changes in ecosystem structure and function, and others interested in land-cover dynamics and continental-scale patterns of North America's changing environment.

20.2 NORTH AMERICAN LAND-CHANGE MONITORING SYSTEM

20.2.1 OVERVIEW

The collective need for a harmonized land-cover monitoring system across North America's political boundaries is a motivating factor for establishing the NALCMS collaboration. The goal is to provide information that simultaneously meets needs at the continental scale, while also providing information for each country to complement existing country-specific monitoring programs. Such an international collaboration offers a number of benefits such as the following:

- Improved mapping accuracy achieved through engagement of local expertise, resources, and reference data
- Increased use of products specifically designed to meet each country's national requirements, which also facilitates continental applications and monitoring
- Standardized and consistent products across North America, generated from the same data source using the same or a similar mapping methodology

At present, land-cover maps of North America at 1-km spatial resolution are available from National Oceanic and Atmospheric Administration–Advanced Very High Resolution Radiometer (NOAA–AVHRR) data as a part of global products IGBP (Loveland and Belward, 1997) and from SPOT-VEGETATION GLC 2000 (Latifovic et al., 2004), at 0.5 km from Terra/Aqua-MODIS global land-cover product (Friedl et al., 2010), and at ~0.3 km from ENVISAT-MERIS GlobCover

(Arino et al., 2008). The most recent land-cover database of North America produced by NALCMS at 0.25-km spatial resolution is presented in this paper. Existing national land-cover products at 30-m spatial resolution, including efforts across Canada (Wulder et al., 2003), the United States (Homer et al., 2004), and Mexico (INEGI, 2005), are designed to address country-specific needs. Thus, a North American compilation of national land-cover products at 30 m was not pursued owing to lack of consistency across national borders, selected classification systems, legends, thematic resolutions, and mapping objectives (Jones, 2008). The same reasons also precluded cross-country land-cover change analysis and environmental assessments that used national land-cover data. It is important to clarify that the land-cover products generated by the NALCMS collaboration do not intend to substitute national land-cover characterization efforts. In fact, NALCMS aims for an approach to convey national land-cover information in a consistent way to a common land-cover product at the continental scale.

Generating annual updates of existing national land-cover products at 30 m is not currently feasible at a continental scale owing to technical and resource constraints. In many cases, the time required for producing and delivering land-cover data at 30-m spatial resolution is a significant limitation for applications where annual land cover is required. It often takes 5 or more years to collect complete cloud-free satellite observations over large areas (United States, Canada, or Mexico), to produce a national land cover, and to make it available. One way to reduce delivery time is to update land cover over change areas using imagery as it becomes available.

Considering the land-cover information needs and opportunities, the general objective of NALCMS was to contribute a common framework to monitor land surface at the continental scale. The proposed approach selected by the involved parties combines annual 250-m spatial resolution satellite observations and 5-year interval 30-m spatial resolution satellite observations, offering products relevant at both spatial and temporal scales of land-cover change (Smith, 2008). The two scales will be provided to users to investigate, confirm, calibrate, and assess 250-m resolution change products with 30-m product support for local areas. The outputs will be products that enable users to identify a greater variety of land-cover change, find change across much smaller land-cover patches, and eventually identify more types of change, for example, gradual change over time. The initial development phase is focused on land-cover and land-cover change prototype products for 2005–2009, which is primarily seen as the period used to develop methodologies, whereas system performance will be evaluated in 2010.

20.2.2 Land-Cover Monitoring

Land-cover monitoring, specifically multitemporal land-cover mapping, over large spatial extents is challenging because of the inconsistencies in satellite measurements. The inconsistencies arise from differences in atmospheric conditions, sun-sensor geometry, geolocation error, variable ground pixel size, sensor noise, vegetation phenology, and surface moisture conditions, which eventually lead to mapping inaccuracy. Existing techniques need to be evaluated and improved to achieve the accuracy and consistency required for continental-scale land-surface monitoring. Manual interpretation can provide excellent results at one point in time as interpreters can make use of contextual and spatial information that cannot be easily incorporated into automated mapping methods. Unfortunately, such interpretations are not easily repeatable, making it difficult to support an automated monitoring effort. To facilitate land-cover monitoring requirements, more sophisticated algorithms based on advances in the fields of pattern recognition and machine learning have emerged. Decision tree (DT) classifiers have often been used for land-cover classification at continental to global scales. Early applications of DTs (Breimann et al., 1984) for remote sensing-based land-cover classification focused on mapping, using coarse-resolution imagery (DeFries et al., 1998; Friedl and Brodley, 1997; Friedl et al., 1999, 2002; Hansen et al., 1996). However, there have been few attempts at producing land-cover time series of sufficient length, consistency, and continuity to study patterns of land-cover variability and change. At high resolution, 30-m land-cover time series generally covers a short period or maintains only a few time steps. Xian et al. (2009) used a DT approach to update changes in 30-m land cover from 2001 to 2006. At coarser resolution (500 m), Moderate Resolution Imaging Spectroradiometer (MODIS) global land-cover version 5 was generated using DTs and prior probabilities to improve classification performance and temporal stability. This product is generally developed for the global modeling community and is not intended to infer change between years (Friedl et al., 2010). Latifovic and Pouliot (2005) produced a 1-km land-cover time series at 5-year intervals from 1985 to 2005, using change detection and evidence-based updating strategies. However, this time series is both temporally and spatially coarse, limiting the change information that can be derived. Currently, there is no spatially extensive, high temporal frequency land-cover product that can be used to analyze land-cover change at the North American scale.

Monitoring land-surface conditions requires satellite data of high temporal frequency to ensure clear-sky composites needed to capture abrupt annual land-cover changes such as forest fire or logging. Currently, two satellite data sources meet the NALCMS requirements: the MODIS on the Terra and Aqua satellites and the Medium Resolution Imaging Spectrometer (MERIS) on ENVISAT. The Visible Infrared Imager Radiometer Suite (VIIRS) on JPSS satellites and Sentinel 2 and 3 will be options for future data continuity. At present, MODIS daily observations are considered the main project input data stream. Data at processing level 1 are readily available through the Distributed Active Archive Center (DAAC) at the Goddard Space Flight Center (GSFC).

NALCMS's initial outputs include development of protocols, products, and partnerships. The following annual products at 250-m cell resolution will be used to demonstrate the monitoring framework:

- North American image composites
- North American thematic land cover
- North American spectral change and land-cover thematic change
- North American fractional vegetation change products

The proposed 30-m resolution products with 5-year repeat frequency will need more time to develop and fund. Leveraging of existing programs that produce 30-m products will be pursued to ensure that costs are kept to the minimum. For example, Canada's Centre for Topographic Information has combined the 30-m sector-based land-cover products and their legends, and an informal Land Cover Community of Practice has been developed to help synergize updates and future multiple 30-m land-cover efforts in the nation. Mexico has a 30-m land-cover mapping program at INEGI, and in the United States, the Multi-Resolution Land Characteristics Consortium coordinates 30-m land-cover production. Potential synergies among these efforts can help support the 30-m requirements of this system and maximize future cost savings.

20.3 METHOD AND DATA

The procedure for generating the North American Land Cover Database 2005 (NALCD2005) has the following steps: selection of a classification system and legend definition, processing of satellite data and ancillary data, classification, and accuracy assessment. The following section describes the design considerations and the country-specific implementation of each step.

20.3.1 CLASSIFICATION SYSTEM AND LEGEND

The NALCMS classification legend is designed at three hierarchical levels, using the Food and Agriculture Organization (FAO) Land Cover Classification System (LCCS). Levels I and II are defined for the North American scale, while level III specifies land-cover information at the national scale. Table 20.1 presents the NALCMS level I legend with 12 classes and level II with 19 classes.

TABLE 20.1

Legend Level I and II of the North American Land-Change Monitoring System

		Level I	Level II	LCCS Basic Classifier
Primarily vegetated areas	Natural and seminatural terrestrial and aquatic	1. Needleleaf forest	1. Temperate or subpolar forest	A3.A10.B2.XX.D2.E1
			2. Subpolar taiga needleleaf forest	A3.A10.B2.XX.D1.E2
		2. Broadleaf forest	3. Tropical or subtropical broadleaf evergreen forest	A3.A10.B2.XX.D1.E2
			4. Tropical or subtropical broadleaf deciduous forest	A3.A11.B2.XX.D1.E2
			5. Temperate or subpolar broadleaf deciduous forest	A3.A14.B2.XX.D1.E2
		3. Mixed forest	6. Mixed forest	A3.A10.B2.XX.D2.E1/ A3.A10.B2.XX.D1.E2
		4. Shrubland	7. Tropical or subtropical shrubland	A4.A20.B3-B9
			8. Temperate or subpolar shrubland	A4.A20.B3-B10
		5. Grassland	9. Tropical or subtropical grassland	A6.A20.B4
			10. Temperate or subpolar grassland	A2.A20.B4.XX.E5
		6. Linchen/ moss	11. Subpolar or polar shrubland-lichen-moss	A4.A11.B3-B10 / A2.A20. B4-B12 / A8.A11-A13
			12. Subpolar or polar grassland-lichen-moss	A2.A20.B4-B12 / A4.A11. B3-B10 / A8.A11-A13
			13. Subpolar or polar barren-lichen-moss	A8.A20-A13 / A4.A11. B3-B10 / A2.A20.B4-B12
		7. Wetland	14. Wetland	A2.A20.B4.C3
	Cultivated/managed terrestrial/aquatic	8. Cropland	15. Cropland	A4-S1
Primarily nonvegetated areas	Terrestrial	9. Barren land	16. Barren land	A1/A2
		10. Urban and built-up	17. Urban and built-up	A4
	Aquatic	11. Water	18. Water	A1
		12. Snow and ice	19. Snow and ice	A2/A3

20.3.2 SATELLITE DATA PROCESSING AND ANCILLARY DATA DESCRIPTION

MODIS observations preprocessed into level 1B data (MOD02QKM and MOD02HKM) are the initial data used to generate 10-day top-of-atmosphere reflectance composites. The L1B data are in 5-min swath granules stored in HDF-EOS (Hierarchical Data Format–Earth Observing System) format. In addition, earth location data fields (latitude, longitude, elevation, etc.) and viewing geometry (sun and satellite zenith and azimuth angles) at the time of acquisition are required. These additional attributes are provided in the MOD03 product for each pixel at 1-km spatial resolution. The level 2 data MOD35 (Cloud Mask) and MODATML2 (atmospheric parameters such as aerosol and water vapor) are used for compositing and atmospheric correction. All MODIS data are acquired online through the LAADS Web site http://ladsweb.nascom.nasa.gov or the EOS Data Gateway Web site http://daac.gsfc.nasa.gov.

Ten-day MODIS composites over North America for 2005 and 2006 were processed at the CCRS following the procedures described by Khlopenkov and Trishchenko (2008) for resampling— Trishchenko et al. (2006) for downscaling of 500-m resolution bands to 250 m and Luo et al. (2008)

for compositing to 10-day intervals. To reduce noise in the processed data, a temporal rank filter was applied. The filter was calculated as the maximum of minimums in a split moving window, where each window contained three composite values. The same filtering approach was found effective in reducing time series data noise used for lake ice-free date detection by Latifovic and Pouliot (2006) and snow detection by Zhao and Fernandes (2009). After rank filtering, a procedure to detect residual spikes in the data was applied based on a neighborhood comparison. Filtered 10-day composites were averaged to generate monthly composites. Initial datasets 2005–2006 contained 12 monthly composites of channels 1 (B1, visible red) and 2 (B2, near-infrared) at 250-m spatial resolution and five channels designed for land applications (bands B3–B7) resampled to 250-m spatial resolution. As an example, monthly composites are shown in Figure 20.1.

In addition to the satellite datasets, a number of other information sources were used to train the classification algorithm and aid the interpretation of specific land-cover classes. Each country used different types of ancillary data depending on their availability at national scale. All ancillary data used in this study are summarized in Tables 20.2 through 20.4. Additional processing was performed for some of the layers, including vector to raster conversion, reprojecting, mosaicing, and resampling to the North American framework.

Over Canada, the main reference data sources were existing land-cover datasets derived from medium-resolution satellite observations (25–60 m), including Satellite Information for Land Cover (SILC), Northern Land Cover of Canada (NLCC), National Land and Water Information System (NLWIS), Agricultural Land Cover Classification (ALCC), and Earth Observation for Sustainable Development (EOSD) in addition to other reference data listed in Table 20.2.

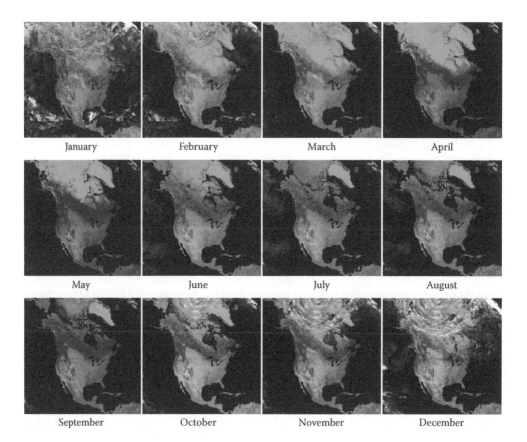

FIGURE 20.1 **(See color insert.)** North America top-of-the-atmosphere reflectance monthly composites from MODIS/Terra 2005 at 250-m spatial resolution.

TABLE 20.2
Ancillary Data Sources for Canada

Title	Source	Use
Water Fraction Map of Canada assembled from the National Topographic Data Base (NTDB), 1:250000 scale	http://geogratis.cgdi.gc.ca, Fernandes et al. (2001)	F
CDED (2000): Canada Digital Elevation Data, Level 1, 1:50000 scale	http://www.geobase.ca	F
NRN (2007). National Road Network, Canada, Level 1, 1:250000 scale	http://www.geobase.ca	F
SILC: Satellite Database for the Land Cover of Canada—a sample of LANDSAT TM/ETM+ scenes (30 m resolution) representing the distribution of land cover across Canada	NRCan, CCRS, SILC, 2000	R
EOSD: Earth Observation for Sustainable Development of Forests Land-Cover Classification, circa 2000 at 30m resolution	Canadian Forest Service, ftp://ftp.ccrs.nrcan.gc.ca/ad/EOSD/ Wulder et al. (2003)	R
NLWIS (2009). National Land and Water Information System, Land Cover for agricultural regions of Canada, circa 2000 at 30 m resolution	Agriculture Agrifood Canada, ftp://ftp.agr.gc.ca	R
NLCC (2008). Circa-2000 Northern Land Cover of Canada at 30 m resolution	NRCan, CCRS, http://geogratis.cgdi.gc.ca, Olthof et al. (2009)	R
ALCC: Classification of Agricultural Lands in Alberta, Saskatchewan and the Peace Region of British Columbia	ALCC (2008). Agriculture Land Cover Classification (ALCC). Classification of Agricultural Lands in Alberta, Saskatchewan and the Peace Region of British Columbia. Digital EnvironmentalTM	R
Agricultural Financial Services Corporation Saskatchewan Crop Insurance Corporation British Columbia Ministry of Agriculture and Lands (Business Risk management Division), circa 2006 at 30 m resolution.		
Treeline	http://data.arcticatlas.org, Timoney et al. (1992)	M
NCEP (2009). NARR: North American Regional Reanalysis-Daily Dataset Degree days	NARR, http://nomads.ncdc.noaa.gov	F
Fire data base	NRCan, CCRS, Zhang et al. (2004a, 2004b); Fraser et al. (2004)	M
Ground truth data	NRCan, CCRS, unpublished	R
MODIS Band1, 2 and 6 minimum and maximum and average reflectance during the peak of the season (July–August)	MODIS/Terra top-of-atmosphere reflectance data at 250 m spatial and 10-day temporal resolution over North America; NRCan/CCRS	F
Average and Integrated NDVI during the peak of the season (July–August)	MODIS/Terra top-of-atmosphere reflectance data at 250 m spatial and 10-day temporal resolution over North America; NRCan/CCRS	F

Note: F—feature, M—mask, R—reference data.

For the United States, the main reference data were the existing National Land Cover Dataset (NLCD) 2001 at 30 m. Several other datasets such as Gap Analysis Program (GAP) and LANDFIRE Existing Vegetation Type (EVT), as well as others listed in Table 20.3, were used to refine training data. The list of ancillary data for Mexico as shown in Table 20.4 includes a digital elevation model and its derivatives slope and aspect; minimum, mean, and maximum temperature; and total precipitation and the number of days with precipitation.

20.3.3 CLASSIFICATION PROCEDURE

The classification procedure used a DT to generate the 2005 NALCMS land-cover map. The C5.0 software was used, which implements a gain ratio criterion in tree development and pruning (Quinlan, 1993). C5.0 also implemented several advanced features that could aid and improve land-cover classification, including boosting and cross-validation. Boosting is a technique that combines multiple classifiers to produce an ensemble result that can be significantly better than that of any single classifier (Bauer and Kohavi, 1999; Freund and Schapire, 1997). Basic classifiers are trained in sequence, and each base classifier is trained using a weighted form of the dataset in which the weighted coefficient associated with each data point depends on the performance of the previous classifier. In particular, points that are misclassified by one of the base classifiers are given greater weight when used to train the next classifier in sequence. Cross-validation can provide an estimate of the land-cover classification quality. In addition, C5.0 can generate a confidence estimate for each classified pixel and a record of the associated classification logic in a text file that can be readily interpreted. DTs have substantial advantages in satellite image classification because of their flexibility, intuitive simplicity, and computational efficiency. Therefore, DT classification algorithms are gaining increased acceptance for land-cover classification, particularly at continental to global scales (Stahler, 1999), and have been employed to generate MODIS global products (Friedl et al., 2010). The design of the classification procedure used to generate the 2005 NALCMS land-cover map included the following steps: stratification by mapping zones, reference data collection and feature-set selection, DT model derivation and classification, and postclassification processing.

20.3.3.1 Mapping Zones

The mapping zone delineation was designed to stratify landscapes across North America into subregions of similar biophysical and spectral characteristics. It was assumed that stratification by spectral patterns delineated different physiological and phenological characteristics of the ecosystem (Homer et al., 2001). Reducing variability of the mapping area increased accuracy and optimized both classification and edge matching (Homer et al., 1997). The other practical reason for using mapping zones was to organize mapping inside national borders to address specific needs, to facilitate the use of national land-cover information for reference and training data not available at the continental scale, and to use local expertise. In delineating mapping zones across borders, a 50-km overlap was applied to ensure better cross-border agreement and easy edge matching (Figure 20.2). Samples inside the cross-border overlap regions were used to generate DT models for both mapping zones.

20.3.3.2 Reference Data and Feature Selection

To carry out the land-cover classification, a large quantity of training data was required for the DT classifier. Different datasets were collected by each country following common general guidelines (Colditz et al., 2008). For each country, several data sources were combined to train classifiers, including visual interpretation of fine and medium spatial resolution satellite imagery, existing ground-truth data, and higher spatial resolution maps. To generate a training set for mapping North America at 250-m spatial resolution, a country-specific sampling procedure governed by data availability and local expertise was implemented. Several additional data sources, as highlighted in Tables 20.2–4, were used to generate masks for postclassification corrections. Other datasets served as features to aid the classifier.

For Canada, the training dataset was derived from the Land Cover Database of Canada 2005 at 250-m spatial resolution produced with a thematic resolution of 39 land-cover classes (Latifovic et al., 2009). The map was produced following the classification approach described by Latifovic et al. (1999) and (2004), which presumed considerable expertise in image interpretation and knowledge of national-scale land-cover distribution. To generate reference data to train the DT, additional features were extracted from ancillary data listed in Table 20.2. The water class was mapped from

FIGURE 20.2 Zones used for mapping North American land cover.

the Water Fraction Map of Canada at 250 m, when the water fraction in a 250-m pixel was greater than 75%. The urban areas were mapped using a road density layer in which low vegetation classes were relabeled as urban if the road density was at least 20% within a 1-km area. Digital elevation data were used to reduce confusion between shadow and water in a complex terrain. A forest fire database was stratified into recent burns within the last 5 years and older burns to differentiate shrub and forest classes.

Training data over Canada were sampled using random stratified sampling. Different sample sizes and feature sets were assessed by cross-validation accuracy and feature usage. The final sample set represented 0.01% of the mapping area and included 15 land-cover types in Canada, each with 3000 samples. The feature set (Table 20.2) was selected based on mutual covariance, theoretical considerations, and exploratory analysis. The satellite-based features were extracted from MODIS red (620–670 nm), NIR (841–876 nm), and SWIR (1628–1652 nm) bands as these contained the majority of the data dimensionality and were least affected by atmospheric contamination (Pouliot et al., 2009). They were extracted only from peak of the growing-season composites. Data during spring and fall were not included because these periods were strongly affected by seasonal changes.

For the United States, training data were collected from several higher resolution sources, including NLCD 2001 land cover, GAP vegetation classification, NOAA Coastal Change and Analysis land cover, LANDFIRE ECV, and others. The data were organized into three mapping zones (eastern, central, and western) to improve accuracy (Figure 20.2). Approximately 120,000 stratified random points were selected for each mapping zone. Because training data were drawn from higher resolution sources, 30-m source data were spatially filtered to identify homogenous patches large enough to serve as training data at the 250-m MODIS scale. Training samples were then systematically drawn from candidate MODIS scale areas using a stratified random-sampling strategy. To ensure adequate training for rare classes, some additional points were collected manually in

TABLE 20.3

Ancillary Data Sources for United States

Title	Source	Use
National Land Cover Data (2001)	USGS	R
National Digital Elevation Model (DEM) Data, 250 m	USGS	F
National Slope Data, 250 m	USGS	F
National Aspect Data, 250 m	USGS	F
National percent tree (0–100) continuous data (2005)	USGS	F
National impervious (0–100) continuous data (2005)	USGS	F
National Existing Vegetation Type (EVT) data (2008)	Landfire, USGS, and USFS 2008	R
NOAA coastal change land cover	NOAA	R
Gap Analysis Program (GAP) data	GAP	R
	Idaho State university	R
Ecological regions of North America	Commission for Environmental Cooperation. "Ecological regions of North America: Toward a common perspective," Commission for Environmental Cooperation, Montreal, Quebec, Canada. 71 pp., 1997.	M
Koppen Climate Classification	University of Melbourne	F
National Isobioclimate data	The University of Montana, Numerical Terradynamic Simulation Group (NTSG)	S

Note: F—feature, M—mask, R—reference data.

targeted areas. For urban classification, the NLCD 2001 percent impervious data were used in conjunction with a road density layer to provide urban training data. LANDFIRE data were also used to help identify wildfire burn areas that affect classification of forests and shrubs. To help differentiate the spatial location of tropical and temperate classes, climatic zone data combined with vegetation classifications from GAP and LANDFIRE were used as training sources. Finally, the NLCD 2001 percent tree canopy layer was used to improve training data quality for forested areas.

For Mexico, the sample data were gathered from various available sources (Table 20.4), and additional samples were generated for rare classes. The total number of sampled pixels was 118,633, of which 80% per class (a total of 94,963 pixels) was selected as potential sites used to train the DT classifier, while the remaining 20% was used for testing. Both potential training and test pixels were evaluated against the INEGI Series III vegetation map (INEGI, 2005) recoded to the NALCMS legend. Only pixels corresponding to the class in the map and with a minimum distance of 500 m to another land-cover class in the reference map were selected to ensure pixel/class purity. This reduced the total number of effective training pixels to 45,548. The classifier was trained for 13 out of the 15 classes existing in Mexico at NALCMS legend level II. Multiple feature sets were generated for the DT. In addition to the ancillary data and the annual time series of monthly 7-band MODIS composites, and Normalized Vegetation Difference Index (NDVI) simple univariate statistics, such as mean, minimum, maximum, standard deviation, and range over temporal periods (annual, 1/2, 1/3, and 1/4 year) were added to the feature set. It was found that the high dimensionality improved the description of the complexity of the Mexican landscape.

20.3.3.3   DT Model Generation and Classification

Land-cover classifications for Canada and the United States were developed using a boosted DT with 10 iterations. In Mexico, a fuzzy approach was used, in which the percentage of each class contained in the individual tree leaves was retained as a class membership. Boosted results were combined

TABLE 20.4
Ancillary Data Sources for Mexico

Title	Source	Use
Digital elevation model of Mexico, 250 m	CONAFOR	F
Monthly average minimum temperature per year in °C (1970–2000)	CONAFOR	F
Monthly average mean temperature per year in °C (1970–2000)	CONAFOR	F
Monthly average maximum temperature per year in °C (1970–2000)	CONAFOR	F
Total days of precipitation per year (1970–2000)	CONAFOR	F
Total of precipitation per year in mm (1970–2000)	CONAFOR	F
Total of evaporation per year in mm (1970–2000)	CONAFOR	F
Land use and vegetation map of Mexico	Dirección General de Geografía - INEGI (2007) (ed.), "Conjunto de Datos Vectoriales de la Carta de Uso del Suelo y Vegetación," Escala 1:250,000, Serie III (CONTINUO NACIONAL), Instituto Nacional de Estadística, Geografía e Informática - INEGI. Aguascalientes, Ags., México	M
Ecological regions of North America	Commission for Environmental Cooperation. "Ecological regions of North America: toward a common perspective," Commission for Environmental Cooperation, Montreal, Quebec, Canada, 71 pp., 1997	M
National forest inventory	Inventario Nacional Forestal, INFyS 2004–2007, CONAFOR, 2007	R
Classification of agricultural lands of Mexico	Colegio de Postgraduados (Colpos) Secretaría de Agricultura, Ganadería, Desarrollo Rural, Pesca y Alimentación (Sagarpa)	R
Urban areas of Mexico	Instituto Nacional de Estadística Geografía e Informática (INEGI) "Localidades de la República Mexicana, 2005," Obtenido de Principales resultados por localidad 2005, II Conteo de población y Vivienda 2005, Editado por Comisión Nacional para el Conocimiento y Uso de la Biodiversidad (CONABIO), 2007, México	R
Agricultural lands of Mexico	Programa de Producción Pecuaria Sustentable y Ordenamiento Ganadero y Apícola 2004–2007	R
Additional data for decision tree training	CONABIO, unpublished	R
Mean, minimum and maximum reflectance in MODIS Band1–7	MODIS/Terra top-of-atmosphere reflectance data at 250 m spatial and 10-day temporal resolution over North America; NRCan/CCRS	F
NDVI mean, minimum, maximum, standard deviation, and range over temporal periods (annual, 1/2, 1/3, and 1/4 year)	MODIS/Terra top-of-atmosphere reflectance data at 250 m spatial and 10-day temporal resolution over North America; NRCan/CCRS	F

Note: F—feature, M—mask, R—reference data.

by summing up these memberships. To improve and stabilize the classification in Mexico, multiple DTs were generated by manipulating features or sample data at the national scale or executing the classification for seven mapping zones. A total of five classifications were combined by averaging the class memberships. To form a North American product, the class estimations were transformed to a discrete map by assigning the dominant class.

20.3.3.4 Postclassification Operations

Postclassification improvement included additional image processing and contextual noise removal to reduce classification inaccuracy caused by spectral mixing, confusion, or ambiguity in the image data. For example, over the Canadian landmass, spectrally similar classes such as low biomass cropland and grassland were confused with each other or with tundra. To clear this confusion, separate tundra and agriculture masks were created. The tundra mask was derived from the treeline mapped by Timoney et al. (1992), whereas the agriculture mask was generated from a winter composite, generated using minimum red reflectance criteria, and an integrated summer NDVI image. The mask was based on the fact that croplands are bright in winter owing to snow cover and have high integrated NDVI relative to tundra grasses in summer. In Mexico, postprocessing included corrections of wetland and urban classes, which were locally overestimated using existing masks of maximum spatial extent. Also, owing to their small spatial extent, water and snow and ice classes were superimposed from stable masks.

For edge matching along the U.S.–Canada border, two main problems were identified from the initial investigation. These were (1) differences in thematic detail across country borders and (2) class discontinuity along edge boundaries. Differences in thematic details were resolved through discussions between Canada and the United States to refine class definitions in order to make them more consistent. For example, in the United States, shrub was interpreted to include both deciduous and evergreen trees smaller than 2 m, whereas in Canada, it was defined to include only small deciduous trees. To address this problem, Canada merged classes for them to be more consistent with those of the United States. Edge discontinuity due to different classes being assigned to the same object along the border was corrected by an object-based reclassification methodology. The first step segmented the pixels along the boundary line into cross-border objects using eCognition. In the second step, each object was assessed to determine the most frequently occurring class, and the pixels for the object were assigned to that class. Figure 20.3a shows the results of object-based edge matching along the west of the Canada–U.S. border. After edge matching, discontinuity visible in the prematched image was corrected and resolved.

Along the U.S.–Mexican border, class probability functions and local manual editing were employed for edge matching. For areas of class disagreement along the border, the second most likely class based on pixel membership was multiplied with a distance layer, and the pixels were relabeled as second class if their probability was higher than those of the dominant class. For instance, around the Falcon Reservoir between Laredo/Nuevo Laredo and McAllen/Reynosa, cropland dominated on the Mexican side, whereas on the U.S. side, tropical and subtropical shrubland and grassland were mapped (Figure 20.3b). Applying the edge-matching technique reduced the amount of cropland on the Mexican side.

Classification noise was removed from the continental map using a "smart eliminate" aggregation algorithm to reduce single pixel noise. This algorithm used eight-corner connectivity from a central pixel to allow nonlinear features to remain intact and accessed a weighting table to allow "smart" decisions on a dissolve protocol that addressed land-cover class similarity. This resulted in a minimum mapping unit of 1 km or 100 ha for the final land-cover map.

20.3.4 Accuracy Assessment Procedure

The accuracy of the 2005 NALCMS land-cover map was evaluated for the level I legend with 12 land-cover classes. The sampling design was a combined stratified random and two-stage cluster

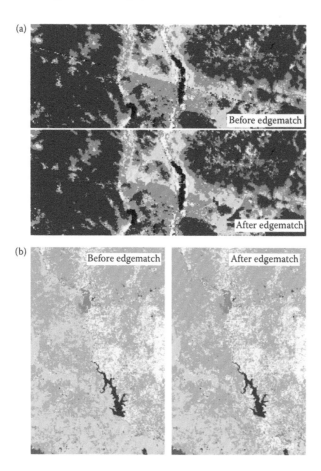

FIGURE 20.3 **(See color insert.)** Examples of matching cross-border land-cover data: (a) the U.S.–Mexico and (b) Canada–U.S. border before and after edge-matching procedure.

sampling approach (Stehman, 2010). In the first stage, the sampling unit was a 15 km × 15 km block, and in the second stage, it was a single pixel. For the first-stage sample, blocks were obtained by selecting 170 of the 89,958 blocks in the combined area of the United States and Canada (0.18898%) and 30 blocks of the 9350 blocks in Mexico (0.32086%). The two samples were selected separately, but in both cases the sampling protocol was simple random sampling. The combined sample of 200 blocks (15 km × 15 km) constituted the full first-stage sample. The first-stage sampling design was constructed to increase the number of blocks sampled in Mexico because Mexico comprises a small proportion of the total area of North America. The United States and Canada are close enough in size, and so it was deemed unnecessary to sample separately in the two countries to equitably distribute the remaining sample blocks.

All pixels within these 200 blocks were sampled using a stratified random sample of 50 pixels per class to yield a final sample of 600 pixels (Figure 20.4). The sample size for each country was not controlled at the second stage of the sample selection, resulting in 55 pixels in Mexico and the remaining 545 pixels being almost equally distributed in the United States and Canada. The sampling design incorporated clustering (the 15 km × 15 km blocks) to reduce the cost of reference data collection by spatially constraining the sample pixels to a smaller search area. Stratification was used in the design to increase the sample size for the rare map classes and reduce the standard errors of users' accuracy estimates. To estimate the error matrix and associated accuracy parameters (overall users' and producers' accuracies), the inclusion probabilities of the sample pixels

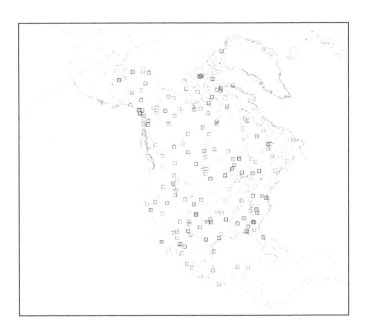

FIGURE 20.4 Stratified random sample used to assess 2005 North American land-cover map accuracy.

needed to be computed and incorporated in the analysis. An inclusion probability was defined as the probability that a particular pixel would be included in the sample. The sampling design was such that pixels had different inclusion probabilities because of a higher proportion of the available blocks in Mexico. The inclusion probability of a pixel was computed as the product of inclusion probabilities from the two stages of sampling (Särndal et al., 1992).

Sample labels were obtained through visual interpretation of high-resolution image data in Google Earth, supported by Landsat data to assist in identification of tree life form in which crown shape was difficult to interpret. Each reference sample was interpreted and assigned a primary label and an alternative label. The primary label referred to the most likely class assigned to a pixel according to the interpreter, whereas the alternative label referred to a second class that could also be considered acceptable in cases where ambiguity existed in the interpretation or where more than one label could be deemed correct. Examples of the latter occurred mainly along transition zones or highly heterogeneous landscapes where two classes co-occurred almost equally within a 250-m pixel.

20.4 RESULTS AND DISCUSSION

20.4.1 NORTH AMERICAN LAND-COVER DATABASE 2005

The NALCD2005 at 250-m spatial resolution NALCD2005 (Figure 20.5) was the first step toward the general objective of NALCMS to contribute, through collective effort, to a harmonized continental-scale land-cover monitoring framework. Country-specific land-cover maps were produced by local experts using the same data preprocessing and information extraction methodologies, though some differences in implementing the DT could not be avoided owing to country-specific requirements and availability of ancillary data. National products were subsequently used to assemble an integrated continental land-cover map. NALCD2005 has 19 classes based on the FAO-LCCS, which ensured its applicability and compatibility with other land-cover mapping projects. The georeferencing parameters provided in Table 20.5 were defined in accordance with the 1:10 million North America Atlas Framework dataset. The latter was designed to lay the foundation for thematic maps under a unique partnership among the National Atlas programs in the United States (USGS),

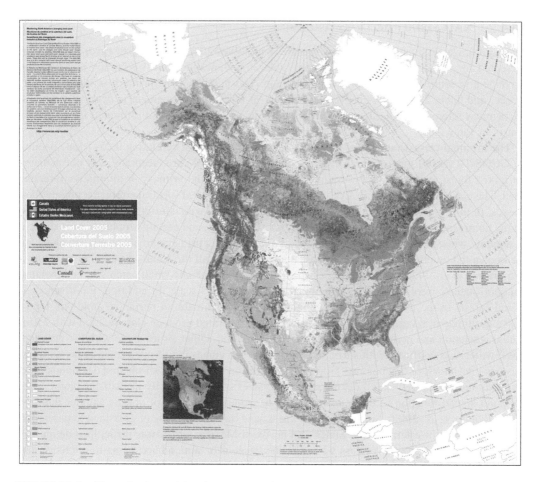

FIGURE 20.5 **(See color insert.)** Land-cover map of North America 2005 at 250-m spatial resolution.

TABLE 20.5
NALCMS Geospatial Framework, Lambert Azimuthal Equal Area Projection Standard Parameters and Boundary Frame

Projection		LAEA				
Longitude of projection center		−100.00				
Latitude of projection center		45.00				
False Easting		0.0				
False Northing		0.0				
Semimajor Axis		6370997.0				
Unit		Meters				
Boundary	ULX	ULY	LRX	LRY	Pixels	Lines
NA	−4418000	4876500	4832000	−3873500	37000	35000
	157°19'39.12"E	52°30'26.94"N	58°29'00.82"	0°09'12.59"S		

Mexico (INEGI), and Canada (NRCan). Existing layers such as base maps (elevation, hydrography, major roads, political boundaries, etc.), terrestrial ecoregions, human influence, and other layers formed a base in which NALCD2005 could be more efficiently integrated with other layers and used by a number of users with different interests.

20.4.2   NALCD2005 ACCURACY ASSESSMENT

The accuracy assessment results are summarized in Tables 20.6 through 20.8. Error matrices are presented for two definitions of agreement. Table 20.6 shows an overall accuracy of 82% when agreement is defined as a match between the map class and either the primary or alternative reference, and Table 20.7 shows an overall accuracy of 68% when agreement is defined using only the primary reference label. Overall users' and producers' accuracies are estimated for each error matrix in Table 20.8.

The minimal class accuracy for land-cover maps with a spatial resolution 0.25–1 km specified by GCOS-T9 (2009) for use as an ECV is 65%. The minimal class accuracy was not achieved for mixed forest, grassland, and wetland classes assessed by the primary or alternative reference label. The classification error matrix for the primary label revealed spectral confusion among these classes. In the case of the mixed forest class, it was confused with conifer or broadleaf depending on the dominant fraction. Shrubland, shrub-covered wetlands, and certain croplands were difficult to separate with spectral data alone owing to all classes being primary broadleaved deciduous and the known radiometric saturation at leaf area index levels for all three classes being in the range of 3–5 (Turner et al., 1999). Wetland classes specifically, with typically small patch size and dynamic temporal nature, were difficult to characterize at the MODIS scale. In addition, the diversity of mixed forests in the transition from temperate to tropical climate, in particular pine-oak forests with a multitude of species as well as cloud forests, complicated accurate classification with 250-m spatial resolution and perhaps resulted in an overestimation. Other issues arose with the grassland class, which was confused with shrub along the tree line consisting of open treed areas with herbaceous understory, or low biomass croplands. Confusion among herb, shrub, and deciduous forest was also due to relatively small disturbance patch sizes of cuts primarily in British Colombia and the point spread function of the sensor, leading to reduced detection of objects smaller than five pixels (Pouliot et al., 2009). Similarly, there was a wide transitional zone of grassland to desert-like shrubland in southwestern United States and northern Mexico, which was difficult to define in a discrete map. The lichen/moss class was either herbaceous or wetland according to reference data

TABLE 20.6
Error Matrix (as Percent of Area)

					Reference Class									
Map	1	2	3	4	5	6	7	8	9	10	11	12	Total	N
1	12.696	0.000	0.577	0.577	0.000	0.000	0.289	0.000	0.000	0.000	0.289	0.000	14.428	50
2	1.195	6.666	0.199	0.199	0.199	0.000	0.000	0.398	0.000	0.199	0.000	0.000	9.056	50
3	0.998	0.111	3.471	0.333	0.000	0.000	0.111	0.000	0.000	0.000	0.111	0.000	5.134	50
4	1.091	0.792	0.000	12.832	0.727	0.364	0.578	0.000	0.000	0.000	0.000	0.000	16.382	50
5	0.197	0.000	0.000	0.591	7.192	0.000	0.313	0.787	0.000	0.197	0.000	0.000	9.277	50
6	0.000	0.000	0.000	0.212	0.000	8.712	0.000	0.000	1.062	0.000	0.637	0.000	10.624	50
7	0.310	0.078	0.078	0.000	0.000	0.078	2.961	0.078	0.078	0.000	0.155	0.000	3.814	50
8	0.000	0.287	0.287	0.000	0.456	0.000	0.338	12.031	0.000	0.000	0.000	0.000	13.398	50
9	0.000	0.000	0.000	0.117	0.000	0.233	0.000	0.000	5.107	0.000	0.069	0.117	5.642	50
10	0.000	0.206	0.000	0.034	0.000	0.000	0.000	0.103	0.034	1.339	0.000	0.000	1.717	50
11	0.000	0.000	0.188	0.188	0.000	0.000	0.000	0.000	0.188	0.188	8.486	0.000	9.238	50
12	*0.000*	*0.000*	*0.000*	*0.000*	*0.000*	*0.000*	*0.000*	*0.000*	*0.026*	*0.000*	*0.000*	*1.264*	*1.290*	*50*
Total	16.487	8.139	4.799	15.083	8.574	9.386	4.589	13.397	6.495	1.923	9.747	1.381	100.000	600
N	67	50	41	52	44	45	47	54	54	42	54	50		

Note: Rows of the error matrix are the map classes, and columns are the reference classes. Here, agreement is defined as a match between the map label and either the primary or alternate reference label.

TABLE 20.7
Error Matrix (as Percent of Area)

Map	1	2	3	4	5	6	7	8	9	10	11	12	Total	N
						Reference Class								
1	10.965	0.000	0.577	0.577	0.000	0.289	0.866	0.000	0.000	0.000	1.154	0.000	14.428	50
2	1.195	5.472	0.398	0.597	0.199	0.000	0.199	0.597	0.000	0.398	0.000	0.000	9.056	50
3	1.109	0.333	2.584	0.554	0.000	0.000	0.333	0.000	0.000	0.000	0.222	0.000	5.134	50
4	1.091	0.792	0.000	11.741	1.818	0.364	0.578	0.000	0.000	0.000	0.000	0.000	16.382	50
5	0.197	0.000	0.000	2.675	4.123	0.000	0.510	1.575	0.000	0.197	0.000	0.000	9.277	50
6	0.212	0.000	0.000	0.212	0.000	7.224	0.850	0.000	1.487	0.000	0.637	0.000	10.624	50
7	1.086	0.078	0.078	0.000	0.000	0.078	2.030	0.078	0.078	0.000	0.310	0.000	3.814	50
8	0.000	0.574	0.287	0.287	1.029	0.000	0.912	9.736	0.000	0.574	0.000	0.000	13.398	50
9	0.000	0.000	0.000	0.117	0.000	0.933	0.000	0.000	3.940	0.000	0.535	0.117	5.642	50
10	0.000	0.343	0.000	0.069	0.000	0.000	0.000	0.137	0.034	1.099	0.034	0.000	1.717	50
11	0.000	0.000	0.188	0.188	0.000	0.000	0.000	0.000	0.188	0.188	8.486	0.000	9.238	50
12	0.000	0.000	0.000	0.000	0.000	0.000	0.000	0.000	0.129	0.000	0.026	1.135	1.290	50
Total	15.854	7.591	4.111	17.018	7.169	8.887	6.276	12.124	5.856	2.455	11.406	1.252	100.00	600
N	73	51	34	66	33	45	47	52	50	38	66	45		

Note: Rows of the error matrix are the map classes, and columns are the reference classes. Here, agreement is defined as a match between the map label and the primary reference label.

TABLE 20.8
Users' and Producers' Accuracies

ID	Class	Primary Label				Primary or Alternate Label			
		Users' (SE)		Producers' (SE)		Users' (SE)		Producers' (SE)	
1	Needleleaf forest	76.0	(6.1)	69.2	(4.3)	88.0	(4.6)	77.0	(4.1)
2	Broadleaf forest	60.4	(7.2)	72.1	(6.6)	73.6	(6.5)	81.9	(5.9)
3	Mixed forest	50.3	(7.2)	62.8	(9.8)	67.6	(6.9)	72.3	(8.8)
4	Shrubland	71.7	(6.6)	69.0	(4.3)	78.3	(6.0)	85.1	(3.9)
5	Grassland	44.4	(7.2)	57.5	(8.6)	77.5	(6.1)	83.9	(6.3)
6	Subpolar, lichen-moss	68.0	(6.7)	81.3	(5.3)	82.0	(5.5)	92.8	(4.0)
7	Wetland	53.2	(7.1)	32.3	(5.8)	77.6	(6.0)	64.5	(8.8)
8	Cropland	72.7	(6.4)	80.3	(4.4)	89.8	(4.1)	89.8	(3.3)
9	Barren lands	69.8	(6.6)	67.3	(6.8)	90.5	(4.1)	78.6	(6.1)
10	Urban	64.0	(6.8)	44.8	(10.5)	78.0	(5.9)	69.6	(12.3)
11	Water	91.9	(3.9)	74.4	(4.9)	91.9	(3.9)	87.1	(4.4)
12	Snow and ice	88.0	(4.6)	90.7	(8.5)	98.0	(2.0)	91.5	(7.7)
		Overall accuracy 68.5% (SE = 2.1%)				Overall accuracy 82.8% (SE = 1.8%)			

owing to the prevalence of both lichen and moss in certain wetlands and the low biomass of both lichen/moss and herbaceous classes. Difficulties of ephemeral wetlands in semiarid regions due to endorheic rivers and streams resulted in inaccuracies, which were corrected by masks in postprocessing. The complexity of the landscape along the frontier of temperate and tropical classes caused some disagreement when assessing classes on NALCMS level II, which was not reflected in this accuracy assessment.

The assessment showed that the overall land-cover map accuracy approached the required standard, but it also indicated the need for methodology improvements and use of additional information to separate spectrally confused classes.

As part of the map assessment, the NALCD2005 was compared to the circa-2005 GLOBCOVER North America regional land cover generated from 300-m MERIS data with a 50-class legend defined by the LCCS. The GLOBCOVER accuracy was evaluated over North America for the 12-class IGBP legend against the same reference data sample used to evaluate NALCD2005. The overall agreement of the current GLOBCOVER over North America when evaluated with the only primary reference labels was 45%, with a kappa of 0.40. Agreement with either the primary or alternative reference label was 59% with a kappa of 0.54, which was significantly lower than the NALCD2005 of 68% with a kappa of 0.59 and 82.8% with a kappa of 0.73. A number of factors could cause the difference between these two products, such as different initial legends, dissimilarity between pixel sizes of MODIS 250 m versus MERIS 300 m, and differences in class description and interpretation. The map-to-map comparison revealed significant disagreement in the Canadian subarctic and arctic regions where GLOBCOVER was highly generalized. The GLOBCOVER had similar difficulties as the NALCD2005 discussed earlier in separating land-cover classes, such as natural grassland and pasture, shrubland and shrub-covered wetlands and certain croplands, as well as lichen/moss and herbaceous classes. The higher accuracy of NALCD2005 was an expected result considering the size of the mapping area of North America compared to global coverage, allowing a better focused legend on the North America biomes and the use of additional auxiliary data and local expertise.

20.4.3 COUNTRY-SPECIFIC ACCURACY

An estimate of the NALCD2005 accuracy over Canada for 12 classes was derived from 250 samples extracted from the North American sampling design shown in Figure 20.4. Accuracy results were estimated for the primary reference label and the primary or alternative label for cases where the primary label did not agree with the pixel label. The confidence level was assigned based on the ease with which a land cover was visually interpreted and on the number of classes present in the 250-m pixel. A high confidence was assigned to pixels where the high-resolution image was acquired at an appropriate time of the year for the class in question (e.g., summer to fall for deciduous) and where one class clearly dominated the land cover within the pixel. For the primary label only, overall accuracy was between 62% and 80% depending on the confidence level, whereas when considering either the primary or the alternative label as being correct, the accuracy was between 77% and 88%. The increase in classification accuracy with high reference data interpretation confidence was a function of greater land-cover homogeneity within the reference data footprint and less ambiguity due to reference image quality.

The accuracy of the U.S. maps was achieved in two ways. First, cross-validation accuracy was estimated from the DT classifier, which was an average of 69% for all zones. Cross-validation can provide relatively reliable estimates for land-cover predictions if the reference data used for cross-validation are collected based on a statistically valid sampling design. The second method was based on comparison with an independent sample of the 250 points for the United States. Error matrices were developed and presented for two definitions of agreement. First, an accuracy of 81% was found when agreement was defined as a match between the map class and either the primary or the secondary reference label. Second, an overall accuracy of 66% was found when agreement was defined using only the primary reference label. Overall users' and producers' accuracies were estimated for each error matrix. This analysis indicated that some proportion of land-cover classes were mixed and highlighted the difficulty in representing mixed pixels in a discrete map.

The accuracy of the Mexican section of the map was estimated using a set of sample data separated before training the DT. The normalized overall accuracy for 13 out of the 15 classes classified for Mexico at Level II yielded 82%. Producers' accuracies ranged between 29% and 92%

(average 68%) and users' accuracies between 36% and 97% with a mean of 75%. Confusion regarding classes was mainly between needleleaf/broadleaved forest and mixed forest and between tropical/subtropical and temperate/subpolar classes of shrubland and grassland.

20.5 CONCLUSIONS

International collaboration on large area land-cover mapping using satellite-based Earth observations offers a number of advantages, such as improved accuracy achieved through engagement of local experts and resources, consistency across the continent, and product standardization. A bottom-up approach assumes compilation of national land-cover maps derived from the same data source using a similar mapping methodology and classification system into a continental product. This approach permits the generation of a harmonized continental scale product that preserves specific national requirements with the use of a multilevel hierarchical legend. Here, the North American 2005 land cover at a spatial resolution of 250 m provides a harmonized view of the physical cover of the earth's surface across the continent. Nineteen land-cover classes were defined using the LCCS standard developed by the FAO of the United Nations. The map is intended for users who require land-cover information at the continental scale. Data are available from http://www.cec.org.

Future research efforts of the NALCMS will focus on developing the change detection component of the monitoring framework. The objective is to develop a methodology for annual updating and delivering continental scale land-cover and land-cover change information at 250 m. The NALCMS collaboration will also contribute to initiatives directed toward finer spatial and temporal resolution land-cover monitoring at continental scales.

ACKNOWLEDGMENTS

The authors are grateful to the trilateral CEC and its Secretariat for support and facilitation of the NALCMS project collaboration. This NALCMS project acknowledges the contribution of Dr. Steve Stehman (State University of New York) who designed the two-stage accuracy assessment sampling framework under CEC contact.

REFERENCES

Arino, O., Bicheron, P., Achard, F., Latham, J., Witt, R., and Weber, J. 2008. The most detailed portrait of Earth. *ESA Bulletin*, 136, 24–31.

Jones, B.K., 2008. North American land cover summit: Summary. In J.C. Campbell, K.B. Jones, J.H. Smith, and M.T. Koeppe (Eds.), *North America Land Cover Summit*. Washington, DC: Associate of American Geographers.

ALCC. 2008. Agriculture Land Cover Classification (ALCC). Classification of Agricultural Lands in Alberta, Saskatchewan and the Peace Region of British Columbia. Digital Environmental TM.

Bauer, E. and Kohavi, R. 1999. An empirical comparison of voting classification algorithms: Bagging, boosting, and variants. *Machine Learning*, 36(1–2), 105–139.

Breimann, L., Friedman, J., Ohlsen, R., and Stone, C. 1984. *Classification and Regression Trees*. Wadsworth, Belmont, CA

CDED. 2000. Canadian Digital Elevation Data. Government of Canada, Natural Resources Canada, Earth Sciences Sector, Centre for Topographic Information.

CEOS. 2006. Satellite observation of the climate system: The Committee on Earth Observation Satellites (CEOS) response to the implementation plan for the Global Observing System for Climate in support of the UNFCCC, 54 pp. Available at: http://www.ceos.org

Colditz, R.R., López, D., Wehrmann, T., Hüttich, C., Crúz, M.I., Muñoz, E., and Ressl, R., 2008. Guideline for automated training data derivation. Working paper in the project: North American Land Change Monitoring System (NALCMS), pp. 18.

DeFries, R., Hansen, M., Townshend, J.R.G., and Sohlberg, R. 1998. Global land cover classifications at 8 km spatial resolution: The use of training data derived from Landsat imagery in decision tree classifiers. *International Journal of Remote Sensing*, 19, 3141–3168.

Fernandes, R.A., Pavlic, G., Chen, W., and Fraser, R. 2001. Canada-wide 1-km water fraction derived from National Topographic Data Base maps. Natural Resources Canada.

Fraser, R., Hall, R., Landry, R., Lynham, T., Raymond, D., Lee, B., and Li, Z. 2004. Validation and calibration of Canada-wide-resolution satellite burned area maps. *Photogrammetric Engineering and Remote Sensing*, 70, 451–460.

Friedl, M.A. and Brodley, C.E. 1997. Decision tree classification of land cover from remotely sensed data. *Remote Sensing of Environment*, 61, 399–409.

Friedl, M.A., Brodley, C.E., and Strahler, A. 1999. Land cover classification accuracies produced by decision trees at continental to global scales. *IEEE Transactions on Geoscience and Remote Sensing*, 37, 969–977.

Friedl, M.A., McIver, D.K., Hodges J.C.F., Zhang X.Y., Muchoney, D., Strahler, A.H., Woodcock, C.E., et al. 2002. Global land cover mapping from MODIS: Algorithms and early results. *Remote Sensing of Environment*, 83, 287–302.

Friedl, M.A., Sulla-Menashe, D., Tan, B., Schneider, A., Ramankutty, N., Sibley, A., and Xiaoman, H. 2010. MODIS Collection 5 global land cover: Algorithm refinements and characterization of new datasets. *Remote Sensing of Environment*, 114, 168–182.

Freund, Y. and Schapire, R.E. 1997. A decision-theoretic generalization of on-line learning and an application to boosting. *Journal of Computer and System Sciences*, 55(1), 119–139.

GCOS. 2006. Systematic observation requirements for satellite-based products for climate. Supplemental details to the satellite-based component of the Implementation Plan for the Global Observing System for Climate in Support of the UNFCCC. GCOS-107, September 2006. Available at: http://www.wmo.int/pages/prog/gcos/Publications/gcos-107.pdf

GTOS. 2008. Terrestrial Essential Climate Variables. Biennial report supplement for Climate Change Assessment, Mitigation and Adaptation. GTOS publication 52. FAp. 2008. 44 pp.

Hansen, M., Dubayah, R., and DeFries, R. 1996. Classification trees: An alternative to traditional land cover classifiers. *International Journal of Remote Sensing*, 17, 1075–1081.

Homer, C.G., Ramsey, R.D., Edwards, Jr., T.C., and Falconer, A. 1997. Land cover-type modeling using a multi-scene Thematic Mapper mosaic. *Photogrammetric Engineering and Remote Sensing*, 63, 59–67.

Homer, C. and Gallant, A. 2001. Partitioning the conterminous United States into mapping zones for Landsat TM land cover mapping. USGS White Paper.

Homer, C., Huang, C., Young, L., Wylie, B., Coan, M. 2004. Development of a 2001 National Landcover Database for the Unites states. *Photogrammetric Engineering and Remote Sensing*, 70(7), 829–840.

INEGI. 2005. Dirección General de Geografía—INEGI (ed.), Conjunto de Datos Vectoriales de la Carta de Uso del Suelo y Vegetación, Escala 1:250,000, Serie III (CONTINUO NACIONAL), Instituto Nacional de Estadística, Geografía e Informática—INEGI. Aguascalientes, Ags., México.

INEGI. 2007. Instituto Nacional de Estadística Geografía e Informática (INEGI) "Localidades de la República Mexicana, 2005," Obtenido de Principales resultados por localidad 2005, II Conteo de población y Vivienda 2005, Editado por Comisión Nacional para el Conocimiento y Uso de la Biodiversidad (CONABIO), 2007, México.

Khlopenkov, K.V. and Trishchenko, A.P. 2008. Implementation and evaluation of concurrent gradient search method for reprojection of MODIS level 1B imagery. *IEEE Transaction on Geoscience and Remote Sensing*, 46, 2016–2027.

Latifovic, R., Cihlar, J., and Beaubien, J. 1999. Clustering methods for unsupervised classification. In Proceedings of the 21st Canadian Remote Sensing Symposium, June 1999, Ottawa, Ont. CD-ROM. Canadian Aeronautics and Space Institute, Ottawa, Ont. Vol. 2, pp. 509–515.

Latifovic, R., Zhu, Z., Cihlar, J., Giri, C., and Olthof, I. 2004. Land cover mapping of North and Central America—Global Land Cover 2000. *Remote Sensing of Environment*, 89, 116–127.

Latifovic, R. and Pouliot, D.A. 2005. Multitemporal land cover mapping for Canada: Methodology and products. *Canadian Journal of Remote Sensing*, 31, 347–363.

Latifovic, R. and Pouliot, D.A. 2006. Analysis of climate change impacts on lake ice phenology in Canada using the historical satellite data record. *Remote Sensing of Environment*, 106, 492–507.

Latifovic, R., Pouliot, D., and Olthof, I. 2009. North America Land Change Monitoring System: Canadian perspective. 30th Canadian Symposium on Remote Sensing, June 22–25. Lethbridge Alberta Canada.

Loveland, T.R. and Belward, A.S. 1997. The IGBP-DIS global 1 km land cover data set, DISCover: Results. *International Journal of Remote Sensing*, 18, 3289–3295.

Luo, Y., Trichtchenko, A.P., and Khlopenkov, K.V. 2008. Developing clear-sky, cloud and cloud shadow mask for producing clear-sky composites at 250-meter spatial resolution for the seven MODIS land bands over Canada and North America. *Remote Sensing of Environment*, 112, 4167–4185.

NCEP. 2009. North American Regional Reanalysis (NARR) Data Sets. Available at: http://nomads.ncdc.noaa. gov/data.php?name=access through the NOAA National Operational Model Archive and Distribution System (NOMADS) of the National Oceanic and Atmospheric Administration (NOAA), National Climatic Data Center (NCDC), Asheville, NC.

NLCC. 2008. Circa-2000 Northern Land Cover of Canada (NLCC). Earth Sciences Sector, Canada Centre for Remote Sensing, Natural Resources Canada.

NLWIS. 2009. National Land and Water Information System (NLWIS) Land Cover for agricultural regions of Canada, circa 2000. Government of Canada/Agriculture and Agri-Food Canada (GC/AAFC).

NRN. 2007. *National Road Network*, 2nd ed. Government of Canada, Natural Resources Canada, Earth Sciences Sector, Geomatics Canada, Centre for Topographic Information.

Olthof, I., Latifovic, R., and Pouliot, D. 2009. Development of a circa 2000 land cover map of northern Canada at 30 m resolution from Landsat. *Canadian Journal of Remote Sensing*, 35, 152–165.

Pouliot, D.A., Latifovic,R., Fernandes, R., and Olthof, I. 2009. Evaluation of annual forest disturbance monitoring using a statistic decision tree approach and 250 m MODIS data. *Remote Sensing of Environment*, 113, 1749–1759.

Quinlan, J.R. 1993. *C4.5: Programs for Machine Learning*. Morgan Kaufmann Publishers Inc., San Francisco, CA.

SILC. 2000. Satellite Information for Landcover of Canada (SILC); Beaubien, J; Blain, D; Chen, J M; Cihlar, J; Fernandes, R; Fraser, R; Latifovic, R; Peddle, D; Tarnocai, C; Trant, D; Wulder, M; Guindon, B; in, Workshop Report, Ottawa, Ontario, August 14–15, 2000.

Särndal, C.E., Swensson, B., and Wretman, J. 1992. *Model-Assisted Survey Sampling*. New York: Springer-Verlag.

Smith, J.H. 2008. North America land cover summit: Introduction. In J.C. Campbell, K.B. Jones, J.H. Smith, and M.T. Koeppe (Eds.), *North America Land Cover Summit*. Washington, DC: Associate of American Geographers.

Stahler. 1999. MODIS Land Cover Product Algorithm Theoretical Basis Document V5.0 MODIS Land Cover and Land-Cover Change.

Stehman, S.V. 2010. Documentation of Sampling Design and Analysis for Assessing the Accuracy of the NALCMS 2005 Land-cover Map. Working paper in the project: North American Land Change Monitoring System (NALCMS), pp. 9.

Timoney, K.P., Laroi, G.H., Zoltai, S.C., and Robinson, A.L. 1992. The high subarctic forest-tundra of northwestern Canada: Position, width, and vegetation gradients in relation to climate. *Arctic*, 45, 1–9.

Trishchenko, A.P., Luo, Y., and Khlopenkov, K.V. 2006. A method for downscaling MODIS land channels to 250m spatial resolution using adaptive regression and normalization, Proceedings of SPIE—The International Society for Optical Engineering v.6366 art. no. 636607. 8 pp.

Turner, P.D., Cohen, W.B., Kennedy, R.E., Fassnacht, K.S., and Briggs, J.M. 1999. Relationships between leaf area index and Landsat TM spectral vegetation indices across three temperate zone sites. *Remote Sensing of Environment*, 70, 52–68.

Wulder, M.A., Dechka, J.A., Gillis, M.A., Luther, J.E., Hall, R.J., Beaudoin, A., and Franklin, S.E. 2003. Operational mapping of the land cover of the forested area of Canada with Landsat data: EOSD land cover program. *Forestry Chronicle*, 79(6), 1075–1083.

Xian, G., Homer, C., and Fry, J. 2009. Updating the 2001 National Land Cover Database landcover classification to 2006 by using Landsat imagery change detection methods. *Remote Sensing of Environment*, 113, 1133–1147.

Zhang, Q., Pavlic, G., Chen,W., Fraser, R., Leblanc, S., and Cihlar, J. 2004a. A semi-automatic segmentation procedure for feature extraction in remotely sensed imagery. *Computers and Geosciences*, 31, 289–296.

Zhang, Q., Pavlic, G., Chen,W., Latifovic, R., Fraser, R., and Cihlar, J. 2004b. Deriving stand age distribution in boreal forests using SPOT/VEGETATION and NOAA AVHRR imagery. *Remote Sensing of Environment*, 91, 405–418.

Zhao, H. and Fernandes., R. 2009. Daily snow cover estimation from advanced very high resolution radiometer polar pathfinder data over North Hemisphere land surface during 1983–2004. *Journal of Geophysical Research*, 114, D05113. doi:10.1029/2008.

21 The Application of Medium-Resolution MERIS Satellite Data for Continental Land-Cover Mapping over South America

Results and Caveats

Lorena Hojas Gascon, Hugh Douglas Eva, Nadine Gobron, Dario Simonetti, and Steffen Fritz

CONTENTS

21.1   INTRODUCTION

In this work, we present a new land-cover map of South America derived from the 300 m resolution MEdium Resolution Imaging Spectrometer (MERIS) sensor, using data from 2009 to 2010. The results are compared to those from similar continental products from SPOT VGT (Global Land Cover 2000, GLC2000) and from MODIS (MODerate Imaging Spectrometer). We used the new map to assess the major land-cover changes that have occurred since the year 2000, using the GLC2000 map (Bartholomé and Belward, 2005) as historical reference. The product is assessed using finer spatial resolution data from Landsat Thematic Mapper.

21.1.1   RATIONALE FOR CONTINENTAL MAPPING

Continental land-cover maps are necessarily made at large mapping scales, generally at around 1:5,000,000. Despite their scale, they have a broad range of applications, such as providing inputs on essential climate variables (ECVs) for global models, showing the distribution of ecosystems, helping with biodiversity priorities (Dinerstein et al., 1995), and outlining species distribution. When updated on a regular basis, they can demonstrate the integrity of protected areas and highlight land-cover change hot spots (Hojas Gascon and Eva, 2011) to focus on more detailed studies (e.g., Achard et al., 2002, 2007).

21.1.2   PREVIOUS LAND-COVER MAPS OF SOUTH AMERICA

A number of land-cover databases, from the 1970s to 1980s, are available for South America, based on climate data and *ad hoc* visual interpretation of satellite data. They represent an inseparable mix of actual and potential land cover, and all describe environmental conditions as they were 30–40 years ago (Holdridge et al., 1971; Hueck and Seibert, 1972; UNESCO, 1981).

Systematic land-cover maps for South America have been produced since the 1990s based on data collected systematically by earth-observing satellites. They benefit from the uniformity of observations across the continent and offer an improved spatial detail; however, they do not have the thematic richness of the earlier products (Eva et al., 2004). They have, nevertheless, introduced a set of advantages. They can provide a relatively up-to-date view of land-cover changes in a dynamic region; provide a synthetic continental view achieved by the same method (unlike the compilation of national maps); exhibit a higher spatial accuracy and precision than conventional maps; demonstrate the major regions of land-cover change since the mid-1970s to date; and are available in digital format, which can be easily updated with new information coming from different sensors; and finally, they can be readily integrated into geographic information systems (GIS) for spatial analysis and query. The International Geosphere Biosphere Programme (IGBP) global land cover, initiated in 1990, was the first effort on global land-cover mapping using data from earth-observing satellites—the 1-km resolution AVHRR (Advanced Very High Resolution Radiometer) sensor. The land-cover classification, with divisions between vegetation cover types, was specifically defined for use in biogeochemical models (Loveland et al., 2000).

In the first phase of the Tropical Ecosystem Environment observation by Satellite project (TREES I), a map was produced, also from the NOAA-AVHRR sensor, for the humid tropical forest of the Amazon basin based on data from 1992. Three main land-cover classes were mapped—forest, nonforest, and fragmented forest—and more specific land-cover types were extracted from the UNESCO vegetation map of South America (1981; Eva et al., 1999). Following the IGBP initiative, data have also been used from the MODIS sensor onboard Aqua and Terra satellites to provide global land-cover mapping in 17 land-cover type classes, originally at 1-km resolution and more recently at 500 m (V005 Land-Cover Type) for the year 2005 (Friedl et al., 2002; USGS, 2009). More recently, a vegetation map for South America (Eva et al., 2004) for the year 2000 was developed as part of the

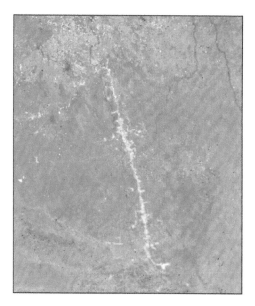

FIGURE 21.1 **(See color insert.)** Trans-Amazonian highway (BR163) at the north of Pará state. Mosaic of SPOT VGT data (left) and mosaic of MERIS FR data (right).

GLC2000 project (Batholomé and Belward, 2005) using data from several satellites, primarily from the 1-km spatial resolution SPOT VEGETATION (VGT) sensor (Eva et al., 2002).

To allow regular land-cover updates and change assessments, the European Space Agency (ESA) generated the global land-cover map GlobCover2005, using an automated processing chain from the full-resolution (FR) 300-m MERIS time series, with both global and regional classification systems (Defourny et al., 2006).

The use of medium spatial resolution satellite data, such as those provided by the MERIS sensor, makes possible considerable improvements in global land-cover mapping (see Figure 21.1). The GlobCover2005 product, however, contains some spatial and thematic inaccuracies, especially in some areas of tropical Latin America, which hinder the application of the data. These inaccuracies may be due, first, to the low number of satellite image acquisitions and valid observations. Spatial data coverage by MERIS FR is lower in the northwest of South America than in other places owing to some programmatic constraints on the acquisition. Moreover, this humid area has persistent cloud cover (Bicheron et al., 2008). Second, some inaccuracies are related to the method used for compositing the mosaic of land-surface reflectance. An automated image preprocessing chain was used using all available data from the MERIS time series, and therefore some images with cloud contamination may have reduced the mosaic quality.

21.2 OBJECTIVES

The objective of this work has been to generate a new land-cover map of South America using selected MERIS data instead of all available images, so as to avoid the problems described above. We have selected a series of MERIS images over the continent from 2008 to 2010 and, with image segmentation, isolated only the cloud-free areas for input into a mosaic composition. The resulting mosaic was classified using an unsupervised technique. A limitation of the method arises from the selection of the images, which invariably come from the dry season. This means that seasonal variations are missing, and therefore class confusions can be introduced, notably between dry forests (when leafless) and savannahs. To overcome this, we have used supplementary information from the Global MERIS FAPAR (Fraction of Synthetically Active Radiation) product at a reduced resolution

(RR) of 1 km. A validation of the final map product has been undertaken using high spatial resolution satellite interpretations produced by the TREES project (Beuchle et al., 2011). Finally, by comparing the map with the GLC2000 map, we have highlighted the major land-cover changes that have occurred since the year 2000.

21.3 MATERIAL AND METHODS

21.3.1 SATELLITE DATA

21.3.1.1 MERIS FR Data

The MERIS instrument was launched in March 2002 by ESA onboard ENVISAT, a polar-orbiting satellite that provides measurements of the earth's atmosphere, ocean, land, and ice surfaces. MERIS is a push-broom imaging spectrometer that measures reflected solar radiation in 15 spectral bands from the visible to the near infrared, at a full ground spatial resolution of 300 m. The MERIS swath is 1150 km wide, nominally obtaining a global coverage in 3 days (ESA, 2000). However, owing to restrictions in the downlink, FR data are not always acquired, resulting in a far lower coverage than desired. The spectral bands selected for our work were the most appropriate for vegetation mapping—681, 708, 753, and 865-nm bands (Dash et al., 2007).

21.3.1.2 MERIS RR Data FAPAR

Although acquisition of FR data is restricted, reduced spatial resolution data at 1 km are globally available. These data are collected and processed by the Joint Research Centre of the European Commission to give a monthly FAPAR (fraction of photosynthetically active radiation) product type (Gobron et al., 2006). This is directly linked to the photosynthetic activity of vegetation and is therefore a good indicator of plant growth and development (Knorr et al., 2007). For this work, we used the monthly average FAPAR data from the year 2009.

21.3.2 DATA PREPROCESSING

21.3.2.1 MERIS FR Data—Data Selection, Calibration, and Cloud Masking

Around 150 suitable preprocessing level 1b MERIS FR images, mainly from the years 2009 and 2010, were identified for the study area with the online MIRAVI ESA catalog (ESA, 2006). The selected MERIS FR data were corrected to top-of-atmosphere reflectance using the onboard calibration coefficients and then geometrically rectified using the tie points provided with the data (Brockmann Consulting, 2010). The clouds and haze from the images were removed, applying the automatically onboard-generated cloud mask and a cloud probability mask using an 80% threshold. This threshold was established by visual interpretation. A buffer of eight pixels around each potential "cloud" was similarly masked to reduce the effect of cloud shadow.

21.3.2.2 MERIS FR Data—Cross-Scan Correction and Forest Normalization

Examination of the MERIS FR data showed a strong cross-scan illumination effect across the forest domain (Figure 21.2), arising from the anisotropic reflectance properties of the target. Depending on the sun azimuth angle, this gave rise to differences in reflectance of over 40% between the east and the west of the image for the same cover type. This effect would produce classification anomalies in the final product, and therefore, a cross-scan illumination correction using a polynomial function was applied, followed by a normalization of the reflectance of the forest pixels to a standard reference value (Hansen et al., 2008).

21.3.2.3 MERIS RR Data

An examination of the MERIS RR data showed that artifacts were present in the monthly composites, probably arising from cloud contamination. Therefore, we performed a second compositing

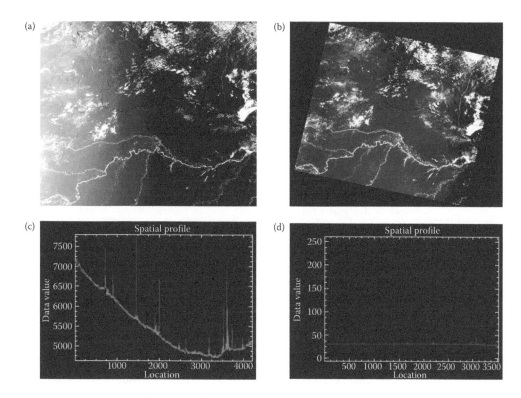

FIGURE 21.2 **(See color insert.)** Top: MERIS image before (a) and after (b) applying cross-track illumination correction. Bottom: spatial profile of the spectral band 2 from a transect of the same image before (c) and after (d) applying cross-track illumination correction.

into four seasonal mosaics, each of 3 months, using the maximum value FAPAR. We also computed the annual mean, maximum, minimum, and standard deviation.

21.4 SATELLITE DATA PROCESSING

21.4.1 MERIS FR DATA—CLASSIFICATION TECHNIQUE

MERIS FR data were composited together into a mosaic using the average reflectance. We used an unsupervised clustering algorithm, ISODATA (ERDAS, 1997), to create 100 spectral classes from the MERIS FR mosaic. These classes were then visually assigned to one of six thematic classes—forest, shrubs, grasslands, agriculture, water, and barren/sparse vegetation—with the use of national vegetation maps and other supporting online data such as Google Earth. A number of class confusions were evident in the data, mainly between "barren" and "agriculture," depending on crop development at the date of acquisition, and between savannahs or shrublands and dry forest, depending on the leaf coverage. To further elaborate and better identify these classes, we examined their FAPAR profiles.

21.4.2 DATA FUSION—COMBINING THE SPATIAL ACCURACY OF THE FR DATA WITH THE SEASONAL INFORMATION FROM THE RR DATA

The FAPAR data from the different thematic classes were clustered into 10 spectral classes. The spectral profile of the resulting clusters were then examined to further discriminate between land-cover classes, such as evergreen and dry forests, grassland and agriculture, or mixed and intensive

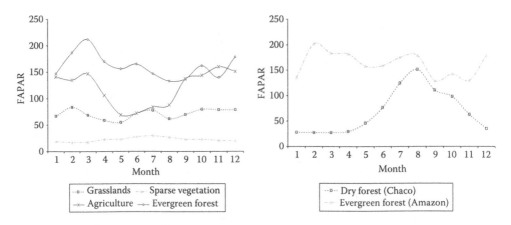

FIGURE 21.3 Monthly FAPAR values extracted from different land-cover types.

agriculture. Figure 21.3 shows an example of spectral profiles of FAPAR for evergreen and dry forests, agriculture, grasslands, and sparse vegetation types.

21.4.3 Validation Dataset

Validation of continental datasets is problematic owing to the extent of the area and the number of classes to be controlled. Recently, experts were employed to validate a series of points distributed across both the area and the thematic spectrum of such data (Mayaux et al., 2006). For this exercise, we used data from the TREES-3 project (Bodart et al., 2011), which comprised 10 × 10 km Landsat TM and ETM subsets located at the confluence of the geographic grid. Most of the Landsat data were obtained from the U.S. Geological Survey's (USGS) National Center for Earth Resources Observation and Science (http://glovis.usgs.gov) at full spatial resolution (30 m). Some 1200 subsets, covering 80% of the land area of South America, were available for the year 2005 (Argentina, Chile, and Uruguay were not covered by the TREES-3 study). The Landsat subsets were classified into forests, shrubland, other land cover (OLC), and water, and were validated by national experts at a series of workshops. The OLC class contained all nonligneous land-cover types (grasslands, agriculture, barren, etc.). We compared the land-cover data from the MERIS classification aggregated into the TREES classes with these data, using regression analysis.

21.5 RESULTS

21.5.1 Results for Land-Cover Mapping

The full continental classification map is shown in Figure 21.4, with the land-cover area per country presented in Table 21.1.

21.5.2 Validation Results

We compared the areas of each land-cover type within the TREES sample sites to those on the MERIS map, using the Pearson correlation coefficient. Forest cover ($R = 0.84$), OLC ($R = 0.7$), and water ($R = 0.9$) compare favorably with the validation data. Shrub, however, is poorly correlated with the validation data ($R = 0.34$). This is not surprising, as it is a transition class between forests and other land-cover types, and apart from this, the "shrub" class in the TREES-3 project also includes forest *regrowth*. Owing to the spatial resolution, this land-cover type is more likely to be

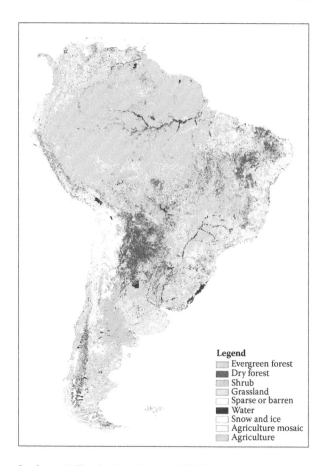

Legend
☐ Evergreen forest
■ Dry forest
☐ Shrub
☐ Grassland
☐ Sparse or barren
■ Water
☐ Snow and ice
☐ Agriculture mosaic
☐ Agriculture

FIGURE 21.4   (See color insert.) The final land-cover MERIS map.

included in the *mosaic of agriculture* in the MERIS map. In Table 21.2, we show the correspondence between permanent agriculture compiled from national statistics by FAO (FAOSTAT, 2011) and that mapped by MERIS. There is a general correspondence between the two datasets, but there are some differences in actual magnitude.

21.5.3   COMPARISON WITH OTHER LAND-COVER CONTINENTAL DATASETS

Comparing our results by country to those produced from GLC2000 and the MODIS product (Figure 21.5), we see a good agreement between the three products in forest areas, but there are major discrepancies in agricultural and natural nonforest vegetation areas (i.e., grasslands and shrublands), primarily when pastures are labeled as agricultural land in the MERIS map and GLC2000 and as savannah in the MODIS map. In general, areas derived from the MERIS map correspond better with the GLC2000 map than with the MODIS map.

21.5.4   LAND-COVER CHANGES SINCE THE YEAR 2000

Comparison of land-cover maps of different resolutions, which are derived from different source materials, poses a number of methodological problems as it is difficult to assess whether changes between the maps are the result of real changes or the result of differences in the input data characteristics (See and Fritz, 2006). Nevertheless, the spatial improvements in the classification between the GLC2000 and the MERIS map are clear (Figure 21.6). It is therefore not possible to accurately

TABLE 21.1
Land-Cover Classes by Country in 1000 ha from the MERIS Land-Cover Map

	Country Area	Evergreen Forest	Dry Forest	Shrub	Grassland	Sparse or Barren	Snow and Ice	Agricultural Mosaic	Agriculture	Water
Argentina	278,040	10,983	27,318	79,767	57,640	43,215	517	27,400	26,814	4386
Bolivia	109,858	44,532	11,338	9826	16,907	15,406	50	8610	1894	1295
Brazil	851,488	356,615	52,969	101,260	28,557	3986	–	200,933	95,327	11,840
Chile	75,610	15,728	1706	11,771	11,294	23,185	2076	4914	3101	1835
Colombia	114,175	54,981	1506	5791	16,192	2473	–	23,049	9768	415
Ecuador	25,637	11,194	258	3005	1275	385	–	7788	1662	70
French Guiana	8353	8040	5	15	77	1	–	143	6	67
Guyana	21,497	19,117	145	975	607	39	–	565	35	15
Paraguay	40,675	8329	11,808	6124	4255	42	–	7258	2239	619
Peru	128,522	70,359	4920	12,235	11,572	18,157	189	7713	2269	1108
Suriname	16,382	15,215	34	107	133	2	–	640	20	230
Uruguay	17,622	603	902	2793	5903	29	–	4079	2905	409
Venezuela	91,205	41,693	3338	7170	14,285	3866	–	13,703	5999	1150
Total	**1,779,064**	**657,390**	**116,247**	**240,840**	**168,697**	**110,787**	**2831**	**306,794**	**152,039**	**23,439**

TABLE 21.2

Areas of Permanent Agriculture (in 1000 ha) Compiled from National Databases and from the MERIS Map

	MERIS	FAOSTAT
Argentina	26,814	33,000
Bolivia	1894	3819
Brazil	95,327	68,500
Chile	3101	1722
Colombia	9768	3461
Ecuador	1662	2500
French Guiana	6	17
Guyana	35	445
Paraguay	2239	4300
Peru	2269	4440
Suriname	20	56
Uruguay	2905	1673
Venezuela	5999	3350
Total	152,039	127,283

quantify changes by such comparisons; however, it is possible to generate general magnitudes and strata of changes if and when those changes are far higher than the satellite spatial resolution. A cross-tabulation of the major aggregated classes from the GLC2000 map and the new MERIS map shows these changes (Table 21.3). The main areas of land-cover change put into evidence by the new map are the Brazilian Amazon (arc of deforestation) and the Chaco region, extending from Bolivia to Paraguay and Argentina. Figure 21.7 shows the extent of the changes that have taken place in Rondônia since 2000. It is clear that such changes are evident as they occur in homogeneous landscapes. Small-scale changes occurring in heterogeneous, fragmented landscapes will be far more difficult to map, as for example land-cover changes occurring in the mountainous regions, or in the *cerrado* and *caatingas* of Brazil. Although the spatial resolution of the MERIS does not allow us to make accurate area estimates, the comparison between the state of the Brazilian Amazon in 2000 (GLC2000) and in the MERIS map gives an increment of the hot spot area of about 19,000 km^2 a year between our two reference dates (Hojas Gascon and Eva, 2011). The official INPE figure for the same period is 17,600 km^2 per year (INPE, 2010).

At the continental level, our data suggest a net annual loss of over 42,000 km^2 of forests from 2000 to 2010 for the continent; this compares with the FAO (2010) estimate of 40,000 km^2 per year. Other changes, however, such as transition from barren to grasslands, can be due to a reinterpretation of the data. We also noted that areas classed as intensive agriculture in the GLC2000 map along the east coast of Brazil are now mapped as mixed agriculture and natural vegetation. This arises from the increased spatial resolution of the MERIS data, allowing a better differentiation between classes.

21.6   DISCUSSION AND CONCLUSIONS

Preparing a new land-cover map of South America from medium spectral resolution MERIS data has required a different approach in data preparation. Major limitations arise from the low number of cloud-free acquisitions across the continent, notably in the equatorial regions. At the same time, major artifacts due to bidirectional reflectance effects need to be removed. Validating such products also poses many problems. In this case, we have used a set of finer spatial resolution classifications

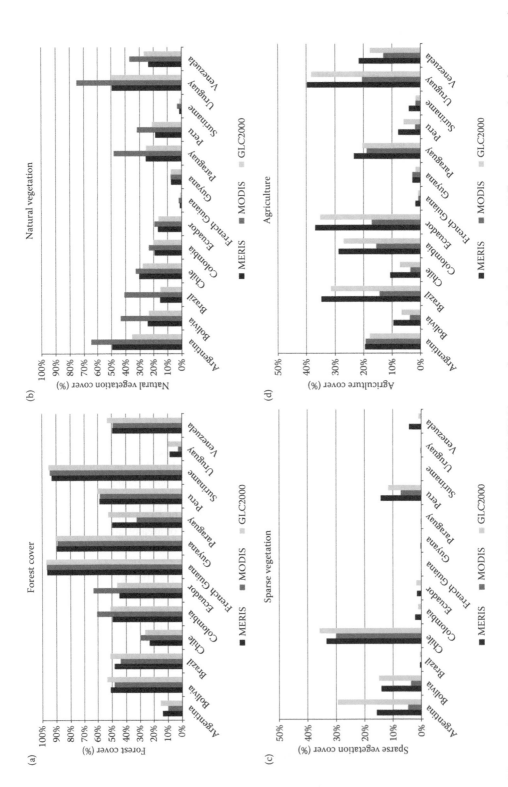

FIGURE 21.5 Comparisons by country from the MERIS, MODIS, and GLC2000 maps of areas of major land-cover type: (a) Forests, (b) natural nonforest vegetation (shrublands and grasslands), (c) sparse vegetation, and (d) agriculture.

TABLE 21.3

The Transitions in 1000 ha of Major Classes between the GLC2000 Map and the New MERIS-Derived Map

	Evergreen Forests	Dry Forests	Shrublands	Grasslands	Barren	Agriculture	Total 2000 (GLC)
Evergreen Forests	679,151	160	153	482	17	28,512	708,476
Dry Forests	222	124,945	281	114	18	9918	135,498
Shrublands	172	261	264,643	303	155	455	265,989
Grasslands	529	81	225	130,315	15,313	231	146,694
Barren	21	16	250	58,665	111,954	73	170,979
Agriculture	477	142	397	201	55	349,888	351,161
Total 2009/2010 (MERIS)	680,572	125,606	265,950	190,080	127,511	389,076	1,778,796

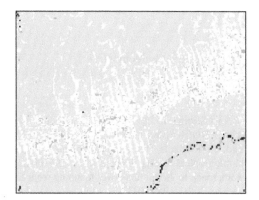
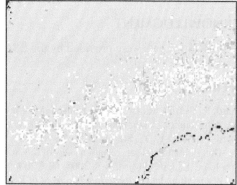

FIGURE 21.6 **(See color insert.)** A 200 km by 150 km extract from the MERIS (left) and GLC2000 (right) maps along the Trans-Amazonian highway in Brazil. Agriculture is represented in gray and light green and forest in darker green.

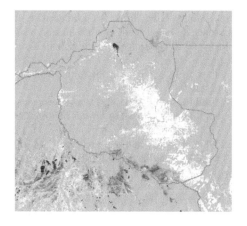
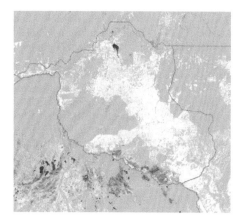

FIGURE 21.7 **(See color insert.)** Extract of Rondônia showing the agricultural expansion in yellow from GLC2000 (left) and MERIS 2009/2010 (right). The forest cover is in green and savannahs in red.

distributed across a large proportion of the study area. These data, along with national statistics, show that while giving a generalized overview of the land cover, major difficulties exist in trying to establish accurate area estimates. Much of this difficulty is due to class definitions. National statistics reflect land use; for example, a savannah may be classed as "agriculture," that is, pastures. Although a comparison of our new map with that of the year 2000 (GLC2000) can indicate major land-cover changes, it will be unwise to use such comparisons for quantitative measures. In our case, the forest changes can be said to give a good indication of where change really has occurred, but other changes may result from either a better interpretation of the data or from the change in spatial resolution between the two datasets. This points to the need for rigorous methodologies when preparing new land-cover maps for assessing changes. In recent years, the availability of data sources and the technical capacity to prepare such maps have increased. Satellite data are easily available through the network for downloading, already geometrically corrected and at various levels of processing. Desktop systems at little or no cost have become available for image processing and GIS. It is clear from our experience in preparing this new map that the input data sources, the spatial and radiometric resolutions and methods used to compile a land-cover map, and the class definitions, can make major differences to the end product. Therefore, comparisons with historical maps are likely to be compromised.

ACKNOWLEDGMENT

The MERIS FR data were provided by the ESA under a Category 1 agreement.

REFERENCES

Achard, F., Defries, R., Eva, H.D., Hansen, M., Mayaux, P., and Stibig, H.J. 2007. Pantropical monitoring of deforestation. *Environmental Research Letters*, 2, 045022. DOI:10.1088/1748-9326/2/4/045022. 11 pp.
Achard, F., Eva, H.D., Stibig, H.J., Mayaux, P., Gallego, J., Richards, T., and Malingreau, J.P. 2002. Determination of deforestation rates of the world's humid tropical forests. *Science*, 297, 999–1002.
Bartholomé, E. and Belward, A.S. 2005. GLC2000: A new approach to global landcover mapping from Earth observation data. *International Journal of Remote Sensing*, 26(9), 1959–1977.
Beuchle, R., Eva, H.D., Stibig, H.-J., Bodart, C., Brink, A., Mayaux, P., Johansson, D., Achard, F., and Belward, A. 2011. A satellite data set for tropical forest change assessment. Accepted for *International Journal of Remote Sensing*, 32(22), 7009–7031.
Bicheron, P., Defourny P., Brockmann, C., Schouten, L., Vancutsem, C., Huc, M., Bontemps, S., Leroy, M., Achard, F., Herold, M., Ranera, F., Arino, O. 2008. Products description and validation report GLOBCOVER. U. C. L., MEDIAS-France, Brockmann Consult, INFRAM, JRC European Commission, GOFC-GOLD, ESA. Toulouse, MEDIAS-FRANCE.
Bodart, C., Eva, H.D., Beuchle, R., Rasi, R., Simonetti, D., Stibig, H-J., Brink, A., Lindquist, E., and Achard, F. 2011. Pre-processing of a sample of multi-scence and multi-date Landsat imagery used to monitor forest cover changes over the tropics. *ISPRS Journal of Photogrammetry and Remote Sensing*, 66(5), 555–563.
Brockmann Consulting. 2010. BEAM Earth Observation Toolbox and Development Platform. Available at: http://www.brockmann-consult.de/cms/web/beam/documentation
Dash, J., Mathur, A., Foody, G.M., Curran, P.J., Chipman, J.W., and Lillesand, T.M. 2007. Land cover classification using multi-temporal MERIS vegetation indices. *International Journal of Remote Sensing*, 28(6), 1137–1159.
Defourny, P., Vancutsem, C., Bicheron, P., Brockmann, C., Nino, F., Schouten, L., and Leroy, M. 2006. GLOBCOVER: A 300m global land cover product for 2005 using ENVISAT MERIS time series. ISPRS Commission Mid-term Symposium, *Remote Sensing: From Pixels to Processes*, Enschede, the Netherlands.
Dinerstein, E., Olson, D.M., Graham, D.J., Webster, A.L., Primm, S.A., Bookbinder, M.P., and Ledec, G. 1995. *A Conservation Assessment of the Terrestrial Ecoregions of Latin America and the Caribbean.* Washington, DC: The World Bank, 129 pp.
ERDAS. 1997. *ERDAS Field Guide*. Atlanta, GA: ERDAS Inc.

ESA. 2000. ESA Missions—ENVISAT—MERIS instrument. Available at: http://envisat.esa.int/instruments/meris/

ESA. 2006. MIRAVI, Envisat MERIS Image Rapid Visualisation. Available at: http://miravi.eo.esa.int/en/

Eva, H.D., Belward, A.S., De Miranda, E.E., Di Bella, C.M., Gond, V., Huber, O., Jones, S., Sgrenzaroli, M., and Fritz, S. 2004. A land cover map of South America. *Global Change Biology*, 10, 731–744. DOI: 10.1111/j.1529-8817.2003.00774.x.

Eva, H.D., Belward, A.S., De Miranda, E.E., Di Bella, C.M., Gond, V., Huber, O., Jones, S., et al. 2002. *A Vegetation map of South America, EUR 20159 EN*. Luxembourg: European Commission.

Eva, H.D., Glinni, A., Janvier, P., and Blair-Myers, C. 1999. *Vegetation Map of Tropical South America at 1:5.000.000*. Ispra, Italy: Join Research Centre European Commission.

FAO. 2010. Global Forest Resources Assessment 2010—Main Report. FAO Forestry Paper 163. Rome, Italy: Food and Agriculture Organization of the UN.

FAO. 2011. FAOSTAT. Available at: http://faostat.fao.org/default.aspx

Friedl, M.A., McIver, D.K., Hodges, J.C.F., Zhang, X.Y., Muchoney, D., Strahler, A.H., Woodcock, C.E., et al. 2002. Global land cover mapping from MODIS: Algorithms and early results. *Remote Sensing of Environment*, 83(1–2), 287–302.

Gobron, N., Pinty, B., Taberner, M., Mélin, F., Verstraete, M.M., and Widlowski, J.-L. 2006. Monitoring the photosynthetic activity vegetation from remote sensing data. *Advances in Space Research*, 38, 2196–2202. DOI: 10.1016/j.asr.2003.07.079.

Hansen, M., Roy, D., Lindquist, E., Adusei, B., Justice, C.O., and Altstatt, A. 2008. A method for integrating MODIS and Landsat data for systematic monitoring of forest cover and change in the Congo Basin. *Remote Sensing of Environment*, 112, 2495–2513.

Hojas Gascon, L. and Eva, H.D. 2011. The application of medium resolution MERIS satellite data for identifying deforestation hotspots over the Brazilian Amazon. 2011. Forestsat 2010, *Operational Tools in Forestry Using Remote Sensing Techniques*. September 7–10, 2011, Lugo, Santiago di Compostella, Spain.

Holdridge, L.R., Grenke, W.C., Hatheway, W.H., Liang, T., and Tosi, J.A. 1971. *Forest Environment in Tropical Life Zones*. Oxford: Pergamon Press, pp. 747.

Hueck, K. and Seibert, P. 1972. Vegetationskarte von Südamerika/Mapa de la Vegetación de America del Sur. Stuttgart: Fischer.

INPE. 2010. Monitoramento da Floresta Amazonica Brasileira por Satelite. (Instituto Nacional De Pesquisas Espaciais, São José dos Campos, 2010). Available at: http://www.obt.inpe.br/prodes/

Knorr, W., Gobron, N., Scholze, M., Kaminski, T., Schnur, R., and Pinty, B. 2007. Impact of terrestrial biosphere carbon exchanges on the anomalous CO_2 increase in 2002–2003. *Geophysical Research Letters*, 34(9), L09703. DOI: 10.1029/2006GL029019.

Loveland, T.R., Reed, B.C., Brown, J.F., Ohlen, D.O., Zhu, J., Yang, L., and Merchant, J.W. 2000. Development of a global land cover characteristics database and IGBP DISCover from 1-km AVHRR data. *International Journal of Remote Sensing*, 21, 1303–1330. DOI:10.1080/014311600210191.

Mayaux, P., Eva, H.D., Gallego, J., Strahler, A.H., Herold, M., Agrawal, S., Naumov, S., et al. 2006. Validation of the global land cover 2000 map. *Geoscience and Remote Sensing, IEEE Transactions on*, 44(7), 1728–1739. DOI: 10.1109/TGRS.2006.864370.

See, L.M. and Fritz, S. 2006. A method to compare and improve land cover datasets: Application to the GLC-2000 and MODIS land cover products. *Geoscience and Remote Sensing, IEEE Transactions on*, 44(7), 1740–1746. DOI: 10.1109/TGRS.2006.874750.

UNESCO. 1981. Carte de la Végétation d'Amérique du Sud—Explicative notes. Paris: UNESCO, pp. 189 + 2 map sheets.

USGS. 2009. Land Cover Type Yearly L3 Global 500 m SIN Grid. Available at: https://lpdaac.usgs.gov/lpdaac/products/modis_products_table/land_cover/yearly_l3_global_500_m/mcd12q1

22 Mapping Land-Cover and Land-Use Changes in China

Xiangzheng Deng and Jiyuan Liu

CONTENTS

22.1 INTRODUCTION

Land-cover and land-use change is being increasingly considered a key subject and an important component of research on global environmental changes and sustainable development (IGBP Secretariat, 2005; Liu et al., 2003a, 2003b, 2010). China, one of the fastest developing countries, is confronted with the challenge of supporting a growing population. Since there is demand for more land to provide food and all kinds of services, the land-cover and land-use patterns in China have dramatically changed during the recent decades (Deng et al., 2010a, 2010b; Liu et al., 2003a, 2003b, 2010). However, data quality and reliability have been the biggest problems in getting a clear picture of the land-cover and land-use changes, and there have been large discrepancies in the estimation of the changes over time and space, which were arrived at by different research institutes (Deng et al., 2010c, 2010d; Liu et al., 2003b; SSB, 1996).

Remote-sensing technology as an efficient investigation method was introduced in China three decades ago to obtain accurate and timely information on land-cover and land-use changes (Zhang and Zhang, 2007). By the end of the 1980s, the China State Land Administration (CSLA; restructured as the Ministry of Land and Resources, MLR, in 1998) sponsored a program to investigate land use in the northwest, using Landsat TM imagery. In 1999, the newly founded MLR launched the National Land Use Change Program especially to monitor, through remote-sensing technology, the scale and pattern of urban land expansion and the decrease in cultivated land. This program has been carried out annually since then and has played an important role in shaping MLR's policies on land management and planning.

To meet the information needs of governments at all levels and promote an understanding of global environmental changes, the Chinese Academy of Sciences (CAS) created a national dataset of land-cover and land-use changes from the late 1980s to the mid-2000s at the scale of 1:100,000; this was based on the Remote Sensing Information Platform of National Resources and Environment (Liu et al., 2003b, 2010).

22.2 DATA AND METHODOLOGY

22.2.1 REMOTE-SENSING DATA

We mapped the land-cover and land-use changes using Landsat TM/ETM scenes at a spatial resolution of 30 m × 30 m. The database included time-series data for four periods: (1) the late 1980s, including Landsat TM scenes acquired from 1987 to 1990; (2) the mid-1990s, including Landsat TM scenes acquired during 1995/1996; (3) the late 1990s, including Landsat TM/ETM scenes acquired during 1999/2000; and (4) the mid-2000s, including Landsat TM/ETM scenes acquired during 2004/2005. For each period, we used 400–500 TM scenes to cover the whole country (514 scenes in the late 1980s, 520 scenes in the mid-1990s, 512 scenes in the late 1990s, and 411 scenes in the mid-2000s). These Landsat TM/ETM images were georeferenced and orthorectified using field-collected ground control points and high-resolution digital elevation models. A hierarchical classification system of 25 land-cover and land-use classes was applied to the final dataset (Table 22.1). The 25 classes of land cover and land use were grouped further into six aggregated classes: cultivated land, forest area, grassland, water area, built-up area, and unused land (Deng et al., 2008; Liu et al., 2003a, 2003b, 2010).

22.2.2 VISUAL INTERPRETATION

Visual interpretation and digitization of TM/ETM images at the scale of 1:100,000 were done to generate the thematic maps. The interpretation process involved preprocessing of digital images, visual interpretation, and detection of land-cover and land-use categories (Deng et al., 2010d), which are summarized in Figure 22.1.

Before visual interpretation, Landsat TM/ETM digital images were preprocessed to remove cloud-fog cover, using the homomorphic filtering method. Then the image distortion brought by radiant errors was cleared through radiometric calibration in which three procedures—remote sensor calibration, atmospheric correction, and topographic correction—were included. Remote sensor calibration procedures handled mainly the incremental correction coefficient and deviation correction (Wu and Cao, 2006).

Atmospheric correction procedures used the empirical model to remove the effects of atmosphere on the reflectance values of images (Zheng et al., 2007). Topographic correction procedures were followed to eliminate illumination effects (Wu et al., 2008). After that, false-color images were fused and produced (Zhang et al., 2002). Next, the fused, false-color digital images were georeferenced and projected into the Albers projection system with reference to the ground control points (Deng et al., 2010d).

Next, the images were visually interpreted using the visual interpretation approach, that is, a manual tracing and on-screen digitization technique to detect land-cover and land-use changes (Liu et al., 2003b). Depending on the sensor resolution, image color, shadow, size, texture, pattern, site, and association, the land-surface features were identified (Deng et al., 2010d). Using ArcGIS (geographic information system) software and vector drawing tools, the land-cover and land-use maps were created by overlaying the images (Deng et al., 2010c). Change detection maps were obtained by combining the four-date land-cover and land-use data. Afterward, land-cover and land-use change maps were produced by overlaying the change detection map, survey data, and secondary data.

Although visual interpretation was not a completely new method, its efficiency reduced the overall classification error as prior knowledge was incorporated into the whole process. The overall interpretation accuracy for land-cover and land-use classification was up to 92.9% for the late 1980s, 97.6% for the late 1990s, and over 95% for the mid-2000s.

TABLE 22.1

Land-Cover and Land-Use Classification in China

First Level of Classification	Second Level of Classification	Descriptions
Cultivated land		Land for agricultural use
	Paddy field	Cultivated area with water resource guarantee and irrigating facilities for rice growing
	Dry land	Land for cultivation without irrigating facilities; dry cropland and land growing vegetables
Forestry area		Land covered by trees including arbor, shrub, bamboo and land for forestry use
	Closed forest	Natural or man-made forests with canopy cover higher than 30%
	Shrub	Land covered by trees less than 2 m high and with canopy cover higher than 40%
	Open forest	Land covered by trees with canopy cover between 10% and 30%
	Other forest	Economic forest cover including tea garden, orchid, etc. and other non-grown-up forest cover
Grassland		Land covered by herbaceous plant with canopy cover higher than 5%, including shrub grass for pasture and woods with canopy cover less than 10%
	Dense grass	Grassland with canopy cover higher than 50%
	Moderate grass	Grassland with canopy cover between 20% and 50%
	Sparse grass	Grassland with canopy cover between 5% and 20%
Water area		Land covered by water bodies or land with facilities for irrigation and water conservation
	Rivers	Land covered by rivers including canals
	Lakes	Land covered by lakes
	Reservoir and ponds	Man-made facilities for water conservation
	Permanent Ice and Snow	Land covered by ice and snow all the year
	Beach and Shore	Land between high tide level and low tide level
	Bottomland	Land between normal water level and flood level
Built-up area		Land used for urban and rural settlements, factories, and transport facilities
	Urban built-up	Land used for cities and counties
	Rural settlements	Land used for settlements in the rural area
	Other built-up area	Land used for factories, quarries, mining, oil-field slattern outside cities and land for special uses such as transportation and airports
Unused land		Land that is not put into practical use or is difficult to use
	Sand	Sandy land with vegetation cover less than 5%
	Gobi	Gravel land with vegetation cover less than 5%
	Salina	Land with saline accumulation and sparse vegetation
	Wetland	Land with a permanent mixture of water and herbaceous or woody vegetation that cover extensive areas
	Bare Soil	Land covered by bare soil and with vegetation cover less than 5%
	Bare Rock	Land covered by bare rocks and with vegetation cover less than 5%
	Other remain area	Other land such as alpine, desert, and tundra

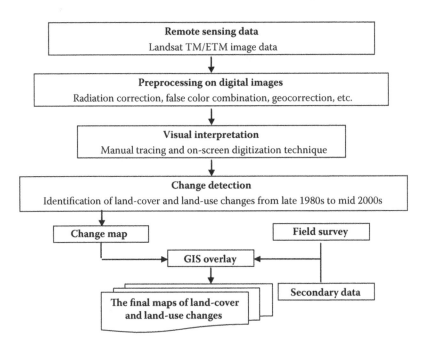

FIGURE 22.1 Diagram of the interpretation of land-cover and land-use changes.

22.2.3 ONE-KILOMETER AREA PERCENTAGE DATASET

We developed an approach to generate 1-km area percentage data to trace land-cover and land-use changes in China (Deng et al., 2010c; Liu et al., 2003a, 2003b). This was based on the map-algebra concept, a data manipulation language designed specifically for geographic cell-based systems.

The procedure has four steps. The first step is to generate a vector map of land-cover and land-use changes during the study period at a scale of 1:100,000 based on the remote-sensing Landsat TM/ETM data. The second is to generate a 1-km FISHNET vector map georeferenced to a boundary map of the study area at a scale of 1:100,000. Each cell of the generated 1-km FISHNET vector map has a unique ID. The third step is to overlay the vector map with the 1-km FISHNET vector map. This is done by aggregating converted areas in each 1-km grid identified by cell IDs of the 1-km FISHNET vector map in the TABLE module of ArcGIS software. Finally, the area percentage vector data are transformed into grid raster data to identify the conversion direction and intensity. The design of the workflow ensures no loss of area information. Without special notification, the statistical area of cultivated land according to the GRID data is the survey area by satellite remote-sensing data, which can be called "gross area." The flowchart depicting generation of the dataset is given in Figure 22.2.

Since the decoded information from Landsat TM/ETM data consists of 25 land-cover and land-use classes, each class of land cover and land use is given an area percentage of that kind of land cover and land use in a grid cell. That is to say, when we aggregate the areas of all the 25 land-cover and land-use classes, we get 100% for each grid cell.

22.3 RESULTS

22.3.1 GENERAL CONDITION OF LAND-COVER AND LAND-USE CHANGES IN CHINA

According to the remote-sensing features of land-cover and land-use changes in China, two phases—from the late 1980s to the late 1990s and from the late 1990s to the mid-2010s—can be identified.

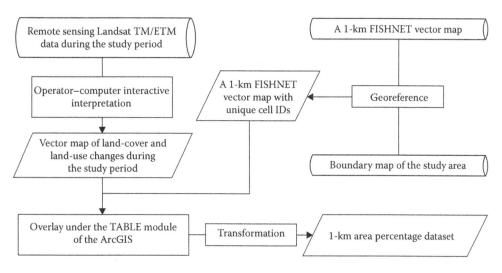

FIGURE 22.2 Flowchart depicting generation of 1-km area percentage dataset.

From the late 1980s to the late 1990s, there was a significant increase in cultivated land, with a more remarkable increase in dry land. The increase in cultivated land occurred in northern China, mostly converted from forest area and grassland, whereas southern China showed a decrease in cultivated land owing to urban land expansion. Both forest area and grassland decreased, whereas the built-up area increased mostly at the expense of cultivated land. From the late 1990s to the mid-2000s, cultivated land decreased, especially paddy field in southern China. Built-up area expanded rapidly and occupied a large area of high-quality cultivated land, especially in the southeastern coastal areas, inland plains, and traditional farming zones. On the contrary, forest area increased because of the "Grain for Green" Project. Grassland decreased because of its conversion to cultivated land.

22.3.2 Land-Cover and Land-Use Changes from the Late 1980s to the Late 1990s

22.3.2.1 Temporal Changes

We estimated that, during the late 1990s, cultivated land was about 141.14 million ha, with paddy field about 35.65 million ha and dry land 105.49 million ha. From the late 1980s to the late 1990s, cultivated land increased by about 2.99 million ha or 2.17% (Table 22.2). Dry land increased by about 2.85 million ha or 2.78%, and paddy field increased by about 0.14 million ha or 0.4%. These changes represented an imbalance between loss and gain of cultivated land. About 3.2 million ha of cultivated land was converted to other land uses, including 1.5 million ha to built-up area (Table 22.3). The conversion from cultivated land to forest area was about 0.52 million ha or 16%, and the conversion to grassland was about 0.64 million ha or 20.2%. However, 6.2 million ha of cultivated land was converted from other land uses, which was much larger than the loss of cultivated lands. As a result, the net changes in cultivated land appeared to increase from the late 1980s to the late 1990s.

Forest area was about 226.74 million ha in the late 1990s, and it decreased to 1.09 million ha or 0.48% from the late 1980s to the late 1990s. The net change of forest area resulted from the imbalance between a loss of 2.7 million ha and a gain of 1.6 million ha. It was estimated that 64% of the loss was converted to cultivated land and 29.8% of the loss to grassland.

There was a loss of 5.6 million ha of grasslands, which were converted to other land uses from the late 1980s to the late 1990s, and more than half of the grasslands (3.4 million ha) was converted to cultivated land. However, about 2.2 million ha of new grassland was obtained by conversion of other land uses. Thus, the net change of grassland decreased by 3.44 million ha or 1.12% during this period.

TABLE 22.2
Land-Cover and Land-Use Changes in China from Late 1980s to the Late 1990s

Land-Cover and Land-Use Types	1990	2000	Change	% Change
Cultivated land	138.15	141.14	2.99	2.17
Paddy field	35.51	35.65	0.14	0.40
Dry land	102.64	105.49	2.85	2.78
Forestry area	227.83	226.74	−1.09	−0.48
Grassland	306.36	302.92	−3.44	−1.12
Water area	32.76	32.92	0.16	0.49
Built-up area	44.67	46.43	1.76	3.94
Unused land	200.50	200.12	−0.38	−0.19

Note: Measured in million hectares.

TABLE 22.3
Land-Cover and Land-Use Conversions in China from Late 1980s to the Late 1990s

From \ To	Cultivated Land	Forestry Area	Grassland	Water Area	Unused Land	Built-up Area
Cultivated land		0.516	0.642	0.364	0.134	1.509
Forestry area	1.747		0.811	0.040	0.028	0.093
Grassland	3.457	1.047		0.150	0.914	0.077
Water area	0.286	0.025	0.092		0.179	0.040
Unused land	0.659	0.040	0.660	0.228		0.049
Built-up area	0.009	0.002	0.004	0.002	0.000	

Note: Measured in million hectares.

The built-up area was about 46.43 million ha in the late 1990s, with an increase of 1.76 million ha or 3.94%. The increase in built-up area was due to conversion, primarily, of cultivated land (85.38%).

22.3.2.2 Spatial Variation

The land-cover and land-use changes during this period showed substantial spatial variations across the country (Figure 22.3). A significant increase in cultivated land occurred in northeast China, northern China, and the Xinjiang oases, whereas there was remarkable decrease in cultivated land in the North China Plains, Yangtze River Delta, Yellow River band in the vicinity of Baotou and Datong section, and Sichuan Basin. Large-scale deforestation was seen in the northeast, and moderate deforestation was seen in the southwest and in the southeast coastal regions. A remarkable decrease in grassland occurred in central China and northwest China. Built-up area expanded widely across China, primarily in North China Plains, the Beijing–Tianjin–Tangshan area, the central part of Gansu, the southeast coastal regions, Sichuan Basin, and the Xinjiang oases.

The rate of land-cover and land-use conversion was obviously different among the subregions. In the northeast, particularly in the four provinces of Inner Mongolia, Liaoning, Jilin, and Heilongjiang, a large area of forest/grassland was converted to cultivated land. In the Northeast China Plain, most of the dry lands were converted to paddy fields because farmers earned more by planting rice. In North China Plain, Yangtze River Delta, and Sichuan Basin, a large area of cultivated land was converted to built-up area because of increasing population. The Northern China Plain and Loess Plateau were characterized by conversion from grassland to cultivated land. Three provinces

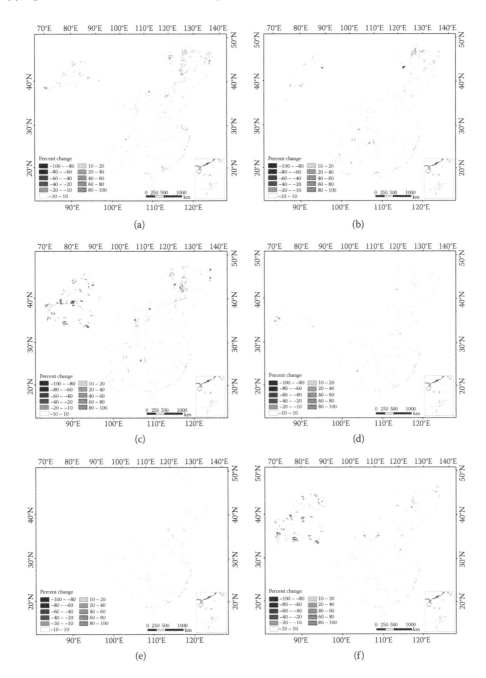

FIGURE 22.3 **(See color insert.)** Spatial patterns of land-cover and land-use changes in China from the late 1980s to the late 1990s: (a) Cultivated land, (b) forest area, (c) grassland, (d) water area, (e) built-up area, and (f) unused land.

Gansu, Ningxia, and Xinjiang showed an obvious increase in cultivated land. Northwest China was characterized by reclamation and abandonment of cultivated land, where half of the decreased grassland changed to cultivated land and the other half to desert. The southeast mountain area showed a conversion from forest area to cultivated land. The southeast coastal area saw conversion from grassland to forest area and from open forest to other forest, pointing to the coexistence of deforestation and afforestation. Southwest China witnessed conversion from forest area to grassland

and from forest area/grassland to cultivated land. The land-use patterns did not obviously change owing to relatively low human disturbance in southwest Qinghai and Tibet.

Urban land expansion was the most dominant feature from the late 1980s to the late 1990s and showed substantial spatial variations across the country. In North China Plain (a traditional agricultural zone) and Yangtze River Delta, built-up area increased significantly, accounting for 50% of the total increased built-up area in China, partly resulting from dense population and more infrastructures. Sichuan Basin, with the highest increasing rate of built-up area, accounted for 18.8% of the increased built-up area, resulting from rapid agricultural development and industrialization. Urban land expansion in the west (including the northwest and southwest) was caused mainly by growth of infrastructure and implementation of the western development strategy. The most remarkable urban land expansion in the 1990s was in Zhujiang Delta and Fujian coastal areas. However, the conversion from cultivated land to built-up area decreased during 1995–2000 partly owing to land management laws. In general, urban land expansion was centralized in the plain cities, economic development zones, and surrounding areas. These economic development zones were Beijing–Tianjin–Tangshan, Shanghai, Nanjing and Hangzhou-Suzhou, Wuxi and Changzhou, coastal and inland areas of Shandong, Xi'an-Xianyang, Chengdu–Chongqing and its peripheral areas, Turpan–Urumqi–Shihezi, and the development axes of Baotou–Lanzhou.

22.3.3 LAND-COVER AND LAND-USE CHANGES FROM LATE 1990s TO MID-2000s

22.3.3.1 Temporal Changes

Our estimate showed that the area of cultivated land decreased by 0.69 million ha from the late 1990s to the mid-2000s, with a decrease of 0.95 million ha of paddy field and an increase of 0.26 million ha of dry land (Table 22.4). The decrease in cultivated land was mainly due to construction, whereas the increase was due to reclamation of unused land, bottomland of rivers, and lakes. Therefore, the overall quality of cultivated land had declined.

Forest area increased by 0.24 million ha, and this included closed forest, shrub, open forest, and other forests. Unrecognizable young plantations on remote-sensing images were excluded from this study. Thirty-five percent of decreased forest area was converted to cultivated land and 36% to grassland.

There was a decrease of 1.19 million ha in grassland. The proportion of grassland reclaimed for cultivation accounted for more than 48% of all the reclamation activities. Grassland expansion was mainly due to the implementation of the "Grain for Green" project.

TABLE 22.4

Land-Cover and Land-Use Changes in China from Late 1990s to Middle 2000s[a]

Land-Cover and Land-Use Types	2000	2005	Change	% Change
Cultivated land	141.14	140.46	−0.69	−0.49
Paddy field	35.65	34.70	−0.95	−2.66
Dry land	105.49	105.76	0.26	0.25
Forestry area	226.74	226.97	0.24	0.10
Grassland	302.92	301.73	−1.19	−0.39
Water area	32.92	33.08	0.16	0.49
Built-up area	46.43	48.14	1.71	3.67
Unused land	200.12	199.93	−0.19	−0.10

[a] Measured in million hectare.

There was an increase of 1.71 million ha in built-up area, which was the most dominant conversion during the first 5 years of this century. The increase in built-up area was mainly derived from cultivated land (about 1.28 million ha), about 75% of the newly developed built-up area.

22.3.3.2 Spatial Variation

The land-cover and land-use changes also showed significant spatial variations across the country (Figure 22.4). In general, traditional farming regions such as the southeastern coastal area and

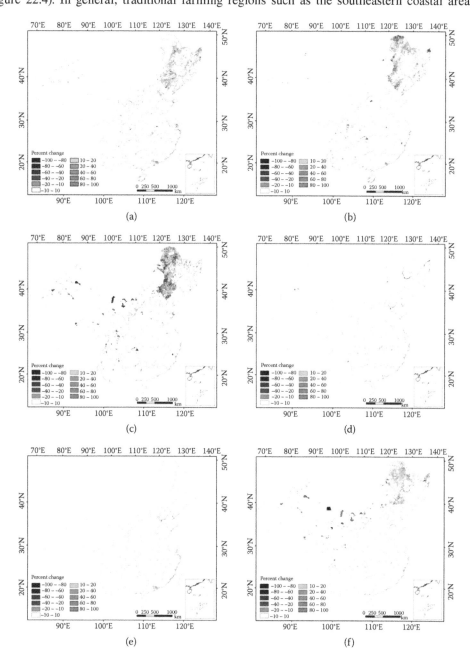

FIGURE 22.4 **(See color insert.)** Spatial patterns of land-cover and land-use changes in China from the late 1990s to the mid-2000s : (a) Cultivated land, (b) forest area, (c) grassland, (d) water area, (e) built-up area, and (f) unused land.

the North China Plain had been shrinking in cultivated land area, and the decrease in paddy field was more remarkable. There was a slight increase in cultivated land in the farming–grazing transitional zone, the farming–forest transitional zone, and the oases across northeast China, northwest China, and north China. Increase in forest area was distributed across Guizhou, Chongqing, Shaanxi, Ningxia, and the southwestern mountains of Inner Mongolia. Forest area decreased in the east, in provinces such as Zhejiang, Fujian, Jiangxi, Guangdong, and Jilin. Decrease in grassland occurred in the central steppe of Inner Mongolia, the oasis in the Xinjiang desert, the farming–grazing transitional zone of the Loess Plateau, and Guizhou and Chongqing in western China. The expansion in grassland was mainly due to the implementation of the "Grain for Green" project in southern Gansu, northern Shaanxi, and the northern part of Sichuan Basin. The expansion of built-up area was concentrated in eastern China. The southeast coastal areas and the plains region in the mainland, such as the North China Plain, Yangtze River Delta, Pearl River Delta, the central area of Gansu Province, Sichuan Basin, and Urumqi–Shihezi region, were the critical areas for urban land expansion.

22.4   CONCLUSION

Land-cover and land-use changes are the most obvious manifestation of the interaction between human activities and natural environment. Accurate information about land-cover and land-use changes in China is of critical importance in assessing environmental and economic sustainability in the future. With the development of remote-sensing technology, we have developed a systematic monitoring system on land-cover and land-use changes to enable long-term and continuous observation and have assessed the impacts of social and economic activities on land-cover and land-use changes.

According to our studies, land-cover and land-use changes in China from the late 1980s to the late 1990s were characterized by highly intense and accelerated changes and significant spatial variations, induced by regional exploitation and rapid socioeconomic development. In the early twenty-first century, land-cover and land-use changes have undergone a transformation from human exploitation such as reclamation and urban land expansion to development and ecological conservation. The ecological environment was recovered to some extent in the mid-western region, where natural land coverage on a regional scale improved significantly—thanks to the implementation of the western development project.

Further research is needed to develop the regional remote-sensing data acquisition, processing, and analyzing systems and combine them to enhance precision and efficiency and to study the effects of land-cover and land-use changes on the environment by linking land-use activities to the natural processes on the earth's surface.

ACKNOWLEDGMENTS

We would like to thank Zengxiang Zhang, Dafang Zhuang, Yimou Wang, Wancun Zhou, Rendong Li, Nan Jiang, Shuwen Zhang, and Shixin Wu for their support on the interpretation of remote-sensing data. The financial support from the National Key Programme for Developing Basic Science of China (2010CB950900) is also appreciated.

REFERENCES

Deng, X., Su, H., and Zhan, J. 2008. Integration of multiple data sources to simulate the dynamics of land systems. *Sensors*, 8, 620–634.

Deng, X, Huang, J., Rozelle, S., and Uchida, E. 2010a. Economic growth and the expansion of urban land in China. *Urban Studies*, 47(4), 813–843.

Deng, X., Jiang, Q., Su, H., and Wu, F. 2010b. Trace forest conversions in northeast China with a 1-km area percentage data model. *Journal of Applied Remote Sensing*, 4, 041893, 1–13. doi:10.1117/1.3491193.

Deng, X., Jiang, Q., Zhan, J., He, S., and Lin, Y., 2010c. Simulation on the dynamics of forest area changes in northeast China. *Journal of Geographical Sciences*, 20(4), 495–509.

Deng, X., Jiang, Q., Ge, Q., and Yang, L. 2010d. Impacts of the Wenchuan Earthquake on the Giant Panda Nature Reserves in China. *Journal of Mountain Sciences*, 2, 197–206.

IGBP Secretariat. 2005. Science Plan and Implementation Strategy. IGBP Report No. 53/IHDP Report No. 19, Stockholm, 64.

Liu, J., Liu, M., Zhuang, D., Zhang, Z., and Deng, X. 2003a. Study on spatial pattern of land-use change in China during 1995–2000. *Science in China Series D*, 46(4), 373–84.

Liu, J., Liu, M., Zhuang, D., and Zhang, Z. 2003b. A study on the spatial-temporal dynamic changes of land-use and driving forces analyses of China in the 1990s. *Geographical Research*, 22(1), 1–12 (in Chinese).

Liu, J., Zhang, Z., Xu, X., Kuang, W., Zhou, W., Zhang, S., Li, R., et al. 2010. Spatial patterns and driving forces of land use change in China during the early 21st century. *Journal of Geographical Sciences*, 20(4), 483–494.

State Statistical Bureau (SSB). 1996. *Statistical Yearbook of China*. Beijing: China Statistical Publisher House.

Wu, J., Bauer, M.E., Wang, D., and Manson, S.M. 2008. A comparison of illumination geometry-based methods for topographic correction of Quickbird images of an undulant area. *ISPRS Journal of Photogrammetry and Remote Sensing*, 63(2), 223–236.

Wu, X. and Cao, C. 2006. Sensor calibration in support for NOAA's satellite mission. *Advances in Atmospheric Sciences*, 23(1), 80–90.

Zhang, J. and Zhang, Y. 2007. Remote sensing research issues of the National Land Use Change Program of China. *ISPRS Journal of Photogrammetry and Remote Sensing*, 62(6), 461–472.

Zhang, Z., Zhang, T., Kang, D., and Yi, J. 2002. False color composite of multi-spectral RS images and its application in environmental geography. *Image Technology*, 1 (in Chinese).

Zheng, W., Liu, C., Zeng, Z., and Long, E. 2007. A feasible atmospheric correction method to TM image. *Journal of China University of Mining & Technology*, 17(1), 112–115.

23 An Approach to Assess Land-Cover Trends in the Conterminous United States (1973–2000)

Roger F. Auch, Mark A. Drummond, Kristi L. Sayler, Alisa L. Gallant, and William Acevedo

CONTENTS

23.1 INTRODUCTION

The resources that human beings depend upon for health and economic well-being are not distributed equally across the landscape. Varying characteristics of climate, geological formations, hydrology, terrain features, soils, and vegetation combine to provide different capacities to support human activities across space and time. Natural deposits of minerals offer potential for some areas to develop and market extractive resources; clement weather, sufficient moisture, and arable soils allow other areas to be farmed for food and fiber; still other areas with adequate moisture and temperature regimes, but steeper terrain or less arable soils, provide forests harvestable for a variety of wood products; extensive tracts of semiarid to arid grasses and shrubs offer open rangeland for grazing livestock; and navigable waterways and terrestrial corridors enable transportation and foster the growth of population centers. Land may have more than one potential use, based on the physical and anthropogenic conditions present. Human interactions with these different land capabilities have resulted in the patterns of land use and land cover (LULC) that we see today.

Availability of local resources, along with technological advances and local to global demands for commodities, shapes the decisions of landowners on how to attain the best and highest economic use of their properties, although not all decisions may involve monetary gains. Because natural resources often have regional-scale distributions, local land-use decisions ultimately add up to regional patterns of LULC, which can affect resources at broader scales and in other regions. For example, decisions of landowners to clear fields for crops have lead to development of agricultural regions in the United States, and the crop varieties, rates, and methods of application of agricultural chemicals, wetland drainage and water use, soil tillage, and crop-rotation practices have altered the quality and quantity of surface water in regions downstream (e.g., Gilliom et al., 2006; U.S. Department of Agriculture, 2003; U.S. Environmental Protection Agency, 2007) and the quality of air, water, and soil in regions downwind (e.g., Daly et al., 2007; Davidson, 2004; Fellers et al., 2004; U.S. Environmental Protection Agency, 2007). Extensive changes in land cover to promote certain land uses have affected local climate and other environmental characteristics such that the land eventually may become less suitable for the intended uses (e.g., Buschbacher et al., 1988; Marshall et al., 2003).

Understanding how land use interacts with other ecosystem services, such as fresh water, flood regulation, nutrient cycling, primary and secondary ecological production, disease and pest regulation, recreation, production of food, fuel, and fiber, and other services, is important for balancing multiple and simultaneous needs often that are in conflict with one another. Monitoring LULC change or stability is one way for resource managers and policymakers to assess which needs are being met and whether government programs and other incentives for land change or stability are successful in meeting goals. Monitoring LULC change can also reveal influences of technological advances, such as increased sizes of crop fields linked with larger and more sophisticated agriculture equipment, development of crop cultivars that withstand colder, drier, and shorter growing-season conditions, or development of various forms of energy (construction of reservoirs to provide consistent sources of hydroelectric power, expansion of land planted in biofuel crops, development of wind farms, etc.). The interplay among local environmental resources and changing global variables including climate, technologies, and economic markets promises to keep LULC highly dynamic over time and requires that we monitor the landscape to assess intended and unintended impacts of, and vulnerabilities to, change.

The U.S. Geological Survey (USGS) Land Cover Trends project was developed in response to the need for a consistent national synthesis of land-cover change at spatial and temporal scales that supported rates of accuracy sufficient for detecting regional change (Loveland et al., 2002; Sohl et al., 2004). Before this project, no single agency or entity had produced such a synthesis for the entire nation. The United States had several important land-use inventory programs that provided a wealth of detailed information. Certain sectors of land use, primarily agriculture (U.S. Department of Agriculture, National Agricultural Statistics Service, 2008) and forest (U.S. Department of Agriculture, U.S. Forest Service, 2010), were regularly assessed, along with the National Resources Inventory program (U.S. Department of Agriculture, Natural Resources and Conservation Service, 2010), which assessed land-use change on private lands within the conterminous United States. Given the different approaches, goals, resource definitions, and spatial monitoring frameworks for understanding land use, the Land Cover Trends project was designed to provide additional information premised on the need for a systematic understanding of national LULC changes that would facilitate the analysis of the consequences of regional LULC-change trends, which is useful in the research on climate, carbon, and natural resources.

The USGS is not bound by any specific land-use, ownership, management, or regulatory mission. This and other advantages made it well suited to undertake a national assessment of LULC change. The agency maintains the Landsat data archive dating back to the early 1970s, which enables a multidecadal analysis of change. The USGS also has extensive experience in producing national and global wall-to-wall land-cover products (e.g., Loveland et al., 2000; Vogelmann et al., 2001). These data products and mapping experiences enabled researchers on the Land Cover Trends project to know that (1) the differences in data sources and mapping objectives and approaches used for previous LULC products would render them not directly comparable for assessing LULC change; (2) time and budget investments for wall-to-wall mapping would be prohibitive for developing a national time series dataset for documenting change; and (3) goals identified as suitable for producing reasonable regional rates of mapping accuracy for past wall-to-wall land-cover maps would be insufficient for detecting the nature of most U.S. LULC change that occurs at local scale resolution. These and related considerations guided the project strategy, from the classification system used to the statistical design, mapping methodology, validation approach, and interpretation of results. The multiyear project was launched with support from the USGS, U.S. Environmental Protection Agency, and the National Aeronautics and Space Administration (NASA), and it demonstrated how aspects of remote sensing of land cover described in this book were brought together for a national multitemporal assessment of land-cover and land-use change.

23.2 METHODOLOGY: STRATEGY AND IMPLEMENTATION

23.2.1 STRATEGIC OVERVIEW

The goals of the Land Cover Trends project were to document types, geographic distributions, and estimated amounts and rates of LULC change on a region-by-region basis from 1973 to 2000 and to determine the key drivers and potential consequences of change. These goals were tempered with the realization that then current wall-to-wall mapping was too costly and time-consuming for mapping nearly 30 years of land change, so a different strategic approach was applied to meet the project assumptions and requirements (Loveland et al., 2002):

- Temporal intervals for assessing change should capture the major types of LULC conversions (Anderson Level I—Anderson et al., 1976) across the country.
- Spatial and statistical LULC change data should include information before and after a change event.
- LULC data must be accurate and consistent.
- Methodology must be extendible to continental and global scales to accommodate future expansion of assessments.

The project design had temporal and spatial elements. The temporal framework consisted of five periods (nominally 1973, 1980, 1986, 1992, and 2000) of Landsat sensor data (Multispectral Sensor [MSS], Thematic Mapper [TM], and Enhanced Thematic Mapper Plus [ETM+]) as the basis for the LULC interpretations. Landsat data, although available to users currently at no cost, were not freely available when the project began, so data needs were met by leveraging existing geoprocessed Landsat datasets. MSS data for 1973, 1986, and 1992 were acquired from the North American Landscape Characterization project (Sohl and Dwyer, 1998). TM and ETM+ data for 1992 and 2000 were obtained from the Multi-Resolution Land Characteristics consortium data collections (Homer et al., 2004; Loveland and Shaw, 1996). The project purchased MSS data for 1980 to meet the desired 6–8 year intervals for assessing change.

Ecoregions (U.S. Environmental Protection Agency, 1999, revised from Omernik, 1987) were used as the spatial strata for estimating amounts and rates, driving forces, and consequences of change (Figure 23.1). The ecoregion framework was developed by synthesizing information on climate, geology, physiography, soils, vegetation, hydrology, and human factors. These regions reflect patterns of LULC potential (Gallant et al., 2004) that could be detected in remotely sensed images. The ecoregions provided the backbone for the probabilistic sampling design adopted by the project.

Experimentation by project researchers found that rectangular landscape blocks of 20 km on a side were large enough to capture even coarse-scale ecoregional landscape patterns but seemingly still met the logistical needs for data-processing effort and the statistical requirements to estimate gross change with a margin of error of 1% at a 0.85 confidence level, given the expected spatial

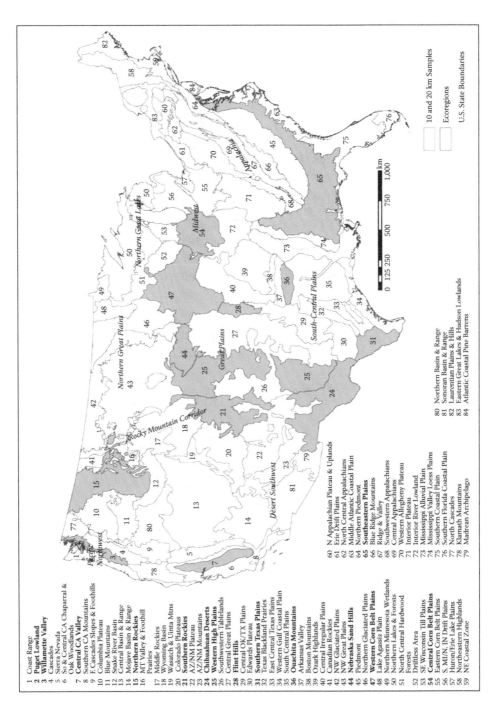

FIGURE 23.1 The 84 Level III ecoregions (U.S. Environmental Protection Agency 1999) used as strata to synthesize information on LULC change for the Land Cover Trends project. Gray-shaded ecoregions and general location names on the map refer to areas highlighted in Section 23.3.

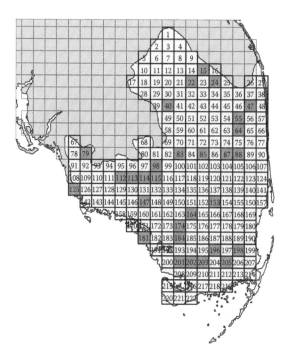

FIGURE 23.2 A 10-km × 10-km grid was laid over the map of U.S. ecoregions in southern Florida. Location of cell center point was used to assign ecoregion membership to each 10-km cell. The numbered cells shown in this example were assigned to the Southern Florida Coastal Plain ecoregion and the magenta-colored cells were those selected at random to be assessed for LULC change from the five mapping periods.

variance of LULC change estimated from the best available data (see Loveland et al., 2002). A grid with 20-km × 20-km cells (sample blocks) was laid over an equal-area projection of the conterminous United States, and each cell assigned membership to one ecoregion so that a stratified random sample could be drawn (Figure 23.2). A statistical evaluation conducted on the results from the initial nine ecoregions revealed that variance in change for some ecoregions exceeded the targeted precision goals. Three to four times as many samples per ecoregion would be needed to meet the desired statistical requirements, but processing this many additional samples could be accommodated logistically only if the geographic area covered by each sample was much smaller. Hence, a compromise solution was to intensify the sample frame but decrease the block size to 10 km × 10 km for the remaining ecoregions in the conterminous United States (Stehman et al., 2003, 2005; Figure 23.3). High changing ecoregions, however, continued to be somewhat problematic, and often exceeded targeted precision goals even with the switch to 10-km sample blocks.

The LULC classification scheme developed for the project focused on two objectives for understanding LULC dynamics. First, the categories had to allow land cover (which is what satellite sensors detect) to serve as a surrogate for land use (which is what analysts infer from the context of the land cover). The USGS Anderson system (Anderson et al., 1976) was designed for that purpose and had a strong history of successful application. Second, the classification scheme had to be applied with consistency and high rates of accuracy. LULC classification schemes that include large numbers or detailed classes tend to incur higher rates of mapping error. Typically, the greater is the number of thematic classes, the more is the opportunity to introduce error in the results (Gallant, 2009). A very general classification scheme, based principally on the Level I hierarchy of the Anderson system, was therefore selected for the project (Table 23.1).

FIGURE 23.3 **(See color insert.)** Ecoregional distribution of the 20-km × 20-km sample blocks selected for the first nine completed ecoregions and the subsequent 10-km × 10-km sample blocks selected for the remaining 75 ecoregions of the conterminous United States.

The classes outlined in Table 23.1 allowed for an "elasticity" that could adapt to different land covers being mapped within the same thematic land-use class while still maintaining national consistency. For example, evidence of harvesting forest stands between image periods ran the gamut of disturbed areas that were a few days to several years old but were still clearly visible as change in the imagery. Thus, "mechanically disturbed" land cover could range in spectral response from that of bare earth to young vegetation regrowth (Figure 23.4).

A second example highlights the elasticity in interpreting land cover versus land use. Wide power-line corridors running through forested land had grassland/shrubland vegetation cover, but represented land managed for development infrastructure. Therefore, power-line corridors were classified as "developed" to remain consistent with class definitions given in Table 23.1. Similarly, wide power lines in croplands, although not visible in Landsat imagery, would be classified as "agriculture" instead of "developed" because the main use of the land beneath the power lines remained agricultural. Grassy areas within multilane highway interchanges might be routinely cut for hay, but the main use of the land was to support transportation; so these areas were classified as "developed." As these and other mapping issues were encountered, a consistent set of decision rules were developed for specific interpretation challenges, allowing analysts to maintain consistency in applying LULC class labels.

Change detection was implemented using LULC mapped for 1992 as a baseline for comparison of successive dates of imagery backward and forward. The project benefitted from a then recent release of the National Land Cover Dataset (NLCD; Vogelmann et al., 2001), a product derived from Landsat TM data for the conterminous United States. NLCD data were extracted for each sample block, and the more detailed classes of the NLCD were collapsed to the 11 classes defined for the Land Cover Trends project. The NLCD land-cover product was designed to provide reasonably

TABLE 23.1

Land-Use/Land-Cover Classifications and Descriptions Used by the Land Cover Trends Project

Land-Cover Class	Description
Open water	Areas persistently covered with water, such as streams, canals, lakes, reservoirs, bays, and oceans
Developed (urban or otherwise built-up)	Areas of intensive use where much of the land is covered with structures or anthropogenic impervious surfaces (residential, commercial, industrial, roads, etc.) or less-intensive use where the land-cover matrix includes both vegetation and structures (low-density residential, recreational facilities, cemeteries, utility corridors, etc.), including any land functionally related to urban or built-up environments (parks, golf courses, etc.)
Agriculture (cropland and pasture)	Land in either a vegetated or an unvegetated state used for the production of food and fiber, including cultivated and uncultivated croplands, hay lands, pasture, orchards, vineyards, and confined livestock operations. Note that forest plantations are considered forests regardless of their use for wood products
Forest and woodland	Nondeveloped land where the tree-cover density is >10%. Note that cleared forestland (i.e., clear-cuts) is mapped according to current cover (e.g., mechanically disturbed or grassland/shrubland)
Grassland/shrubland	Nondeveloped land where cover by grasses, forbs, and/or shrubs predominates and tree-cover density is <10%
Wetland	Land where water saturation is the determining factor in soil characteristics, vegetation types, and animal communities. Wetlands can contain both water and vegetated cover
Mines and quarries	Areas with extractive mining activities that have a significant surface expression, including mining buildings, quarry pits, overburden, leach, evaporative features, tailings, or other related components
Barren	Land comprised of soils, sand, or rocks where <10% of the area is vegetated. Does not include land in transition recently cleared by disturbance
Mechanically disturbed	Land in an altered, often unvegetated transitional state caused by disturbance from mechanical means, as by forest clear-cutting, earthmoving, scraping, chaining, reservoir drawdown, and other similar human-induced changes
Nonmechanically disturbed	Land in an altered, often unvegetated transitional state caused by disturbance from nonmechanical means, as by fire, wind, flood, animals, and other similar phenomena
Snow and ice	Land where the accumulation of snow and ice does not completely melt during summer (e.g., alpine glaciers and snowfields)

accurate estimates for regional and broad-scale applications and was not developed with the level of effort necessary for estimates to be accurate at the local scale (Vogelmann et al., 2001). Therefore, the LULC data were edited manually to improve classification accuracy, based on source Landsat TM data with support from aerial photographs with similar temporal coverage acquired through the National Aerial Photography Program (NAPP). The project opted for a minimum classification unit of 60 m, as land-cover features occupying multiple adjacent pixels (in the 30-m cell resolution provided by TM data) could be interpreted with greater accuracy and consistency than can single-pixel features. The 60-m minimum mapping unit also allowed a more seamless transition to the coarse-scaled MSS data used for mapping the earlier time periods.

MSS February 6, 1973   MSS June 26, 1979

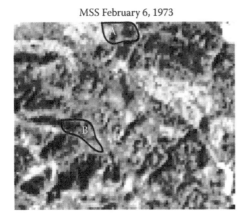 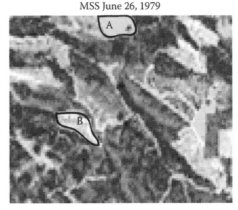

FIGURE 23.4 (See color insert.) The definition for the "mechanically disturbed" LULC class accommodates a range of variance in land-cover conditions to support the conceptual intent of the project. In this sample block from the Ouachita Mountains ecoregion, Areas "A" and "B" were mature forests in 1973. The subsequent image for 1980 era reveals Area A as recently disturbed and unvegetated and Area B as vegetated but obviously altered since 1973.

The revised 1992 LULC product for the sample blocks was investigated using as a baseline for mapping change in successive image dates. The Land Cover Trends project initially used an enhanced change-vector approach (Dwyer et al., 1996) as an automated method to identify pixels representing LULC change between image dates; however, this approach proved inefficient, as seasonal changes in vegetation phenology and landscape moisture often had greater intraannual and interannual differences in spectral responses than exhibited by pixels representing actual LULC conversions (Sohl et al., 2004). Subsequently, manual interpretation of Landsat data, aided by aerial photography and other ancillary information, was employed as the sole means of identifying change.

The manual interpretation approach allowed image analysts to overcome issues of spectral variance that challenged the use of the enhanced change-vector analysis, as well as ambiguity in spectral responses exhibited by similar but different land-cover types. Analysts were able to take advantage of the myriad contextual clues (e.g., patch size, patch shape, shadow effects, texture, pattern, or spatial associations) that varied across the landscape so as to improve LULC classification accuracy (Sohl et al., 2004). They were able to readily distinguish grass cover in school athletic fields, mowed parks, cemeteries, and golf courses (all "developed") from grass cover in hay fields and highly maintained pastures ("agriculture") and from the grass cover of open rangeland ("grassland/shrubland"). They sometimes encountered unusual or unknown land uses, requiring a search for additional information to yield a classification outcome consistent with the project goals as well as with the decisions applied in other ecoregions or by other analysts.

To map LULC through time, the 1992 baseline LULC map for each sample block was duplicated to initiate mapping for the 1986 period. The 1986 classification product was then modified only for pixels identified as locations of change between 1992 and 1986, as identified from Landsat data. Likewise, LULC for 1980 was mapped based on areas that had changed relative to the 1986 product, and LULC for 1973 was mapped based on areas that had changed relative to the 1980 product. Comparable processing was followed to develop the 2000 LULC map from the 1992 baseline map. On completion of the mapping of all periods, the analysts generated a series of change matrices to check for illogical LULC conversions, such as developed land changing back to agriculture, which would signal a need to revisit the imagery and interpreted LULC map. The sample block was then turned in for further quality-control procedures.

23.2.2 Data Quality

Several strategies were evaluated for ensuring that the LULC map results met the project's data-quality objectives. The Land Cover Trends project had examples of accuracy assessment approaches designed for national or more extensive land-cover mapping efforts (e.g., Scepan et al., 1999; Zhu et al., 2000), but it did not have resources to conduct such an assessment. More importantly, potential sources of data available nationally, such as historical aerial photography, which could be applied towards a formal assessment of accuracy, were being largely used to interpret Landsat imagery for classifying LULC in the sample blocks. An alternative approach was to validate mapping results from independent datasets.

The Land Cover Trends project investigated the potential for applying information collected by national inventory programs, such as the U.S. Forest Service Forest Inventory and Analysis Program (U.S. Department of Agriculture, U.S. Forest Service, 2010) and the Natural Resources and Conservation Services National Resources Inventory (U.S. Department of Agriculture, U.S. Natural Resources and Conservation Service, 2010), but these programs, although national in scope, collected data only for lands under certain ownership and land-use categories. Moreover, the data were confidential because of private land ownership and accessibility was severely restricted.

The project also explored validation of classification results by summarizing sample block information at the level of ecoregions for comparison with survey or census data aggregated from independent county-level databases. For example, analysts compared LULC change results for ecoregions with time series data from the U.S. Census of Agriculture (U.S. Census Bureau, 1977–1994; U.S. Department of Agriculture, National Agricultural Statistics Service, 1999, 2004) on the extent of land in various agricultural categories. These analyses were informative, but were challenged by county boundaries not aligning with the boundaries of ecoregion strata (particularly for ecoregions in the western United States that contained few or no counties in entirety) as well as by differences in agricultural class definitions used by the Census of Agriculture and by the project. Additionally, similar county databases were not available for validating changes in most other LULC types.

A more feasible solution for the project was a quality-control approach. One such method was to assign an independent analyst to classify a proportion of block samples already classified by other analysts. The drawbacks of this method were that it was time-consuming and identified disagreements between interpretations without resolving them, thereby requiring more time to be spent on it. The project adopted a more efficient method employing a group review process for all sample blocks. It conducted a review upon completion of LULC mapping for all sample blocks within an ecoregion. In this procedure, the analyst responsible for mapping a particular sample block briefed the rest of the team on the LULC change results and relayed additional information to provide the context for LULC change characteristics and drivers in and around the sample block (e.g., if the block was near a major source of commercial activity, or a drought period was accompanied by occurrences of wildfires, or observed LULC changes were consistent with reports for participation in new programs encouraging alteration of land management, etc.).

A helpful device for highlighting LULC change in a block was for the analyst to create an animation that cycled through the sequence of LULC maps for the five periods mapped. The team, as a group, would view the animations to note areas of change, then view more carefully the LULC maps for the individual periods, as well as the source Landsat imagery. This helped the group identify areas of questionable results and raise discussion on any new issues of LULC change that may not have been encountered in previous ecoregions. The analyst responsible for the block would document the issues raised, such as mapping errors, consistency with other blocks within the ecoregion, consistency with mapping decisions for the national effort, or other reasons, and revise the sample block accordingly. This group review process was a major tool for maintaining consistency in LULC interpretations across the nation and over the approximately 10 years required to complete the assessment. Several dozen interpreters contributed to the project over time, and the group review was the means by which the collective knowledge base was retained and shared. The review process was assisted largely by web-enabled sharing of computer screens, allowing image analysts at remote locations to participate in the reviews.

23.3 RESULTS

Results from the Land Cover Trends project show the variability of amounts, rates, types, and, ultimately, causes of LULC conversion that occurred across the nation. Here, we highlight some results of and insights into U.S. LULC change as examples of the types of information available from project outputs (see also the recent publications by Drummond and Loveland, 2010; Napton et al., 2010; Sleeter et al., 2010).

23.3.1 RATES AND FOOTPRINT OF CHANGE

Basic insights into the characteristics of U.S. LULC change come from an understanding of the pace and extent of ecoregion change. LULC change across ecoregions of the conterminous United States was highly variable, related to interactions among different economic, social, and environmental processes. For example, the average annual rate of change from 1973 to 2000 for the Chihuahuan Deserts ecoregion (Figure 23.1), where limited expansion of mining and developed land cover were the main pathways of change, was quite low at 0.02% ± 0.01% of the ecoregion, whereas the average annual rate for the Ouachita Mountains ecoregion (Figure 23.1), caused by cycles of timber harvest and regrowth, was rapid at 2.35% ± 0.38%. The overall spatial extent, or footprint, of change between 1973 and 2000 ranged from 0.5% ± 0.2% in the Chihuahuan Deserts to 33.9% ± 5.2% in the Ouachita Mountains. The footprint of change was an estimate of the total area of an ecoregion that had undergone some type of LULC change at least once during the 27-year study period, though it might have changed more than once, such as in the clearance and regrowth of forest after wood harvest. In general, the footprint of change was the highest in the southeastern and northwestern United States, where 13 ecoregions had greater than 10% change, although there were also high rates of change in the south-central United States, the Great Plains, and the Northern Great Lakes regions. The footprint was the lowest (less than 5%) in the ecoregions of the southern Rocky Mountains, southwestern Deserts, and in several ecoregions in the eastern Great Plains, the Midwest, and the Appalachians (Figures 23.1 and 23.5).

23.3.2 NET CHANGE IN ECOREGIONS

The net outcome of gross LULC conversion in ecoregions can indicate a number of different trends that may be broadly characterized as an acceleration of change, a relatively consistent or level trend, a punctuated spike in a particular sector, or a regional shift in the direction of change. Results from the Central Corn Belt Plains ecoregion (Figure 23.6a), for example, indicated an expansion and acceleration of developed land cover between 1973 and 2000. Approximately 85% of the gain in developed land was at the expense of agriculture, and another 8% was converted from forest cover. The Atlantic Coastal Pine Barrens ecoregion (Figure 23.6b) exhibited a relatively consistent pattern of change across all four time intervals of increased development and decline in forest and agriculture, with a notable amplitude change during 1980–1986. Such consistency was not common, indicating the nonlinearity of most LULC changes. For example, the forested Western Allegheny Plateau ecoregion (Figure 23.6c) that includes coal mining, urban centers (Pittsburgh, PA), and valley agriculture had a mix of consistent and varied trends across the four time intervals.

Significant shifts in net gains of LULC types could also occur, often reflecting changes in anthropogenic driving forces, such as changes in governmental policy. A large spike in grassland/shrubland in the Western High Plains ecoregion (Figure 23.6d) coincided with a decline in agricultural land cover. Most of the gain occurred during 1986–1992, which was associated with enrolment in the newly created federal Conservation Reserve Program (CRP). Initiated in 1985 (Food

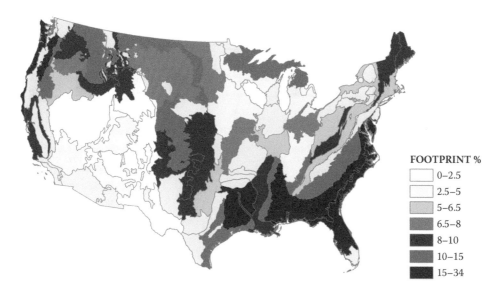

FOOTPRINT %

	0–2.5
	2.5–5
	5–6.5
	6.5–8
	8–10
	10–15
	15–34

FIGURE 23.5 **(See color insert.)** Estimates of the total spatial extent, or footprint, of land-cover change for the 84 ecoregions of the conterminous United States.

Security Act of 1985, Public Law No. 99–196, 99 Statute 1354), the CRP encouraged landowners to take marginal cropland out of production to reduce soil erosion. A previous expansion in agriculture from 1973 to 1980 and smaller gains from 1980 to 1986 were followed by the large directional change. The decline in agricultural land cover associated with CRP enrolment underscores the important role of government policy in certain ecoregions for LULC decisions. Some changes appear cyclical or repeatable. For example, the Ouachita Mountains ecoregion (Figure 23.6e) had a substantial net reforestation from 1986 to 1992, which slowed but did not reverse a nearly 7% forest loss during the 27-year study period. Forest harvest through mechanical disturbance and subsequent forest regrowth remained the major story of change. Nonmechanical disturbance, primarily from wildland fires, also caused spikes in the rate of annual net change, for example, in the forested Sierra Nevada ecoregion (Figure 23.6f).

23.3.3 COMMON LAND CONVERSIONS

Some of the most useful information for understanding the underlying causes of ecoregion change comes from data on LULC conversions. Observations at the ecoregion scale showed wide variations in the types, rates, and net results of conversion. In the conterminous United States, the most common conversion by area was "forest" to "mechanically disturbed" (Figure 23.7), primarily driven by the economic demand for wood-based products such as paper and building materials. Three of the top five conversions were related to forest harvest, including regrowth of forest (mechanically disturbed to forest or mechanically disturbed to grassland/shrubland as a transitional cover where regrowth was slower and had not returned to forest cover by the next period). The other two most common LULC conversions involved fluctuations between agriculture and grassland/shrubland, especially the influence of the federal CRP during the third time interval. Increases in the extent of developed land were often prominent through time, though they were not the most common types of LULC conversion for the nation as a whole. Changes in developed land, however, were different than most other LULC types because they were almost always permanent and many times unidirectional. Agricultural land cover was the leading source of new developed land, followed by forest, grassland/shrubland, or wetland land covers (Figure 23.8). In

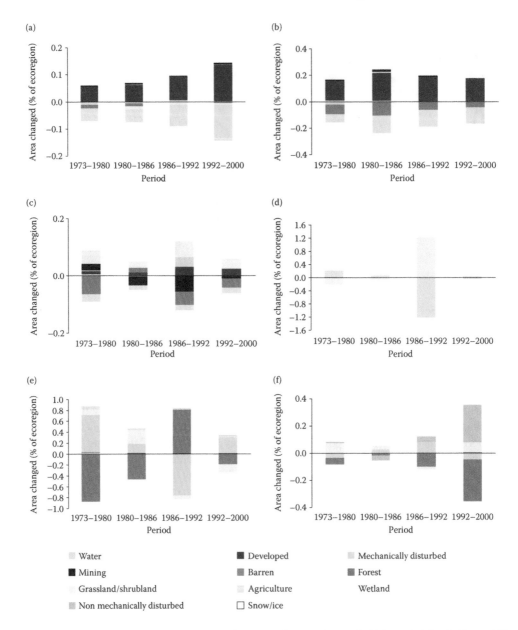

FIGURE 23.6 The average annual net change for the (a) Central Corn Belt Plains, (b) Atlantic Coastal Pine Barrens, (c) Western Allegheny Plateau, (d) Western High Plains, (e) Ouachita Mountains, and (f) Sierra Nevada ecoregions.

many ecoregions, the temporal circumstances and mix of different LULC conversion types were perhaps as important as the extent of any particular conversion alone.

23.3.4 HUMAN–ENVIRONMENT INTERACTIONS

How do the amounts, rates, and types of change relate to specific human–environment interactions? The southeastern United States has become an important region for pulp and timber production, which has contributed to extensive changes in several ecoregions. The footprint of change

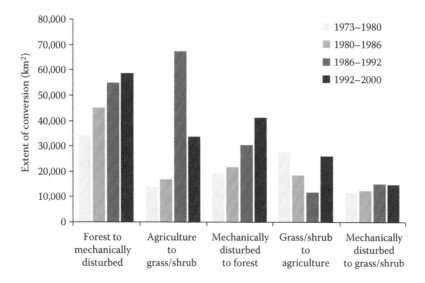

FIGURE 23.7 The five most extensive LULC conversions by area.

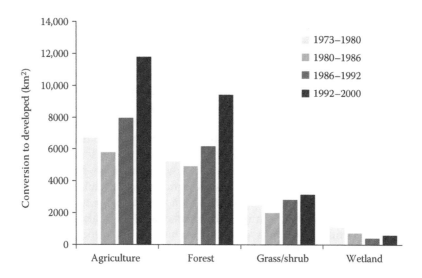

FIGURE 23.8 The most common conversions to developed land.

in the Southeastern Plains ecoregion was an estimated 20.4% ± 3.9%. More than half the change involved forest harvest (forest to mechanically disturbed) and regrowth (mechanically disturbed to forest). Industrial pine plantations tended to have relatively fast-cutting and replanting cycles that contributed to a large footprint of change and high rates of gross change. Economic opportunities to produce pulp and wood products in the region's favorable climate with fast growing native pine species converged with a move away from marginal agricultural use of the land. The rapid cycles of change in the Southeast contrasted with the slower regrowth rates and associated longer intervening grassland/shrubland transitions in the Northeast ecoregions where intensive forestry was practiced.

In the Great Plains, the characteristics of change were tied to climate variability, land quality, government farm policy, and other socioeconomic factors including globalization of trade. Several ecoregions in the drier western plains exhibited higher rates of LULC change than eastern plains ecoregions where more favorable climate and extensive areas of stable agriculture were found.

Environmental conditions and water resources, coupled with economic and policy concerns, drove these LULC fluctuations. Still, individual ecoregions in the Great Plains had dissimilar profiles of change resulting from differences in water availability for crop irrigation, precipitation increases (in the northern plains), encroachment of woody shrubs and woodlands in the south, and urbanization (in the east and southeast). For example, the high footprint of change in the Western High Plains ecoregion at 11.6% ± 2.4% was driven by agricultural expansion in the 1970s, which was followed by even greater amounts of conversion back to grassland/shrubland, much of it in response to economic and conservation incentives of the CRP to convert large amounts of marginal and environmentally sensitive cropland back to grassland/shrubland (Figure 23.6d). Landowner enrolment in the CRP was high in many ecoregions and was a major cause of higher annual rate of change between 1986 and 1992 and continued, although lower, into the 1992–2000 interval.

Low rates and small footprints of change were observed in several relatively stable agricultural ecoregions in the eastern Great Plains and the Upper Midwest as the ability to grow high-value row crops with little to no irrigation on highly productive soils favored consistent land use. Cropland area remained mostly unchanged throughout the Western Corn Belt Plains and Central Corn Belt Plains ecoregions, although conversion from agriculture to developed land was one of the main types of change documented. Little change was observed in the hilly rangelands of the Nebraska Sand Hills and Flint Hills ecoregions, lands largely unsuited for crop production.

Ecoregions comprising the Rocky Mountain corridor showed a wide range of rates and extent of change. The Northern Rockies ecoregion had a large footprint of change (13.9% ± 3.9%) driven by forest harvest (mechanically disturbed) and fire disturbance (nonmechanically disturbed), which was particularly notable between 1992 and 2000. Grassland/shrubland cover increased by nearly 14% because of the slow recovery of forests following clear-cutting and, to a lesser degree, stand-replacing fire. The Southern Rockies ecoregion had a small footprint of change (1.0% ± 0.3%), consistent with other southwestern ecoregions, although it received pressure from recreational and urban development. Forest harvest was also more limited in the Southern Rockies and other mountainous ecoregions of the Southwest because of fewer commercially valuable tree species and slower growth rates because of different climatic conditions when compared to the Northern Rockies.

The footprints of change for desert and forested ecoregions of the southwestern United States were consistently small. This was due in part to the large extent of low-productivity public lands, low population density, and lack of industrial scale agriculture. Particularly since 2000, forested ecoregions there and in other parts of the West had been affected by insect damage and forest dieback linked to drought and climate warming (Breshers et al. 2005; Carroll et al. 2004). Some of these changes were starting to be observed in 2000 era imagery and helped boost the nonmechanical disturbance LULC change during the last time interval.

High rates of change in the Pacific Northwest were related to cycles of forest harvest, grassland/shrubland transitional land cover, and forest recovery. The Puget Lowland ecoregion had a 10% decline in forest cover between 1973 and 2000, with conversion from forest to developed land being a leading type of change in addition to forest harvest. The Cascades ecoregion had heavy forest cutting earlier in the study period that accounted for most of the overall 26.6% ±3.7% LULC change, but constant replanting of trees, favorable climate for regrowth, and policy shifts dealing with forest cutting on public land in the 1990s lead to nearly the same percentage of forest land cover in 2000 as in 1973. The western valley agricultural ecoregions (Central California Valley and Willamette Valley) also had high rates of change, although a substantial amount of change in the Willamette Valley ecoregion was related to wood harvest and regrowth in the forested foothills along the eastern and western edges of the ecoregion. The large footprint of change in the Central California Valley was caused primarily by various expansions and contractions of agriculture and grassland/shrubland, which resulted in a small net expansion of agriculture. A nearly 40% increase in developed land there contributed to a decline in grassland/shrubland cover and perhaps diminished the expansion of agriculture, although some grassland/shrubland conversion to agriculture might have resulted from other agricultural loss to urbanization.

23.4 CONCLUSIONS

The ability to generalize the story of change across 84 individual ecoregions, indicated by the preceding survey of LULC change results emerging from the project, is one of the strengths of the approach. Clearly, there are multiple pathways of change, driven by numerous interacting forces. The individual ecoregion stories are much more complex than can be examined here, and future work should focus on explaining that complexity, its drivers, and its consequences.

The USGS Land Cover Trends project illustrates a robust approach for using remotely sensed data to document types and rates of fine-scaled LULC change for a national assessment. It has provided the first multiyear assessment across the conterminous United States, not restricted by particular thematic land sectors or jurisdiction. The project methodology is both durable and flexible enough to accommodate an extended time frame and considerable geographic variability. This systematic analysis provides a foundation of LULC-change knowledge to better inform future investigations and monitoring efforts as well as regional and national policy and program decisions. Land-Cover Trends data have supported a variety of block- to regional-scale studies, including studies of topographic change (Gesch, 2006), effects on LULC change on animal habitat (e.g., Price et al., 2006), radiative forcings (Barnes and Roy, 2010), and carbon cycling (e.g., Liu et al., 2004; Tan et al., 2005). The database of LULC change compiled during this project provides a strong foundation and rich heritage for continued monitoring of landscape dynamics in the United States.

ACKNOWLEDGMENTS

We would like to thank all the members of the Land Cover Trends project who readied the Landsat data and interpreted the land cover of the 2755 sample blocks used in the project. We are grateful for research funding provided by the U.S. Geological Survey's Geographic Analysis and Monitoring and Office of Global Change programs, the U.S. Environmental Protection Agency's Office of Research and Development, and NASA's Land-Cover and Land-Use Change Program.

REFERENCES

Anderson, J.R., Hardy, E.E., Roach, J.T., and Witmer, R.E. 1976. A Land Use and Land Cover Classification System for Use with Remote Sensor Data, Professional Paper 964, U.S. Geological Survey.

Barnes, C.A., and Roy, D.P. 2010. Radiative forcing over the conterminous United States due to contemporary land cover land use change and sensitivity to snow and inter-annual albedo variability. *Journal of Geophysical Research*, 115: G04033.

Breshears D.D., Cobb, N.S., Rich, P.M., Price, K.P., Allen, C.D., Balice, R.G., Romme, W.H. et al. 2005. Regional vegetation die-off in response to global-change-type drought. *Proceedings of the National Academy of Sciences, U.S.A.* 102: 15144–15148.

Buschbacher, R., Uhl, C., and Serrao, E.A.S. 1988. Abandoned pastures in eastern Amazonia. II. Nutrient stocks in the soil and vegetation. *Journal of Ecology*, 76: 682–299.

Carroll, A.L., Taylor, S.W., Regniere, J., and Safranyik, L. 2004. Effects of climate change on range expansion by the mountain pine beetle in British Columbia. In T.L. Shore, J.E. Brooks, and J.E. Stone (Eds.), *Mountain Pine Beetle Symposium: Challenges and Solutions* (pp. 223–232.). Kelowna, BC: Natural Resources Canada, Canadian Forest Service, Pacific Forestry Centre.

Daly, G.L., Lei, Y.D., Teixeira, C., Muir, D.C.G., Castillo, L.E., and Wania, F. 2007. Accumulation of current-use pesticides in neotropical montane forests. *Environmental Science and Technology*, 41: 1118–1123.

Davidson, C. 2004. Declining downwind: Amphibian population declines in California and historical pesticide use. *Ecological Applications*, 14(6): 1892–1902.

Drummond, M.A. and Loveland, T.R. 2010. Land-use pressure and a transition to forest-cover loss in the Eastern United States. *BioScience*, 60: 286–298.

Dwyer, J.L., Sayler, K.L., and Zylstra, G.J., 1996, Landsat pathfinder datasets for landscape change analysis. In R.E. MacIntosh, C.T. Swift, and S.J. Frasier (Eds.), *Remote Sensing for a Sustainable Future* (pp. 547–550), International Geoscience and Remote Sensing Symposium, Lincoln, NE, May 27–31, 1996, Proceedings, v. 1: Piscataway, NJ: Institute of Electrical and Electronics Engineers (IEEE).

Fellers, G.M., McConnell, L.L., Pratt, D., and Datta, S. 2004. Pesticides in mountain yellow-legged frogs (*Rana muscosa*) from the Sierra Nevada mountains of California, USA. *Environmental Toxicology and Chemistry*, 23(9): 2170–2177.

Gallant, A.L. 2009. What you should know about land-cover data. *Journal of Wildlife Management*, 73(5): 796–805.

Gallant, A.L., Loveland, T.R., Sohl, T., and Napton, D. 2004. Using a geographic framework for analyzing land cover issues. *Environmental Management*, 34(S1): 89–110.

Gesch, D.B. 2006. An Inventory and Assessment of Significant Topographic Changes in the United States, Ph.D. dissertation, South Dakota State University, Geospatial Science and Engineering Program.

Gilliom, R.J., Barbash, J.E., Crawford, C.G., Hamilton, P.A., Martin, J.D., Nakagaki, N., Nowell, L.H., et al. 2006. The Quality of Our Nations Waters—Pesticides in the Nations Streams and Ground Water, 1992–2001, Circular 1291, U.S. Geological Survey.

Homer, C., Haung, C., Yang, L., Wylie, B., and Coan, M. 2004. Development of a 2001 National Landcover Database for the United States. *Photogrammetric Engineering and Remote Sensing*, 70: 929–840.

Liu, S., Loveland, T.R., and Kurtz, R.M. 2004. Contemporary carbon dynamics in terrestrial ecosystems in the southeastern plains of the United States. *Environmental Management*, 33(Supplement 1): S442–S456.

Loveland, T.R., Reed, B.C., Brown, J.F., Ohlen, D.O., Zhu, Z., Yang, L., and Merchant, J.W. 2000. Development of a global land cover characteristics database and IGBP DISCover from 1 km AVHRR data. *International Journal of Remote Sensing*, 21(6–7): 1303–1330.

Loveland, T.R. and Shaw, D.M. 1996. Multi-resolution land characterization: Building collaborative partnerships.

Loveland, T.R., Sohl, T.L., Stehman, S.V., Gallant, A.L., Sayler, K.L., and Napton, D.E. 2002. A strategy for estimating the rates of recent United States land-cover changes. *Photogrammetric Engineering and Remote Sensing*, 68(10): 1091–1099.

Marshall, C.H., Pielke, Sr., R.A., and Steyaert, L.T. 2003. Crop freezes and land-use change in Florida. *Nature*, 426: 29–30.

Napton, D.E., Auch, R.F., Headley, R., and Taylor, J.L. 2010. Land changes and their driving forces in the southeastern United States. *Regional Environmental Change*, 10: 37–53.

Omernik, J.M. 1987. Ecoregions of the conterminous United States. *Annals of the Association of American Geographers*, 77: 118–125.

Price, S.J., Dorcas, M.E., Gallant, A.L., Klaver, R.W., and Willson, J.D., 2006. Three decades of urbanization: Estimating the impact of land-cover change on stream salamander populations. *Biological Conservation*, 133: 436–441.

Scepan, J., Menz, G., and Hansen, M.C. 1999. The DISCover validation image interpretation process. *Photogrammetric Engineering and Remote Sensing*, 65(9): 1075–1081.

Sleeter, B.M., Wilson, T.S., Soulard, C.E., and Liu, J. 2010. Estimation of late twentieth century land-cover change in California. *Environmental Monitoring and Assessment*, 173(1–4): 251–266.

Sohl, T.L., and Dwyer, J.L. 1998. North American landscape characterization project: The production of a continental scale three-decade landsat data set. *Geocarto International*, 3(3): 43–51.

Sohl, T.L., Gallant, A.L., and Loveland, T.R. 2004. The characteristics and interpretability of land surface change and implications for project design. *Photogrammetric Engineering and Remote Sensing*, 70(4): 439–448.

Stehman, S.V., Sohl, T.L., and Loveland, T.R. 2003. Statistical sampling to characterize recent United States land-cover change. *Remote Sensing of Environment*, 86: 517–529.

Stehman, S.V., Sohl, T.L., and Loveland, T.R. 2005. An evaluation of sampling strategies to improve precision of estimates of gross change in land use and land cover. *International Journal of Remote Sensing*, 26(22): 4941–4957.

Tan, Z., Liu, S., Johnston, C.A., Loveland, T.R., Tieszen, L.L., Liu, J., and Kurtz, R. 2005. Soil organic carbon dynamics as related to land use history in the northwestern Great Plains. *Global Biogeochemical Cycles*, 19(3): 1-10.

U.S. Bureau of the Census. 1977. *1974 Census of Agriculture*. Washington, DC: U.S. Government Printing Office.

U.S. Bureau of the Census. 1981. *1978 Census of Agriculture*. Washington, DC: U.S. Government Printing Office.

U.S. Bureau of the Census. 1984. *1982 Census of Agriculture*. Washington, DC: U.S. Government Printing Office.

U.S. Bureau of the Census. 1989. *1987 Census of Agriculture*. Washington, DC: U.S. Government Printing Office.

U.S. Bureau of the Census. 1994. *1992 Census of Agriculture*. Washington, DC: U.S. Government Printing Office.

U.S. Department of Agriculture. 1999. *1997 Census of Agriculture*, National Agricultural Statistics Service, Washington, DC.

U.S. Department of Agriculture. 2003. *Agricultural Resources and Environmental Indicators,* Agriculture Handbook No. AH-722.

U.S. Department of Agriculture. 2004. *2002 Census of Agriculture*, National Agricultural Statistics Service, Washington, DC.

U.S. Department of Agriculture. 2008. The census of agriculture–About the census. National Agricultural Statistics Service. Available from: http://www.agcensus.usda.gov/About_the_Census/index.asp (accessed February 1, 2011).

U.S. Department of Agriculture. 2010. U.S. Forest Service. Forest Inventory and Analysis National Program-home page. Available from: http://fia.fs.fed.us/ (accessed February 1, 2011).

U.S. Department of Agriculture. 2010. Natural Resources and Conservation Service. 2010. National Resources Inventory-home page. Available from: http://www.nrcs.usda.gov/technical/NRI/ (accessed February 1, 2011).

U.S. Environmental Protection Agency. 1999. Level III Ecoregions of the Continental United States, National Health and Environmental Effects Research Laboratory, Corvallis, Oregon (1:7,500,000-scale map).

U.S. Environmental Protection Agency. 2007. National Water Quality Inventory: Report to Congress, 2002 Reporting Cycle, EPA 841-R-07-001, Washington, DC.

Vogelmann, J.E., Howard, S.M., Yang, L., Larson, C.R., Wylie, B.K., and van Driel, N. 2001. Completion of the 1990s national land cover data set for the conterminous United States from landsat thematic mapper data and ancillary data sources. *Photogrammetric Engineering and Remote Sensing*, 67(6): 650–662.

Zhu Z., Yang L., Stehman S.V., and Czaplewski R.L. 2000. Accuracy assessment for the U.S. Geological Survey regional land cover mapping program: New York and New Jersey region. *Photogrammetric Engineering and Remote Sensing* 66:1425–1435.

24 Is Africa Losing Its Natural Vegetation?

Monitoring Trajectories of Land-Cover Change Using Landsat Imagery

Andreas Bernhard Brink, Hugh Douglas Eva, and Catherine Bodart

CONTENTS

24.1 INTRODUCTION

Sub-Saharan Africa represents nearly 20% of the earth's surface. The landscape includes many biologically rich and unique ecoregions, such as tropical forests, montane forests, and wood and grass savannas. The region is also home to 13% of the world's population, and since it has the highest birth rate among all continents—2.4 % compared with a world average of 1.3%—its population is projected to grow to 2 billion by 2050. Currently, 60% of the population lives in rural areas, and 55% of the economically active population depends on agriculture; however, the annual urban growth rate at nearly 4% is the most rapid in the world and nearly twice the global average (FAOSTAT, 2006).

Sub-Saharan Africa has the lowest total gross domestic product (GDP) and GDP per capita in the world (ERS/USDA, 2008). The economy of the region is based mainly on primary products or natural resources. The exploitation of these is often related to loss and degradation of forests and woodlands, loss of animal and plant species, degradation of land, increase in water shortage, and decline in water quality. The pressure on the environment and its degradation are mainly the result of high population growth, which has exceeded the capacity of natural resources to meet increasing human needs with the current technology. Population increase and economic growth are linked to increased use of fuel energy—mostly in the form of firewood and charcoal—which accounts for over 75% of the energy consumption in sub-Saharan nations (EIA, 1999). The conversion of natural vegetation to agriculture, associated with poor land management practices, causes degradation and erosion of land. It is estimated that about 25% of the land is subject to erosion by water and 22% to erosion by wind, and desertification affects over 45% of the land area of which 55% is at high risk to very high risk of desertification (UNEP, 2005).

The effects of land-cover changes on natural vegetation, biodiversity, socioeconomic stability, and food security may have a long-term impact on natural resources such as freshwater and forest resources, sustainable food production, climate and, last but not the least, human welfare (Foley et al., 2005). Therefore, assessing the dynamics of land-cover and land-use changes and understanding its underlying causes have been recognized as key areas of research on regional and global environmental change.

The following study aims at using an independent method to assess and quantify main land-cover changes in sub-Saharan Africa over a 25-year period (from 1975 to 2000) by using earth-observing satellites. Four broad land-cover classes—forests, natural nonforest vegetation, agriculture, and barren areas—are analyzed, and the driving forces of land-cover changes are discussed.

24.2 METHOD

The study is based on the mapping capacity of earth-observing satellites of high spatial resolution, which have been operating since the early 1970s. Although these types of data are appropriate for mapping, they have a restricted coverage both in time and space. To cover the full sub-Saharan region, a huge number of scenes are required, which would increase the cost for both image acquisition and processing. A standard technique of land resource inventories is, therefore, to use a sampling strategy across the target area (e.g., Achard et al., 2002). Based on the White/UNESCO vegetation map of Africa, which aggregates similar land-cover types in so-called ecoregions, a stratified random sample of 57 study sites was selected from a hexagonal-based grid across sub-Saharan Africa (Figure 24.1). For each of these sites, remotely sensed satellite images from 1975 to 2000 were acquired.

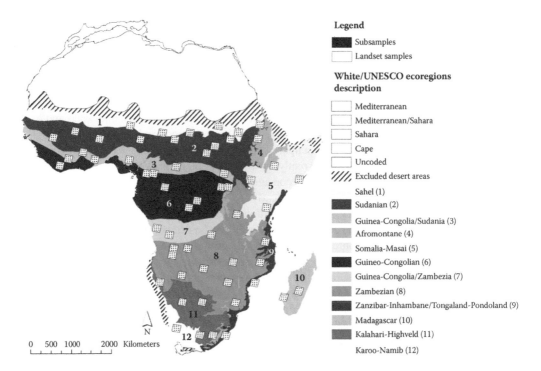

FIGURE 24.1 Ecoregion stratification and sample sites. Ecoregions as described in White's (1983) vegetation map of Africa. The striped areas represent the permanent deserts not included in the study. Under Sections 24.3 and 24.4, the ecoregions of Sahel, Sudanian, and Guinea-Congolia/Sudania are included when referring to the Sahel belt or area.

Historical satellite images were obtained from Landsat Multi-Spectral Scanner (MSS), which has a spatial resolution of 80 m, and recent satellite images were obtained from the Landsat Thematic Mapper and Enhanced Thematic Mapper, which have a finer spatial definition of 30 m.

Furthermore, to improve the variance of the estimates and reduce the errors that occur in full-scene classification, in each of the 57 sites nine subscenes of 20 × 20 km were extracted and were used to assess the land cover. The resulting 511 subscenes that were analyzed covered about 1% of sub-Saharan Africa (Brink and Eva, 2009).

The subscenes were independently classified using an unsupervised classification algorithm and then visually interpreted into four main land-cover classes (forests, nonforest natural vegetation, agriculture, and barren), along with water and "no data" (clouds/missing area). The *forest* class included closed evergreen, semievergreen, and dry deciduous forests with at least 40% of coverage. Deciduous open and degraded forests, wood- and shrublands, and grass savannas were associated with the *nonforest natural vegetation* class. The *agriculture* class was related to the agricultural domain including irrigated and rain-fed croplands, smallholdings, and plantations. In certain cases, man-made pastures were classified under this heading, predominantly in the Sahel and Sudanian regions, where arid shrublands had been cleared for grazing. Natural pastures (in fact, a land-use term) were classified under *nonforest natural vegetation*. The class *barren* included very sparse grasslands (or pseudo-steppe), bare soil, and rocks. Although the class *water* was included in the interpretation, it was not reported in the assessment, as a special sampling scheme would be required to assess the current area of water bodies.

For each subscene and each reference year, the proportions of land cover were calculated for every class, and the changes between the two dates were assessed. The results were then extrapolated by direct expansion (Gallego and Delincé, 1991a and 1991b) to the different strata and to the whole of sub-Saharan Africa.

24.3   RESULTS

The estimates of our study show that in the year 2000 sub-Saharan Africa was covered by slightly more than 17% of agriculture, almost 20% of forests, 60% of nonforest natural vegetation such as wood- and shrublands and savannas, and 2.5% of barren land, excluding the permanent deserts. These findings correspond closely to the land-cover map of the continent made by Mayaux et al. (2004) for the same year. The differences range from 1% in the forest class to a maximum of 2.5% in the agriculture class. The latter map was derived using a different methodology based on full wall-to-wall coverage of SPOT VGT satellite data of low spatial resolution.

Analysis of the results over the 1975–2000 period (Table 24.1) reveals that sub-Saharan Africa has lost some 130 million hectares (Mha) of natural vegetation (forest and nonforest natural

TABLE 24.1
Land-Cover Changes in Sub-Saharan Africa between 1975 and 2000

	Forest	NF vegetation	Agriculture	Barren
	(000 ha)	(000 ha)	(000 ha)	(000 ha)
Land cover 1975	438,917	1,247,980	215,274	42,912
Land cover 2000	367,592	1,189,085	338,687	49,477
Total change	−71,325	−58,894	123,413	6565
Total (%) change	−16	−5	57	15
Average annual change	−2853	−2356	4937	263
Average annual (%) change	−0.7	−0.2	2.3	0.6

vegetation) over the last 25 years. Forest and nonforest natural vegetation have decreased by 16% (71 Mha) and 5% (60 Mha), respectively. On the other hand, at the expense of natural vegetation, sub-Saharan Africa has gained about 120 Mha of new agriculture land, from just over 220 Mha in 1975 to nearly 340 Mha in 2000, which represents a 57% increase over the 25-year period. Barren areas have increased by 15%, which amounts to 6.5 Mha. Another Landsat sample-based study confirms this general trend of expansion of agriculture in Africa, estimating a nearly 60% increase in agricultural areas at the expense of natural vegetation (Gibbs et al., 2010). Assuming a linear change over time, the yearly deforestation rate has been 0.7%, which means that the whole region has been losing nearly 3 Mha of forests every year. The yearly deficit in nonforest natural vegetation has been 0.2%, which is equal to more than 2 Mha lost every year. This amounts to over 5 Mha of natural vegetation lost per year, which is about the size of a country like Togo. On the other hand, the annual gain of agriculture land has been almost 5 Mha, which means an average annual change rate of 2.3%. Barren areas have increased by a yearly rate of 0.6%, which means over 0.26 Mha every year.

The general trend of natural vegetation loss on the one hand and increase in agriculture on the other hand is evident throughout the sub-Saharan African region. However, considerable differences are apparent in the geographic distribution and intensity of the land-cover changes in the region. Furthermore, the sources of natural vegetation (either forests or nonforest natural vegetation) converted to agricultural land vary substantially throughout the region (Figure 24.2).

The most important loss of natural vegetation during the 25-year period has occurred in the Sahel belt—including the ecoregions of Sahel, Sudanian, and Guinea-Congolia/Sudania—as described in Figure 24.1. This major loss is visible in the whole of West Africa and to a lesser extent in East Africa and Angola. The Sahel region has the largest proportion of loss of natural vegetation (about 40% of the total), and therefore it accounts for over one quarter of the total increase in agricultural land. In contrast, relatively little change has appeared in central and southern Africa. Forest-cover change accounted for the highest proportion of loss of land cover. We noted that most of the forest-cover changes in sub-Saharan Africa have occurred outside the humid forest domain of Central Africa, which accounts for less than 15% of the total forest loss, despite accounting for two-thirds of the subcontinent's forest area. Deforestation has occurred mainly in West Africa, in the southern part of the Sahel belt, in some areas in the dry forest domain of East Africa, and in Angola. Losses in nonforest natural vegetation are predominantly in the northern part of the Sahel region and in some areas in East Africa, which together account for over 80% of loss in wood- and shrubland. Large-scale conversion of intact forest into agricultural land has taken place primarily in West Africa and to a lesser extent in some parts of East Africa, Madagascar, and Angola. Forest degradation and fragmentation—from closed forest to open and degraded forest and woodlands that are included in the nonforest natural vegetation class—are most evident not only in Sudan and Angola, but also to some extent in Chad, Central African Republic, Ghana, and southern Tanzania. In areas where long-lasting civil disturbance has occurred, land abandonment associated with an increase of nonforest natural vegetation areas is observed (i.e., Angola and some parts of Sudan). In the northern part of the Sahel region where trees are replaced by smaller species owing to lower rainfall, agricultural land has been converted largely from nonforest natural vegetation such as wood- and shrubland. In total, the Sahel area accounts for over 60% of loss in nonforest natural vegetation over sub-Saharan Africa. The remaining decline in nonforest natural vegetation is accounted for by the wood- and shrub savannas of eastern Africa and the miombo woodlands of southern Africa. Very little conversion of natural vegetation to agricultural land is detected in South Africa and the coastal areas of Mozambique. Here, in 1975, vast areas were already dominated by agriculture, and therefore little land was available for further cropland expansion. No change between 1975 and 2000 could be identified in the Kalahari zone of Namibia and Botswana, which is probably due to the adverse living conditions and the resultant low population density. Regeneration and regrowth of forest has occurred in Cameroun and to some extent in eastern Zambia.

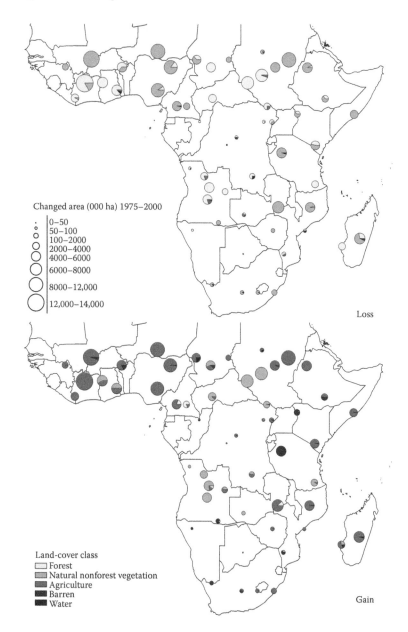

FIGURE 24.2 (See color insert.) Distribution and proportion of land-cover changes between 1975 and 2000. The top image represents the loss of land cover, whereas the bottom image shows the gain in land cover. The size of the pie chart corresponds to the extent of area changed.

24.4 DISCUSSION AND CONCLUSIONS

Making use of a stratified sample of high-resolution Landsat data, we were able to report on major land-cover changes in sub-Saharan Africa between 1975 and 2000. We estimate the loss of natural vegetation to be over 130 Mha in this period, caused primarily by expansion of agriculture, which has increased by 57% from just over 220 Mha in 1975 to nearly 340 Mha in 2000, and by other forms of degradation caused by human activities, such as timber harvesting and wood collection. Geist and Lambin (2002) described the human driving forces of land-cover changes as an interlinking of three key variables: expansion of agriculture, extraction of wood, and development of infrastructure—the last one being difficult to identify by the applied sampling methodology and being less important in the study. Expansion of agriculture is further subdivided into shifting cultivation, permanent cultivation, and cattle ranching, while variable wood extraction is further subdivided into commercial wood extraction (clear-cutting, selective harvesting), fuelwood extraction, polewood extraction, and charcoal production. The loss of natural vegetation in sub-Saharan Africa can be explained by the first two variables proposed by Geist and Lambin (2002). This is a direct response to the demand generated by a fast-growing population in the African continent. In 2007, sub-Saharan Africa's population was about 800 million, and since it has the highest birth rate among all continents, its population is projected to grow to 2 billion by 2050. Although 60% of the population lives in rural areas and 55% of the economically active population depends on agriculture, the urban population has increased from 22% to 36% between 1980 and 2005 (The World Bank, 2007).

The driving force behind deforestation in sub-Saharan Africa has been usually seen as the clearing of trees for agriculture (Brink and Eva, 2009; Foley, 1985). The available evidence from our study supports this hypothesis, but also highlights the fact that forest degradation due to wood extraction is an additional key factor throughout the region, especially in the Sahel belt, the miombo woodland zone of Tanzania, Zambia, southern Democratic Republic of Congo, and Angola. In sub-Saharan Africa, the main use of extracted wood is for production of energy (Kebede et al., 2010; Mugo and Ong, 2006; Mwampamba, 2007). This means that degradation of forest and woodland for production of firewood and charcoal is not just a by-product of bigger forces such as logging for timber and expansion of agriculture, but a specific need to meet energy needs—there is even evidence of export of charcoal from Tanzania and Somalia to the Middle East (SWALIM, The Guardian, 2005). Although the region is endowed with a huge diversity of energy sources such as oil, gas, coal, uranium, and hydropower, the local infrastructure and use of these commercial energy sources are very limited. Traditional sources of energy in the form of firewood and charcoal account for over 75% of the total energy use in the region (Mwampamba, 2007; Kebede et al., 2010). These are used mainly for cooking and also for agriculture, and in the rural industry for brick-making, food processing, baking, tobacco curing, etc. (Kebede et al., 2010).

There is a direct relationship between population growth and energy demand and between expansion of agriculture and energy demand. With the current development, population and economic growth impose a dependence on the remaining natural vegetation to meet further expansion of agriculture and energy needs. Analyzing the population-increase figures and relating them to the land-cover-change figures of our study, we see that the average amount of agricultural land per head of rural population has fallen by 20% from 1.09 to 0.91 ha—however, in certain zones it is far more dramatic. At the same time, the extent of land available for future exploitation (i.e., forests, savannas, and woodlands) has necessarily reduced, emphasizing the pressure on the remaining forests and woodlands. To meet the population and economic demands in sub-Saharan Africa without overexploiting natural vegetation, technological development of agriculture is needed on the one hand, and energy-efficient technologies and a diversification of energy sources are needed on the other hand. The use of modern technologies such as mechanization, fertilization, and irrigation in agriculture and a more efficient use of biomass energy sources linked to the development of a modern energy infrastructure are a prerequisite for a sustainable use of natural vegetation in sub-Saharan Africa.

REFERENCES

Achard, F., Eva, H., Stibig, H.J., Mayaux, P., Gallego, J., Richards, T., and Malingreau, J.P. 2002. Determination of deforestation rates of the world's humid tropical forests. *Science*, 297, 999–1003.
Brink, A.B. and Eva, H.D. 2009. Monitoring 25 years of land-cover change dynamics in Africa: A sample based remote sensing approach. *Applied Geography*, 29, 501–512.

Energy Information Agency (EIA). 1999. Major African environmental challenge: Use of biomass energy. Available from: /http://www.eia.doe.gov/ emeu/cabs/chapter7.htmlS

ERS/USDA. 2008. Real historical gross domestic product (GDP) and growth rates of GDP for baseline countries/regions (in billions of 2005 dollars), 1980–2005. Available from: http://www.ers.usda.gov/Data/Macroeconomics/Data/HistoricalRealGDPValues.xls.

FAO. (2006). *FAOSTAT—2006 Database*. Rome: Food and Agriculture Organization. Available from: http://faostat.fao.org/ (accessed July 5, 2007).

Foley, G. 1985. Woodfuel, deforestation and tree growing in the developing world. *Energy Policy*, 3(2), 190–192.

Foley, J.A., Defries, R., Asner, G.P., et al. 2005. Global consequences of land use changes. *Science*, 309, 570–574.

Gallego, F.J. and Delincé, J. 1991a. Stratification for acreage regression estimators with remote sensing. In A. Annoni, F. Dicorato, and J. Stakenborg (Eds.), *Manual for the Use of Software for Agricultural Statistics Using Remotely Sensed Data*. Ispra, Varese, Italy: Commission of the European Communities, Joint Research Centre.

Gallego, F.J. and Delincé, J. 1991b. Crop area estimation through area frame sampling and remote sensing. In A. Annoni, F. Dicorato, and J. Stakenborg (Eds.), *Manual for the Use of Software for Agricultural Statistics Using Remotely Sensed Data*. Ispra, Varese, Italy: Commission of the European Communities, Joint Research Centre.

Geist, H.J. and Lambin, E.F. 2002. Proximate causes and underlying driving forces of tropical deforestation. *BioScience*, 52(2), 143–150.

Gibbs, H.K., Ruesch, A.S., Achard, F., Clayton, M.K., Holmgren, P., Ramankutty, N., and Foley, J.A. 2010. Tropical forests were the primary sources of new agricultural land in the 1980s and 1990s. *Proceedings of the National Academy of Science, USA*, 107(38), 16732–16737.

Kebede, E., Kagochi, J., and Jolly, C.M. 2010. Energy consumption and economic development in sub-Sahara Africa. *Energy Economics*, 32, 532–537.

Mayaux, P., Bartholomé, E., Fritz, S., and Belward, A. 2004. A new land-cover map of Africa for the year 2000. *Journal of Biogeography*, 31, 861–877.

Mugo, F. and Ong, C. 2006. Lessons of eastern Africa's unsustainable charcoal trade. ICRAF Working Paper no. 20. Nairobi, Kenya. World Agroforestry Centre.

Mwampamba, T.H. 2007. Has the woodfuel crisis returned? Urban charcoal consumption in Tanzania and its implications to present and future forest availability. *Energy Policy*, 35, 4221–4234.

SWALIM—Somalia Water and Land Information Management. http://www.faoswalim.org/ (accessed December 14, 2010).

The Guardian. 2005. Bleak future as charcoal burning flourishes. Available from: www.ippmedia.com (accessed July 5, 2005).

The World Bank. 2007. World Development Indicator 2007, World Bank, Washington D.C.

UNEP. 2005. *One Planet Many People, Atlas of Our Changing Environment*. Division of Early Warning and Assessment (DEWA), United Nations Environment Programme (UNEP), Nairobi.

White, F. 1983. *Vegetation Map of Africa*. Paris: UNESCO.

Section IV

Looking Ahead

25 The NASA Land-Cover and Land-Use Change Program
Research Agenda and Progress (2005–2011)

Garik Gutman, Chris Justice, and LeeAnn King

CONTENTS

25.1 BACKGROUND

Land-cover and land-use change are perhaps the most immediate and visible aspects of global change. Land cover can change through the natural processes of succession, natural disturbances such as fire, or from changing climatic conditions. At the local scale, land-use change occurs because of a decision by individual farmers, ranchers, landowners, or managers. Change in land use at the regional scale occurs through processes summarized as agricultural intensification, extensification, abandonment, or through a change in policy. Most land-use changes are associated with a change in land cover. Changes in land cover can be monitored directly from space and airborne-sensing systems.

The NASA Land-Cover/Land-Use Change (LCLUC) Program initiated in 1996 leant heavily on the intellectual formulation of the International Geosphere–Biosphere Programme (IGBP), International Human Dimensions Programme (IHDP), Land-Use and -Cover Change (LUCC) program, and the related global data initiatives developed by the IGBP Data and Information System (IGBP–DIS; Lambin et al., 1995; Turner et al., 1995). The NASA LCLUC program has been at the intersection of several discipline programs at the NASA Headquarters, primarily because it includes studies on the impacts of change in land cover and land use on the carbon cycle, climate, hydrology, ecology, and biodiversity (Gutman et al., 2004). The LCLUC program is distinct from other discipline programs within the NASA science program, as it directly addresses societal aspects of land processes, which requires the involvement of social scientists. Thus, the integration of physical and social sciences and the use of satellite observations in an interdisciplinary framework has been the thrust of the LCLUC program. The program has a special place in NASA's earth sciences in developing interdisciplinary science with a high degree of societal relevance.

The long-term objectives of the LCLUC program are to develop the capability to perform repeated global inventories of land-use and land-cover change from space and to improve the scientific understanding of land-cover and land-use processes and the models necessary to simulate the processes taking place from local to global scales. The program aims at modeling current land-use and land-cover change, developing projections of future changes and their direct and indirect impacts, and evaluating the societal consequences of the observed and predicted changes. It helps to establish the operational provision of land-use and land-cover data and information products, services, models, and tools for multiple users, including scientists, resource managers, and policymakers. The key science questions addressed by it are as follows: (1) Where are land cover and land use changing, what is the extent of the change, and over what time scale? (2) What are the causes and consequences of LCLUC? (3) What are the projected changes of LCLUC and their potential impacts? and (4) What are the impacts of climate variability and change on LCLUC and the associated feedbacks?

LCLUC is one of five programs comprising the Carbon Cycle and Ecosystems Focus Area within NASA's earth sciences and has developed links to other Focus Research Areas and NASA Instrument and Data programs. For example, to address the question of the drivers of land-cover change in relation to carbon and water cycles, a close partnership has been developed between LCLUC and the NASA Terrestrial Ecology, Surface Hydrology, and Radiation Science programs. The Terrestrial Ecology and LCLUC programs have had several jointly funded projects, for example, supporting studies of land-use change as part of the Large-Scale Biosphere-Atmosphere Experiment in Amazonia (LBA program; Keller et al., 2009). The LCLUC program, jointly with the Radiation Sciences and the Terrestrial Hydrology programs, supported studies of the impacts of LCLUC on the atmosphere, such as the effect of biomass burning, and on water resources and the hydrological cycle, respectively. The NASA Interdisciplinary Science (IDS) program and the Applied Sciences Program have also supported LCLUC-related research. In addition to internal NASA solicitations, the LCLUC program has developed research solicitations jointly with the U.S. Department of Agriculture.

In the early days of the program, the role of land-use change in the global carbon cycle was an emerging topic (Dixon et al., 1994; Janetos and Justice, 2000; Watson et al., 2000) and, as a result, several of the funded research topics addressed aspects of carbon and forestry. Carbon cycle science has remained a strong part of the program, which has expanded to address aspects of the water cycle, the associated land–atmosphere interactions, urban environments, agricultural land-use change and climate impacts on land use, as well as vulnerability and adaptation of land use to environmental changes. Gutman et al. (2004) presented the history of the LCLUC program and research directions taken while transitioning from Phase 1 (1996–2004) to Phase 2 (2005–2011) of the program. In this chapter, we outline the status of the program, its structure and goals, focusing on recent trends and future directions.

25.2 OBSERVATIONS AND DATA FOR LCLUC RESEARCH

The LCLUC program aims at developing and using NASA remote-sensing technologies, as well as other U.S. and non-U.S. satellite data sources, to improve our understanding of human interactions with the environment, and thus provide a scientific foundation for understanding the sustainability, vulnerability, and resilience of human land use and terrestrial ecosystems. In doing so, its major goal is to further the understanding of the consequences of land-cover and land-use change on environmental goods and services, carbon and water cycles, and management of natural resources (Janetos et al., 1996).

Starting with the AVHRR (Advanced Very High Resolution Radiometer) and Landsat Pathfinder programs in the early 1990s, NASA has been developing procedures for generating regional to global datasets on land cover and change (Justice and Townshend, 1994; Justice et al., 1995). These activities are continuing through the LCLUC Program and other NASA data-oriented initiatives such as Advancing Collaborative Connections for Earth System Science (ACCESS) and Making Earth Science Data Records for Use in Research Environments (MEaSUREs).

25.2.1 MODERATE-RESOLUTION OBSERVATIONS

The workhorse of the program has been the U.S. Landsat moderate spatial resolution sensor. Other non-U.S. Landsat-like sensors have also been launched into space during the last two decades, but limited accessibility to data from these sensors remains an issue for the LCLUC science team. Thus, Landsat has been the sensor of choice, providing the necessary science-quality observations at spatial resolutions well suited to mapping land cover and monitoring change (Goward et al., 2009, 2011; Warner et al., 2009). Most of the early applications of Landsat data for land-cover mapping were undertaken on local areas, within an individual Landsat scene (185 km × 185 km). The focus of LCLUC science during the recent years has been on regional to continental scale studies, requiring multiple Landsat scenes. Perhaps the earliest example of regional mapping of forest-cover change using Landsat was provided by the NASA Landsat Pathfinder Program for the Amazon Basin (Skole and Tucker, 1993). At that time, Landsat data were purchased on a per-scene basis, and an investment of several million dollars was needed by NASA to purchase multitemporal regional datasets. This "proof of concept" set the stage for wall-to-wall regional change analysis at 30 m, demonstrating the means by which to quantify local changes at the regional scale, which is essential for the LCLUC program.

For the tropics, the biomes of greatest interest in terms of deforestation and the carbon budget, the low frequency of Landsat cloud-free acquisitions resulted in the need for combining data from successive years to obtain a cloud-free mosaic of the land surface. This "epoch" approach to land-cover mapping has been a central theme of regional Landsat mapping through the last decade. Following this approach, in 2005, the LCLUC program initiated and provided expertise to help design, assemble, and distribute to the scientific community the Global Land Survey (GLS) datasets (Gutman et al., 2008). Through this initiative, Landsat data were assembled and processed to generate cloud-free, orthorectified global datasets for epochs centered around 1990, 2000, 2005, and 2010. These datasets contributed to the development of a number of new regional derived products quantifying land cover and land-cover change. The current GLS approach aims at providing users with one clear image during leaf-on conditions for every location of the global land area. While the current GLS framework does not include seasonal coverage for the globe (i.e., multiple images throughout the growing season), multiple Landsat-7 and Landsat-5 images are provided for selected areas in the tropics, where it is not generally possible to obtain a single cloud-free image during the growing season. More details on the design and processing of GLS datasets can be found in Gutman et al. (2008).

During the past decade, the LCLUC program has supported a number of regional land-cover mapping initiatives in support of science studies of the Amazon (Skole et al., 2004), Central Africa

(Hansen et al., 2008), the United States (Huang et al., 2009a; Loveland et al., 1991), European Russia (Potapov et al., 2009), the Americas (Huang et al., 2009a, 2009b), S.E. Asia (Samek et al., 2004), the Black Sea Region (Olofsson et al., 2009), and Monsoon Asia (Xiao et al., 2009). More recently, the program has expanded to include regional thematic mapping of specific land-cover types, for example, mangroves (Giri et al., 2007; Simard et al., 2006), agriculture (Ozdogan et al., 2006), and urban areas (Schneider et al., 2003).

25.2.2 TOWARD GLOBAL-SCALE LANDSAT PRODUCTS

The opening up of the Landsat archive by the USGS for free access has fueled the development of a new generation of Landsat products. In areas of frequent cloud cover, such as in the tropics, all Landsat scenes available within a year for a given location are being analyzed to generate regional cloud-free mapping of forest-cover change (Broich et al., 2011; Hansen et al., 2009; Lindquist et al., 2008). Dense stacks of time series from the 40-year Landsat record are also being used to understand the land-cover changes (Huang et al., 2009a, 2009b; 2010). The NASA MEaSUREs Web Enabled Landsat Data (WELD) Project now provides 30-m mosaics of composited Landsat data at weekly, monthly, and annual periods for the United States and Alaska (Roy et al., 2010). The approach developed by the NASA MODIS (Moderate Resolution Imaging Spectroradiometer) Land Team, whereby atmospherically corrected surface reflectance data provide the basis for a number of higher order products (Justice et al., 2002), has now been applied to Landsat data by the Landsat Ecosystem Disturbance Adaptive Processing System (LEDAPS; Masek et al., 2006) and the WELD Project (Hansen et al., 2011). At the same time, LCLUC is supporting research to develop a fully automated mapping of land-cover type from calibrated satellite products, preparing the way for automated land-cover mapping (Baraldi et al., 2010). With free access to Landsat data and the various processing methods recently developed and the availability of high-performance computing, the development of a complete 40-year record of global land-cover change at 30-m resolution within the next several years is foreseeable.

25.2.3 COARSE-RESOLUTION OBSERVATIONS

Starting with the NOAA AVHRR, land-cover mapping was developed using time series data (Townshend et al., 1991; Tucker et al., 1985). More recently, the suite of land products generated from MODIS established a milestone in land remote sensing (Justice and Tucker, 2009). These products and the associated quality assessment and validation provide a more than 10-year record of science-quality data with which to monitor land-surface changes (Ramachandran et al., 2011). Through systematic quality assessment and validation by the MODIS science team, the MODIS land products have been incrementally improved (Masuoka et al., 2011; Morisette et al., 2002; Roy et al., 2002). The products have been reprocessed completely four times, and a fifth is underway. The product suite includes land cover (Friedl et al., 2011), vegetation continuous fields (Carroll et al., 2011), and fire and burned areas (Justice et al., 2011). All these have been used by LCLUC research projects, and their value increases as the time series is extended.

25.2.4 OBSERVATION CONTINUITY FOR LCLUC RESEARCH

Our ability to quantify land-cover change using satellite observation requires data continuity. NASA is participating in missions that will continue the systematic observations from Landsat and MODIS. The Landsat Data Continuity Mission (LDCM), being jointly developed by NASA and USGS, is to be launched in January 2013 (Irons and Masek, 2006). The OLI instrument, a solid-state linear array, will have an enhanced spectral capability and will continue the dynamic data continuity of the Landsat series. A separate two-band thermal instrument (TIRS) will be collocated on the platform (Reuter et al., 2010). While the USGS is responsible for the LDCM ground segment, resource commitments for the routine generation of science-quality land-cover products from LDCM are yet to be made. Some of these products are being prototyped by the LCLUC program. The future of the U.S. Landsat Program beyond LDCM is currently under discussion (Goward et al., 2011). There is a strong demand from the LCLUC science and applications community that Landsat become a truly operational system.

The Joint Polar Satellite System's VIIRS instrument is designed to continue the coarse-resolution MODIS observational record (Justice et al., 2011). As the bridge to an operational series of instruments, the first VIIRS instruments will provide near-daily global coverage with an early afternoon overpass. For the land community, the band selection and spatial resolution is similar to that of MODIS. Although there is a clear need, a firm commitment for generating a suite of science-quality VIIRS land-cover products to provide continuity with MODIS is yet to be made.

25.2.5 INTERNATIONAL ASSETS FOR LAND-COVER AND LAND-USE CHANGE RESEARCH

The failure of the Landsat 7 Scan-Line Corrector in 2003 highlighted the fragility of the U.S. observing systems needed for monitoring land cover. Owing to the possibility of a significant Landsat data gap, the LCLUC program considered the potential role of various international space-borne assets with Landsat-like capabilities (Goward et al., 2009). Countries with such assets include France, China, Brazil, India, Japan, and the U.K.; however, each instrument has slightly different characteristics, and the associated agencies have different data policies. Even so, all these systems are used in various ways around the world for land-cover mapping. It is worth noting that the increased temporal frequency of the Indian Resource Satellite AWiFS instruments provides new capabilities for agricultural land-use monitoring and points to the advantage of a constellation approach for moderate-resolution sensing systems (Goward et al., 2011). With the increasing demand for timely access to global cloud-free satellite data and the evolving observation requirements, international coordination of satellite land observations needs to be given more attention (Gutman et al., 2008; Townshend et al., 2011). However, removal of restrictive or inequitable data accessibility and pricing policies will be an important prerequisite for developing international moderate-resolution data partnerships.

25.3 LCLUC RESEARCH COMPONENTS

The LCLUC program was designed initially around a number of regional case studies representing different typologies of land-cover change, complemented by methodological studies that explore the production and validation of particularly important regional and global remote-sensing land-use and land-cover related datasets (Gutman et al., 2004). In the first phase of the LCLUC program (1996–2004), the focus was on funding a small number of regional activities in areas where important land-use changes had recently taken place, for example, in Central Africa (Laporte et al., 2004), South Africa (Prince, 2004), Central America (Sader et al., 2004), South East Asia (Samek et al., 2004), Amazon (Skole et al., 2004), North American high latitudes (McGuire et al., 2004), Northern Eurasian boreal forests (Krankina et al., 2004), the Middle East (Hole and Smith, 2004), and Central United States (Wessman et al., 2004).

The emphasis of the second phase (2005–2011) was on expanding local process, case studies, and synthesis studies at the regional scale. Funding was directed toward continuing studies on the impacts of land-use change on carbon and hydrological cycles; on predictive modeling of future land use as well as improved regional datasets; field process and parameterization studies; and improved modeling of land-use dynamics. During the last 3 years, the program has expanded to include studies of rapid urban expansion, the impacts of LCLUC on climate, and climate-change impacts on land use. It also undertook a pilot initiative on aspects of land-use vulnerability and sustainability.

Below, we present some results of research studies from the second phase of the LCLUC program (2005–2011). Information on all the projects can be found at http://lcluc.hq.nasa.gov.

25.3.1 LCLUC Detection and Monitoring

Quantifying rates of forest-cover change is important for improved carbon accounting and climate-change modeling, management of forestry and agricultural resources, and biodiversity monitoring. Matt Hansen (South Dakota State University) and his team in collaboration with USGS established a global forest-monitoring capability with multiresolution and multitemporal remotely sensed observations. They extended previous research on global forest-cover dynamics and land-cover change estimation to establish a robust, operational forest monitoring and assessment system. Satellite-based monitoring of forest regions was conducted at annual and interannual intervals using an approach that integrated both MODIS and Landsat data to provide timely biome-scale forest-change estimation. This was achieved by using annual MODIS change indicator maps to stratify biomes into low-, medium-, and high-change categories. Landsat image pairs were then sampled within these strata and analyzed for estimating the area of forest cleared. Results for the humid tropics reveal that 27.2 million hectares of forest were cleared from 2000 to 2005, with nearly 50% of this change occurring in Brazil (Hansen et al., 2008). Indonesia was a distant second in forest loss, accounting for 12% of the biome total. The approach enabled regional intercomparisons and could be implemented repeatedly in a monitoring context. For example, a national-scale study of Indonesia, using the above method with AVHRR forest loss indicator maps and Landsat sample blocks for the 1990 to 2000 epoch, estimated average annual clearing to be 1.78 million hectares (Hansen et al., 2009). Clearing from 2000 to 2005 averaged 0.71 million hectares per year. This dramatic downturn might be related to the drivers of forest clearing having changed at the turn of the century, including political and social upheaval, an economic downturn, and the occurrence of widespread, human-induced fires during the ENSO event of 1997 and 1998 (Hansen et al., 2010).

Boreal forest clearing from 2000 to 2005 actually exceeded that of the humid tropics, totaling 35.1 million hectares. The proportion of forest lost in 2000 was 4.02%, compared to 2.36% for the humid tropics. In the boreal biome, fire was a major cause of forest-cover loss and was estimated to account for nearly 60% of the total. As a percentage of forest cover, forest loss in North America for 2000 was nearly twice that of Eurasia (5.63%–3.00%; Hansen et al., 2010). Overall, the method enables global, biome, and targeted national/regional-scale quantification of forest-cover change. The method requires less effort than exhaustive mapping approaches, includes a measure of uncertainty, and through regression estimation provides a spatial depiction of the estimated change.

Land-cover changes in the extreme north, such as the Yamal peninsula region in northwest Siberia, represent the type of changes that are likely to become much more common in the tundra areas of Russia and the circumpolar region within the next decade. Skip Walker with his team at University Alaska, Fairbanks, and in collaboration with University of Virginia, NASA Goddard Space Flight Center, and Finnish and Russian institutions, used space-based technologies and models to address land-cover/land-use change problems in the Yamal peninsula (Walker et al., 2009). The existing oil and gas activities, and changes in the Russian political-economic structure over the past 30+ years, have had profound impacts on the socioecological systems of the local Nenets people. The prospect of a rapidly expanding infrastructure network and changes in climate further threaten their way of life (Walker et al., 2011). The team examined the cumulative effects of resource development, climate change, and traditional land use. Quickbird, in combination with ASTER and Landsat data, was used to evaluate the extent of land-cover change in the target area with a gas field. The area where change was detected amounts to 33.3 km² of the total 448 km² perimeter. Indirect impacts of roads and pipelines are seen in migration corridors and have the greatest effect on Yamal's resource development. There is also a high potential for extensive landscape effects due to unstable sandy soils and extremely ice-rich permafrost near the surface. Land withdrawals by industry, increasing Nenets population, and larger reindeer herds are all increasing pressure on the rangelands. While for Yamal, satellite data have shown only modest summer land-surface warming and only slight greening changes during the past two-and-a-half decades, trends are much stronger in the other parts of the Arctic (Comiso et al., 2008). The value of an analysis of cumulative effects on the Yamal peninsula will be in the lessons learned and the applications of those lessons to other areas of potential development in the Arctic. Current research by the team requires the development of tools to better predict future changes by combining scientific and traditional knowledge of the landscapes, detailed field observations, socioeconomic analyses, remote sensing, climate change analyses, vegetation-change models as well synthesis of large-scale studies across the Arctic.

25.3.2 Carbon and Biogeochemical Cycle Impacts

Fires represent one of the largest disturbances in land cover affecting the earth's system carbon cycle. The Langley Research Center (LaRC) LCLUC project led by Amber Soja, in collaboration with scientists from the Russian Academy of Sciences, examined the relationships between weather, extreme fire events, and fire-induced land-cover change in the changing climate of Siberia. The research focused on elucidating the factors that force the dominant driver of land-cover change in Siberia, fire, which is shaped by human and climate dynamics. It is predicted that warming in Northern Eurasia will exceed 40% of the global mean (Groisman et al., 2006). Their investigation shows that January temperatures in the Sayani mountain range of south central Siberia have already exceeded the 2090 Atmosphere Ocean General Circulation Model predictions (Soja et al., 2007). The team concluded that 7 of the 9 years under investigation had witnessed extreme fire seasons, which implies that the definition of a "normal" fire year may already be changing (Groisman and Soja, 2009). According to this research, fire, under the influence of weather, climate, and human management, is a mechanism to maintain vegetation stability and diversity in equilibrium with the climate and a mechanism by which land cover moves more quickly toward a new equilibrium with changing climate. The research demonstrates that the effects of fire and weather are regional and are particularly evident in ecotones at upland and lowland treelines of mountainous regions and at the far southern and northern reaches of Siberia. The strong influence of fire weather in Sakha, northern Siberia, provides information that can be used to predict future vegetation change, driven by fire, weather, and climate (Soja et al., 2004). In Tuva, a region in southern Siberia, the team revealed the disappearance of the relic *Pinus sylvestris* forests due to the combination of fire and the absence of weather conditions conducive for germination and survival. The research provides substantial evidence of warming- and fire-induced land-cover change across Northern Eurasia, which suggests a potential nonlinear, rapid response to climate change, as opposed to the predicted slow linear change. These results corroborate the early suggestion by Weber and Flannigan (1997) that an "altered fire regime may be more important than the direct effects of climate change in forcing or facilitating species distribution changes, migration, substitution and extinction."

25.3.3 Water and Energy Cycle Impacts

To study interactions of edaphic and land-use factors on water resources of the Cerrado region of Brazil, Eric Davidson's team at the Woods Hole Research Center joined with scientists of the Carnegie Institution at Stanford University and with Brazilian and Argentinean colleagues. Research has shown that deforestation changes the hydrological, geomorphological, and biochemical states of river systems by decreasing evapotranspiration and increasing the run-off and river discharge across all spatial scales. Increased run-off and decreased vegetative cover increase erosion and altered river and floodplain morphology, as sediments are deposited inside channels and bars. However, detection of such changes in large rivers is difficult because deforestation often takes place before instrumentation begins and coincides with other human alterations of the river channel, such as the construction of dams and levees. The team showed that the deforestation that began in the 1960s in the savannah region of central Brazil (locally known as Cerrado) had altered 62% of the landscape and had significantly altered the morphological and hydrological characteristics of a 120,000 km^2 watershed of the otherwise unmodified Araguaia River (Coe et al., 2011).

Fieldwork and satellite image analysis show a 28% increase in sediment transport, 188 million tons of stored sediment, and an increase in the number of sandy bars but a decrease in the number of islands since the 1960s. Observed discharge increased by 25% from the 1970s to the 1990s, and simulations with a land-surface vegetation and hydrology model indicated that about 2/3 of the increase might have been from deforestation. These results provide an unequivocal quantification of human alterations of the hydrology and geomorphology of a large tropical river. Further, they suggest that similar changes have occurred throughout the 2,000,000 km^2, hydrologically important Cerrado region and that many other large tropical rivers are similarly affected by ongoing deforestation (Coe et al., 2011).

25.3.4 ECOSYSTEMS AND BIODIVERSITY IMPACTS

The LCLUC program collaborated with the NASA biodiversity program in supporting studies on conservation or biodiversity as related to land-use changes. For example, the program supported the study of the interactive effects of conservation and development policies on land cover and panda habitat in the Sichuan Giant Panda Sanctuary of China; the study was done by Jianguo (Jack) Liu and his team at Michigan State University, in collaboration with scientists from the State University of New York and the Chinese Academy of Sciences. Government policies shape human activities that drive land-cover changes and impact wildlife habitats (Liu et al., 2007). They can be put into two broad categories: development policies and conservation policies. Development policies seek to improve human well-being, whereas conservation policies seek to protect and restore natural ecosystems. This research project investigated the interactive effects of the concurrent implementation of two conservation policies and a development policy on land cover and panda habitat dynamics across the Sichuan Giant Panda Sanctuary—a recently created World Heritage Site of UNESCO. Annual rate of forest decline fluctuated between 0.6% and 1.8% from 1994 to 2001 (from 2001 to 2007 the annual rate of forest recovery varied between 0.5% and 1.9%), though some townships toward the east are still experiencing overall forest-cover losses (Viña et al., 2007). The Wenchuan earthquake of 2008 induced drastic losses in forest cover, particularly in the townships in the eastern portion of the sanctuary near the epicenter. This resulted in a reversal of the net gains experienced from 2001 to 2008. The combined effects of development and earthquake-induced landslides would have drastically reduced the forest cover without the implementation of the conservation policies (Viña et al., 2010). Further evaluation of the socioeconomic influences on the sanctuary and habitat conservation policies is on going.

25.3.5 PREDICTIVE LAND-USE MODELING

To explore modeling strategies for adaptation to coupled climate and land-use change in the United States, Scott Goetz and his team from Woods Hole Research Center, with scientists from NASA, used a set of coupled ecosystem and hydrology models that evaluate the combined impacts of climate and land-use change. These models also simulate the influence of potential mitigation and adaptation strategies by predicting land-use change scenarios that incorporate alternative practices. This research spatially predicts future land-use change and incorporates those predictions under different scenarios and land-use management options that can mitigate additional climate warming by increasing carbon sequestration via changes in primary productivity. Land-use change is modeled at a combination of spatial scales. At fine-grain resolution (30 m), the regional urban modeling system, SLEUTH (Slope, Land cover, Exclusion, Urbanization, Transportation, and Hillshade) 3d, has been used to forecast urban growth up to the year 2030 for the Chesapeake Bay Watershed and adjacent counties (Goetz et al., 2007; Jantz et al., 2010a). This effort is based on the SLEUTH urban growth model but has been modified to include new functionality and fit metrics, besides enhancing performance and applicability (Jantz et al., 2010b). The model was calibrated for subregions within the watershed, based on satellite mapping of urbanization for 1990 and 2000, accurately matching urban change within each subregion. At the national scale, the Spatially Explicit Regional Growth Model (SERGoM) was used to produce housing density and impervious surface-area scenarios for the entire United States (Theobald et al., 2009). Four housing density and impervious surface scenarios are now being generated for the period from 2010 to 2100. These are being used to determine changes in soil infiltration properties, which are then used as a direct input to TOPS (the Terrestrial Observation and Prediction System). Results from initial TOPS simulations for the Delaware and Chesapeake watersheds for 2000–2030 show an increase in annual run-off, a significant increase in run-off per storm event and an overall decrease in annual gross primary productivity associated with increasing impermeable surfaces associated with urbanization.

25.3.6 Climate Variability and Change

Irina Sokolik and her team from Georgia Institute of Technology with Chinese and Russian collaborators have tried to improve our understanding of how and to what extent land-cover/land-use changes and varying dust loadings and their interactions have been affecting climate of dry lands in Central Asia over the past 50 years. Growing evidence suggests that changes in land use and land cover and atmospheric dust loadings are among the key drivers of climate change in the dry land regions of Central Asia. Desiccation of the Aral Sea, conversion of the steppe in Kazakhstan to agricultural fields, and severe desertification of northeast China are just a few examples of land-use changes that have altered the source and emission of dust. The impacts of dust storms are not only regional, but may affect areas thousands of kilometers away from their source, making interactions between climate change, land use, and dust aerosols globally relevant (Sokolik et al., 2001).

To improve the ability to predict impacts of dust on the climate and environment, the team developed a regional masking derived from MODIS surface albedo for Central and East Asia. This involved the development and implementation of a new dust module DuMo in the NCAR Weather Research and Forecasting (WRF) model. The dust module included two different state-of-the-art schemes that explicitly accounted for land properties (including vegetation and soil moisture) and meteorology, providing a new modeling capability for studying land–atmospheric dust interactions (Darmenova et al., 2009). Another component of this project was the development of the Asian Dust Databank by integrating the diverse satellite and ground-based data on land use/land cover, atmospheric mineral dust, and climatic variables in Central and East Asia.

Dust emission was prevalent in the Taklamakan, Badain Jaran, and Gurbantungut Deserts of northwestern China, which was chosen as a study site for this research. The analysis of World Meteorological Organization (WMO) data from 1950 to 2006 revealed complex patterns of spatial and temporal distributions of dust outbreaks. The newly developed WRF DuMo regional modeling system in conjunction with the Asian Dust Databank is being used to study the effect of dust and LCLUC on the climate of dry lands in Central Asia. The analysis demonstrates that land-cover roughness index is a dynamic characteristic changing both with season and on much shorter time scales, and interactions between soil and atmosphere are important for climate systems analysis and earth studies, though climate model representation of dust poses significant challenges (Waggoner and Sokolik, 2010).

25.3.7 Land Use, Vulnerability, and Adaptation

This theme is relatively new in the LCLUC program, and the projects that address this issue are in their early stages. For example, a project by Kirsten de Beurs (University of Oklahoma) in collaboration with scientists from South Dakota State University and Radford University is investigating recent trends in land abandonment in Russia, assessing vulnerability and adaptation to changing climate and population dynamics.

Russia's population is projected to shrink by 29% by 2050. Differential dynamics among rural populations are correlated with ethnicity, natural condition, and remoteness of large cities, and they are key drivers in the spatial disintegration of rural Russia (Ioffe et al., 2006). Currently, Russia is slowly transitioning into a country with an internal "archipelago" of islands of productive agriculture around cities set within a matrix of much less productive and abandoned croplands. This heterogeneous spatial pattern is mainly driven by depopulation of the least favorable parts of the countryside, where "least favorable" is a function of lower fertility of land, higher remoteness from urban markets, or both. The project investigates potential sustainable productivity of the remaining croplands under climatic and demographic changes. The team aims at improving current understanding of the interactions of climate change and the spatiotemporal impacts of agricultural reform in European Russia. The project includes modeling land abandonment based on past abandonment estimates retrieved from satellite imagery, age-structured population models, and spatially structured metapopulation models. This will involve using sociodemographic data, distance to major population centers, and bioclimatic potential derived from a combination of current temperature and moisture regimes retrieved from space-borne sensors and predicted future regimes from IPCC AR4 models. The modeling approach will predict how possible future climates can influence abandonment patterns in Russia and how adaptive strategies can affect rural recolonization and recultivation patterns.

25.4 SYNTHESIS STUDIES

The LCLUC program has supported a number of synthesis studies aimed at advancing the conceptual underpinning of LCLUC science. Such studies are based on summarizing state-of-the-art knowledge, and synthesizing results and findings from available relevant datasets and research studies to advance our understanding of the processes, drivers, and impacts of LCLUC. Such analysis inevitably leads to the identification of data and research gaps and the ways proposed to fill the gaps. This process is aimed at developing new understanding and conceptual frameworks through development of theory and hypothesis testing, compilation and comparative analyses, data, and model integration. The findings are then articulated by publishing a refined or new conceptual framework for some aspect of LCLUC. Some reviews and synthesis studies are presented by Gutman et al. (2004), and more synthesis studies based on Phase 2 of the program are expected in the next 3 years. New understanding leads to new community research priorities. The LCLUC program aims at being responsive to such changing research priorities.

25.5 U.S. INTERAGENCY COORDINATION IN THE STUDY OF LCLUC

NASA is a major contributor to the U.S. Global Change Research Program (USGCRP), which is the interagency forum for coordinating global change research (USGCRP, 2010). The USGCRP Land Use Interagency Working Group (LUIWG) coordinates federal land-use research. It takes advantage of the complementary roles of the different federal agencies, including USGS, USDA, USFS, EPA, and USAID, and their particular emphases on land-use research, to expand the scope of research on land-use and land-cover change. Strengthening interagency collaboration to complement NASA's research is a priority for the LCLUC program.

Although NASA instruments collect systematic, time series observations in support of its science mission, the responsibility for operational monitoring within the U.S. belongs to the operational agencies, for example, NOAA USGS, USDA, and USF. As a result, there is often a disconnect in the transition of proven and tested research to routine operations, which more often than not comes down to issues of funding. The NASA Science Applications program working with other federal agencies can create a bridge between science and operational applications.

25.6 INTERNATIONAL CONTEXT OF THE NASA LCLUC PROGRAM

Important land-use questions related to global food supply, regional water resources, carbon sources and sinks, biodiversity loss, and population growth necessitate working in a number of different regions, which results in a strong international component to the LCLUC program. The program emphasizes studies where land-use change is rapid or where there are significant regional or global implications. In most countries, scientists are undertaking land-use research that reflects national priorities. In developing countries, such research is focused on the pressing issues associated with improving land, resource management, human health, and livelihoods. The LCLUC program promotes collaboration with in-country scientists and regional science networks, increasing their accessibility to NASA space-borne assets and in turn helping NASA scientists access international data and analyze and understand complex local and regional land-use issues and practices.

NASA has a history of coordinating their activities around regional scientific initiatives or campaigns (e.g., FIFE: Sellers et al., 1992; BOREAS: Sellers et al., 1997). During its second phase, the LCLUC program contributed to the following large regional science programs: Central African Regional Program for the Environment (CARPE), the Land-Biosphere-Atmosphere (LBA), the Northern Eurasia Earth Science Partnership Initiative (NEESPI), and the Monsoon Area Integrated Regional Study (MAIRS). NASA LCLUC contributed to these programs by soliciting, selecting, and funding research projects, supporting NASA scientists to attend regional science workshops and undertake LCLUC research aligned with the objectives of these programs.

NEESPI and MAIRS are regional initiatives under the international programs IGBP, IHDP, and WCRP. NEESPI seeks to develop a comprehensive understanding of the Northern Eurasian terrestrial ecosystem dynamics, biogeochemical cycles, surface energy and water cycles, and human activities and how they interact with and alter the biosphere, atmosphere, and hydrosphere of the earth. The LCLUC program has been supportive of NEESPI since its inception and has been instrumental in improving accessibility to satellite datasets for the region from NASA and SCANEX, a regional data provider in Moscow, Russia. MAIRS focuses on human monsoon system interaction and seeks to understand to what extent human activities modulate the Asia monsoon climate and how the changed climate will further affect Asia's socioeconomic development. The LCLUC Program is strengthening the land-use research component of MAIRS. CARPE contributes to the broader Congo Basin Forest Partnership (CBFP) and aims at promoting sustainable natural resource management in the Congo Basin. CARPE works to reduce the rate of forest degradation and loss of biodiversity by supporting increased local, national, and regional natural resource management capacity. LBA includes a focus on how tropical forest conversion, regrowth, and selective logging influence carbon storage, nutrient dynamics, hydrologic processes, trace gas fluxes, and the prospect for sustainable land use in Amazonia.

The IGBP–IHDP Global Land Project (GLP) followed on from the international LUCC program. Communication with the GLP is usually done through the informal channels of scientific exchanges and participation in research. There are five LCLUC projects on the current list of GLP-endorsed projects.

With a primary emphasis on satellite observations, the LCLUC program recognizes the importance of establishing the global earth-observing systems needed for long-term monitoring of land cover and land use. Through its research program, LCLUC is developing, prototyping, and demonstrating the methods required for such a system. However, although the LCLUC program can contribute to the design and development of these systems, NASA is not an operational agency and cannot put the long-term operational monitoring systems in place. NASA primarily aims at testing aerospace technology, developing and launching instruments, providing observations, generating data products, and developing the associated modeling and analysis methods to address earth science questions.

In recent years, with decreasing budgets, more attention has been paid to international cooperation between space research agencies. To meet the requirements of global change research, the global monitoring systems will need to be international, conforming to internationally accepted standards of data quality, product accuracy, and data continuity (Townshend et al., 2011).

The LCLUC program is a major supporter of the Global Observation of Forests and Land Cover Dynamics (GOFC-GOLD) program, which is a component of the Global Terrestrial Observing System (GTOS) and which aims at establishing operational monitoring systems through international cooperation (Townshend et al., 2004). GOFC-GOLD focuses on forest and land cover and fire and has a number of coordinating initiatives underway, including the development of a sourcebook for Reducing Greenhouse Gas Emissions from Deforestation and Degradation (REDD) in developing countries, and it has created regional networks of scientists doing research on land cover, land use, and fire. LCLUC supports much of the GOFC-GOLD regional capacity-building activities and contributes to developing regional information scientific networks. The LCLUC program also supports the Project Office for the Fire Implementation Team of GOFC-GOLD. Through GOFC-GOLD, LCLUC contributes to a number of land cover-related tasks of the Global Earth Observing System of Systems (GEOSS), including the global 30-m land-cover initiative, the Land Surface Imaging (LSI) constellation, and the agricultural land-use change component of the GEOSS Agricultural Monitoring Task (Becker-Reshef et al., 2010). GEOSS provides an important international coordinating mechanism with focus on earth observations for societal benefits.

25.7   FUTURE DIRECTIONS OF THE LCLUC PROGRAM

With the increasing and competing demand for land for producing food, animal feed, and fuel from a growing human population and economic development, more attention will have to be paid to understanding the associated issues of land-use change. The trade-off between competing demands for land and the potential outcomes of different management strategies will need to be modeled in ways that can inform policies and the associated decision-making processes (Lambin and Meyfroidt, 2011). With global markets and economic development driving land-use change, continued focus on the distal economic drivers of land-use change will be needed (DeFries et al., 2010). The challenge of global sustainability, initiated at the United Nations Conference on Environment and Development (UNCED) in 1992, remains relevant today (Turner et al., 2007). With the twentieth anniversary of UNCED approaching, the international community will inevitably renew its focus on issues such as sustainable use of land, reducing environmental degradation, and maintaining the delivery of environmental goods and services, including preservation of biodiversity and improvement of livelihoods. Changes in land-use policy and subsidies aimed at improving resource production may need to be reconsidered in the light of the adverse impacts of land-use change on regional climate (McAlpine et al., 2009). In addition to the aforementioned socioeconomic drivers of change, climate variability and change are already leading to changes in land cover and land use. Studies that have examined the distribution of land use primarily related to changing physical climatic variables will need to be refined as the global and regional climate models are improved. Models of land-use change that address economic scenarios, constraints, and opportunities will need to be linked to these climate projections (e.g., Fischer et al., 2001). New integrated assessment models will be needed to provide a realistic coupling of human and natural systems and quantification of land-use transitions (e.g., Hurtt et al., 2002, 2006). These models will benefit from an improved set of data products from the satellite record providing more nuanced information on land-cover characteristics, land-use and land-management practices. Land-use practices, which mitigate climate change, are being promoted, subsidized, and incentivized, and these include reduced deforestation, no-till agriculture, agroforestry, etc. Large tracts of land are being converted from food crops to crops for fuel (e.g., corn and sugarcane for ethanol) or from woodland to agriculture, to reduce dependence on fossil fuel. However, such land-management changes have implications for food supply, water quality, and biodiversity. Addressing individual demands for land in isolation ignores the proximate and distal impacts of land-use change, and a more holistic view of land use is needed to mitigate climate change (DeFries and Rosenzweig, 2010).

The science community is expanding its focus from understanding the physical climate system and modeling the climate system to addressing the impacts of climate change and adaptation (e.g., USGCRP, 2011). Changes in land use will be a primary adaptation to climate change, and issues of vulnerability of land-use systems will need to be addressed.

Given the various trends and research issues associated with land-use change, the question arises as to the role and focus of the NASA LCLUC program for future research. The program will continue to support cutting-edge research, and our researchers will contribute to some of the important land-use research issues associated with economic development and climate variability and change. Given the nature of land-use decisions and practices, it is important to continue the integrated approach to land-use science, combining physical and social science. Emphasis will be on issues of regional to global significance and of societal benefit.

The primary emphasis will remain on the use of remote sensing. The forthcoming near-term NASA and USGS missions, which include LDCM and VIIRS, will provide new observations and capabilities for land-use science that will need developing. The focus will be on developing new land-use and land-cover datasets to meet science needs and consistent long-term data records to quantify change. New technologies and capabilities will be encouraged but will be implemented by other parts of the NASA program. Within the LCLUC program, fusion of data from various sensors and various parts of electromagnetic spectrum will be emphasized.

Land-use change is pervasive, and the global connections of land use and the global importance of regional land-use changes require a global approach. The international dimension of the program will continue, enhancing collaborations with international scientists and supporting international programs that provide regional expertise and meet the goals of the LCLUC program.

ACKNOWLEDGMENTS

The authors would like to thank NASA LCLUC P.I.'s Matt Hansen, Skip Walker, Amber Soja, Eric Davidson, Jack Liu, Scott Goetz, Irina Sokolik, and Kirsten De Beurs for the information on their respective research projects.

REFERENCES

Baraldi, A., Durieux, L., Simonetti, D., Conchedda, G., Holecz, F., and Blonda, P. 2010. Automatic spectral-rule-based preliminary classification of radiometrically calibrated SPOT-4/-5/IRS, AVHRR/ MSG, AATSR, IKONOS/ QuickBird/ OrbView/ GeoEye, and DMC/ SPOT-1/-2 Imagery—Part I: System design and implementation. *IEEE Transactions on Geoscience and Remote Sensing*, 3: 1326–1354.

Becker-Reshef, I., Justice, C.O., Sullivan, M., Vermote, E., Tucker, C., Anyamba, A., Small, J., et al. 2010. Monitoring global croplands with coarse resolution earth observations: The Global Agriculture Monitoring (GLAM) project. *Remote Sensing*, 2: 1589–1609.

Broich, M., Hansen, M.C., Potapov, P., Adusei, B., Lindquist, E., and Stehman, S.V. 2011. Time-series analysis of multi-resolution optical imagery for quantifying forest cover loss in Sumatra and Kalimantan, Indonesia. *International Journal of Applied Earth Observation and Geoinformation*, 13: 277–291.

Carroll, M., Townshend, J., Hansen, M., DiMiceli, C., Sohlberg, R., and Wurster, K. 2011. MODIS Vegetation cover conversion and vegetation continuous fields. *Land Remote Sensing and Global Environmental Change*, 11: 725–745.

Coe, M.T., Latrubesse, E.M., Ferreira, M.E., and Amsler, M.L. 2011. The effects of deforestation and climate variability on the streamflow of the Araguaia River, Brazil. *Biogeochemistry*, 105: 119–131.

Comiso, J.C., Parkinson, C.L., Gersten, R., and Stock, L. 2008. Accelerated decline in the Arctic sea ice cover. *Geophysical Research Letters*, 35: L01703. doi:10.1029/2007GL031972.

Darmenova, K., Sokolik, I.N., Shao, Y., Marticorena, B., and Bergametti, G. 2009. Development of a physically-based dust emission module within the WRF model: Assessment of dust emission parameterizations and input parameters for source regions in Central and East Asia. *Journal of Geophysical Research*, 114: D14201. doi: 10.1029/2008JD011236.

DeFries, R. and Rosenzweig, C. 2010.Toward a whole-landscape approach for sustainable land use in the tropics. *PNAS*, 107: 19627–19632.

DeFries, R., Rudel, T., Uriarte, M., and Hansen, M. 2010. Deforestation driven by urban population growth and agricultural trade in the twenty-first century. *Nature Geosciences*, 3: 178–181.

Dixon, R.K., Brown, S., Houghton, R.A., Solomon, A.M., Trexler, M.C., and Wisniewski, J.1994. Carbon pools and flux of global forest ecosystems. *Science*, 263: 185–190.

Fischer, G., Shah, M., van Velthuizen, H., and Nachtergaele, F.O. 2001. Global agro-ecological assessment for agriculture in the 21st century. IIASA Research Report 02–02, International Institute for Applied Systems Analysis, Laxenburg, Austria, 119.

Friedl, M.A., Zhang, X., and Strahler, A. 2011. Characterizing global land cover type and seasonal land cover dynamics at moderate spatial resolution with MODIS data. *Land Remote Sensing and Global Environmental Change*, 11: 709–724.

Giri, C., Pengra, B., Zhu, Z., Singh, A., and Tieszen, L.L. 2007. Monitoring mangrove forest dynamics of the Sundarbans in Bangladesh and India using multi-temporal satellite data from 1973 to 2000. Elsevier Ltd. *Estuarine Coastal and Shelf Science*, 73: 91–100.

Goetz, S.J., Prince, S.D., and Jantz, C.A. 2007. Satellite maps of the Chesapeake watershed help to monitor urban sprawl. *Chesapeake Bay Journal*, 17: 20.

Goward, S.N., Arvidson, T., Williams, D.L., Irish, R., and Irons, J.R. 2009. Moderate spatial resolution optical sensors. In T.A. Warner, M.D. Nellis, and M.D. Foody (Eds.), *Handbook of Remote Sensing* (pp. 123–138). London: SAGE Publications.

Goward, S., Williams, D., Arvidson, T., and Irons, J. 2011. The future of Landsat-class remote sensing. In B. Ramachandran, C.O. Justice, and M.J. Abrams (Eds.), *Land Remote Sensing and Global Environmental Change: NASA's Earth Observing System and the Science of ASTER and MODIS* (pp. 873). Series: Remote Sensing and Digital Image Processing 11. New York: Springer.

Groisman, P.Y., Knight, R.W., Razuvaev, V.N., Bulygina, O.N., and Karl, T.R. 2006. State of the ground: Climatology and changes during the past 69 years over Northern Eurasia for a rarely used measure of snow cover and frozen land. *Journal of Climate*, 19: 4933–4955.

Groisman, P. and Soja, A.J. 2009. Ongoing climatic change in Northern Eurasia: Justification for expedient research. *Environmental Research Letters*, 4: 1–7.

Gutman, G., Janetos, A., Justice, C., Moran, E., Mustard, J., Rindfuss, R., Skole, D., and Turner II, B.J. (Eds.). 2004. *Land Change Science: Observing, Monitoring, and Understanding Trajectories of Change on the Earth's Surface*. New York: Kluwer Academic Publishers.

Gutman, G., Byrnes, R., Masek, J., Covington, S., Justice, C., Franks, S., and Headley, R. 2008. Towards monitoring land cover and land use changes at a global scale: The Global Land Survey 2005. *Photogrammetric Engineering and Remote Sensing*, 74: 6–10.

Hansen, M.C., Stehman, S.V., Potapov, P.V., Loveland, T.R., Townshend, J.R.G., Defries, R.S., Pittman, K.W., et al. 2008a. Humid tropical forest clearing from 2000 to 2005 quantified by using multitemporal and multiresolution remotely sensed data. *Proceedings of the National Academy of Sciences USA*, 105: 9439–9444, doi:10.1073/pnas.0804042105.

Hansen, M.C., Roy, D., Lindquist, E., Justice, C.O., and Altstaat, A. 2008b. A method for integrating MODIS and Landsat data for systematic monitoring of forest cover and change in the Congo Basin. *Remote Sensing of Environment*, 112: 2495–2513.

Hansen, M., Stehman, S.V., Potapov, P.V., Arunarwati, B., Stolle, F., and Pittman, K. 2009. Quantifying changes in the rates of forest clearing in Indonesia from 1990 to 2005 using remotely sensed datasets. *Environmental Research Letters*, 4: 034001.

Hansen, M.C., Stehman, S.V., and Potapov, P.V. 2010. Quantification of global gross forest cover loss. *Proceedings of the National Academy of Sciences USA*, 107: 8650–8655, doi:10.1073/pnas. 0912668107.

Hansen, M.C., Egorov, A., Roy, D.P., Potapov, P., Ju, J., Turubanova, S., Kommareddy, I., and Loveland, T. 2011. Continuous fields of land cover for the conterminous United States using Landsat data: First results from the Web-Enabled Landsat Data (WELD) project. *Remote Sensing Letters*, 2: 279–288.

Hole, F. and Smith, R. 2004. Arid land agriculture in Northeastern Syria: Will this be a tragedy of the commons? In G. Gutman, A.C. Janetos, C.O. Justice, E.F. Moran, J.F. Mustard, R.R. Rindfuss, D. Skole, B.L. Turner II, and M.A. Cochrane (Eds.), *Land Change Science: Observing, Monitoring, and Understanding Trajectories of Change on the Earth's Surface* (pp. 209–222). Dordrecht, the Netherlands: Kluwer Academic Publishers.

Huang, C., Goward, S.N., Schleeweis, K., Thomas, N., Masek, J.G., and Zhu, Z. 2009a. Dynamics of national forests assessed using the Landsat record: Case studies in eastern United States. *Remote Sensing of Environment*, 113: 1430–1442.

Huang, C., Kim, S., Altstatt, A., Song, K., Townshend, J.R.G., Davis, P., Rodas, O., et al. 2009b. Assessment of Paraguay's forest cover change using Landsat observations. *Global and Planetary Change*, 67: 1–12.

Huang, C., Goward, S.N., Masek, J.G., Thomas, N., Zhu, Z., and Vogelmann, J.E. 2010. An automated approach for reconstructing recent forest disturbance history using dense Landsat time series stacks. *Remote Sensing of Environment*, 114: 183–198.

Hurtt, G.C., Pacala, S.W., Moorcroft, P.R., Caspersen, J., Shevliakova, E., and Houghton R.A. 2002. Projecting the future of the U.S. carbon sink. *Proceedings of the National Academy of Sciences USA*, 99: 1389–1394.

Hurtt, G.C., Frolking, S., Fearon, M.G., Moore, B., Shevliakova, E., Malyshev, S., Pacala, S.W., and Houghton, R.A. 2006. The underpinnings of land-use history: Three centuries of global gridded land-use transitions, wood-harvest activity, and resulting secondary lands. *Global Change Biology*, 12: 1–22.

Ioffe, G., Nefedova, T., and Zaslavsky, I. 2006. *The End of Peasantry? The Disintegration of Rural Russia*. Pittsburgh, PA: University of Pittsburgh Press.

Irons, J. and Masek, J. 2006. Requirements for a Landsat continuity mission. *Photogrammetric Engineering and Remote Sensing*, 70: 1102–1110.

Janetos, A.C. and Justice, C.O. 2000. Land cover and global productivity: A measurement strategy for the NASA program. *International Journal of Remote Sensing*, 21: 1491–1512.

Janetos, A.C., Justice, C.O., and Harriss, R.C. 1996. Mission to planet Earth: Land cover and land use change. In J.S. Levine (Ed.), *Biomass Burning and Global Change* (pp. 1–13). Cambridge, MA: MIT Press.

Jantz, C.A., Goetz, S.J., Claggett, P., and Donato, D. 2010a. Modeling regional patterns of urbanization in the Chesapeake Bay watershed. *Computers, Environment and Urban Systems*, 34: 1–16.

Jantz, C.A., Goetz, S.J., Donato, D., and Claggett, P. 2010b. Designing and implementing a regional urban modeling system using the SLEUTH cellular urban model. *Computers, Environment and Urban Systems*, 34: 1–16.

Justice, C.O. and Townshend, J.R.G. 1994. Data sets for global remote sensing: Lessons learnt. *International Journal of Remote Sensing*, 15: 3621–3639.

Justice, C.O. and Tucker, C.J. III. 2009. Coarse resolution optical sensors. In T.A. Warner, M.D. Nellis, and G.M. Foody (Eds.), *Handbook of Remote Sensing* (pp. 139–150). London: SAGE Publications Inc.

Justice, C.O., Bailey, G.B., Maiden, M.E., Rasool, S.I., Strebel, D.E., and Tarpley, J.D. 1995. Recent data and information system initiatives for remotely sensed measurements of the land surface. *Remote Sensing of the Environment*, 51: 235–244.

Justice, C.O., Townshend, J.R.G., Vermote, E.F., Masuoka, E., Wolfe, R.E., El Saleous, N., Roy, D.P., and Morisette, J.T. 2002. An overview of MODIS Land data processing and product status. *Remote Sensing of Environment*, 83: 3–15.

Justice, C.O., Vermote, E., Privette, J., and Sei, A. 2011. The evolution of U.S. moderate resolution optical land remote sensing from AVHRR to VIIRS. In B. Ramachandran, C.O. Justice, and M.J. Abrams (Eds.), *Land Remote Sensing and Global Environmental Change: NASA's Earth Observing System and the Science of ASTER and MODIS* (pp.873). Series: Remote Sensing and Digital Image Processing, 11. Springer Verlag, New York, Dordrecht, Heidelberg, London.

Keller, M., Bustamante, M., Gash, J., and Silva Dias, P. (Eds). 2009. *Amazonia and Global Change*, Geophysical Monograph Series, 186, 576. Washington DC: AGU.

Lambin, E.F. and Meyfroidt, P. 2011. Global land use change, economic globalization and the looming land scarcity. *PNAS*, 108: 3465–3472.

Lambin, E.F., Baulies, X., Bockstael, N., Fischer, G., Krug, T., Leemans, R., Moran, E.F., et al. 1995. Land Use and Land Cover Change (LUCC) implementation strategy. IGBP Report 48, IHDP Report 10, IGBP Stockholm. 125.

Laporte, N.T., Lin, T.S., Lemoigne, J., Devers, D., and Honzák, M. 2004. Towards an operational forest monitoring system for Central Africa. In G. Gutman, A.C. Janetos, C.O. Justice, E.F. Moran, J.F. Mustard, R.R. Rindfuss, D. Skole, B.L. Turner II and M.A. Cochrane (Eds.), *Land Change Science: Observing, Monitoring, and Understanding Trajectories of Change on the Earth's Surface* (pp. 97–110). Dordrecht, the Netherlands: Kluwer Academic Publishers.

Lindquist, E.J., Hansen, M.C., Roy, D.P., and Justice, C.O. 2008. The suitability of decadal image datasets for mapping tropical forest cover change in the Democratic Republic of Congo: Implications for the global land survey. *International Journal of Remote Sensing*, 29: 7269–7275.

Liu, J.G., Dietz, T., Carpenter, S.R., Alberti, M., and Folke, C. 2007. Complexity of coupled human and natural systems. *Science*, 317: 1513–1516.

Loveland, T.R., Merchant, J.W., Ohlen, D.O., and Brown, J.F. 1991. Development of a land-cover characteristics database for the Conterminous U.S. *Photogrammetric Engineering and Remote Sensing*, 57: 1453–1463.

Masek, J.G., Vermote, E.F., Saleous, N.E., Wolfe, R., Hall, F.G., Huemmrich, K.F., Gao, F., Kutler, J., and Lim, T.K. 2006. A Landsat surface reflectance dataset for North America 1990–2000. *IEEE Geoscience and Remote Sensing Letters*, 3: 68–72.

Masuoka, E., Roy, D., Wolfe. R., Morisette, J., Sinno, S., Teague, M., Saleous, N., Devadiga, S., Justice, C.O., and Nickeson, J. 2011. MODIS land data products: generation, quality assurance and validation. In B. Ramachandran, C.O. Justice, and M.J. Abrams (Eds.), *Land Remote Sensing and Global Environmental Change: NASA's Earth Observing System and the Science of ASTER and MODIS* (pp. 873). Series: Remote Sensing and Digital Image Processing. New York: Springer.

McAlpine, M.A., Syktus, J., Ryan, J.G., Deo, R.C., McKeon, G.M., McGowan, G.M., and Phinn, S.R. 2009. A continent under stress: interactions, feedbacks and risks associated with impact of modified land cover on Australia's climate. *Global Change Biology*, 15: 2206–2223.

McGuire, A.D., Apps, M., Chapin, F.S. III, Dargaville, R., Flannigan, M.D., Kasischke, E.S., Kicklighter, D., et al. 2004. Land cover disturbances and feedbacks to the climate system in Canada and Alaska, Chapter 9. In G. Gutman, A.C. Janetos, C.O. Justice, E.F. Moran, J.F. Mustard, R.R. Rindfuss, D. Skole, B.L. Turner II, M.A. Cochrane (Eds.), *Land Change Science: Observing, Monitoring, and Understanding Trajectories of Change on the Earth's Surface* (pp. 139–161). Dordrecht, the Netherlands: Kluwer Academic Publishers.

Morisette, J.T., Privette, J.L., and Justice, C.O. 2002. A framework for validation of the MODIS land products. *Remote Sensing of Environment*, 83: 77–96.

Olofsson, P.Y., Woodcock, C., Baccini, A., Houghton, R.A., Ozdogan, M., Gancz, V., Blujdea, V., Torchinava, P., Tufekcioglu, A., and Baskent, E. 2009. The effects of land use change on terrestrial carbon dynamics in the Black Sea Region. In P. Groisman and S.V. Ivanov (Eds.), *Regional Aspects of Climate-Terrestrial-Hydrologic Interactions in Non-boreal Eastern Europe* (pp. 175–182). Netherlands: Springer.

Ozdogan, M., Woodcock, C.E., Salvucci, G.D., and Demir, H. 2006. Changes in summer irrigated crop area and water use in Southeastern Turkey: Implications for current and future water resources. *Water Resources Management*, 20: 467–488.

Potapov, P., Hansen, M.C., Stehman, S.V., Pittman, K., and Turubanova, S. 2009. Gross forest cover loss in temperate forests: biome-wide monitoring results using MODIS and Landsat data. *Journal of Applied Remote Sensing*, 3: 1–23.

Prince, S. 2004. Mapping desertification in Southern Africa. In G. Gutman, A.C. Janetos, C.O. Justice, E.F. Moran, J.F. Mustard, R.R. Rindfuss, D. Skole, B.L. Turner II, and M.A. Cochrane (Eds.), *Land Change Science: Observing, Monitoring, and Understanding Trajectories of Change on the Earth's Surface* (pp. 163–185). Dordrecht, the Netherlands: Kluwer Academic Publishers.

Ramachandran, B., Justice C.O., and Abrams, M.J. (Eds.). 2011. *Land Remote Sensing and Global Environmental Change: NASA's Earth Observing System and the Science of ASTER and MODIS* (pp. 873). Series: Remote Sensing and Digital Image Processing. 11, New York: Springer.

Reuter, D., Richardson, C., Irons, J., Allen, R., Anderson, M., Budinoff, J., Casto, G., et al. 2010. The thermal infrared sensor on the Landsat data continuity mission. *Geoscience and Remote Sensing Symposium (IGARSS), IEEE International*, Honolulu Hawaii, 754–757.

Roy, D.P., Borak, J.S., Devadiga, S., Wolfe, R.E., Zheng, M., and Descloitres, J. 2002. The MODIS land product quality assessment approach. *Remote Sensing of the Environment*, 83: 62–76.

Roy, D.P., Ju, J., Kline, K., Scaramuzza, P.L., Kovalskyy, V., Hansen, M.C., Loveland, T.R., Vermote, E.F., and Zhang, C. 2010. Web-enabled Landsat Data (WELD): Landsat ETM+ composited mosaics of the conterminous United States. *Remote Sensing of Environment*, 114: 35–49.

Sader, S.A., Chowdhury, R.R., Schneider, L.C., and Turner, B.L. 2004. Forest change and human driving forces in Central America. In G. Gutman, A.C. Janetos, C.O. Justice, E.F. Moran, J.F. Mustard, R.R. Rindfuss, D. Skole, B.L. Turner II, and M.A. Cochrane (Eds.), *Land Change Science: Observing, Monitoring, and Understanding Trajectories of Change on the Earth's Surfaces* (pp. 57–76). Dordrecht, the Netherlands: Kluwer Academic Publishers.

Samek, J.H., Xuan Lan, D., Silapathong, C., Navanagruha, C., Syed Abdullah, S.M., IGunawan, I., Crisostomo, B., Hilario, F., Hien, H.M., and Skole, D.L. 2004. Land-use and land-cover change in Southeast Asia. In G. Gutman, A. Janetos, C. Justice, E. Moran, J. Mustard, R. Rindfuss, D. Skole, and B.J. Turner II (Eds.), *Land Change Science: Observing, Monitoring, and Understanding Trajectories of Change on the Earth's Surface*. New York: Kluwer Academic Publishers.

Schneider, A., Friedl, M.A., McIver, D.K., and Woodcock, C.E. 2003. Mapping urban areas by fusing multiple sources of coarse resolution remotely sensed data. *Photogrammetric Engineering and Remote Sensing*, 69: 1377–1386.

Sellers, P.J., Hall, F.G., Asrar, G., Strebel, D.E., and Murphy, R.E. 1992. An overview of the first international satellite land surface climatology project (ISLSCP) field experiment (FIFE). *Journal of Geophysical Research*, 97: 18345–18371.

Sellers, P.J., Hall, F.G., Kelly, R.D., Black, A., Baldocchi, D., Berry, J., Ryan, M., et al. 1997. BOREAS in 1997: Experiment overview, scientific results, and future directions. *Journal of Geophysical Research*, 102: 28731–28769.

Simard, M., Zhang, K.Q., Rivera-Monroy, V.H., Ross, M.S., Ruiz, P.L., Castaneda-Moya, E., Twilley, R.R., and Rodriguez, E. 2006. Mapping height and biomass of mangrove forests in Everglades National Park with SRTM elevation data. *Photogrammetric Engineering and Remote Sensing*, 72: 299–311.

Skole, D. and Tucker, C.J. 1993. Tropical deforestation and habitat fragmentation in the Amazon: Satellite data from 1978 to 1988. *Science*, 25: 1905–1910.

Skole, D.L., Cochrane, M.A., Matricardi, E., Chomentowski, W.H., Pedlowski, M., and Kimble, D. 2004. Pattern to process in the Amazon region: Measuring forest conversion, regeneration, and degradation. In G. Gutman, A. Janetos, C. Justice, E. Moran, J. Mustard, R. Rindfuss, D. Skole, and B.J. Turner, II (Eds.), *Land Change Science: Observing, Monitoring, and Understanding Trajectories of Change on the Earth's Surface*. New York: Kluwer Academic Publishers.

Soja, A.J., Cofer, W.R., Shugart, H.H., Sukhinin, A.I., Stackhouse Jr., P.W., and McRae, D.J. 2004. Estimating fire emissions and disparities in boreal Siberia 1998 through 2002. *Journal of Geophysical Research*, 109: D14S06.

Soja, A.J., Tchebakova, N.M., French, N.H.F., Flannigan, M.D., Shugart, H.H., Stocks, B.J., Sukhinin, A.I., Varfenova, E.I., Chapin, F.S., and Stackhouse, P.W. Jr. 2007. Climate-induced boreal forest change: Predictions versus current observations. *Global and Planetary Change*, 56(3–4): 274–296.

Sokolik, I.N., Winker, D., Bergametti, G., Gillette, D., Carmichael, G., Kaufman, Y., Gomes, L., Schuetz, L., and Penner, J. 2001. Introduction to special section on mineral dust: Outstanding problems in quantifying the radiative impact of mineral dust. *Journal of Geophysical Research*, 106: 18,015–18,027.

Theobald, D.M., Goetz, S.J., Norman, J., and Jantz, P. 2009. Watersheds at risk to increased impervious surface cover in the coterminous United States. *Journal of Hydrologic Engineering*, 14: 362–368.

Townshend, J.R.G., Justice, C.O., Li, W., Gurney, C., and McManus, J. 1991. Global land classification by remote sensing: Present capabilities and future prospects. *Remote Sensing of the Environment*, 35: 248–256.

Townshend, J.R.G., Justice, C.O., Skole, D.L., Belward, A., Janetos, A., Gunawan, I., Goldammer, J., and Lee, B. 2004. Meeting the goals of GOFC: An evaluation of progress and steps for the future. In G. Gutman, A.C. Janetos, C.O. Justice, E.F. Moran, J.F. Mustard, R.R. Rindfuss, D. Skole, B.L. Turner II, and M.A. Cochrane (Eds.), *Land Change Science: Observing, Monitoring, and Understanding Trajectories of Change on the Earth's Surface* (pp. 31–52). Dordrecht, the Netherlands: Kluwer Academic Publishers.

Townshend, J.R.G., Latham, J., Justice, C.O., Janetos, A., Conant, R., Arino, O., Balstad, R., et al. 2011. International coordination of satellite land observations: Integrated observations of the land. In B. Ramachandran, C.O. Justice, and M.J. Abrams (Eds.), *Land Remote Sensing and Global Environmental Change: NASA's Earth Observing System and the Science of ASTER and MODIS*. Series: Remote Sensing and Digital Image Processing. 11, New York: Springer.

Tucker, C.J., Townshend, J.R.G., and Goff, T.E. 1985. African land cover classification using satellite data. *Science*, 227: 369–375.

Turner, B.L. III, Lambin, E.F., and Reenberg, A. 2007. The emergence of land change science for global environmental change and sustainability. *PNAS*, 104: 20666–20671.

Turner, B.L. II, Skole, D., Sanderson, S., Fischer, G., Fresco, L., and Leemans, R. 1995. Land-use and Land-cover Change. *Science/Research Plan*. IGBP Report 35, IHDP Report 7. IGBP Stockholm, 132.pp

USGCRP. 2010. Our Changing Planet: The US Global Change Research Program for FY 2010. Washington DC, 164.

USGCRP. 2011. Our Changing Planet: the US Global Change Research Program for FY 2010. Washington DC, 84.

Viña, A., Bearer, S., Chen, X., He, G., Linderman, M., An, L., Zhang, H., Ouyang, Z., and Liu, J. 2007. Temporal changes in giant panda habitat connectivity across boundaries of Wolong Nature Reserve, China. *Ecological Applications*, 17: 1019–1030.

Viña, A., Chen, X., McConnell, W.J., Liu, W., Xu, W., Ouyang, Z., and Liu, J. 2010. Effects of natural disasters on conservation policies: The case of the 2008 Wenchuan Earthquake, China. *Ambio*, 40: 274–284.

Waggoner, D.G. and Sokolik, I.N. 2010. Seasonal dynamics and regional features of MODIS-derived land surface characteristics in dust source regions of East Asia. *Remote Sensing of Environment*, 114: 2126–2136.

Walker, D.A., Leibman, M.O., Epstein, H.E., Forbes, B.C., Bhatt, U.S., Raynolds, M.K., Comiso, J.C., et al. 2009. Spatial and temporal patterns of greenness on the Yamal Peninsula, Russia: Interactions of ecological and social factors affecting the Arctic normalized difference vegetation index. *Environmental Research Letters*, 4: 16.

Walker, D.A., Forbes, B.C., Leibman M.O., Epstein H.E., Bhatt, U.S., Comiso, J.C., Drozdov, D.S., et al. 2011. Cumulative effects of resource development, reindeer herding, and climate change on the Yamal Peninsula, Russia and contrasts with the Alaska North Slope. In G. Gutman and A. Reissell (Eds.), *Arctic Landcover and Land-Use in a Changing Climate: Focus on Eurasia*. New York: Springer.

Warner, T.A., Almutairi, A., and Lee, J.Y. 2009. Remote sensing of land cover. In T.A. Warner, M.D. Nellis, and M.D. Foody (Eds.), *Handbook of Remote Sensing* (pp. 123–138). London: SAGE Publications.

Watson, R.T., Noble, I.R., Bolin, B., Ravindranath, N.H., Verardo, D.J., and Dokken, D.J. (Eds.). 2000. *Land Use and Land–Use Change and Forestry*. IPCC Special Report. London: Cambridge University Press.

Weber, M.G. and Flannigan, M.D. 1997. Canadian boreal forest ecosystem structure and function in a changing climate: Impact on fire regimes. *Environmental Reviews*, 5:145–166.

Wessman, C.A., Archer, S., Johnson, L.C., and Asner, G.P. 2004. In G. Gutman, A.C. Janetos, C.O. Justice, E.F. Moran, J.F. Mustard, R.R. Rindfuss, D. Skole, B.L. Turner II, and M.A. Cochrane (Eds.), *Land Change Science: Observing, Monitoring, and Understanding Trajectories of Change on the Earth's Surface* (185–208). Dordrecht, the Netherlands: Kluwer Academic Publishers.

Xiao, X., Biradar, C., Czarnecki, C., Alabi, T., and Keller, M. 2009. A simple algorithm for large-scale mapping of evergreen forests in Tropical America, Africa and Asia. *Remote Sensing*, 1: 355–374.

26 Building Saliency, Legitimacy, and Credibility toward Operational Global and Regional Land-Cover Observations and Assessments in the Context of International Processes and Observing Essential Climate Variables

Martin Herold, Lammert Kooistra, Annemarie van Groenestijn, Pierre Defourny, Chris Schmullius, Vasileios Kalogirou, and Olivier Arino

CONTENTS

26.1   INTRODUCTION

Reliable observations of the earth's land cover are crucial to the understanding of climate change and its impacts, to sustainable development, to natural resource management, to conserving of biodiversity, and to the understanding of ecosystems and biogeochemical cycling. Land-cover change is an issue with far-reaching policy implications, internationally, nationally, or locally. The United Nations Conference on Environment and Development's Agenda 21, the World Summit on Sustainable Development in Johannesburg 2002, and the existing UN conventions, most prominently the United Nations Framework Convention on Climate Change (UNFCCC), have further emphasized the importance of sustained land-cover assessments in their implementation plans (GCOS, 2004). The recently adopted implementation plan of the Group on Earth Observation (GEO) highlights in particular the importance of land cover for all areas of societal benefits.

Despite such emphasis, it is to be noted that land observations are not as operational as other major earth observation domains (oceans and atmosphere). It has further been shown that simply providing technically and scientifically sound data sets is often not sufficient for operational land-cover observations and assessments. Such activities have to be part of a process of engaging with user organizations, policy makers, and political processes in order to ensure that mapping and monitoring products are seen as legitimate and salient, apart from their being technically valid (Clark et al., 2006). Thus, one objective is to ensure interaction with, and technical support for, high-level political processes, such as GEO and UN conventions, with increasing need for land-cover earth observations and assessments. This requires an advocacy role for participating in international mechanisms to enhance the importance of land-cover observations, understand the requirements for specifying land observation strategies, and provide technical support and feedback to the political processes and policy discussions.

The additional objective is to build technical credibility, and it is twofold. The first component focuses on the area of international consensus building on technical implementation guidelines. Moving from research to operation for global monitoring has to integrate and synthesize scientific progress for defining the most suitable and internationally accepted approaches. Such detailed technical protocols specify the current best practices and lay the foundation for actual mapping and monitoring efforts. These include identifying best practices for land-cover validation, evolving standards for land-cover characterization, and setting guidelines for monitoring deforestation and associated carbon emissions in developing countries. The second component deals with implementation activities and further technical studies to address remaining critical issues in order to support the development of international standards and showcasing of their successful application. Different case studies of land-cover harmonization and validation, as well as land-change analysis, are presented to evaluate and assess technical guidelines and foster operational land-cover assessment activities in practice. In addition, implementation of the strategies and technical guidelines developed as part of this study is addressed by participating in specific global mapping activities.

The present work has been carried out as part of the Global Observation of Forest Cover and Land Dynamics (GOFC-GOLD) in conjunction with the ongoing European efforts to monitor land cover globally. The activities, linked to the Global Terrestrial Observing System (GTOS), provide the essential platform for linking scientific research with multiple activities and actors, such as data producers, political processes, and information users in global land observations.

26.2   LEARNING FROM GLOBAL ASSESSMENTS

Global environmental assessments have become a key element in international, national, and local policy and decision making. They are an important means for scientists to provide inputs to address environmental problems at the policy level, apart from the more traditional ways of peer-reviewed publications, popular media, or private advice to relevant actors (Clark et al., 2006). The global nature of such efforts emphasizes their role in addressing problems that require cooperation among different countries, between scientists and policy makers, and across different areas. There are several such assessments, including the Millennium Ecosystem Assessment, the Intergovernmental Panel on Climate Change (IPCC) assessment reports, United Nations Environment Programme's Global Environmental Outlook, or the Forest Resources Assessments conducted by the Food and Agriculture Organization (FAO). There is no doubt that good scientific information is essential for environmental decision making. However, mechanisms to link scientific research to policy-level discussions and decisions are not an easy matter. For example, there is temporal dependence and an evolutionary cycle that links scientific findings, creation of an observation and monitoring program, and related policy and public awareness and action; this is referred to as the "issue attention cycle." Furthermore, global assessments require a social communication process in which scientists, decision makers, advocates, and the media interact and interpret findings in particular ways. Thus, it is to be noted that the impact of global assessment depends not merely on the science being robust and technically believable (credibility). Users must view the assessment as being "salient," "legitimate," and "credible." When potential users believe that the information generated by an assessment process is relevant to their decision making, it can be considered salient. Legitimacy is provided if the process is perceived as being fair and taking into account the concerns and insights of the relevant stakeholders. In this context, five general principles have been advocated for practitioners to produce efficient global and regional assessments (Clark et al., 2006):

- Focus on the process, not the report
- Focus on salience, legitimacy, and credibility
- Assess with multiple audiences in mind
- Involve stakeholders and connect with existing networks
- Develop influence over time

These principles are also relevant for global land monitoring and assessment efforts, which, besides providing the data and observations, should engage in international processes to make the findings and results more relevant and acceptable. Adoption of this concept in this work and in global land-cover observations in general is presented in Figure 26.1. The ultimate determinants of historical context, key user characteristics, and assessment characteristics lay the foundation by specifying activities to ensure progress in the three areas. To improve saliency, three specific political processes have been chosen for engaging with and analyzing user requirements. Interaction with these processes improves saliency and thereby the relevance, usefulness, and acceptance of land-cover observations. Global land-cover observations and assessments have become more legitimate by establishing a long-term interaction process among relevant actors, developing technical consensus, and implementing projects jointly. Setting standards for land-cover characterization and validation and showcasing their application in global land-cover and land-change assessment studies have in particular ensured technical credibility.

26.3 PROGRESS IN ENHANCING SALIENCY AND LEGITIMACY

A high-level political process was the UNFCCC discussions and negotiations on reducing emissions from deforestation and forest degradation in developing countries (REDD); it was the key mitigation option for the post-Kyoto climate agreement. Related policy negotiations have been suffering from a lack of technical understanding of whether the historical and future pace of forest loss and associated carbon emissions in developing countries can be estimated

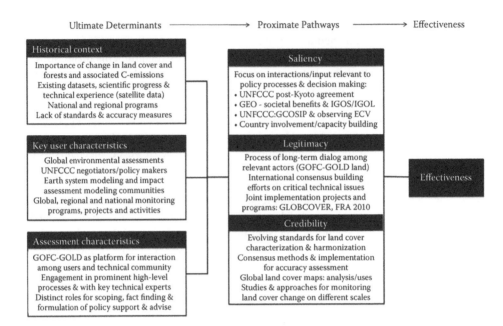

FIGURE 26.1 Adoption of the conceptual framework for considering effective environmental assessments (Clark et al., 2006), emphasizing the approaches to improving saliency, legitimacy, and credibility for global land-cover observations.

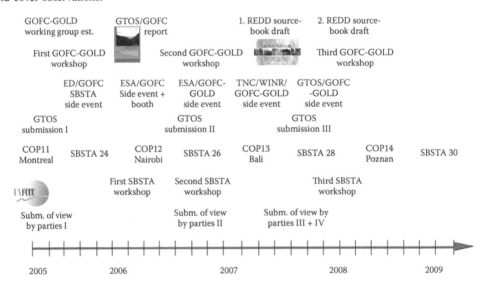

FIGURE 26.2 Timeline showing the major REDD events as part of the UNFCCC process and the contributing activities of the GOFC-GOLD working group to assist in and respond to the UNFCCC requirements (COP = Conference of the Parties, SBSTA = Subsidiary Body of Science and Technical Advice).

and accounted for in an operational, verifiable, transparent, and efficient manner. Thus, the user requirements are rather specific. A direct effort to provide technical input to this process was initiated when the issue was identified in 2005 (Figure 26.2). Dedicated technical input has been developed to assist the negotiations and build country capacity (Herold and Johns, 2007), including a GOFC-GOLD REDD sourcebook. It describes, in a user-friendly format, the consensus reached and the transparent methods adopted to produce estimates of changes in forest area and the production of carbon stocks resulting from deforestation and degradation (www. gofc-gold.uni-jena.de/redd).

The second prominent area in the UNFCCC that calls for progress in global land-cover observation relates to research and systematic observation (GCOS, 2004). The aim is to continuously monitor essential climate variables (ECVs) so as to reduce uncertainties in understanding the global climate system; this includes land cover as one such variable. The related Global Climate Observing System (GCOS) implementation plan (GCOS, 2004) mentions a number of specific tasks to improve global observation of land cover as an ECV, including the creation of international standards, consensus methods for map accuracy assessment, continuity for fine-scale satellite observations, development of an *in situ* reference network, implementation of an operational validation framework, generation of annual global land-cover products, and development of a high-resolution global land-cover change data set. As requested by the UNFCCC Subsidiary Body of Science and Technical Advice, reporting guidelines and standards are being developed for each ECV, including land cover. Work on this issue is documented at www.fao.org/gtos/topcECV.html.

GEO is now the most prominent political process concerned specifically with earth observation. It resulted from three ministerial-level earth observation summits (2003–2005) and started implementation activities in 2006. A dedicated task in the GEO 2006–2011 work plan was global land-cover observation, and the first step toward this was the process of linking GEO requirements (nine areas of societal benefits) with land-cover observation variables (Figures 26.3 and 26.4).

26.4 BUILDING TECHNICAL CREDIBILITY

The UNFCCC and GEO requirements have been studied and applied to develop comprehensive observation strategies and define implementation priorities. Key strategies derived as part of the

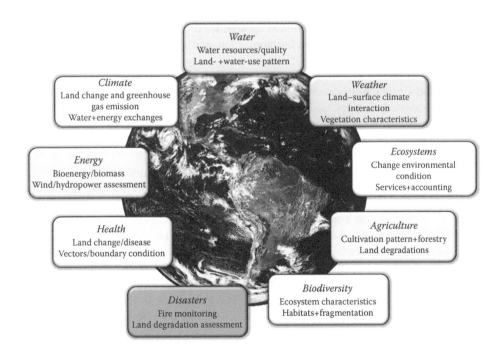

FIGURE 26.3 (See color insert.) GEO areas of societal benefits and key land-cover observation needs emphasize the multitude of services from continuous and consistent global terrestrial observations.

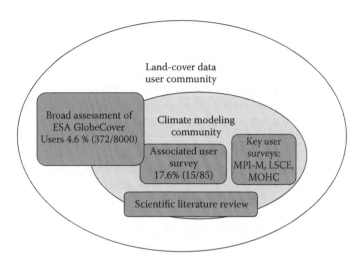

FIGURE 26.4 Concept of user communities and activities within user consultation plan to assess land-cover requirements of general land-cover user community and climate modeling community.

research are reflected in the Integrated Global Observations for Land (IGOL; Townshend et al., 2007) and in the GEO land-cover tasks (Herold et al., 2008).

An operational land-cover observation framework integrates information from different scales, that is, moderate- and high-resolution satellite data, and *in situ* observations (or high-resolution satellite data). These measurements have contributed to global monitoring in terms of spatial and thematic detail and require temporal updates. An integrated system combines the advantages to provide worldwide consistency and link the local and global observation levels. The concept assumes that there is observation continuity on all observation scales and that the data and information products are consistent and compatible. This assumes the availability of coordinated operational global satellite and *in situ* observations and common approaches to characterize, describe, and compare land-cover information (standardization, harmonization, and validation) and to facilitate the joint application of mapping products.

In the context of harmonization and evolving standards for semantic land-cover characterization, the UN Land-Cover Classification System (LCCS) currently provides the most comprehensive, internationally accepted, and flexible framework (Herold et al., 2006). Within LCCS, land-cover classifiers provide standardization of terminology and include descriptions of vegetation life form, leaf type, leaf longevity, and percent cover, as well as characterization of nonvegetation cover types (such as terrestrial vs. aquatic/regularly flooded areas).

Progress in the statistically robust accuracy assessment of land-cover data focuses on the development of standard methods (Strahler et al., 2006) for operational efforts and specific validation exercises. Dedicated case studies for assessing the accuracy of global land-cover maps (GLC2000, Mayaux et al., 2006), and the comparative assessment of different global maps linking harmonization and validation, have resulted in a thorough understanding of the existing uncertainties and strategies so as to assess and reduce them in the future.

Based on the global land-cover products with 1-km spatial resolution and a better understanding of their inconsistencies, this research has contributed to refinement, better utilization, and updating of the products. This includes the development of synergy maps with thematic characterizations aimed at specific applications (SYNMAP, Jung et al., 2006), their application in earth system models, and the exploration of linkages and heterogeneities between existing global and regional land cover. The new GLOBCOVER 2005/06 product (Arino et al., 2007) provides the next generation of global land-cover data based on available international standards, with more detailed spatial resolutions and with the aim of generating them annually.

An operational global land-cover observing system must provide land-cover change estimates to fully deliver the benefits. A combined approach using coarse-resolution and fine-scale satellite observations and *in situ* observations seems most suitable for global and regional scales. Case studies using coarse-resolution satellite data investigating land-cover dynamics emphasize the value of studying long-term trends, interannual versus intraannual dynamics, and the indication of hot spots and cumulative land change. However, fine-scale (i.e., Landsat-type) satellite data are currently the most suitable data source for observing with confidence a large array of land-cover/land-use change processes. The need for such operational approaches is now emphasized in starting national forest monitoring activities in many developing countries in order to build capacities for participation in the post-2012 climate agreement. In general, successful implementation and technical credibility of a global land-cover change assessment requires involvement, agreement, and capacity building among countries and their regional and national experts and networks. In that context, FAO's Forest Resources Assessment for 2010 will use operational satellite-based approaches for a regional and global monitoring of historical forest change processes.

26.5 REQUIREMENTS FOR MONITORING LAND-COVER ECV

For observing land cover as an ECV, several areas require attention:

- The need to address UNFCCC requirements
- The need for product specifications to be driven by the climate user communities
- The need for implementation to focus on a truly global system and process, including:
  - coordinated observations
  - integrated and standardized mapping
  - independent quality assessment

Any ECV monitoring efforts have to ensure saliency and legitimacy besides technical credibility. An international coordination mechanism among key actors worldwide (users, producers, science, regional/national experts) is essential to ensure that land-cover products are accepted internationally and by UNFCCC. The recent GTOS report (Herold et al., 2009) summarizes the level of standardization and desired observations as a general assessment of the UNFCCC requirements and needs (Table 26.1).

Besides, the analysis emphasizes the need for coordinated observations. An operational global land-cover monitoring integrates information from different observation scales, that is, integrating coarse- and fine-scale satellite data and *in situ* data. ECV monitoring assumes the use of all useful data sources—from historical archives, present assets, and future monitoring programs—in a seamless and consistent manner. Acquisitions and the derivation of standard products should be coordinated among space agencies (e.g., with the support of GEO, Committee of Earth Observation Satellites).

"Integrated and standardized mapping and monitoring" refers to the need for both static and updated maps and dynamic monitoring products at different spatial and temporal scales (Table 26.1). These outputs require different sets of observations and monitoring approaches. The development and derivation of the mapping products need consistency in land-cover characterization in order to be interoperable as part of an integrated global observing system. The broad areas and topics requiring international consensus are outlined in this document. There is also a need to ensure synergy with other ECV observation products (i.e., fire, biophysical parameters, snow cover) that are directly related to land-cover characteristics.

The issue of independent quality assessment follows the need to ensure that the required standards are met and that uncertainties are quantified and reduced as far as practicable. Considering the suite of important land-cover information (Table 26.1), a diversity of products contributing to

TABLE 26.1

Characteristics of Land-Cover Mapping and Monitoring Products Useful for Observing Land-Cover as an ECV

Name	Spatial Resolution	Frequency of Product Update	Maturity
Mapping of Land Cover			
Land-cover maps	250 m–1 km	Annual	Preoperational
Fine-scale land-cover and land-use maps	10–30 m	3–5 years	Preoperational (for land cover)
Global land-cover reference sample database	*in situ*/1 m	1–5 years	Preoperational (CEOS, GOFC-GOLD)
Monitoring of Dynamics and Change			
Global land-cover dynamics and disturbances	250 m–1 km	Intraannual/long-time series	Preoperational (for several processes)
Fine-scale land-cover and land-use change	10–30 m	1–5 years	Preoperational (for land cover)
Monitoring areas of "rapid change"	1–30 m	1–2 years or less	Preoperational (for some change processes)

Source: Adapted from Herold, M., et al. Assessment of the status of the development of the standards for the terrestrial essential climate variables: Land cover, FAO/GTOS ECV Report T9. 2009. Available at http://www.fao.org/gtos/doc/ECVs/T09/T09.pdf

ECV monitoring is to be expected. While diversity and redundancy are useful for building a sustained global land-cover monitoring system and for ensuring flexibility in incorporating new technologies, an independent assessment mechanism led by the international community is also needed. This mechanism should provide comparative assessment and validation of individual products and work toward synergy to ensure that a common framework is used for global assessments and that the "best global estimates" are made available on the basis of current knowledge, data, and information. The basis for such efforts is a sustained global network of calibration and validation sites, international agreement, and standards and approaches for land-cover characterization and validation, and an internal coordination mechanism.

26.6 DETAILED CLIMATE USER ASSESSMENT—THE EUROPEAN SPACE AGENCY LAND-COVER CLIMATE CHANGE INITIATIVE PROJECT EXPERIENCES

Despite the general requirements of the UNFCCC and its subsidiary bodies, the ECV monitoring implementation and the definition of more detailed product specifications should be driven by a more consolidated climate user assessment. There are different levels at which users reflecting the climate research community can be involved in this process. Some of the key climate users will represent particular research groups and fields, and their specific requirements will have a key impact on product specification. However, the user consultation should further address a broad range of issues related to the nature of the interactions of land cover and climate. Climate determines the distribution of natural vegetation; so changes in vegetation indicate climate change. Land-cover changes also occur because of changes in land management practices and land use (e.g., agricultural intensification or forest clearance for cropland). Changes in land cover force climate change by modifying water and energy exchanges with the atmosphere and by changing greenhouse gas and aerosol sources and sinks. Global land observations are used in climate, carbon, and ecosystem models that provide predictions and scenarios for use by the parties negotiating development of UNFCCC; observations of land variables have to be reported by the parties to the convention in order to document their own overall contribution to changes in the earth's atmospheric constituents, including greenhouse gas concentrations.

One of the key requirements is that current IPCC assessment reports are based on an uncertain understanding of the land surface dynamics and related processes; these issues call for continuous monitoring systems and data.

For an adequate modeling of processes at the land surface boundary to the atmosphere, an accurate representation of the land surface is necessary. A climate model used to simulate these processes requires proper determination of the land surface characteristics used in its parameterizations as boundary conditions. These parameters include background surface albedo, surface roughness length due to vegetation, fractional vegetation cover, leaf area index (LAI), forest ratio, plant-available soil water holding capacity, and volumetric wilting point.

Three major communities—Global Circulation Modeling (GCM), Earth System Modeling (ESM), and Integrated Assessment Modeling (IAM)—play an important role in understanding and quantifying earth and climate system analysis and, specifically, in understanding the role of land-use and land-cover change. They have a common global scope of some kind but focus on different objectives.

A variety of approaches to addressing land-use and land-cover change have been considered by these communities. GCM includes a rather coarse level of ecological and biogeochemical process representation and uses land cover as a generic and fixed boundary condition. ESM modelers take an approach that stems from a combination of basic ecosystem (e.g., carbon cycle) and dynamic global vegetation models (DGVMs) and incorporate different plant functional types (PFTs) into their structures. These aspects of ESMs are increasingly being used for impact assessments, both for ecosystems and for the impacts on hydrology, which are modified by ecosystem responses. The ESM approach is derived from a tradition of using complex models to analyze the different components and interactions of the physical system. The focus has mainly been on the climate system, with an initial description of coupled ocean–atmosphere systems and, more recently, the carbon cycle and dynamic vegetation. By enlarging its focus, the ESM approach is increasingly coupling climate with hydrology, agriculture, and urban systems as integral components of the earth system.

The IAM approach comes largely from a tradition of modeling human behavior explicitly and the interaction of human activities, decision making, and the environment, including economic production and consumption, energy systems, greenhouse gas emissions, and land use. This community has also recognized the importance of land use as a critical factor in socioeconomic decision making, for example, for food and timber production, the state of ecosystems and their services, and, increasingly, as a response to the demand for biofuels for the electricity and transportation sectors. Whereas many IAMs have focused strongly on energy–economy systems and included land-use emissions only as exogenous factors, this is now changing with the development and implementation of coupled socioeconomic and climate modeling strategies.

In addition, improved land-cover observations might lay the foundation for fostering the next-generation climate model concepts and applications. There is already a debate within the climate modeling community on new and revised concepts to better parameterize land-cover characteristics for better process representation. In this context, the initial activities of the Terrabites EU Cost action (http://www.terrabites.net/) observed that there are scale limits to the second generation of DGVMs. PFTs need an improved representation, taking into account the dynamics in both space and time. Most DGVMs are "area-based" models in which grid cell fractions occupied by homogeneous populations of PFTs exist without any real age or size structure and do not mechanistically simulate the process of vegetation succession or competition for light resources between PFTs. Second-generation DGVMs are already using a more advanced spatial representation of vegetation (e.g., LPJ-GUESS using populations of individual plants). Dynamic changes in PFTs can be represented by characterization of changes and variation in plant traits, for example, phenology, R/K strategy. This also links closely to ECV variables like fraction of absorbed photosynthetically active radiation and LAI. As many plant traits vary much more within species than between species within PFTs, PFTs are still a good base to represent the inherent variability. A new definition of PFTs should be able to represent vegetation coexistence caused by vertical and spatial ecosystem heterogeneity.

Thus, land-cover user requirements for climate modeling and climate research are expected to diversify.

26.6.1 User Survey Overview

As part of the requirement analysis, a user consultation mechanism (Figure 26.4) was set up to involve different climate modeling groups by conducting surveys for different types of users: (1) a group of key users, most of them also participating in the Climate Modelling User Group (CMUG) of the European Space Agency Climate Change Initiative (ESA CCI); (2) associated climate users involved in and leading the development of relevant key climate models and applications; and (3) the broad land-cover data user community reflected in scientific literature and represented by users of the ESA GlobCover product. The surveys were carried out in September and October 2010 and focused on the three major ways in which land-cover observations are used in climate models:

1. As proxy for a set of land surface parameters assigned on the basis of PFTs
2. As proxy for human activities in terms of natural versus anthropogenic and tracking human activities, that is, land use affecting land cover (land-cover change as a driver of climate change)
3. As data sets for validation of model outcomes (i.e., time series) or to study feedback effects (land-cover change as a consequence of climate change)

The growth of requirements for the three aspects from the current models to new modeling approaches was specifically taken into account. Next to the surveys, requirements from the GCOS Implementation Plan 2004 and 2010 and associated strategic earth observation documents for land cover (GTOS, IGOL, Integrated Global Carbon Observation, and CMUG) were considered and reviewed. Finally, a detailed literature review was carried out with special attention to innovative concepts and approaches in order to better reflect land dynamics in the next-generation climate models.

26.6.2 Analysis of User Requirements

The outcome of user requirement assessment shows that although the range of requirements of the climate modeling community is broad, the requirements coming from different user groups and the broader requirements derived from GCOS, CMUG, and other relevant international panels are well matched. As a starting point of the Land-Cover CCI project, activities have been closely aligned with specific land-cover tasks listed in the GCOS Implementation Plan of 2004 and 2010 (Table 26.2). For example, LCCS should be adopted as an approach to thematic characterization of land-cover classes (Action T22), particularly because of its compatibility with the PFT concept. The project will further address critical tasks that have not made much progress to date, that is, on the implementation of an operational reference network and validation (Action T25) and creation of annual maps of global land cover (Action T26).

26.6.2.1 Spatial Detail

The users provided information on the level of spatial details they required, and the results are summarized in Figure 26.5. First, there is not one spatial resolution that fits all purposes; on an average, climate models run on broad spatial levels of detail, and a resolution of 300 m or coarser is sufficient to meet the modeling requirements of most users. However, for some, and in particular for

TABLE 26.2

Key Tasks for Land-Cover Theme from GCOS Implementation Plan (2004 and 2010), Progress Reported in 2009 (for COP 15), and How These Tasks Are Taken Up by Land-Cover CCI Project

GCOS Implementation Plan Task (2004 and 2010)	Status Reported in Recent Progress Report (2009)	Issues Addressed by Land-Cover CCI
Action T22 international standards for land-cover maps; in IP 2010, T22 was removed	The UN LCCS (under ISO) provides the required standards and specifications (good progress)	LCCS classifiers, generic classes, and related legends targeted at user requirements will be used to develop the product
Action T23 methods for land-cover map accuracy assessment; in IP 2010, defined as T26	Standard validation protocols, methods, and best practices have been developed by the CEOS Working Group on Calibration and Validation (WGCV), working with GOFC-GOLD (good progress)	The project uses a comprehensive validation approach that is independent, internationally agreed, and repeatable
Action T25 development of *in situ* reference network for land cover; in IP 2010, T22 is reflected in ecosystem observing network	As a start, GOFC-GOLD and CEOS WGCV have developed the framework for an *in situ* reference network for operational global land-cover validation (low progress)	For product validation, a comprehensive approach, making best use of existing resources and aiming at developing an operational reference network, is applied
Action T26 annual land-cover products; in IP 2010, defined as T27	There are several global land-cover products at the requested resolution, including GlobCover and MODIS (moderate progress)	The activities build upon the GlobCover heritage, cooperating with the MODIS team and aiming at annual global products
Action T27 regular fine-resolution land-cover maps and change; in IP 2010, defined as T28	No concerted action toward a global product at the required fine resolution (10–30 m) has been achieved (low progress)	The issue of fine-scale land cover/land-cover change is not specifically addressed here, while some methodological steps could be extended to higher resolution dataset

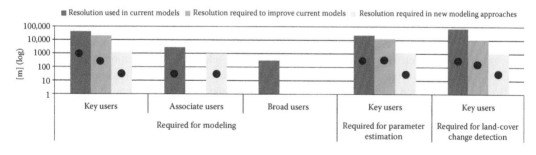

FIGURE 26.5 (See color insert.) Spatial resolution median requirements (note *y*-axis in log-scale) from user surveys. The orange points indicate the minimum requirement.

the future (Figure 26.5), more detailed resolutions are required. This would mean that in the coming years land-cover observations for estimating model parameters and for describing change would need to develop toward fine-scale satellite observations coming from Landsat-type observations (e.g., Sentinel-2). This would also require prioritization of regions for which this level of detail is most relevant.

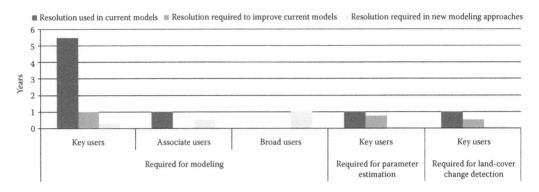

FIGURE 26.6 Overview of temporal resolution requirements (median of responses) from user surveys.

26.6.2.2 Temporal Resolution

Many users rely on annual updating of parameters initially derived from land-cover data. While annual data are currently not available for land cover, the modeling community is using interpolation and ancillary data (i.e., from the literature or models) to provide the temporal detail required.

The need for increased temporal resolution data is pertinent to all user groups. Particularly for the future, moving into intraannual and monthly dynamics of land cover is considered essential (Figure 26.6). While any addition to the temporal resolution of the currently often static land-cover data is useful, the need to explore the potential of dense remote-sensing time-series signals is vital. In terms of the temporal range, models use periods beyond the remote-sensing era back in time, and this range is expected to widen further in the future.

26.6.2.3 Land-Cover Characterization and Land-Change Requirements

Whereas almost all major land categories in current maps are important, the surveys particularly highlighted the need for increasing detail in forest, herbaceous, and agriculture classes in the current models. Considering all users, the need for wetland and urban classes is expected to increase in future models and other land-cover applications. Forests and some other vegetation classes (i.e., shrubs) are commonly separated by leaf type and phenology. Since users require a suite of different types of land-cover categories (or PFTs) for model parameterization that varies with type of model and modeling approach, any land-cover product will need to offer some flexibility in responding to these different thematic needs. Broad surveys have shown that more than 90% of the users find the UN LCCS suitable for thematic characterization—an approach that is also compatible with the PFT concept of many models. The surveys stressed the need for additional information on the separation of C3/C4 grasses and crops and consideration of human activities and land management practices. For example, "disturbed fraction" has been advocated as one such requirement. While some information is commonly not available from remote-sensing data sets (i.e., C3/C4 separation), the use of external products or nonsatellite-derived data may be needed if it improves accuracy and parameter estimation procedures.

A fair amount of users (in particular, key users) do not utilize any change or dynamic products from land-cover remote sensing in their modeling. However, as stated in Figure 26.7, the need for more dynamic information and land-cover/land-use changes in the future is pertinent. Important information is required for vegetation phenology, agricultural expansion, forest loss/deforestation, and urbanization. In addition, the need to monitor wetland dynamics, fire, land degradation, and long-term vegetation trends is highlighted by the community of associated users.

It is also important to note that about half of the broad user community and one-fifth of the associated users did not mention the need for any change/dynamic information. This reemphasizes the need for both stable and accurate land cover and dynamic components reflecting time series and changes.

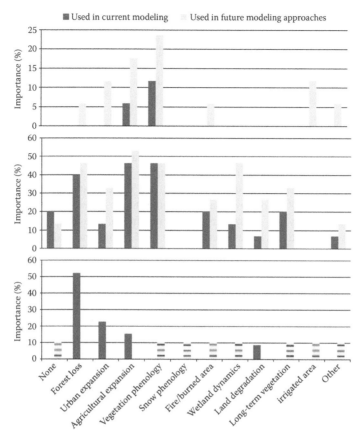

FIGURE 26.7 Classes of interest for land dynamic monitoring for key users (upper figure), associate users (middle figure), and broad users (lower figure, grey striped classes were not asked).

26.6.2.4 Accuracy Requirements

There are three types of quantitative requirements for the accuracy of the CCI land-cover products coming from GCOS, CMUG, and CCI. Since available land-cover maps have an overall area-weighted accuracy of around 70%, it can be assumed that the accuracy requirements for the land-cover CCI should be higher. Second, GCOS requirements mention a maximum of 15% omission/commission per class and those from CMUG and the CCI an error of 5%–10%. CMUG further requires stability in accuracies over time of less than 10%. Those requirements can be understood as quantitative guidelines; however, in the current knowledge of global land-cover mapping, there are two main problems in using such statements for the upcoming land-cover mapping efforts:

- Errors of 5%–10% either per class or as overall accuracy are rare and hard to achieve in any land-cover mapping effort with more than—two or three categories
- The accuracy of the product depends on its actual use in the model

In particular, the analysis of the model parameters versus land-cover types emphasizes that the relative importance of different class accuracies varies heavily depending on which parameter is estimated (Section 26.6.3). This is an important implication that cannot be considered by using a standard overall accuracy reporting. Any accuracy analysis should provide flexibility to account for such differences in the way land-cover data are used in models and the related impact on the uncertainty of the input data. In addition, the need for stability in the accuracy should be reflected in implementing a multidate accuracy assessment.

Finally, the users stressed the need for quality flags and controls, the probability for the land-cover class or anticipated second class or even probability distribution function for each class (coming from the classification algorithm), and the need for accuracy numbers for land-cover classes (potentially also with regional estimates).

26.6.3 ANALYSIS OF MODEL PARAMETERS VERSUS LAND-COVER

The relationship between land-cover types and model parameters is one of the most important issues determining the accuracy and relevancy of land-cover data for parameterization, calibration, and validation of climate-related models. We used the climate model parameterization as described by Hagemann (2002) to provide a better quantitative understanding of why and on what level of detail and accuracy the climate users require thematic land-cover information. In this study, 75 different land surface classes according to the Olsson land and ecosystem map (Olson, 1994a, 1994b) were parameterized for nine land surface parameters, using literature data and expert analysis. We used these data to analyze the relative importance of different land-cover classes for estimating model parameters. The importance of each differentiating two land-cover classes is reflected in the relative similarity for each actual land surface parameter value. Thus, for each pair of classes (x, y), the similarity (Sim_{xy}) can be calculated by relating the specific parameter values for each class (Par_x, Par_y) to the overall range of parameter values across all classes ($\text{Par}_{max} - \text{Par}_{min}$):

$$\text{Sim}_{xy} = \frac{\left|\text{Par}_x - \text{Par}_y\right|}{\text{Par}_{max} - \text{Par}_{min}}$$

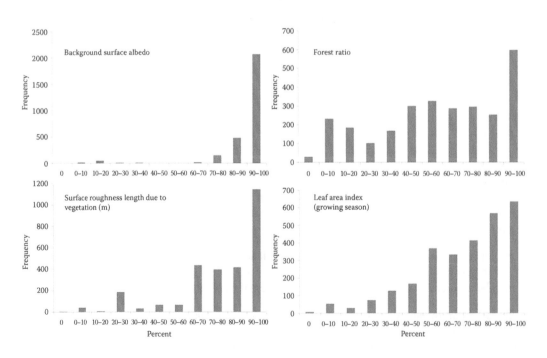

FIGURE 26.8 Histogram of similarity values (in percent, x-axis) representing the relative difference in values among two land-cover classes for four land surface parameters. Note: the scale of y-scale varies for different diagrams.

The similarity value is reported as a percentage, with 100% representing the same parameter value for this pair of classes. The result is a symmetric matrix of 75 × 75 classes for each of the nine surface parameters; the aggregated results are presented here.

Figure 26.8 shows the histograms of the distribution of the similarities among all class pairs for four parameters. For different land surface parameters, the patterns of class similarity are somewhat different. For albedo, there are a few classes that obviously have a value (snow, ice, water) very different from those of many of the other classes that are relatively similar. For the forest ratio, there is more of an equal distribution among the range of similarity values. A more varied distribution is indicated for vegetation surface roughness and the LAI. This highlights the fact that the relative importance, and thereby the accuracy, of land-cover categories for model parameterization varies depending on what model parameter is estimated from the data. A single overall accuracy for a land-cover map value will not be able to provide information on how accurate a specific map is for parameter estimation.

Table 26.3 shows the average similarity from nine parameter values for 12 land-cover classes aggregated from 75 Olsson map classes. The areas with pink table cells have the highest average

TABLE 26.3

Matrix of the Similarity between the 12 Generalized Land-Cover Classes as Average for Nine Land Surface Model Parameters

Generalized Land-Cover Legend		1 Evergreen Needleleaf Trees	2 Evergreen Broadleaf Trees	3 Deciduous Needleleaf Trees	4 Deciduous Broadleaf Trees	5 Mixed/Other Trees	6 Shrubs	7 Herbaceous Vegetation	8 Cultivated and Managed Vegetation	9 Urban/Built-Up	10 Snow and Ice	11 Barren	12 Open Water
1	Evergreen needleleaf trees												
2	Evergreen broadleaf trees	87											
3	Deciduous needleleaf trees	74	75										
4	Deciduous broadleaf trees	67	78	85									
5	Mixed/other trees	70	73	89	89								
6	Shrubs	54	59	78	80	78							
7	Herbaceous vegetation	45	50	70	71	72	89						
8	Cultivated and managed vegetation	52	61	75	82	80	92	87					
9	Urban/built-up	34	33	53	48	55	58	65	55				
10	Snow and ice	21	15	39	32	42	50	58	45	69			
11	Barren	36	30	54	47	57	65	73	60	78	85		
12	Open water	30	24	48	40	50	57	65	53	75	88	89	

similarities among all land surface parameters. They tend to be located near the diagonal of the matrix, reflecting somewhat the ranking of classes 1–12 from forests to barren and water areas. Most dissimilar are the nonvegetation and vegetation classes. A misclassification and confusion between two classes with large similarity will cause a much lower error in the quantitative parameter estimation than uncertainties among very dissimilar classes.

26.7 SUMMARY OF AND RECOMMENDATIONS FOR LAND-COVER ECV EFFORTS

In summary, the requirements coming from different user groups and the broader requirements derived from relevant international panels are well matched. The findings highlight that

- There is need for both stable land-cover data and a dynamic component in the form of time series and changes in land cover
- Consistency among the different model parameters is often more important than accuracy of individual data sets, and it is important to understand the relationship between land-cover classifiers with the parameters and the relative importance of different land-cover classes
- Providing information on natural versus anthropogenic vegetation (disturbed fraction) and tracking human activities and defining history of disturbance is of increasing relevance, in particular for land use affecting land cover, with more details needed for focus areas with large anthropogenic effects
- Land-cover products should provide flexibility to serve different scales and purposes in terms of both spatial and temporal resolution
- The relative importance of different class accuracies varies significantly depending on the surface parameter that is estimated, and the need for stability in accuracy should be reflected in implementing a multidate accuracy assessment
- Future requirements for temporal resolution refer to intraannual and monthly dynamics of land cover, including remote-sensing time-series signals
- More than 90% of the general land-cover users find LCCS (Herold and Johns, 2007) suitable for thematic characterization, and this approach is also quite compatible with the PFT concept of many models
- Quality of land-cover products needs to be transparent by using quality flags and controls and should include information on the probability for the land-cover class or anticipated second class or even the probability distribution function for each class (coming from the classification algorithm).

As a next step within the Land-Cover CCI project (2010–2013), the outcome of this user requirement analysis will be used as input for the product specification of the next-generation Global Land Cover data set that will be developed within this project.

ACKNOWLEDGMENTS

The authors gratefully acknowledge the contributions of the European Space Agency's Data User Element (ESRIN) for supporting related activities, in particular for funds provided to the GOFC-GOLD land-cover project office and the Land-Cover CCI project.

REFERENCES

Arino, O., Leroy, M., Ranera, F., Gross, D., Bicheron, P., Nino, F., Brockman, C., et al. 2007. GLOBCOVER—A global land cover service with MERIS, *Proceedings of Envisat Symposium 2007*, on CD Rom.

Clark, W.C., Mitchell, R.B., and Cash, D.W. 2006. Evaluating the influence of global environmental assessments. In R.B. Mitchell, W.C. Clark, D.W. Cash, and N.M. Dickson (Eds.), *Global Environmental Assessments: Information and Influence*. Cambridge, MA: MIT Press 1–28.

GCOS. 2004. Implementation plan for the Global Observing System for Climate in support of the UNFCCC, October 2004, GCOS–92, WMO Technical Document No. 1219, (WMO; Geneva), p.153. Available at: http://www.wmo.int/pages/prog/gcos/index.php?name=AboutGCOS

Hagemann, S. 2002. An improved land surface parameter dataset for global and regional climate models. MPI-M Report 336.

Herold, M., Latham, J.S., Di Gregorio, A., and Schmullius, C.C. 2006. Evolving standards on land cover characterization. *Journal of Land Use Science*, 1(2–4), 157–168.

Herold, M., and Johns, T. 2007. Linking requirements with capabilities for deforestation monitoring in the context of the UNFCCC-REDD process. *Environmental Research Letters*, 2, 045025 (7 pp), Available at: erl.iop.org.

Herold, M., Woodcock, C.E., Loveland, T.R., Townshend, J., Brady, M., Steenmans, C., and Schmullius, C. 2008. Land Cover Observations as part of a Global Earth Observation System of Systems (GEOSS): Progress, activities, and prospects. *IEEE Systems*, 2(3), 414–423.

Herold, M., Woodcock, C., Cihlar, J., Wulder, M., Arino, O., Achard, F., Hansen, M., et al. 2009. Assessment of the status of the development of the standards for the terrestrial essential climate variables: Land cover, FAO/GTOS ECV Report T9. Available at http://www.fao.org/gtos/doc/ECVs/T09/T09.pdf

Jung, M., Henkel, K., Herold, M., and Churkina, G. 2006. Exploiting synergies of global land cover products for carbon cycle modeling. *Remote Sensing of Environment*, 101(4), 534–553.

Mayaux, P., Eva, H., Gallego, J., Strahler, A.H., Herold, M., Agrawal, S., Naumov, S., et al. 2006. Validation of the Global Land-Cover 2000 map. *IEEE Transactions on Geoscience and Remote Sensing*, 44, 1728–1739.

Olson, J.S. 1994a. Global ecosystem framework-definitions. USGS EROS Data Center Internal Report.

Olson, J.S. 1994b. Global ecosystem framework-strategy. USGS EROS Data Center Internal Report.

Strahler, A., Boschetti, L., Foody, G.M., Friedl, M.A., Hansen, M.C., Herold, M., Mayaux, P., Morisette, J.T., Stehman, S.V., and Woodcock, C. 2006. Global land- cover validation: Recommendations for evaluation and accuracy assessment of global land cover maps, Report of Committee of Earth Observation Satellites (CEOS)—Working Group on Calibration and Validation (WGCV), JRC report series.

Townshend, J.R., Latham, J., Arino, O., Balstad, R., Belward, A., Conant, R., Elvidge, C., et al. 2007. Integrated global observation of land: An IGOS-P theme. Available at: www.fao.org/gtos.

Index

A

ACCESS, *see* Advancing Collaborative Connections for
    Earth System Science (ACCESS)
Advanced Land Imager (ALI), 115
Advanced very high resolution radiometer (AVHRR), 17,
    69, 92, 155, 187, 266, 381
    AVHRR GIMMS NDVI product, 96
    AVHRR LAI, 93, 94
    for large-scale forest clearing quantification, 192
Advancing Collaborative Connections for Earth System
    Science (ACCESS), 381
Afforestation, 292, 297
Africa, sub-Saharan, 369
    deforestation, 274
    ecoregion stratification, 370
    forest degradation and fragmentation, 372
    GDP in, 369
    land-cover changes, 371–372, 373
    MSS use, 371
    natural vegetation loss, 372
    sample sites, 370
    subscene classification, 371
Agricultural Land Cover Classification (ALCC), 308
Agriculture
    extensification of, 292, 296
    intensification of, 296
Airborne Visible InfraRed Imaging Spectrometer
    (AVIRIS), 158
ALI, *see* Advanced Land Imager (ALI)
Ancillary data sources, 308
    for Canada, 309
    for Mexico, 313
    for United States, 312
Anderson Classification System (ACS), 15, 43, 67,
    69, 70, 77
    agricultural land, 78
    barren land, 79
    forestland, 78–79
    perennial snow or ice, 80
    rangeland, 78
    tundra, 79–80
    urban or built-up, 77–78
    water, 79
    wetland, 79
ANN, *see* Artificial neural network (ANN) model
Anthropogenic land-use/land-cover change, 3
AO, *see* Arctic Oscillation (AO)
AOI, *see* Area of Interest (AOI)
Arctic Oscillation (AO), 102
Area of Interest (AOI) tool, 142
Artificial neural network (ANN) model, 233
ASCR, *see* Automatic scattergram-controlled regression
    (ASCR)
Asia, tropical, 275
    climatic conditions, 275, 276

forest-cover monitoring program, 279, 280
JAXA, 281
land cover, 275, 276
    northeast India, 280
land-cover mapping, 276, 277, 278
    Borneo, 281
    Indonesia, 280–281
    Lower Mekong Basin, 279
    Malesia vegetation map, 277
    Southeast Asia, 280, 281
    subregional, 279
    tropical vegetation regional, 277
    NOAA AVHRR imagery, 277–278
    regional forest, 276, 277
Atmospheric-Vegetation Interactive Model
    (AVIM), 92
At-sensor spectral radiance, 120–121
Autoassociative memory, 139
Automated GlobCover classification chain, 51, 52
Automatic scattergram-controlled regression (ASCR),
    157, 164
AVHRR, *see* Advanced very high resolution radiometer
    (AVHRR)
AVIRIS, *see* Airborne Visible InfraRed Imaging
    Spectrometer (AVIRIS)

B

Bidirectional reflectance distribution function (BRDF),
    158, 122, 249
Bitemporal change detection, 161–162
Bitemporal classification approach, 194

C

C5, *see* Collection 5 (C5)
Calibration and validation (Cal/Val), 114
Canonical correlation analysis (CCA), 102–103
Canonical factor (CF), 103
CAR, *see* Central African Republic (CAR)
CARPE, *see* Central African Regional Program for the
    Environment (CARPE)
CAS, *see* Chinese Academy of Sciences (CAS)
Cascades ecoregion, 364
CBD, *see* UN Convention on Biological Diversity (CBD)
CBERS, *see* China–Brazil Earth Resources Satellite
    (CBERS)
CBFP, *see* Congo Basin Forest Partnership (CBFP)
CBI, *see* Composite burn index (CBI)
CCA, *see* Canonical correlation analysis (CCA)
C-CAP, *see* Coastal Change Analysis Program
    (C-CAP)
CCD, *see* UN Convention to Combat Desertification
    (UNCCD)